Electrostatic Effects in Soft Matter and Biophysics

D1591369

NATO Science Series

A Series presenting the results of scientific meetings supported under the NATO Science Programme.

The Series is published by IOS Press, Amsterdam, and Kluwer Academic Publishers in conjunction with the NATO Scientific Affairs Division

Sub-Series

I. Life and Behavioural Sciences	IOS Press
II. Mathematics, Physics and Chemistry	Kluwer Academic Publishers
III. Computer and Systems Science	IOS Press
IV. Earth and Environmental Sciences	Kluwer Academic Publishers

The NATO Science Series continues the series of books published formerly as the NATO ASI Series.

The NATO Science Programme offers support for collaboration in civil science between scientists of countries of the Euro-Atlantic Partnership Council. The types of scientific meeting generally supported are "Advanced Study Institutes" and "Advanced Research Workshops", and the NATO Science Series collects together the results of these meetings. The meetings are co-organized bij scientists from NATO countries and scientists from NATO's Partner countries – countries of the CIS and Central and Eastern Europe.

Advanced Study Institutes are high-level tutorial courses offering in-depth study of latest advances in a field.
Advanced Research Workshops are expert meetings aimed at critical assessment of a field, and identification of directions for future action.

As a consequence of the restructuring of the NATO Science Programme in 1999, the NATO Science Series was re-organized to the four sub-series noted above. Please consult the following web sites for information on previous volumes published in the Series.

http://www.nato.int/science
http://www.wkap.nl
http://www.iospress.nl
http://www.wtv-books.de/nato-pco.htm

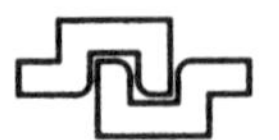

Series II: Mathematics, Physics and Chemistry – Vol. 46

Electrostatic Effects in Soft Matter and Biophysics

edited by

Christian Holm

Max-Planck-Institut für Polymerforschung,
Mainz, Germany

Patrick Kékicheff

Institut Charles Sadron,
C.N.R.S., Strasbourg, France

and

Rudolf Podgornik

Department of Physics,
University of Ljubljana, Ljubljana, Slovenia

Kluwer Academic Publishers

Dordrecht / Boston / London

Published in cooperation with NATO Scientific Affairs Division

Proceedings of the NATO Advanced Research Workshop on
Electrostatic Effects in Soft Matter and Biophysics
Les Houches, France
1-13 October 2000

A C.I.P. Catalogue record for this book is available from the Library of Congress.

ISBN 1-4020-0196-7 (HB)
ISBN 1-4020-0197-5 (PB)

Published by Kluwer Academic Publishers,
P.O. Box 17, 3300 AA Dordrecht, The Netherlands.

Sold and distributed in North, Central and South America
by Kluwer Academic Publishers,
101 Philip Drive, Norwell, MA 02061, U.S.A.

In all other countries, sold and distributed
by Kluwer Academic Publishers,
P.O. Box 322, 3300 AH Dordrecht, The Netherlands.

Printed on acid-free paper

All Rights Reserved
© 2001 Kluwer Academic Publishers
No part of the material protected by this copyright notice may be reproduced or utilized in any form or by any means, electronic or mechanical, including photocopying, recording or by any information storage and retrieval system, without written permission from the copyright owner.

Printed in the Netherlands.

Preface

Recently theoretical interest in soft matter has finally caught up with its undisputed practical and technological relevance (high-polymers and elastomers, polyelectrolyte hydrogels, food processing and storage, waste management, new tailored materials, etc.). But even more importantly charged soft matter appears to be at the heart of biotechnologies that are widely believed to be the most important technologies of this century and that already figure in many countries' technological strategies. Recent impetus in the elucidation of the structure of synthetic genosomes, complexes of negatively charged DNA and positively charged lipids, reveals all the subtleties that one has to master in order to be able to study and understand materials that are both soft and charged.

Soft charged matter is dominated by the interplay of thermal fluctuations and long-range electrostatic effects on different length and time scales which results in complicated phase diagrams. The range of expertise that one needs to study charged soft matter is so wide, encompassing continuum mechanics, statistical mechanics, field theory, molecular simulations on the theoretical side and sophisticated experimental techniques such as direct force measurements, laser tweezers, light, X-ray and neutron scattering, atomic force microscopy, that it is quite unusual for any university department to be proficient in all its facets.

The idea to organize the NATO Advanced Study Institute "Electrostatic Effects in Soft Matter and Biophysics" grew out of a workshop organized by W. Gelbart, A. Parsegian and P. Pincus in Santa Barbara at the ITP in 1998 with a similar theme. While there it was devoted to active researchers, mainly theorists, already familiar with the subject, we felt that it was time to do our share in educating young people to foster the development of this exciting and rapidly growing field.

Our aim was to bring together theorists, computer simulators and experimentalists from research areas as diverse as macromolecular charged colloids and biological materials like DNA. We tried to cover most basic investigative tools on soft charged materials: analytical methods and approaches, computational tools and tricks of the trade, as well as experimental methods in their different ranges of applicability. The lectures were divided into three thematic blocks:

The first block consisted of modern analytical methods used to describe electrostatic interactions between charged soft bodies with fixed charges immersed in a solution of mobile charge carriers. H. Wennerström introduced us to the intermolecular interactions specifically concentrating on the dielectric description in the condensed phase and the Lifshitz theory. R. Kjellander and R. Netz covered in depth the introduction to integral equation theories for the electric double layer and the mean-field Poisson-Boltzmann equation and all its varied consequences respectively. They also introduced us to the various modern theoretical approaches that go beyond the mean-field approximation. B. Shklovskii introduced us to his approach to electrostatic correlations based on the Wigner crystal model and showed us how different physical problems can be treated with this theory. R. Podgornik reviewed the work on the equation of state in charged macromolecular arrays motivated by recent experimental findings on ordered DNA arrays and ordered multilamellar lipid arrays. B. Gelbart and A. Khokhlov reviewed some recent experimental and theoretical insights into DNA condensation and complexation, cationic histone proteins, and chromatin, and told us about new analytical and experimental approaches to understand these biologically important structures. J.-F. Joanny showed us how many problems in the physics of polyelectrolytes and polyampholytes in dilute and semi-dilute solutions can be treated with simple scaling theories.

The second block consisted of an introduction to simulational methods of molecular dynamics and Monte Carlo approaches to soft matter systems with long-range Coulombic interactions. B. Jönsson introduced us to protein electrostatics and compared theoretical approaches to Monte Carlo simulations. We also learned about correlation effects between plates and spheres and the collapse of DNA-like systems due to correlation interactions. K. Kremer prepared a grand tour of polymer simulations, showed us how poor solvent modifies the behavior of polyelectrolytes and gave examples for deviations between theory and simulations for stiff rod polyelectrolytes.

The last thematic block treated experimental methods. In depth coverage of light, neutron, and X-rays scattering was performed by M. Caffrey, M. Rawiso, and C. Williams. They emphasized new type of experiments like real-time measurements, neutron spin echo, standing X-ray waves, reflectivity, photon correlation spectroscopy, X-ray speckles, and use of X-ray microbeams. A comparison between the main techniques of direct force measurements was presented by P. Kékicheff. These include techniques based on osmotic equilibria, surface force balances, surface force apparatus, atomic force microscope and force measurement techniques on individual freely moving particles (total internal reflection microscopy, etc). The advantages but also the limitations of each technique were underlined and discussed in different systems. In addition new methods applicable to charged systems (colloidal dispersions, polyelectrolytes, DNA assemblies) were presented (video microscopy, fluorescent microscopy, etc) by D. Grier in conjunction with structural and chemical inhomogeneities, defects and microphases either already present or induced upon confinement or under stress.

The chapters in this book do not necessarily follow the courses presented at the school. The authors rather focus on a specific part of the lectures' content, giving adequate background material, trying to provide a reference chapter for further reading. Exactly for this purpose we also added an introductory chapter on the Poisson-Boltzmann description of the cell model, because it lies at the basis of many theoretical advances, and which was not treated in this way in the school. Unfortunately other urgent activities did not allow M. Rawiso to prepare a full version of his contribution in time, but he included a guide to the literature of his lectures.

We think that all participants will agree that the Les Houches School provided a unique learning environment for all of us. We had enjoyable mountain trips with the daring experience of how long a return trip through the French Alps can be, if the guide gets lost, and a great cultural experience by visiting the van Gogh exhibition in Martigny. We spent a whole day without running water, thus testing the dedication and passion for science of the students as well as the lecturers. The students discussed eagerly their work at the poster sessions, and were willing to ask for more at the tutoring sessions that we organized after the lectures. We obtained a total of 140 applications for the 57 available positions from people of more than 19 different countries. We agreed only to accept a very limited number of senior scientists, and gave full priority to young scientists. For all of those that could not make it to Les Houches, and also because there is no single one textbook covering all the different facets of soft charged matter, it is our hope to service the science community with these proceedings.

We would like to thank the people who helped us finish this project. At various stages those were T. Caramico, M. Karttune, and M. Richter. Many thanks go also to the staff at Les Houches, the members of the pep group at Mainz, and of course to all authors who accepted the heavy workload of contributing to this volume.

Finally we gratefully acknowledge generous financial support by the NATO as ASI PST.ASI.976196 and the European Community under grant HPCFCT -19 99-0130, with supplementary funds by the Max-Planck-Institut für Polymerforschung which made this meeting possible.

Christian Holm, Patrick Kékicheff, Rudi Podgornik

Mainz, Strasbourg, Ljubljana, August 2001.

List of Participants

Directors

Holm, Christian
Max–Planck–Institut für Polymerforschung, Ackermannweg 10, 55128 Mainz, Germany. E-Mail: holm@mpip-mainz.mpg.de

Kékicheff, Patrick
Institut Charles Sadron, C.N.R.S., 6 rue Boussingault, 67083 Strasbourg, France. E-Mail: kekichef@ics.u-strasbg.fr

Podgornik, Rudolf
Department of Physics, University of Ljubljana, Jadranska 19 SI–1000 Ljubljana, Slovenia. E-Mail: rudolf.podgornik@fiz.uni-lj.si

Lecturers

Caffrey, Martin
Ohio State University, Department of Chemistry, 100 West 18th Avenue Columbus, OH 43210–1173, USA. E-Mail: caffrey@chemistry.ohio-state.edu

Gelbart, William
Department of Chemistry, U.C.L.A. University of California, Los Angeles, CA 90095–1569, USA. E-Mail: gelbart@chem.ucla.edu

Grier, David G.
University of Chicago, James Franck Institute, 5640 South Ellis Ave. Chicago, IL 60637–1433, USA. E-Mail: grier@fafnir.uchicago.edu

Joanny, Jean–François
Physicochimie Curie, Institut Curie Section Recherche, 11 rue P.et M.Curie, 75231 Paris Cedex 05. E-Mail: jean-francois.joanny@curie.fr

Jönsson, Bo
Physical Chemistry Depart., Lund University Chem. Center, POB 124, 22100 Lund, Sweden. E-Mail: bo.jonsson@teokem.lu.se

Khokhlov, Alexei R.
Moscow State University, Physics Department Moscow, 117234 Russia. E-Mail: polly.phys.msu.su

Kjellander, Roland
Physical Chemistry Dep., Göteborg University Fysikalisk kemi, 41296 Göteborg, Sweden. E-Mail: rkj@phc.chalmers.se

Kremer, Kurt
Max–Planck–Institut für Polymerforschung, Ackermannweg 10, 55128 Mainz, Germany. E-Mail: kremer@mpip-mainz.mpg.de

Netz, Roland R.
Max–Planck–Institut für Kolloid– und Grenzflächenforschung, 14424 Potsdam, Germany. E-Mail: netz@mpikg-golm.mpg.de

Rädler, Joachim O.
Max–Planck–Institut für Polymerforschung, Ackermannweg 10, 55128 Mainz, Germany. E-Mail: raedler@mpip-mainz.mpg.de

Rawiso, Michel
Institut Charles Sadron, C.N.R.S., 6 rue Boussingault, 67083 Strasbourg, France. E-Mail: rawiso@cerbere.u-strasbg.fr

Shklovskii, Boris I.
University of Minnesota School of Physics and Astronomy, 116 Church St. SE, Minneapolis, MN 55455, USA. E-Mail: shklovskii@physics.spa.umn.edu

Wennerström, Håkan
Department of Physical Chemistry, 1 Box 124 Chemical Center, 22100 Lund, Sweden. E-Mail: Hakan.Wennerstrom@fkem1.lu.se

Williams, Claudine E.
CNRS–Collège de France, 11 Place Marcellin–Berthelot, 75321 Paris, France. E-Mail: williams@ext.jussieu.fr

ASI Students

Baigl, Damien, CNRS–College de France, Physique de la Matiere Condensee, 11, place Marcelin Berthelot, 75005 Paris, France.
E-Mail: damien.baigl@college-de-france.fr

Barbosa, Marcia, Instituto de Fisica, UFRGS, Caixa Postal 15051, RS 91501–970 Porto Alegre, Brazil. E-Mail: barbosa@if.ufrgs.br

Behrens, Sven Holger, University of Chicago, The James Franck Institute, 5640 S. Ellis Avenue, Chicago, IL 60637, USA.
E-Mail: sbehrens@uchicago.edu

Bergenholtz, Johan, University of Gothenburg, Dept. of Physical Chemistry, 41296 Göteborg, Sweden. E-Mail: jbergen@phc.chalmers.se

Betterton, Meredith, Curie Institute UMR CNRS/IC, 1b8 11 rue Pierre et Marie Curie, 75248 Paris Cedex 05, France.
E-Mail: betterton@post.harvard.edu

Borukhov, Itamar, UCLA, Dept. of Chemistry and Biochemistry, 607 C. Young Dr. East, Los Angeles, CA 90095–1569, USA.
E-Mail: itamar@chem.ucla.edu

Burak, Yoram, Tel Aviv University, School of Physics and Astronomy, Tel Aviv, 69978, Israel. E-Mail: yorambu@post.tau.ac.il

Castelnovo, Martin, Institut Charles Sadron, 6 Rue Boussingault, 67083 Strasbourg, France. E-Mail: castel@cerbere.u-strasbg.fr

Cepic, Mojca, Jozef Stefan Institute, Jamova 39, 1000 Ljubljana, Slovenia. E-Mail: mojca.cepic@ijs.si

Deserno, Markus, UCLA, Dept. of Chemistry, 405 Hilgard Ave. Los Angeles, CA 90095–1569, USA. E-Mail: markus@chem.ucla.edu

Dzeshkouski, Aliaksandr, University of North Carolina, Chemistry Dept., CB♯3290, Chapel Hill, NC 27599–3290, USA. E-Mail: ad@unc.edu

Fleck, Christian, Universität Konstanz, Fachbereich Physik, 78457 Konstanz, Germany. E-Mail: christian.fleck@uni-konstanz.de

Forsberg, Björn, University of Gothenburg, Dept. of Physical Chemistry, 41296 Göteborg, Sweden. E-Mail: bjofo@phc.chalmers.se

Häußler, Wolfgang, Max–Planck–Institut für Polymerforschung, Ackermannweg 10, 55128 Mainz, Germany.
E-Mail: haeussle@mpip-mainz.mpg.de

Hagberg, Daniel, Lund University, Theoretical Chemistry Chem. Center, POB 124, 22100 Lund, Sweden. E-Mail: daniel@signe.teokem.lu.se

Hansen, Per Lyngs, National Institutes of Health, LPSB/NICHD, Bldg. 12A Rm. 2041, 12 South Drive, Bethesda, MD 20892–5626, USA. E-Mail: lyngs@helix.nih.gov

Jonsson, Marie, Physical Chemistry 1, Box. 124, 22100 Lund, Sweden. E-Mail: marie.jonsson@fkem1.lu.se

Jusufi, Arben, Heinrich–Heine–Universität Düsseldorf, Institut für Theoretische Physik II, Universitätsstr. 1, 40225 Düsseldorf, Germany. E-Mail: jusufi@thphy.uni-duesseldorf.de

Karttunen, Mikko, Helsinki University of Technology, Laboratory of Computational Engineering, P.O. Box, 9400 02015 Hut Helsinki, Finland. E-Mail: karttune@lce.hut.fi

Khan, Malek, Lund University, Theoretical Chemistry, Box 124, 22100 Lund, Sweden. E-Mail: malek.khan@teokem.lu.se

Kramarenko, Elena, Moscow State University, Physics Dept. Moscow, 117234, Russia. E-Mail: kram@polly.phys.msu.su

Kranenburg, Marieke, University of Amsterdam, Dept. Chemical Engineering Nieuwe, Achtergracht 166, 1018 WV Amsterdam, The Netherlands. E-Mail: marieke@its.chem.uva.nl

Kunze, Karl–Kuno, Max–Planck–Institut für Kolloid– und Grenzflächenforschung, 14424 Potsdam, Germany. E-Mail: kunze@mpikg-golm.mpg.de

Ligoure, Christian, CNRS GDPC CC26, University Montpellier, 2 Montpellier, 34095 CEDEX 05, France. E-Mail: ligoure@gdpc.univ-montp2.fr

Limbach, Hans Jörg, Max–Planck–Institut für Polymerforschung, Ackermannweg 10, 55128 Mainz, Germany. E-Mail: limbach@mpip-mainz.mpg.de

Lobaskin, Vladimir, Physics Dept. Chelyabinsk, State University, 454136 Chelyabinsk, Russia. E-Mail: vladimir.lobaskin@unifr.ch

Loison, Claire, CECAM, ENS Lyon, 46 Allee d'Italie, 69364 Lyon Cedex 07, France. E-Mail: loison@mpip-mainz.mpg.de

Lopes, Antonio, ITQB–Institute for Techonological Chemistry and Biology, New University of Lisbon Ap. 127, 2781–901 Oeiras, Portugal. E-Mail: alopes@itqb.unl.pt

Lukatsky, Dmitry, Department of Materials and Interfaces, Weizmann, Institute of Science, Rehovot, 76100, Israel.
E-Mail: lukatsky@wicc.weizmann.ac.il

Martín–Molina, Alberto, Dept. Of Applied Physics, Campus de Fuentenueva Univ. of Granada, 18071 Granada, Spain. E-Mail: almartin@ugr.es

Messina, René, Max–Planck–Institut für Polymerforschung, Ackermannweg 10, 55128 Mainz, Germany. E-Mail: messina@mpip-mainz.mpg.de

Migliorini, Gabriele, Max–Planck–Institut für Polymerforschung, Ackermannweg 10, 55128 Mainz, Germany. E-Mail: migliori@mpip-mainz.mpg.de

Moldakarimov, Samat, Institute of Polymer Materials and Technology, Satpaev Str. 18a, 480013 Almaty, Kazakhstan.
E-Mail: samat.ipmt-kau@usa.net

Moreira, Andre Guerin, Max–Planck–Institut für Kolloid– und Grenzflächenforschung, 14424 Potsdam, Germany.
E-Mail: amoreira@mpikg-golm.mpg.de

Nguyen, The Toan, University of Minnesota, School of Physics, 116 Church Street S.E., Minneapolis, MN 55455, USA.
E-Mail: ttnguyen@physics.spa.umn.edu

Oksana, Ismailova, Heat Physics Dept., Uzbek Academy of Sciences, Katartal 28, 700135 Tashkent, Uzbekistan. E-Mail: shmi@janus.silk.org

Patkowski, Adam, A. Mickiewicz University, Umultowska 85, 61–614 Poznan, Poland. E-Mail: patkowsk@amu.edu.pl

Pinchuk, Anatoliy, Kyiv Shevchenko University, Eugene Potje 9/43, 03057 Kyiv, Ukraine. E-Mail: onmf@serv.biph.kiev.ua

Ramirez, Rosa, CECAM. ENS–Lyon 46, Allee d'Italie, 69007 Lyon, France.
E-Mail: rosa@cecam.fr

Rastegar, Abbas, University of Nijmegen, Research Institute for Materials, Toernooiveld 1, 6525ED Nijmegen, The Netherlands. E-Mail: rastegar@sci.kun.nl

Rescic, Jurij, University of Ljubljana Faculty of Chemistry and Chemical Technology, Askerceva 5, 1000 Ljubljana, Slovenia. E-Mail: jurij.rescic@uni-lj.si

Schiessel, Helmut, Max–Planck–Institut für Polymerforschung, Ackermannweg 10, 55128 Mainz, Germany. E-Mail: heli@mpip-mainz.mpg.de

Sens, Pierre, Institut Charles Sadron, 6 Rue Boussingault, 67083 Strasbourg, France. E-Mail: sens@ics.u-strasbg.fr

Smith, Benjamin A., McGill University–Condensed matter Physics 3600 University st. Montreal, PQ H3A 2TB Canada.
E-Mail: bsmith@physics.mcgill.ca

Squires, Todd, Harvard Physics Department, Cambrigde, MA 02138, USA.
E-Mail: squires@physics.harvard.edu

Tchoukov, Plamen Hristov, Institute of Physical Chemistry, Bulgarian Academy of Science Acad. G. Bonchev Str. bl. 11, 1113 Sofia, Bulgaria.
E-Mail: tchoukov@ipchp.ipc.bas.bg

Tsori, Yoav, Tel Aviv University, School of Physics and Astronomy, Tel Aviv, 69978, Israel. E-Mail: tsori@post.tau.ac.il

Vilfan, Andrej, Cavemdish Laboratory, TCM Group, Madingley Road, CB3 0HE Cambridge, UK. E-Mail: avilfan@ph.tum.de

Volk, Nicole Helena, Universität Mainz , Institut für Physikalische Chemie, Welder Weg 11, 55099 Mainz, Germany. E-Mail: volk@mail.uni-mainz.de *von Gruenberg, Hans–Henning*, Universität Konstanz, PF 5560, 78457 Konstanz, Germany. E-Mail: hennig.vongruenberg@uni-konstanz.de

Whiting, Carole Jo, University of Leeds, Polymer IRC, Physics Department Leeds, LS2 9JT West Yorkshire, UK. E-Mail: phycw@phys-irc.leeds.ac.uk

Wilk, Agnieszka, Adam Mickiewicz University Poznan, Institute of Physics, ul. Umultowoska 85, 61–614 Poznan, Poland.
E-Mail: agniwilk@friko5.onet.pl

Zakharova, Svetlana, Leiden University, HB 325, LIC 7, Gorlaeus Laboratories, Einteinweg 55, RA 2300 Leiden, The Netherlands.
E-Mail: s.zakharova@chem.leidenuniv.nl

Zeldovich, Konstantin, Moscow State University, Physics Dept. Moscow, 117234, Russia. E-Mail: zeld@polly.phys.msu.su

Zhumadilova, Gulmira, Institute of Polymer Materials and Technology, Satpaev str. 18a, 480013 Almaty, Kazakhstan. E-Mail: ipmt-kau@usa.net

Ziherl, Primoz, University of Pennsilvania, 209 South 33rd Street, Philadelphia, PA 19104–6396, USA. E-Mail: primoz@physics.upenn.edu

Contents

STRUCTURE AND DYNAMIC PROPERTIES OF MEMBRANE LIPID AND PROTEIN*

MARTIN CAFFREY
Biochemistry, Biophysics, Chemistry
The Ohio State University, Columbus, Ohio 43210. USA

*Parts of this chapter are reproduced with permission from Annu. Rev. Biophys. Biophys. Chem. Vol 24, 1995 by Annual Review, Inc., from Topics Curr. Chem. Vol 151, 1989 by Springer-Verlag and from Curr. Opin. Struct. Biol. 10, 2000 by Elsevier Science Ltd.

The prodigious x-ray flux and tuneability of synchrotron radiation has proved an enormous boon in studies of the structural biology and dynamic properties of membranes and their components. Over the past several years, this bright x-ray source has been used to advantage in investigations of the dynamics and mechanism of lipid phase transitions in bulk systems, lyotrope transport between and within bulk mesophases, x-radiation damage of lipid, lipid-lipid and lipid-lyotrope miscibility, Langmuir-Blodgett lipid films, and of the crystallization of membrane proteins in lipidic mesophases. A review of these disparate but related investigations is presented here.

1. Kinetics of Lipid Phase Transitions Using TRXRD

There are many vital cellular processes, such as membrane fusion, which must involve changes in lipid phase state. To date, little attention has been paid to the dynamics of lipid phase interconversions. If such changes are physiologically relevant, they must occur on a time scale comparable with those taking place *in vivo*. Thus, there is a need to establish the kinetics of lipid phase transitions. In addition to the kinetics, little is known about the mechanism of lipid phase transitions. Such information is integral to our understanding of the structural and compositional requirements for transitions that occur in biological, reconstituted and formulated systems.

C. Holm et al. (eds.), Electrostatic Effects in Soft Matter and Biophysics, 1–26.
© 2001 *Kluwer Academic Publishers. Printed in the Netherlands.*

A method, referred to as time-resolved x-ray diffraction (TRXRD), has been developed which facilitates direct and quantitative measurement of bulk lipid phase transitions [1, 2]. Mechanistic insights are provided by the ability to detect transition intermediates. The method makes use of a focused, monochromatic synchrotron derived x-ray beam and an x-ray imaging device to continuously monitor diffracted x-rays. Kinetic and mechanistic studies are performed by using TRXRD in conjunction with relaxation measurements following the imposition of a rapid perturbation in one or more thermodynamic variables.

1.1. MOTIVATION

Despite the fact that life is dynamic, surprisingly little attention has been devoted to the study of the dynamics of vital processes at the level of macromolecular structure and assembly. Fortunately, the situation is changing. Interest has recently been focused on the cell membrane, itself an extremely lively structure. Consider for a moment the structural rearrangements it undergoes in membrane fusion during the fertilization event, vesicle mediated transport or cellular locomotion. Such events integrate enormous structural change. It is the involvement of biological liquid crystals, or lipids, in these processes that motivates studies of lipid phase transition kinetics.

The stability of a given lipid phase depends on a range of biologically relevant thermodynamic quantities, both environmental and compositional [3, 4, 5]. These include temperature, pressure, pH, salt and water concentration, and overall lipid composition. A change in any one of these can effect phase change, which in turn can profoundly influence the behavior, performance, and in some cases, survival of the system as a whole. Lipid phase transitions and the rates at which they occur are important for a host of other reasons. At the physiological level, they have been implicated in a range of processes including fat digestion, development of disease, regulation of membrane protein activity, membrane permeability and penetrability by macromolecules such as proteins, transbilayer lipid movement, visual transduction, and anaesthetic action. It should be borne in mind also that most membrane lipids in isolation have a variety of phases accessible to them at or close to physiological conditions, that many phase transitions can be effected isothermally, and that nonbilayer phases can accumulate in biological systems under conditions of stress.

From a commercial standpoint an understanding of lipid phase behavior is also important. For example, the organoleptic properties, shelf-life and behavior during processing, formulation and reconstitution of foods, feeds and pharmaceuticals are sensitive functions of the phase relations existing among the various lipid components. Cryopreservation and anhydrobiosis are other areas where membrane lipid phase behavior is important [6].

Failure to survive freezing and thawing protocols or extreme desiccation relates to mesomorphic stability and to the dynamics and mechanism of phase changes taking place in the membrane lipid.

1.2. LIPID MESOMORPHISM

Membrane lipids can exist in a number of intermediate states or mesophases more ordered than the liquid but less so than the crystalline solid (3-5). Such materials are called liquid crystals and their multiple intermediate phase character is termed mesomorphism. There are four types of mesomorphic behavior: lyotropic, thermotropic, barotropic and ionotropic. These refer to the expression of the disparate liquid crystal phases by manipulating solvent (lyotrope) content, temperature, pressure, and salt concentration, respectively. Most membrane lipids exhibit the four types of mesomorphism.

Phase refers to that part of a system which is chemically and physically homogeneous (at a macroscopic level) throughout. The different lyotropic phases are distinguished by the relative spatial conjunction of lipid and solvent (Figure 1). In the case of the hydrated lamellar phase, for example, lipid molecules pack to form extensive planar bilayers which stack one atop the other each separated by a layer of water. This phase has long been used as a model for the biomembrane. Lipids can also adopt a number of nonbilayer configurations, the most commonly encountered of which are the hexagonal and cubic phases. The hexagonal phase consists of a hexagonally packed array of parallel rods of either water or lipid. In the normal hexagonal (HI) phase, water constitutes the continuous medium supporting hydrocarbon rods, whereas in the HII, or inverted hexagonal phase, hydrocarbon forms the continuous medium. The cubic phase has many modifications that arise as a result of the three-dimensional packing of rodlike or lamellar aggregates some of which have the property of forming infinite periodic minimal surfaces. Nonbilayer phases are not simply physiochemical or crystallographic curiosities but are phases accessible to most membrane-derived lipids at or close to physiological conditions.

1.3. TIME-RESOLVED X-RAY DIFFRACTION

1.3.1. *The Principle*

A limitation of many of the methods for studying the dynamics of bulk lipid mesomorphic transitions, with the exception of x-ray diffraction, is that the measured parameter is oftentimes not a unique or a direct reflection of the phase under investigation. In contrast, x-ray diffraction provides direct phase identification and microstructure characterization in well-ordered systems. However, to record such a pattern can take from minutes to days depending on the x-ray source, sample, phase type and detector. According-

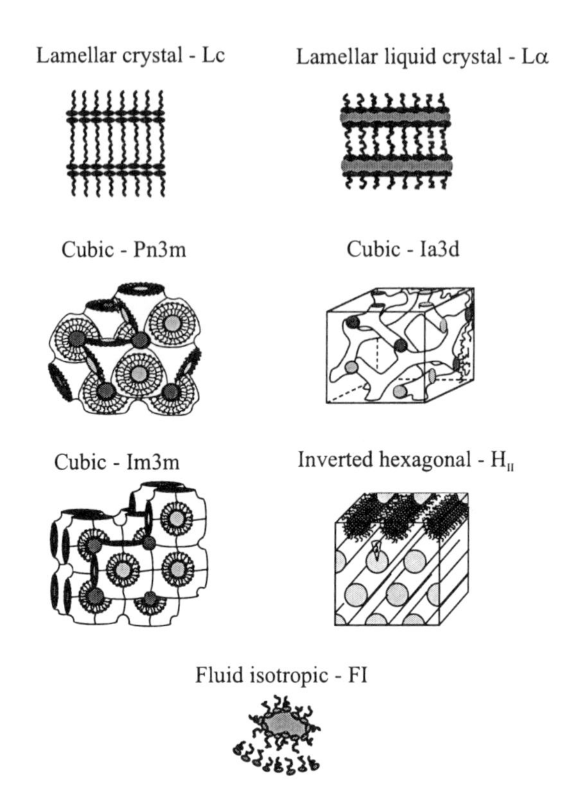

Figure 1. Cartoon representation of the various solid, mesophase and liquid states adopted by lipids.

ly, conventional x-ray diffraction techniques are not suited for examining the dynamics of lipid mesomorphism where phase changes can occur on a milliseconds time scale. This is where the method of time-resolved x-ray diffraction comes to the fore. The ability to image x-rays continuously comes about as a result of the combination of two essential elements: on the one hand, a bright source of (monochromatic) x-rays as is available at a synchrotron and, on the other, a live-time x-ray imaging device. This combination provides a means for direct phase identification and characterization continuously in time.

1.3.2. *Information Content*

Insofar as the dynamics and mechanism of phase transitions are concerned, TRXRD provides two important pieces of information: 1) transition kinetics obtained by following the relaxation of the system in response to an applied perturbation and 2) details of intermediates that form in the process. These data provide a basis for formulating, evaluating and refining transition mechanisms (see reference [7] as an example).

1.3.3. *Experimental Aspects*

The experimental arrangement used by the author in making TRXRD measurements is shown schematically in Figure 2. The system includes a synchrotron source, monochromating and focusing optics, the sample, the detector and recording and image analyzing devices. The former includes an image intensifier tube, while the latter incorporates a charge coupled device (CCD), video camera, video cassette recorder and monitor and a digital image processor. Sample manipulation devices have been omitted from the figure for sake of clarity.

1.3.4. *Selected Results*

A summary of selected results pertaining to lipid phase transition kinetics established by TRXRD has been presented [1, 2]. Some general observations will be made in regards to these measurements, as follows: i. Some transitions are fast, occurring on a time-scale of milliseconds, while others are extremely sluggish requiring weeks to complete. ii. Certain transitions are equally fast in the forward and reverse direction. Others have rates that differ by many orders of magnitude. iii. Many of the transitions studied appear to be two-state in the sense that at any point during the transition only two phases co-exist. In contrast, other transitions are continuous, displaying a pronounced second order character while others again reveal the presence of transition intermediates. iv. While some transitions are readily reversible and non-hysteretic in phase type, lattice parameters and transition temperature, others are extremely hysteretic. v. For certain lipid systems, transition time is independent of hydration, salt concentration, lipid identity, changes in the periodicity of the transforming phases and in long- and short-range order but is dependent on the magnitude and direction of the T-jump. vi. Long-range order is preserved throughout many lipid phase transitions. This suggests that the transforming units remain coupled and undergo the transition cooperatively and that long-range order is established rapidly in the nascent phase. In contrast, other transitions involve an intermediate state devoid of long-range order. In such cases, the precipitous loss of the (ordered) phase undergoing change is accompanied or followed by the emergence of diffuse scatter and/or line broadening which sharpens up with time and eventually gives way to the ordered nascent phase. vii. Depending on the system, temporal correlations of changes in long- and short-range order may or may not exist. viii. In many of the transitions studied, the interconverting phases appear to be incommensurate. ix. In certain systems, changes occur in short-range order without a corresponding adjustment in long-range order of the interconverting phases.

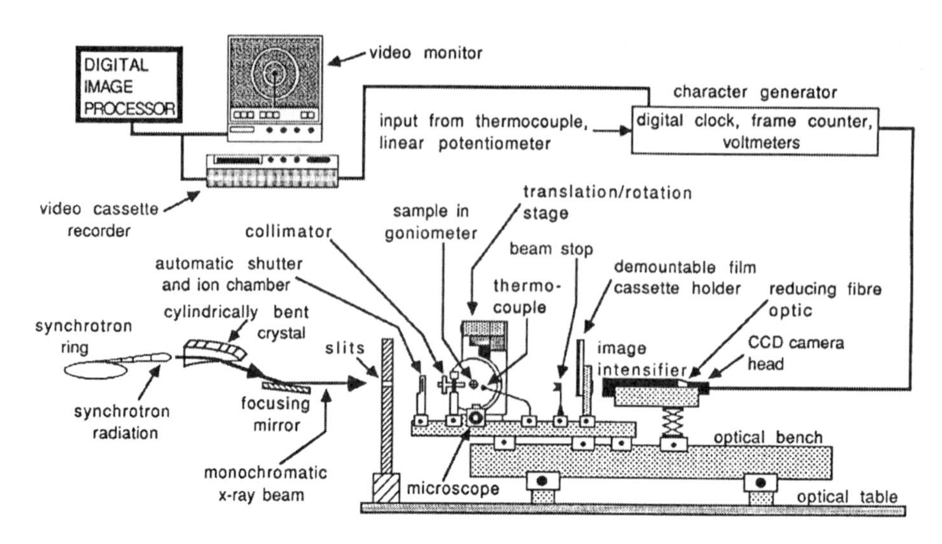

Figure 2. Schematic diagram of the experimental arrangement for monitoring x-ray diffraction in live-time using synchrotron radiation (not drawn to scale).

x. In addition to rearrangements occurring during a phase transition, quite extensive and rapid microstructure and volume changes take place within single phase regions.

An example of recently recorded data on an acid-induced transition in large unilamellar lipid vesicles is shown in Figure 3. A stop-flow apparatus was used to rapidly adjust sample pH during the course of the TRXRD measurement performed using synchrotron radiation as part of a study of membrane fusion [8]. The pH change triggers an initial and rapid formation of a bulk lamellar phase (d001, 48.6 Å) which in time transforms to a bulk inverted hexagonal phase (d10, 61.1 Å).

It is clear that a sizeable body of information has been gathered concerning the dynamics and mechanism of lipid phase transitions based upon the TRXRD method. Further, the data serve to illustrate the potential of this approach in revealing the molecular structure and compositional dependence of mesomorph stability and interconvertibility and, ultimately, the underlying transition mechanism.

2. X-Radiation Damage

The call for brighter synchrotron x-radiation sources for use in structural biology research is barely audible as we embark on a new millennium.

Our brightest sources are already creating havoc when used at design specifications because of radiation damage. The problem is particularly severe in studies involving kinetics and mechanism where cryotechniques are not always viable. Accordingly, we need to understand the very nature of radiation damage and to devise means for minimizing it. This is the thrust of an ongoing study as applied to lipid membranes and mesophases. Experiments were performed at the most brilliant beamlines at the ESRF (Grenoble, France) and the APS (Argonne, IL). In these investigations we found two very different types of radiation damage. One involved a dramatic phase transformation and the other a disordering of lamellar stacking. The work highlights the nature of the damage process and the need for additional studies if we are to make most efficient use of an important resource, synchrotron radiation [9, 10, 11].

3. Two Novel Approaches for Studying the Mesomorphic Properties of Lipids

The stability of liquid crystal phases depend on temperature and composition and each lipids' pattern of dependency is conveniently described in the form of an isobaric temperature-composition phase diagram. Two related methods of collecting mesomorphic phase information which are less time-consuming and more efficient and informative than conventional techniques have been developed. By incorporating a range of conditions into each sample preparation - a temperature gradient in the first method [12] and a lyotrope gradient [13] in the second - and utilizing the method of TRXRD, the time required to collect phase information for a complete diagram is reduced to minutes. Temperature-composition phase diagrams constructed using these methods compare well with those constructed by conventional means.

In the first approach, a temperature gradient is imposed on samples of constant composition so that each sample in an x-ray capillary represents a vertical line, or isopleth, in the corresponding phase diagram. In the second approach, the sample contains a solvent gradient and is raised to progressively higher temperatures. At each temperature the sample represents a horizontal line, or isotherm, in the phase diagram. These techniques are referred to, respectively, as the temperature gradient and lyotrope gradient methods.

To analyze samples in both the temperature and lyotrope gradient methods, capillaries are passed through a synchrotron-derived x-ray beam while the image-intensified diffraction pattern is recorded in live-time continuously along the length of a sample.

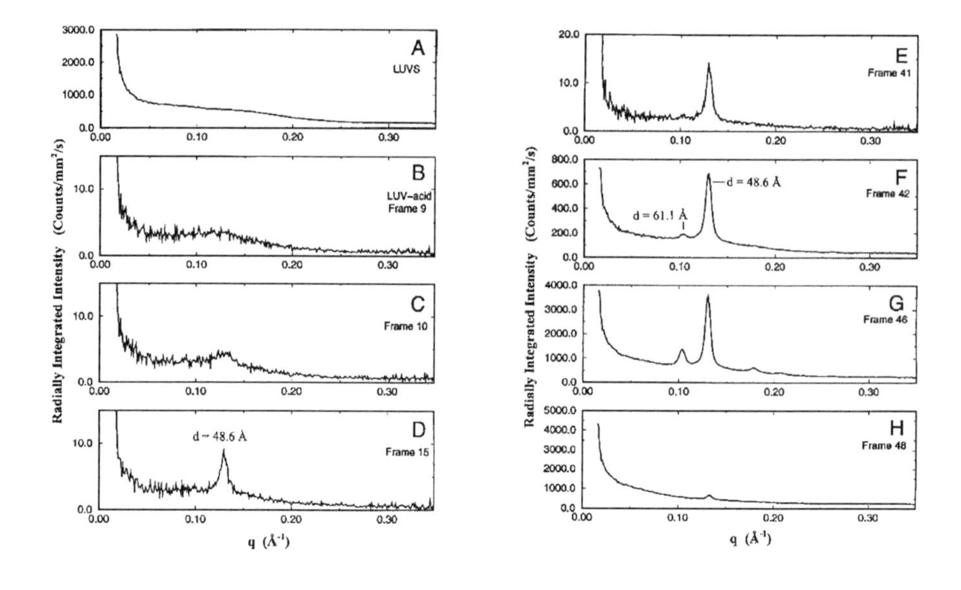

Figure 3. TRXRD measurements of changes in low-angle diffraction from a suspension of large unilamellar DiPoPE vesicles induced by a pH-jump from high to low pH [8]. Data collected at X12B (NSLS) with a multiwire detector were obtained by averaging results from 19 successive mixing/flushing cycles at 39.2°C. (A) LUV suspension before mixing, 60 s integration. Integrated 0-30 ms (B), 30-60 ms (C), 180-210 ms (D), 930-960 ms (E), 1-2 s (F), 5-6 s (G) after mixing. (H) Integrated 0-5 s after mixing and flushing the cell.

4. Water Transport in Lyotropic Mesophases

It was realized at an early stage in the development of the lyotrope gradient method described above that the transport properties of water in a contiguous series of lipid mesomorphs should be accessible. This provided the impetus for the development of a computer algorithm based on a finite-difference method for determining the mutual diffusion for water in the contiguous mesomorphs of the monoolein/water system. The FORTRAN program based on this algorithm searches for diffusion coefficients across a phase region such that calculated phase boundary positions coincide with moving boundary positions determined experimentally by TRXRD. From the derived diffusion coefficients the phase development in time can be predicted. Another practical application is the prediction of the amount of time and lyotrope required to prepare samples in a defined mesomorphic state without mechanical mixing.

Results for the moolein/water system give mutual diffusion coefficients ranging from 0.7-1.5 x 10^{-6} cm^2.s in the fluid isotropic, lamellar liquid crystal and cubic phases [14].

5. Membrane Structure Studies Using X-Ray Standing Waves

Understanding membrane structure and how this relates to the biological function of membrane components represents, today, one of the grand challenges in structural biology research. By knowing how structure impacts on function, we can hope to manipulate structure to our collective advantage. Membrane structure is considered here from the point of view of membrane lipid and protein topology as well as the aqueous region at and next to the membrane surface. Studying such disparate aspects of membrane topology is a non-trivial undertaking. The problem arises primarily for lack of suitable methods of membrane characterization. By suitable, we refer to a method that provides direct structure information with atomic resolution in an element-specific manner on samples that may take the form of an isolated membrane or lipid monolayer and that can be used with intrinsically ordered and disordered systems. In other words, we seek a method that can work with molecularly thin films where the amount of material interrogated in the course of measurement is extremely small. The x-ray standing wave (XSW) technique introduced here is such a method [15].

The XSW technique offers many of the advantages of a surface technique provided we can suitably dispose the membrane to be analyzed on a solid substrate for interrogation. The structure information provided by way of the XSW measurement is the location of a marker atom in the profile structure of the sample. By this we mean the one-dimensional projection of the spatially averaged in-plane structure along the normal to the supporting solid surface. The method offers element-specificity and spatial resolution on an atomic scale. While it does not provide direct in-plane structure information, such can be obtained by using data collected normally in the course of an XSW measurement. Another attractive feature of the method is that it can be applied with equal ease to systems that are inherently ordered and disordered. An example of the former is a two-dimensional crystalline array of bacteriorhodopsin in the purple membrane or a Langmuir-Blodgett (LB) film of calcium stearate. Ion or protein distribution in the aqueous medium bathing the membrane is an example of the latter. In addition to being useful in determining the structure of systems at equilibrium, the method can also be used to advantage in monitoring membrane-related dynamic processes. In this regard, it has found application in studies of membrane lipid phase transitions and of ion movement in membrane systems. Future applications include protein folding, membrane protein insertion, surface potential driven rearrangements in membrane-associated ion, lipid and/or protein distribution, and surface binding.

Herein is a summary of progress made in developing and in applying the XSW approach for studying membrane topology. It chronicles a series of measurements made on relatively simple model systems by way of demonstrating definitively the utility and versatility of this new structural tool. What emerges is an information-rich technique for investigating systems that do not naturally or easily lend themselves to detailed structure investigations.

5.1. PROGRESS TO DATE

5.1.1. *Background*

An XSW can be generated by the interference between two coherently related x-ray beams [15]. Conventional XSW measurements use dynamical Bragg diffraction from perfect single crystals for generating XSW with d-spacing periods ranging from 1 to 4 Å. This technique was first applied to the study of implanted heavy atom layers. It reached a new level of interest when it was realized that the standing wave extended above the surface, and could therefore be used for examining adsorbed surface layers. Such measurements have proven to be very precise in determining where, and how well, heavy atom distributions register with respect to the perfect crystal diffraction planes. This information is obtained by observing the modulation in the impurity atom x-ray fluorescence yield as the standing wave antinodes shift inward by one-half of a d-spacing during a scan in angle through the Bragg reflection. This same approach would also be useful for studying the layered arrangement of atoms in LB films. However, the characteristic modulo-d length scale of a few ångströms, which makes the XSW technique so appropriate for studying bond length distances between atom layers at single crystal surfaces, is too short for measuring the spacing between heavy atom layers in LB films. For this reason, XSW with periods at and above 20 Å to better match the thickness of LB films of biologically relevant materials are needed. In what follows, we show how such long-period XSW generated by Bragg diffraction from layered synthetic microstructures (LSM) or by total external reflection from mirror surfaces can be used.

5.1.2. *A Molecular Yardstick for Model Biomembranes*

LB films have been used extensively as models for biomembranes. In this study, we developed and implemented an approach that uses XSWs for determining the position and width of heavy atom layers in LB films with atomic resolution [16]. Specifically, structural information on an atomic scale was obtained for an LB trilayer system by means of long-period XSWs. The LB trilayer of zinc and cadmium arachidate was deposited on an LSM consisting of 200 tungsten/silicon layer pairs with a 25 Å period. A 30 Å thermally induced inward collapse of the zinc atom layer that was initially located in the LB trilayer at 53 Å above the LSM surface was observed. The mean position and width of the zinc atom layer was determined with a precision of ± 0.3 Å. This study represented a first in terms of using XSW with a probing length scale comparable to the dimensions of the biological membrane. It set the stage for the following series of definitive demonstration experiments.

5.1.3. *Structure Studies of Membranes up to 1,000 Å Thick with Ångström Resolution*

Based on the success of the measurements described above, we speculated regarding the utility of the XSW approach in structure studies of supramolecular aggregates and assemblies such as prevail, for example, in membrane receptor - ligand interactions and in multimembrane stacks. Measurements of this type would require necessarily that the probing XSW be well defined at distances from the standing wave generating surface far greater than was demonstrated previously for such systems. Given the biological import of such structure studies and the enormous potential afforded by the method we set out to establish that it is possible to generate an XSW that is well defined at close to a thousand ångströms above a mirror surface. This length scale was chosen in light of the biological systems likely to be investigated using XSW in the foreseeable future. The samples examined in this study consisted of an octadecyl thiol (ODT) coated gold mirror on top of which a variable number (0, 14) of ω-tricosenoic acid (ωTA, C23) bilayers followed by a single, upper inverted bilayer of zinc arachidate (ZnA, C20) was deposited by the LB technique. The results demonstrated convincingly that *the XSW field is well defined at close to a thousand ångströms above the mirror surface and can be used to establish heavy atom layer mean position and width with ångström resolution* [17]. The quality of the data demonstrates the enormous potential of XSW as structure probes of membranes and of thin film and interface related phenomena.

5.1.4. *Membrane Protein Topology*

The elucidation of the structure and function of membrane proteins represents one of the grand research challenges in structural biology. Based on the demonstrated utility of XSW in studying model membranes, we proposed to use the XSW technique as a means for uncovering key features of membrane protein structure with the protein still membrane bound. The thrust is to determine how membrane proteins such as receptors, transport, light harvesting and signal and energy transducing proteins are arranged in and on their supporting lipid bilayer membrane and, by extension, how they function in such an environment. To address these fundamental issues, it is necessary to prepare a series of lipids and proteins specifically labeled with appropriate 'heavy' atoms. Many membrane proteins contain heavy atoms naturally. In other cases, heavy atom substitutions can be made that do not perturb the structure or the activity of the protein. For example, selenomethionine can be site-selectively substituted for methionine, a sulfur-containing amino acid, using modern semisynthetic chemistry and/or recombinant DNA technology. Our objective was to apply these and other methodologies in establishing the membrane topological relationship of the electron transfer protein, cytochrome c.

The principle upon which the membrane cytochrome c topology measurements are based is as follows. With cytochrome c bound as a monolayer to a flat supported lipid adlayer on an XSW generating surface, the plane containing the heme-iron (or zinc in the case of zinc cytochrome c) serves as an internal reference benchmark plane from and to which distance measurements are made. A series of cytochromes c is prepared each member of which is singly labeled with selenomethionine at a defined site in the protein. Each cytochrome c preparation is in turn adsorbed onto a lipid adlayer deposited on an appropriate LSM or x-ray mirror. By suitably tuning the energy of the incident x-ray beam, the objective is to stimulate (and to record) simultaneously the x-ray fluorescence of both the reference iron (or zinc) and the reporter selenium atoms. Fluorescence and reflectivity data are collected in the appropriate angular range and analyzed to calculate the coherent position and fraction of each heavy atom in the bound layer of cytochrome c. For each cytochrome c, the selenium-to-iron (or zinc) distance is calculated. Such measurements in conjunction with the known 3-D structure of the protein and assuming it does not change, will suggest a unique orientation for the protein at the membrane surface. From the orientation information the site of interaction and thus, the amino acids in contact with the membrane can be identified.

An initial round of XSW and reflectivity measurements on cytochrome c deposited directly on a freshly prepared silver surface and on lipid films has been made. The results of the XSW measurements on a sample where cytochrome c is situated directly on the mirror surface, are shown in Figure 4. The data show that the selenium in the protein is suspended as a monolayer with a mean position of 11.5 ± 2 Å above the mirror surface and with a spread (half-width at half-height, HWHH) of 6 Å [18]. This result suggests that the protein retains its globular shape while adsorbed at the mirror surface. In support of this statement, we have shown independently that the surface density of cytochrome c in this sample is 1,040 ± 10 $Å^2$/molecule. This is based on a comparison of the fluorescence yield values recorded above the critical angle with a similar fluorescence measurement on a reference sample of known selenium surface density. Since the average diameter of cytochrome c is 34 Å, in an hexagonally close-packed arrangement, each

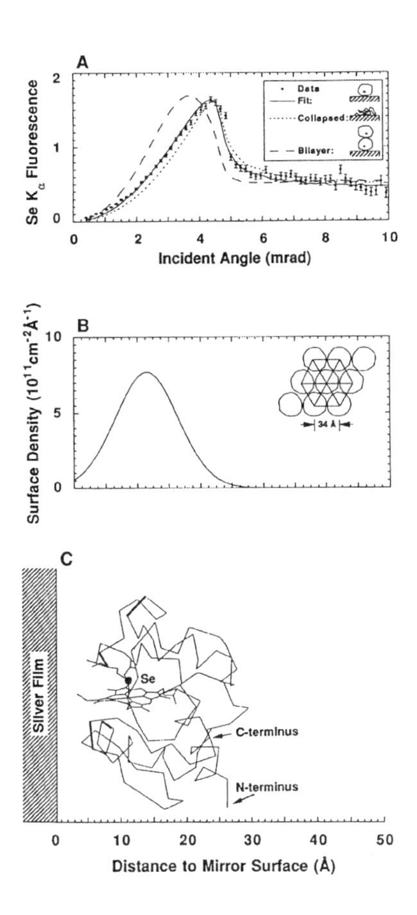

Figure 4. Experimental (circles) and theoretical (lines) angular dependence of the selenium K_α fluorescence yield (**A**) from SeMet^{80} cytochrome c at a silver mirror surface at 13.3 keV. The calculated selenium distribution above the mirror that best fits the fluorescence yield data (solid line in **A**) is shown in **B**. The inset in **B** shows an hexagonally close-packed arrangement of cytochrome c molecules with an average diameter of 34 Å. Separate theoretical fluorescence yield profiles have been generated simulating an unfolded protein with the selenium marker atom at the mirror surface (dotted line in **A**) and a film consisting of a protein bilayer (dashed line in **A**). The α-carbon trace of horse heart cytochrome c is shown (**C**) with the protein oriented in such a way as to position the selenium marker atom (assumed to replace isomorphously the sulfur in the native protein) 11.5 Å above the silver mirror surface. This particular arrangement has the heme plane perpendicular to and the exposed heme edge facing the mirror surface. The α-carbon backbone of several lysines have been emboldened. The N- and C-termini are indicated. Note, that the side chains of the amino acid residues fill the 4 Å space between the mirror surface and the α-carbon backbone of the protein in contact with the surface. This is just one of the possible, but physically reasonable, orientations for cytochrome c at the mirror surface.

cytochrome c molecule occupies 1,000 Å^2. This suggests, therefore, that the protein maintains its globular shape and is hexagonally close-packed at the silver mirror surface. The data are consistent with four positively charged lysine residues, located at positions 13, 27, 72 and 79, forming a ring around the exposed heme edge and a platform, of sorts, for the protein to dock on the metal surface. This model has the heme plane oriented perpendicular to and the exposed heme edge facing the mirror surface.

5.1.5. *Diffuse-Double Layer at the Membrane-Aqueous Interface*
The properties of the interface formed by an electrolyte in contact with a charged surface govern a large number of processes, including electrodeposition, colloidal suspension, and ion transport to and through biological membranes. Due to the separation of charge at such an interface, a gradient in the electrostatic potential arises across the interface which polarizes the solution and affects ion distribution in the interfacial region (Figure 5A). Since the interface resides below a relatively thick aqueous overlayer, high-resolution structure-determining techniques, which rely on charged particle beams or vacuum, cannot be used facilely to profile the ion distribution in the solution immediately above a charged surface. As a consequence, the treatment of this interfacial region has, up until now, been idealized and based on models such as those given by Helmholtz, Gouy-Chapman, and Stern.

X-Rays, which can penetrate through millimeters of water, should, in principle, be ideally suited for solving this long-standing problem in membrane electrostatics and electrochemistry. In the study to be described, we measured directly the ion distribution in an electrolyte solution in contact with a charged polymerized phospholipid membrane by using long-period XSWs. The 27Å thick lipid monolayer was supported on a tungsten/silicon mirror. XSW were generated above the mirror surface by total external reflection of a 9.8 keV x-ray beam from a synchrotron undulator. The membrane surface, which contained negatively charged phosphate headgroups, was bathed in a dilute zinc chloride solution (Figure 5B). The concentration of zinc ions in the condensed layer at the membrane surface and the zinc ion distribution in the diffuse layer were measured as a function of headgroup charge [19]. The Debye length of the diffuse layer varied by an order of magnitude depending on the pH of the solution above the phospholipid monolayer. The results agree qualitatively with the Gouy-Chapman-Stern model which predicts that the charged surface can be partially neutralized by a condensed (or adsorbed) layer of counterions and that the ion distribution in the solution will form a diffuse layer with an exponential decay functional form (Figure 5C). The thermodynamic reversibility of the pH induced changes in the diffuse double layer profile has been demonstrated

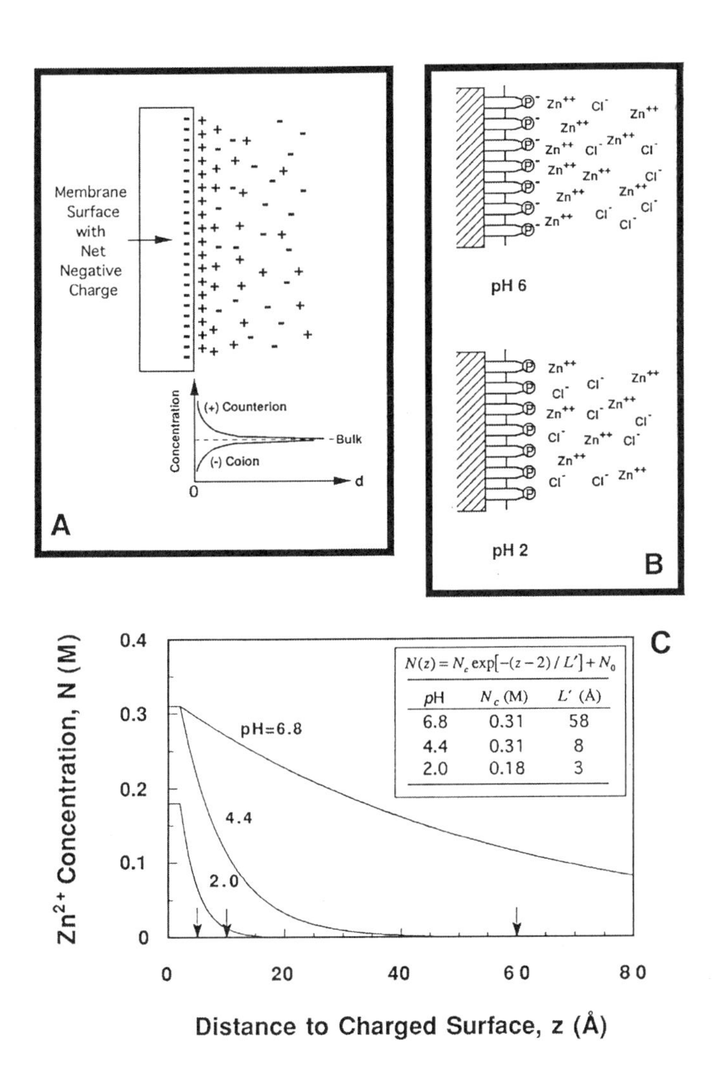

Figure 5. Ion distribution at the film/aqueous interface. A. Schematic representation of the counterion and coion distribution in the aqueous phase next to a surface bearing a net negative charge. B. Schematic of the cross-linked phospholipid membrane deposited on a silanated W/Si LSM. C. Zinc ion distribution in the aqueous phase next to a cross-linked phospholipid membrane at different pH values as measured using XSW generated under specular conditions. The equation describing the exponential decay distribution is shown in the inset where N is the zinc concentration at distance Z from the surface, N_c is the excess surface zinc concentration, L' is the ion Debye length and N_o is the zinc concentration in bulk solution.

also [15, 32].

This study demonstrates how long-period x-ray standing waves can be used to directly measure the ion distribution profile in an electrolyte solution in contact with a charged membrane surface. With this new method in hand and its utility established we have ahead of us almost a century worth of theory to set about testing experimentally.

6. Membrane Protein Crystallization in Lipidic Mesophases

A major impasse on the route that eventually leads to membrane protein structure through to activity and function is found at the crystal production stage. Diffraction quality crystals, with which structure is determined, are particularly difficult to prepare currently when a membrane source is used. The reason for this is our limited ability to manipulate proteins with hydrophobic/amphipathic surfaces that are usually enveloped with membrane lipid. More often than not, the protein gets trapped as an intractable aggregate in its watery course from membrane to crystal. As a result, access to the structure and thus function of tens of thousands of membrane proteins is limited. In contrast, a veritable cornucopia of soluble proteins have offered up their structure and valuable insight into function, reflecting the relative ease with which they are crystallized. There exists therefore an enormous need for new ways of producing crystals of membrane proteins.

The more traditional methods for crystallizing membrane proteins involve solubilization in surfactants and subsequent treatment with 'precipitants' as for soluble proteins. I refer to this as the *in surfo* method. The Rosenbusch group at the Biozentrum in Basel has been a hotbed of activity in this area from the beginning and has contributed many important membrane protein structures through the years [20]. Little wonder then that it was this same group that introduced yet another method for growing crystals of membrane proteins which uses the lipidic mesophase as an incubus [21]. From a phase science point of view, it represents an extension of their previous work.

The cubic mesophase (Figure 1) figures prominently in this, what is sometimes called the *in cubo*, method. We know very little about how and why the method works. Nor indeed do we know the identity of the structure or phase that feeds *directly* the growing crystal surface. Accordingly, I prefer to refer to it as the *in meso* method, where *meso* stands for the more general, non-committal middle or liquid crystal phase. Mesophase and liquid crystal are used synonymously.

The recipe for growing crystals *in meso* follows [21]: I. Combine 2 parts protein solution/dispersion with 3 parts lipid (monooein). II. Mix. The cubic phase forms spontaneously. III. Add precipitant/salt, incubate and crystals form in hours to weeks. IV. All operations are at 20 °C. It is that simple, and it works with soluble and membrane proteins.

6.1. MINING CRYSTALS *IN MESO*

We really know very little about how *in meso* crystallization comes about. However, crystals of a host of materials have been grown from what starts out as a doped cubic phase [23]. Given that we know the initial and end states, it is reasonable to speculate as to how we get from one to the other. If we use bacteriorhodopsin (BR) as an example, the end state is a crystal with proteins arranged in sheets. The starting condition has BR dispersed in an aqueous micellar solution to which residual purple membrane (PM) lipid remains bound. The protein presumably is cummerbunded with solubilizing alkyl glycoside (AG) detergent around its hydrophobic midsection. The dispersion is then combined with lipid (monoolein, MO) by mechanical mixing to produce a mixture with the following approximate overall composition: water/MO/AG/PM lipid/BR = 300,000/15,000/700/10/1 by mole.

Let us now contemplate the fate of the major components in the system up to crystallization. The detergent has high aqueous solubility. But it is also amphipathic with a proclivity for hydrophobic spaces and surfaces. At the ratio of monoolein-to-water used typically (40 %(w/w) water), the cubic phase is stable (A in Figure 6B) [24]. It incorporates a network of cubic membranes (lipid bilayers) which separate two interpenetrating but non-contacting aqueous networks (Figure 1). Both the polar aqueous and the apolar bilayer interior serve as compartments into which the detergent will partition. But the detergent is also drawn to the hydrophobic surface of the protein on which it piggybacked into the mix in the first place (Figure 7a-d).

Typically, the aqueous protein/detergent solution is combined with dry monoolein at 20°C. Upon initial contact, water will migrate from the solution into the monoolein. In so doing, it will establish a water activity gradient along which a series of phases will form. The sequence of phases is the same as found on the 20°C isotherm in the monoolein/water phase diagram (Figure 6B). At low hydration levels, the first liquid crystal phase to form is of the lamellar type (D in Figure 6B). This gives way to the cubic phase with increasing hydration (C in Figure 6B, Figure 7d).

As water leaves the aqueous protein solution for the dry lipid, there is a corresponding increase in the concentration of detergent and protein (etc.) in the residual aqueous solution. This will favor the formation of a lamellar phase by reference to the AG/water phase diagram [25]. At the same time, the slightly soluble monoolein will partition into it, facilitated no doubt by the detergent. In this way, the mixed detergent/protein micelle will acquire monoolein. With reference to the monoolein/AG/water phase diagram [26],

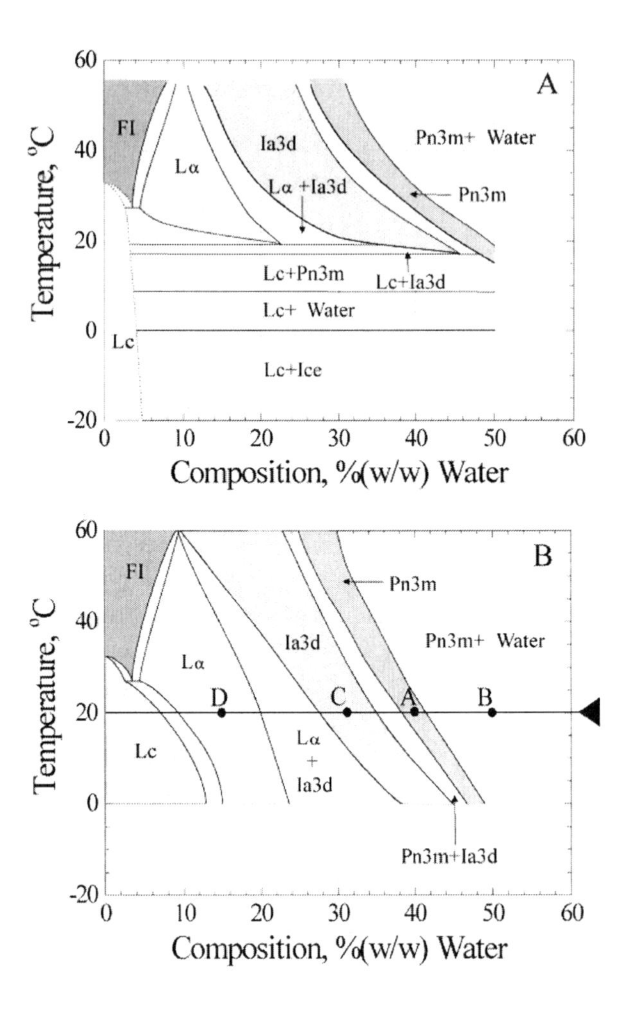

Figure 6. Temperature-composition phase diagram for the monoolein/water system [24]. A. Equilibrium phase diagram. B. Metastable phase diagram. In B, points along the 20°C isotherm identified by letters are referred to in the text.

we see that high concentrations of AG again favor the lamellar phase and it is likely that the protein now finds itself in a monoolein-enriched micelle with strong lamellar tendencies (Figure 7e). Close by, bulk monoolein is giving way to a local lamellar phase and it is possible that the two fuse. This produces a bilayer which is continuous into the bulk lipid and in which the protein is now reconstituted (Figure 7e). Vectoral orientation and oligomerization of the protein within a given layer may be imposed at this stage in the process. As the system approaches equilibrium, defined by the overall composition of the sample, the lamellar phase transforms to the bulk cubic phase. Since the lipid bilayer is continuous throughout, and the protein and detergent can diffuse within it, both are likely to distribute in the cubic phase (Figure 7e,f). Consistent with this is the observation that in the case of BR, the phase adopts a homogeneous purple hue at this stage in the process.

In this state (Figure 7f), the protein has been removed from the potentially hostile environment of a detergent micelle. It is stabilized in a lipid bilayer with physico-chemical properties more akin to the native membrane. We now turn our attention to the protein crystallization process. For nuclei to form and for crystals to grow, requires that the system be perturbed in some way. Not unlike crystallization protocols for soluble proteins, typically, this involves adding salt (Figure 7g) [27]. What is it that salt does to trigger crystal nucleation and growth? We know that the salt will compete with the lipid (and the detergent and the protein) in the cubic phase for available water. This is accompanied on occasion by deliquescence and the creation of a salt-saturated liquid [22]. The water-withdrawing effect of the salt causes the cubic lattice to contract and for bilayer curvature to rise [28]. This may be perturbation enough to herd the proteins into a lipidic corral where they crowd together, associate and eventually organize into a crystal lattice (Figure 7h). All the while, the salt dissolves in the aqueous medium. By shielding charges on the protein, the elevated ionic strength should also facilitate close protein contact, nucleation and crystal growth. The dehydrating effect of the salt, if sufficient in degree, can induce transient and local lamellar phase formation (D in Figure 6B). Whether nucleation and growth takes place directly in/from the cubic phase and without the involvement of any other, possibly local, intermediate structure or phase is not known. This is what I refer to as the postulated portal question.

As crystals grow, the mesophase loses color [21]. This is a relatively slow process happening over a period of days to weeks in the case of the purple colored BR. Presumably, during the course of growth the crystal face is being fed by protein diffusing in the lipid bilayer through the mesophase up to the crystal face to be ratcheted into place. This suggests the existence of a tethering portal structure between the bulk mesophase and the crystal surface.

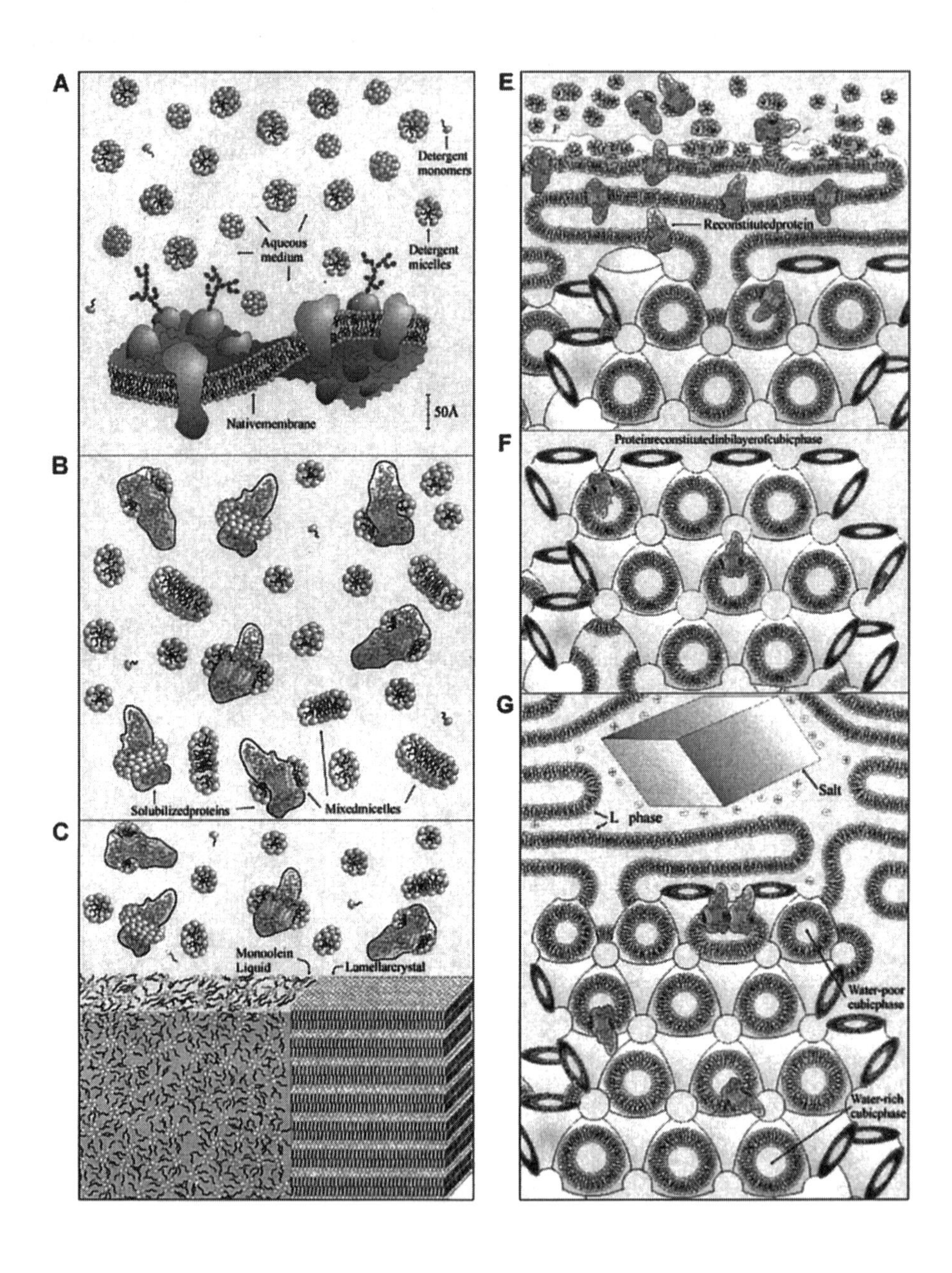
A
Detergent monomers
Aqueous medium
Detergent micelles
Nativemembrane
50Å
B
Solubilizedproteins
Mixedmicelles
C
Monoolein Liquid
Lamellarcrystal
E
Reconstitutedprotein
F
Proteinreconstitutedinbilayerofcubicphase
G
Salt
L phase
Water-poor cubicphase
Water-rich cubicphase

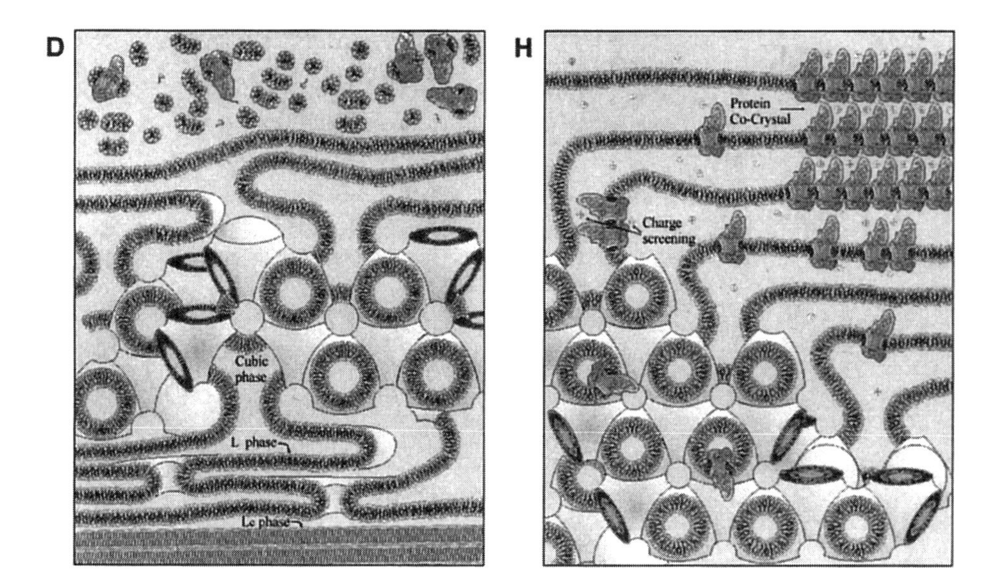

Figure 7. Solubilization, reconstitution and *in meso* crystallization of membrane proteins. Cartoon representation of the events taking place during solubilization (a, b), reconstitution (c-f), and *in meso* crystallization (g, h) of an integral membrane protein as outlined in the text. As much as possible, the dimensions of the monoolein, detergent, native membrane lipid, protein (RCV, PDB 1PRC [31]), bilayer, and aqueous channels have been drawn to scale. a. Isolated biological membrane in the presence of detergent micelle solution. b. Solubilized membrane proteins exist as mixed micelles in solution. c. Combining solubilized protein solution with monoolein in the fluid isotropic and lamellar crystal phases. d. Hydration of monoolein to form contiguous Lα and cubic phases. Limited detergent and monoolein exchange occurs. e. Protein reconstitution and dispersion in mesophases. f. Reconstituted protein in bilayer of cubic phase. g. Addition of precipitant (salt) to initiate crystallization by water-withdrawing and charge screening effects. Bilayer curvature in cubic phase increases as water content drops. h. Reversible crystallization of protein (and bound lipid, in the case of co-crystallization) from cubic phase through lamellar portal.

6.2. ADDITIONAL CONSIDERATIONS

6.2.1. *Salt*

It may be that the initial salt-induced dehydration is what provides the driving force for nucleation and crystal growth. In addition to the lipid and detergent, the protein too must give up some of its bound water. Protein-protein and other types of contacts may be favored to compensate for the lost water. These serve to recruit more proteins from within the mesophase. Further, since the protein is designed to span a bilayer and does not normally encounter regions of high curvature, the natural inclination is for the accreting domain to grow laterally to produce a layered structure (Figure 7h). If the same is happening in adjacent layers, the proteins will establish contacts between layers and in so doing, set up the three-dimensional lattice of the crystal. The stacking of sheets is likely favored by local dehydration since it frees up additional water for sequestration by the salt.

6.2.2. *Detergents*

Non-ionic detergents have found extensive use in solubilizing membrane proteins for subsequent characterization, reconstitution and crystallization studies. They are likely to accompany the protein into the *in meso* crystallization mix. As surfactants, they can wreak havoc on the lipidic mesophase, which we assume is integral to growing crystals. Molecular geometry considerations suggest that the complimentary molecular shapes of the popular AG detergents and monoolein should lead to a destabilization of the highly curved cubic phase in favor of a lamellar type structure in the presence of the detergent [26]. Using x-ray diffraction, we have shown that relatively high concentrations of the AG are tolerated by the cubic phase, which is good news for the *in meso* method. However, further additions bring about a complete transformation from the cubic to the lamellar phase [26]; As noted, this is relevant to the mechanism of crystallization in the presence of detergents.

6.2.3. *Screen Compatibility*

Typically, a precipitant is added to trigger nucleation and growth of membrane protein crystals in the *in meso* method. The commercially available screen solution series are convenient for use in such crystallization trails [27]. However, they contain an array of components, many of which could destroy the cubic phase. We have determined which of the Hampton Screen (50 solutions) and Hampton Screen 2 (48 solutions) series of solutions support cubic phase formation by means of x-ray diffraction [29]. The data show that over 90 % of the screen solutions produced the cubic phase at 20°C. In contrast, about half of the screens were cubic phase compatible at 4°C.

6.2.4. *Equilibrium vs. Metastability*

The liquid crystalline phases are notorious in their capacity to undercool - in the same way that water remains liquid below 0°C when cooled appropriately. This property is responsible for the differences seen below 20°C between the two monoolein/water phase diagrams in Figure 6 where one represents equilibrium (Figure 6A) and the other, metastable (Figure 6B) behavior [24]. In the latter, the metastable cubic and lamellar phases can persist down to 0°C (Figure 6B) for years. What makes long-lived metastability so attractive from the point of view of crystallization *in meso* is the possibility of retaining the cubic phase to low temperatures. The proteins crystallized *in meso* thus far are robust and were crystallized at 20°C. For more labile proteins, it might be desirable to work at lower temperatures. With the monoolein system, it should be possible to do the *in meso* crystallization in the $0-4^{\circ}C$ range where the cubic phase persists under these conditions (Figure 6B).

6.2.5. *In Meso en Masse*

Crystallization trials involve vast numbers of samples. We have described the construction of an inexpensive mixing/delivery device mostly from commercially available parts that lends itself nicely to such high-throughput applications [30]. It was developed for working with milligram quantities of lipid (micrograms of protein), to facilitate the handling of highly viscous cubic mesophases and to have precise control over sample composition. The device is simple to use and is being implemented in *in meso* crystallization trials and other applications in labs worldwide. Large volume syringes can be employed to prepare gram quantities of cubic phase/protein dispersions in a single step. One to three milligram quantities of the dispersion are then placed in small tubes, screen solutions or solid precipitants are added and the samples stored for crystallization. In the hands of an experienced user, hundreds of crystallization trials can be set up daily. With a few obvious and simple modifications, the device can be adapted for miniaturization, automation and robotization for use in high-throughput applications.

6.3. WHY LIPIDIC PHASES SHOULD PRODUCE GOOD CRYSTALS

6.3.1. *Microgravity Analogy*

The mechanism responsible for the reported improvement in crystal and diffraction quality has not been established but much is made of the convection-free environment microgravity affords [27]. In solution, when the depletion zone forms next to the growing crystal on earth, a gradient in protein concentration exists between the crystal surface and the bulk of the solution. The associated density gradient creates a continuous, gravity-driven convective current and the depletion zone is disturbed as the heavier protein-rich layers fall through the lighter layers onto the crystal surface. In a microgravity environment, the depletion zone is stabilized which allows for the slow and orderly addition of protein to the growing crystal.

It is likely that growing crystals *in meso* is akin to growing them in microgravity for the following reasons. Upon nucleation and as the crystals begin to grow, a depletion zone is created in the surrounding mesophase. However, *in meso*, the crystal presumably is tethered to and embedded in a highly viscous medium which will not support convection of the type described above for solution. Further, protein diffusion in the supporting phase is even slower than it is in solution. Thus, a depletion zone is stabilized and the crystal face is fed by slowly diffusing protein from the bulk. This increases the likelihood of growing larger and more ordered crystals. It is likely too that impurities are 'filtered out' by the lipid bilayer, producing a higher grade crystal.

Settling of newly formed crystals onto others in the mix is commonly encountered in solution for earth grown crystals [27]. It produces defects and limits crystal growth. Under conditions of microgravity and *in meso*, for obvious but different reasons, sedimentation related impairments to crystal growth are not an issue.

6.3.2. *Epitaxy.*
Epitaxial nucleation is a process whereby crystal seed formation is facilitated when proteins in solution align themselves on an exposed lattice whose size approaches an integral multiple of the nascent crystal lattice. For protein crystals grown from solution, a variety of materials have served this purpose [27]. The growth of membrane proteins from a mesophase may represent a variation on this theme. Thus, in the case of membrane proteins crystallized *in meso* for which structures are available, the crystal consists of stacked planar sheets of protein. The simple-minded scheme presented above for how the lipidic phase feeds the crystal surface invokes a planar stack of lipid bilayers that extend from the bulk cubic phase to the crystal surface. There may be an epitaxial relationship between the crystal lattice and the conducting lamellar phase.

7. Acknowledgments

I am especially grateful to the members of my research group, past and present, and to colleagues who have contributed in their own unique ways to this review. Work included in this review was supported by NIH (DK36849, DK45295, GM56969, GM61070) and NSF (DIR9016683, DBI9981990).

References

1. Caffrey, M. (1989) The study of lipid phase transition kinetics by time-resolved x-ray diffraction. *Ann. Rev. Biophys. Biophys. Chem.* **18**, 159–186.
2. Caffrey, M. (1989) Structural, mesomorphic and time-resolved studies of biological liquid cryatals and lipid membranes using synchrotorn x-radiation. *Topics Curr. Chem.* **151**, 75–109.
3. Luzzati, V. (1968), in D. Chapman (ed.), *Biological Membranes, Physical Fact and Function*, Academic Press, New York, Vol. 1. pp 71–123.
4. Shipley, G.G. (1973), in D. Chapman and D.F.H. Wallach (eds.), *Biological Membranes*, Academic Press, New York, Vol 2, pp 1–89.
5. Small, O.M. (1987) *Handbook of Lipid Research: The Physical Chemistry of Lipids from Alkanes to Phospholipids* Vol 4. Plenum Press, NY. 672 pp.

6. Caffrey, M. (1986) The influence of water on the phase properties of membrane lipids: Relevance to anhydrobiosis, in A.C. Leopold (ed.), *Membranes, Metabolism and Dry Organisms*. Comstock Publ. Assoc., Ithaca, New York, pp. 242–258.
7. Caffrey, M. (1985), Kinetics and mechanism of the lamellar gel/lamellar liquid crystal and lamellar/inverted hexagonal phase transition in phosphatidylethanolamine: A real-time x-ray diffraction study using synchrotron radiation. *Biochemistry* **24**, 4826–4844.
8. Siegel, D.P., Capel, M. and Caffrey, M. (1999). Using time-resolved x-ray diffraction to search for fusion-related intermediates in an Lα/Hii phase transition., Biophys. J. **76**, A438.
9. Cherezov, V., Cheng, A., Petit, J.M., Diat, O. and Caffrey, M. (2000) Biophysics and synchrotron radiation. Where the marriage fails. X-Ray damage of lipid membranes and mesophases. *Cell. Mol. Biol.* **46**, 1133–1145.
10. Cheng, A-c., Hogan, J.L. and Caffrey, M. (1993) X-Rays destroy the lamellar structure of model membranes. *J. Mol. Biol.* **229**, 291–294.
11. Cheng, A. and Caffrey, M. (1996) Free radical mediated x-ray damage of model membranes. *Biophys. J.* **70**, 2212–2222.
12. Caffrey, M. and Hing, F.S. (1987) A temperature-gradient method for lipid phase diagram construction using time-resolved x-ray diffraction. *Biophys. J.* **51**, 37–46.
13. Caffrey, M. 1989. A lyotrope gradient method for liquid crystal temperature-composition-mesomorph diagram construction using time-resolved x-ray diffraction. *Biophys. J.* **55**, 47–52.
14. Gerritsen, H. and Caffrey, M. (1990) Water transport in lyotropic liquid crystals and lipid-water systems: Mutual diffusion coefficient determination. *J. Phys. Chem.* **94**, 944–948.
15. Caffrey, M. and Wang, J. (1995) Membrane-structure studies using x-ray standing waves. *Annu. Rev. Biophys. Biomol. Struct.* **24**, 351-378.
16. Bedzyk, M., Bilderback, D. H., Bommarito, M., Caffrey, M. and Schildkraut, J. (1988) X-ray standing waves: A molecular yardstick for biological membranes. *Science* **241**, 1788-1791.
17. Wang, J., Bedzyk, M. J., Penner, T. L. and Caffrey, M. (1991) Structural studies of membranes and surface layers up to 1,000 Å thick using x-ray standing waves. *Nature* **354**, 377-380.
18. Wang, J. , Wallace, C., Clark-Lewis, I. and Caffrey, M. (1994) Structure characterization of membrane bound and surface adsrobed protein. *J. Mol. Biol.* **237**, 1-4.
19. Bedzyk, M., Bommarito, M., Caffrey, M. and Penner, T. L. (1990) Diffuse-Double Layer at a Membrane/Aqueous Interface Measured with X-ray Standing Waves. *Science* **248**, 52-56.
20. Rosenbusch, J.P., Lustig, A., Grabo, M., Zulauf, M. and Regenass, M. (2001) Approaches to determining membrane protein structures to high resolution: do selections of subpopulations occur? *Micron* **32**, 75-90.
21. Landau, E.M. and Rosenbusch, J..P. (1996) Lipidic cubic phases: a novel concept for the crystallization of membrane proteins. *Proc Natl Acad Sci USA* **93**, 14532-14535.
22. Caffrey, M. (2000) A lipid's eye view of membrane protein crystallization in mesophases. *Curr. Opin. Struct. Biol.* **10**, 486-497.
23. Landau, E.M., Rummel, G., Cowan-Jacob, S.W. and Rosenbusch, J..P. (1997) Crystallization of a polar protein and small molecules from the aqueous compartment of lipidic cubic phases. *J Phys Chem B* **101**, 1935-1937.

24. Qiu, H. and Caffrey, M. (2000) Phase diagram of the monoolein/water system: Metastability and equilibrium aspects. *Biomaterials* **21**, 223-234.
25. Warr, G, Drummond, C. and Greiser, F. (1986) Aqueous solution properties of nonionic n-dodecyl □-D-maltoside micelles. *J Phys Chem* **90**, 4581-4586.
26. Ai, X. and Caffrey, M. (2000) Membrane protein crystallization in lipidic mesophases. Detergent effects. *Biophys. J.* **79**, 394-405.
27. McPherson, A. (1999) *Crystallization of Biological Macromolecules.* CSHL Press.
28. Chung, H. and Caffrey, M. (1994) The curvature elastic energy function of the cubic mesophase. *Nature* **368**, 224-226.
29. Cherezov, V., Fersi, H. and Caffrey, M. (2001) Crystallization screens. Compatibility with the lipidic cubic phase for *in meso* crystallization of membrane proteins. *Biophys. J.* **81**, 225-242.
30. Cheng, A., Hummel, B., Qiu H. and Caffrey, M. (1998) A simple mechanical mixer for small viscous lipid-containing samples. *Chem Phys Lipids* **95**, 11-21.
31. Deisenhofer, J., Epp, O., Sinning, I. and Michel, H. (1995) Crystallographic refinement at 2.3 Å resolution and refined model of the photosynthetic reaction centre from Rhodopseudomonas viridis. *J Mol Biol* **246**, 429.
32. Wang, J., Caffrey, M., Badzyk, M., Penner, T. (2001) Direct profiling and reversibility of ion distribution at a charged membrane/aqueous interface: An X-ray standing wave study. *Langmuir* **17**, 3671-3681.

CELL MODEL AND POISSON-BOLTZMANN THEORY: A BRIEF INTRODUCTION

MARKUS DESERNO[1], CHRISTIAN HOLM[2]
[1] *Department of Chemistry and Biochemistry, UCLA, USA*
[2] *Max-Planck-Institut für Polymerforschung, Mainz, Germany*

The cell-model and its treatment on the Poisson-Boltzmann level are two important concepts in the theoretical description of charged macromolecules. In this brief contribution we provide an introduction to both ideas and summarize a few important results which can be obtained from them. Our article is organized as follows: Section 1 outlines the sequence of approximations which ultimately lead to the cell-model. Section 2 is devoted to two exact results, namely, an expression for the osmotic pressure and a formula for the ion density at the surface of the macromolecule, known as the contact value theorem. Section 3 provides a derivation of the Poisson-Boltzmann equation from a variational principle and the assumption of a product state. Section 4 applies Poisson-Boltzmann theory to the cell-model of linear polyelectrolytes. In particular, the behavior of the exact solution in the limit of zero density is compared to the concept of Manning condensation. Finally, Section 5 shows how a system described by a cell model can be coupled to a salt reservoir, *i.e.*, how the so-called Donnan equilibrium is established.

Our main motivation is to compile in a concise form a few of the basic concepts which form the arena for more advanced theories, treated in other lectures of this volume. Many of the basic concepts are discussed at greater length in a review article by Katchalsky [1], which we warmly recommend.

1. The cell model

1.1. THE NEED FOR APPROXIMATIONS

Solutions of charged macromolecules are tremendously complicated physical systems, and their theoretical treatment from an "ab initio" point of view is surely out of question. The standard solvent itself – water – already poses formidable problems. Adding the solute requires additional

C. Holm et al. (eds.), Electrostatic Effects in Soft Matter and Biophysics, 27–52.
© 2001 *Kluwer Academic Publishers. Printed in the Netherlands.*

understanding of the solvent-solute interaction, the degree of dissociation of counterions, the conformation of the macroions, its intricate coupling with the distribution of the counterions and many further complications. How can one ever hope to achieve even some qualitative predictions about such systems?

Many of the interesting features of polyelectrolytes are ultimately a consequence of the presence of charges. One may thus hope that a theoretical description focusing entirely on a good treatment of the electrostatics and using crude approximations for essentially all other problems will unveil why these systems behave the way they do. In a first important step any quantum mechanical effects are ignored by using a classical description. A second simplification is to treat the solvent as a dielectric continuum and consider explicitly only the objects having a monopole moment. This "dielectric approximation" is motivated by the long-range nature of Coulomb's law and works surprisingly well [2]. Since a classical system of point charges having both signs is unstable against collapse, a short-range repulsive interaction is required, which is most commonly modeled as a hard core. For a simple electrolyte this approximation is called the "restricted primitive model".

In 1923 Debye and Hückel studied such a system using the linearized Poisson-Boltzmann theory [3]. Their treatment accounted for the fact that ions tend to surround themselves by ions of opposite charge, which reduces the electric field of the central ion when viewed from a distance. While the exponentially screened Coulomb potential is one of the most prominent results, it must be noted that the authors computed the free energy of the electrolyte. Its electrostatic contribution scales as the $3/2$ power of the salt density, which explains why a virial expansion must fail. A good textbook account is given in Ref. [4].

Though approximate, Debye-Hückel-theory works very well for 1:1 electrolytes. However, perceptible deviations are already much larger in the 1:2 case. This is not merely a consequence of the presence of multivalent ions, namely that the increased strength of electrostatic interaction may correlate ions more strongly. Rather, the *asymmetry* of the situation itself is a key source of the problem – see Kjellander's lecture for more details [5]. It is for this very reason that in the highly asymmetric case of a charged macromolecule surrounded by small counterions the standard Debye-Hückel theory cannot be applied.

1.2. DECOUPLING THE MACROIONS

The cell model is an attempt to turn this situation into an advantage: If the situation is highly asymmetric, there is no reason to pursue a symmetric treatment. Since the macroions all have the same charge, their mutual

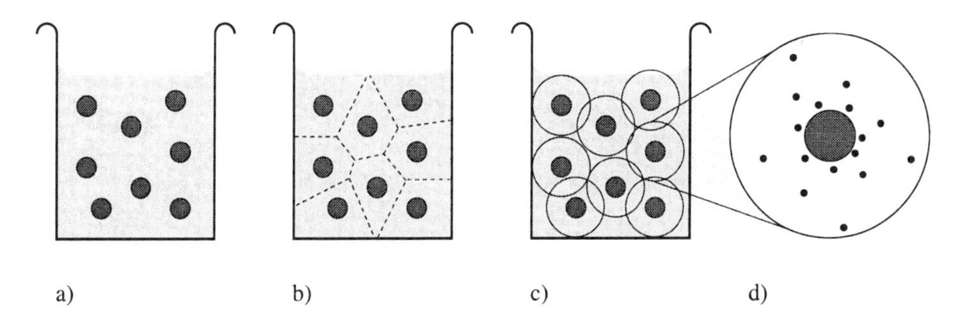

Figure 1. Approximation stages of the cell model. The full solution (a) is partitioned into cells (b), which are conveniently symmetrized (c). Subsequently the attention is restricted to just one such cell (d) and the counterion distribution within it.

pair interaction is repulsive. Unless there are effects which cause them to attract, aggregate and ultimately fall out of solution, *i.e.*, unless the effective pair potential is no longer repulsive, the macroions will organize so as to keep themselves as far apart as possible. The total solution can now be partitioned into cells, each containing one macroion, the right amount of counterions to render the cell neutral, and possibly salt molecules as well. As a consequence of the assumed homogeneous distribution of macroions, these cells will all have essentially the same volume, equal to the total volume divided by the number of macroions. Observe that different cells do not have strong electrostatic interactions, since they are neutral by construction.

The cell model approximation consists in restricting the theoretical description of the total system to just one cell. While the interactions between the small ions with "their" macroion as well as with small ions in the same cell are explicitly taken into account, all interactions across the cell boundary are neglected. Note that the existence of cells, all of which have essentially the same size, requires correlations to be present between the macroions. However, these correlations are no longer the subject of study. The cell model can thus be viewed as an approximate attempt to factorize the partition function in the macroion coordinates, *i.e.*, replacing the many polyelectrolyte problem by a one polyelectrolyte problem. Figure 1 (a) $\rightarrow$ (d) illustrates this process for the spherical case.

The remaining effect of all other macroions is to determine the *volume* of the cell, but so far nothing has been said about its *shape*. It is conveniently chosen so as to simplify further progress. For instance, in computer simulations the cell could be identified with the replicating unit of periodic boundary conditions. This requires space-filling cells, for instance, cubes. In an analytical treatment one usually tries to maximize the symmetry of the problem. Hence, spherical colloids are centered in spherical cells, see Fig. 1 (c). The main advantage of this strategy is that a density-functional approach neglecting symmetry-breaking fluctuations becomes a one-dimensional problem. Linear polyelectrolytes are enclosed in cylindrical cells, but it is more difficult to give a precise meaning to these cylinders – they may not even exist in the solution. For instance, even if the polyelectrolyte is very stiff and thus locally straight (like, *e.g.*, DNA), the whole molecule can be fairly coiled on larger scales. In this case one may look at the cylindrical tube enclosing the molecule and simply neglect the fact that it is bent on scales larger than the persistence length of the polyelectrolyte, provided that the tube diameter is small compared to the latter. We thereby pretend that a locally rod-like object is also globally straight. This point of view is justified as long as the observables that we set out to calculate are dominated by local, effectively short-ranged, interactions—as is the case for ion profiles close to the macroion or the osmotic pressure of the counterions. Care must be exercised for observables that may depend on the actual global shape of the macroion, like for instance the viscosity of the solution.

In the cylindrical case a very common further approximation is to neglect end-effects at the cylinder caps by assuming the cylinders to be infinitely long. This additional approximation can of course only be good if the actual finite cylinders are much longer than they are wide. Obviously, the aim of all these approximations is to capture the dominant effect of a locally cylindrical electrical field that an elongated charged object generates.

2. Some exact results

Although the partition function for systems with interacting degrees of freedom can only be evaluated in very special cases, it is frequently possible to derive rigorous relations which the exact solution has to satisfy. For restricted primitive electrolytes there exist *e.g.* the Stillinger-Lovett moment conditions [6], which pose restrictions on the integral over the ion-ion correlation functions and its second moment, or extensions of these conditions to non-uniform electrolytes [7]. Such results are of great theoretical interest, since they can be used as a consistency test for approximate theories. They can also be used to check simulations and may provide very direct ways for analyzing them.

In this section we give the derivation of two exact results which are particularly relevant for the cell-model. The first is an exact expression for the osmotic pressure in terms of the particle density at the cell boundary. The second is known as the "contact value theorem" and provides a relation between the osmotic pressure, the particle density, and the surface charge density at the point of contact between the macroion and the electrolyte.

It has first been derived by Henderson and Blum [8] within the framework of integral equation theories and later by Henderson *et. al* [9] using more general statistical mechanical arguments.

Wennerström *et. al* [10] give a transparent proof of both results based on the fact that derivatives of the free energy with respect to the cell boundaries can be expressed in terms of simple observables. Their argument goes as follows: Let A_R denote the area of the outer cell boundary and R its position, such that infinitesimal changes $\mathrm{d}R$ change the cell volume by $A_R \mathrm{d}R$. The free energy is given by $F = -k_\mathrm{B}T \ln Z$, where $Z = \mathrm{Tr}\,\{\mathrm{e}^{-\beta H}\}$ is the canonical partition function, H is the Hamiltonian, $\beta \equiv 1/k_\mathrm{B}T$, and $\mathrm{Tr}\,\{\cdot\}$ is the integral ("trace") over phase space. The pressure is then given by

$$P = -\frac{\partial F}{\partial V} = \frac{k_\mathrm{B}T}{A_R Z}\frac{\partial Z}{\partial R}. \tag{1}$$

It is easy to see that the energy of the system is independent of the location of the outer cell boundary, since it is hard and carries no charge. Hence, R enters the partition function only via the upper boundaries in the configuration integrals, and the derivative of Z with respect to R can be transformed according to

$$\begin{aligned}
\frac{\partial Z}{\partial R} &= \frac{\partial}{\partial R}\,\mathrm{Tr}\,\exp\Big\{-\beta H(\boldsymbol{r}_1,\ldots,\boldsymbol{r}_N)\Big\} \\
&= A_R \sum_{i=1}^{N} \mathrm{Tr}_{\mathrm{not}\ \boldsymbol{r}_i} \exp\Big\{-\beta H(\boldsymbol{r}_1,\ldots,\boldsymbol{r}_i \to R,\ldots,\boldsymbol{r}_N)\Big\} \\
&= A_R\, N\, \mathrm{Tr}_{\mathrm{not}\ \boldsymbol{r}_1} \exp\Big\{-\beta H(\boldsymbol{r}_1 \to R,\boldsymbol{r}_2,\ldots,\boldsymbol{r}_N)\Big\} \\
&= A_R\, Z\, n(R).
\end{aligned} \tag{2}$$

The trace over phase space is an N-fold volume integral over all particle coordinates, each containing a radial integration from r_0 to R. As a consequence of the product rule, the derivative with respect to R is the sum of N terms, in which the integral from r_0 to R over the radial coordinate of particle i is differentiated with respect to R, *i. e.*, the integration is omitted and the radial coordinate is set to R. The integration over the two remaining coordinates now yield a prefactor A_R. Since all particles are identical, these N terms are all equal. In the last step we used the fact that the trace over all particles but the first one is equal to Z times the probability distribution of the first particle, so a multiplication by N gives the ion density $n(R)$ at the cell boundary. Combining Eqns. (1) and (2) we obtain the pressure:

$$\beta P = n(R). \tag{3}$$

In words: The osmotic pressure in the cell-model is exactly given by $k_{\rm B}T$ times the particle density at the outer cell boundary. If more than one species of particles are present, $n(R)$ is replaced by the sum $\sum_i n_i(R)$ over the boundary densities of these species. Note that despite its "suggestive" form Eqn. (3) does by no means state that the particles at the outer cell boundary behave like an ideal gas. Even if the system is dense and the particles are strongly correlated, Eqn. (3) is valid, since it is completely independent of the pair interactions entering H.

In a similar fashion one can compute the derivative of the free energy with respect to the inner cell boundary, *i.e.*, the location of the surface of the macroion. In this case the geometry enters the problem, since in the non-planar case a change in the location of this surface necessarily also changes the surface charge density, if the total charge is to remain the same. For simplicity we will restrict ourselves to the planar case here and defer the reader to Ref. [10] for the other geometries.

Let A_{r_0} be the area of the inner cell boundary and r_0 the position of the surface of the macroion, such that an infinitesimal positive change $\mathrm{d}r_0$ makes the macroion larger, but reduces the volume available for the counterions by $A_{r_0}\mathrm{d}r_0$. The pressure is thus given by

$$P = -\frac{\partial F}{\partial V} = -\frac{k_{\rm B}T}{A_{r_0}Z}\frac{\partial Z}{\partial r_0}. \tag{4}$$

The key difference with the previous case is that the energy of an ion also depends on the location of the wall, since the latter is charged. Hence, our calculation leading to Eqn. (2) must be supplemented by an additional term $\mathrm{Tr}\big[\frac{\partial}{\partial r_0}\mathrm{e}^{-\beta H}\big]$, which leads to the expression

$$\frac{\partial Z}{\partial r_0} = -A_{r_0}\,Z\,n(r_0) - \frac{Z}{k_{\rm B}T}\left\langle\frac{\partial H}{\partial r_0}\right\rangle. \tag{5}$$

It is easy to see that $\partial H/\partial r_0 = -2\pi\ell_{\rm B}\tilde{\sigma}^2 A_{r_0}$, independent of the ion coordinates. Here, $\ell_{\rm B} = \beta e^2/4\pi\varepsilon_0\varepsilon_{\rm r}$ is the Bjerrum length, *i.e.*, the distance at which two unit charges have interaction energy $k_{\rm B}T$, and $\tilde{\sigma}$ is the number density of surface charges. Combining this with Eqns. (4) and (5) finally gives

$$\beta P = n(r_0) - 2\pi\ell_{\rm B}\tilde{\sigma}^2. \tag{6}$$

This equation is known as the contact value theorem, since it gives the contact density at a planar charged wall as a function of its surface charge density and the osmotic pressure. The occurrence of the second term is related to the presence of an electric field, which vanishes at the outer cell boundary and which contributes its share to the total pressure via the

Maxwell stress tensor [11]. Observe finally that by subtracting Eqns. (3) and (6) we obtain a relation between the ion density at the inner and outer cell boundary. Taking into account different ion species, it reads

$$\sum_i n_i(r_0) - \sum_i n_i(R) \;=\; 2\pi\ell_{\rm B}\tilde{\sigma}^2. \tag{7}$$

This is a rigorous version of an equation which has been derived on the level of Poisson-Boltzmann theory by Grahame [12]. Note that since the densities $n_i(R)$ are bounded below by 0, the contact density is at least $2\pi\ell_{\rm B}\tilde{\sigma}^2$.

We would like to emphasize that Eqns. (6) and (7) only apply to the planar case. For a cylindrical or spherical geometry the contact density for the same values of $\tilde{\sigma}$ and P is lower [10]. We will briefly return to this point in Sec. 4.4.

3. Poisson-Boltzmann theory

What makes the computation of the partition function so extremely difficult? It is the fact that all ions interact with each other, implying that their positions are mutually correlated. Stated differently, the many-particle probability distribution does not factorize into single-particle distributions, and hence the partition function does not factorize in the ion coordinates. Poisson-Boltzmann theory is the mean-field route to circumventing this problem. Its following derivation demonstrates this point in a particularly clear way. We largely follow the lines of Ref. [13, Ch. 4.8].

Quite generally, the free energy F can be bounded from above by [14]:

$$F \;\leq\; \langle H\rangle_0 - T\,S_0, \tag{8}$$

where $\langle H\rangle_0 \;=\; \mathrm{Tr}\,\{p_0 H\}$ is the expectation value of the energy in some arbitrary state existing with probability p_0 and $S_0 \;=\; -k_{\rm B}\,\mathrm{Tr}\,\{p_0 \ln p_0\}$ is the entropy of that state. This relation is sometimes referred to as the Gibbs-Bogoliubov-inequality and provides a general and powerful way of deriving mean-field theories from a variational principle [13]. Its equality version holds if and only if p_0 is the canonical probability $\mathrm{e}^{-\beta H}/\,\mathrm{Tr}\,\{\mathrm{e}^{-\beta H}\}$.

Assume we have a system of N point-particles of charge ze and mass m within a volume V, and additionally some fixed charge density $en_{\rm f}(\boldsymbol{r})$. The Hamiltonian H is the sum of the kinetic energy $K(\boldsymbol{p}_1,\ldots,\boldsymbol{p}_N)$ and the potential energy $U(\boldsymbol{r}_1,\ldots,\boldsymbol{r}_N)$, and up to an irrelevant additive constant it is given by

$$\begin{aligned} H &= \sum_{i=1}^{N} \frac{\boldsymbol{p}_i^2}{2m} + \sum_{i<j=1}^{N} \frac{z^2 e^2}{4\pi\varepsilon_0\varepsilon_{\mathrm{r}}|\boldsymbol{r}_i - \boldsymbol{r}_j|} + \int_V \mathrm{d}^3 r \sum_{i=1}^{N} \frac{z n_{\mathrm{f}}(\boldsymbol{r}) e^2}{4\pi\varepsilon_0\varepsilon_{\mathrm{r}}|\boldsymbol{r}_i - \boldsymbol{r}|} \\ &= \sum_{i=1}^{N} \frac{\boldsymbol{p}_i^2}{2m} + ze \sum_{i=1}^{N} \Big(\frac{1}{2}\,\psi(\boldsymbol{r}_i) + \psi_{\mathrm{f}}(\boldsymbol{r}_i) \Big), \end{aligned} \tag{9}$$

where ψ and ψ_{f} are the electrostatic potentials originating from the ions and from the fixed charge density, respectively. In a classical description position and momentum are commuting observables, so the momentum part of the canonical partition function factorizes out. Since this is just a product of N identical Gaussian integrals, its contribution to the free energy is readily found to be

$$\beta F_{\boldsymbol{p}} = -\ln \mathrm{Tr}_{\boldsymbol{p}} \left[\mathrm{e}^{-\beta K} \right] = \ln(N!\, \lambda_{\mathrm{T}}^{3N}) \simeq N \Big[\ln(N\lambda_{\mathrm{T}}^3) - 1 \Big] \tag{10}$$

where $\lambda_{\mathrm{T}} = h/\sqrt{2\pi m k_{\mathrm{B}} T}$ is the thermal de Broglie wavelength, and where Stirling's approximation $\ln N! \simeq N \ln N - N$ has been used in the last step.

The complication comes from the $\boldsymbol{r}$-part of the partition function, specifically from the fact that the couplings between the positions $\boldsymbol{r}_i$ appearing in U render the N-particle distribution function $p_N(\boldsymbol{r}_1, \dots, \boldsymbol{r}_N) \equiv \mathrm{e}^{-\beta U} / \mathrm{Tr}_{\boldsymbol{r}} \left[\mathrm{e}^{-\beta U} \right]$ essentially intractable. The purpose of any mean-field approximation is to remove these correlations between the particles. One way of achieving this goal is by replacing the N-particle distribution function by a product of N identical one-particle distribution functions:

$$p_N(\boldsymbol{r}_1, \dots, \boldsymbol{r}_N) \stackrel{\text{mean-field}}{\longrightarrow} p_1(\boldsymbol{r}_1)\, p_1(\boldsymbol{r}_2) \cdots p_1(\boldsymbol{r}_N). \tag{11}$$

Of course, this *product state* is different from the canonical state, but if used as a trial state in the Gibbs-Bogoliubov-inequality (8) it yields an upper bound for the free energy. The electrostatic contribution $\langle U \rangle_0$ is then given by

$$\begin{aligned} \langle U \rangle_0 &= \int_V \mathrm{d}^3 r_1 \cdots \int_V \mathrm{d}^3 r_N \, p_1(\boldsymbol{r}_1) \cdots p_1(\boldsymbol{r}_N)\, ze \sum_{i=1}^{N} \Big(\frac{1}{2}\,\psi(\boldsymbol{r}_i) + \psi_{\mathrm{f}}(\boldsymbol{r}_i) \Big) \\ &= N \int_V \mathrm{d}^3 r \, p_1(\boldsymbol{r})\, ze \Big(\frac{1}{2}\,\psi(\boldsymbol{r}) + \psi_{\mathrm{f}}(\boldsymbol{r}) \Big). \end{aligned} \tag{12}$$

Similarly, the entropy S_0 is found to be

$$\begin{aligned} S_0 &= -k_{\mathrm{B}} \int_V \mathrm{d}^3 r_1 \cdots \int_V \mathrm{d}^3 r_N \, p_1(\boldsymbol{r}_1) \cdots p_1(\boldsymbol{r}_N) \, \ln\left(p_1(\boldsymbol{r}_1) \cdots p_1(\boldsymbol{r}_N) \right) \\ &= -N k_{\mathrm{B}} \int_V \mathrm{d}^3 r \, p_1(\boldsymbol{r}) \ln\left(p_1(\boldsymbol{r}) \right). \end{aligned} \tag{13}$$

Notice the key effect of the factorization assumption (11): It reduces an N-dimensional integral to N identical one-dimensional integrals, thereby making the problem tractable. Observe also that by definition the one-particle distribution function is proportional to the density:

$$p_1(\boldsymbol{r}) \;\equiv\; \frac{n(\boldsymbol{r})}{\int_V \mathrm{d}^3r\; n(\boldsymbol{r})} \;=\; \frac{n(\boldsymbol{r})}{N} \tag{14}$$

Combining Eqns. (8), (10), (12), (13) and (14), we arrive at the following bound for the free energy:

$$F \;\leq\; F_{\mathrm{PB}}[n(\boldsymbol{r})], \tag{15}$$

where the Poisson-Boltzmann density functional is given by

$$F_{\mathrm{PB}}[n(\boldsymbol{r})] \;=\; \int_V \mathrm{d}^3r \,\Big\{ zen(\boldsymbol{r})\Big[\frac{1}{2}\,\psi(\boldsymbol{r}) + \psi_{\mathrm{f}}(\boldsymbol{r})\Big] \\ +\; k_{\mathrm{B}}T\, n(\boldsymbol{r})\,\Big[\ln\big(n(\boldsymbol{r})\lambda_{\mathrm{T}}^3\big) - 1\Big]\Big\}. \tag{16}$$

Clearly, we aim for the best – *i.e.* lowest – upper bound; we therefore want to know which density $n(\boldsymbol{r})$ minimizes this functional. Hence, the mean-field approach has led us to the variational problem of *minimization of a density functional.* Setting the functional derivative $\delta F_{\mathrm{PB}}[n]/\delta n$ to zero and requiring that (*i*) the charge density and the electrostatic potential be related by Poisson's equation (see below) and (*ii*) the total number of particles be N leads finally to the Poisson-Boltzmann equation.[1]

In order to fulfill the first constraint, we add a term $\mu_0\,(n(\boldsymbol{r}) - N/V)$ to the integrand in Eqn. (16), where μ_0 is a Lagrange multiplier. The second constraint is automatically satisfied if we rewrite $\psi(\boldsymbol{r})$ in terms of $n(\boldsymbol{r})$. The functional derivative then gives:

$$0 \;=\; \frac{\delta F[n(\boldsymbol{r})]}{\delta n(\boldsymbol{r})} \;=\; \mu(\boldsymbol{r}) \;=\; \mu_0 \,+\, ze\,\psi_{\mathrm{tot}}(\boldsymbol{r}) \,+\, k_{\mathrm{B}}T\,\ln\big(n(\boldsymbol{r})\lambda_{\mathrm{T}}^3\big), \tag{17}$$

where $\psi_{\mathrm{tot}} = \psi + \psi_{\mathrm{f}}$ is the total electrostatic potential. We could have "guessed" this equation right away from a close inspection of the free energy density in Eqn. (16), which consists of only two simple terms: The first is the electrostatic energy of a charge distribution $en(\boldsymbol{r})$ in the potential created by itself and by an additional external potential $\psi_{\mathrm{f}}(\boldsymbol{r})$; the second is the entropy of an ideal gas with density $n(\boldsymbol{r})$. Stated differently, apart

[1] Note the mathematical analogy in classical mechanics, where the Euler-Lagrange differential equations correspond to Hamilton's variational principle of a stationary action functional.

from the fact that the particles are charged, we are effectively dealing with an ideal gas. In fact, the right hand side of Eqn. (17) is just the (local) electrochemical potential of a system of charged particles in the "ideal gas approximation", *i.e.*, assuming that the activity coefficient is equal to 1.

The condition $\mu(\boldsymbol{r}) = 0$ can be written in a more familiar way:

$$n(\boldsymbol{r}) \;=\; \lambda_{\mathrm{T}}^{-3} \mathrm{e}^{-\beta(ze\,\psi_{\mathrm{tot}}(\boldsymbol{r})+\mu_0)} \;=\; n_0\,\mathrm{e}^{-\beta ze\,\psi_{\mathrm{tot}}(\boldsymbol{r})} \tag{18}$$

where μ_0 or n_0 are fixed by the equation $\int \mathrm{d}^3r\; n(\boldsymbol{r}) \;=\; N$, *i.e.*, by the requirement of particle conservation. Note that n_0 is the particle density at a point where $\psi_{\mathrm{tot}} = 0$. Eqn. (18) states that the ionic density is locally proportional to the Boltzmann factor. Although this appears to be a very natural equation, which is taken for granted in most "derivations" of the Poisson-Boltzmann equation, the above derivation shows it to be the result of a mean-field treatment of the partition function.

Combining Eqn. (18) with Poisson's equation $\Delta\psi_{\mathrm{tot}}(\boldsymbol{r}) = -e(zn(\boldsymbol{r}) + n_{\mathrm{f}}(\boldsymbol{r}))/\varepsilon_0\varepsilon_{\mathrm{r}}$ yields the Poisson-Boltzmann equation

$$\Delta\psi_{\mathrm{tot}}(\boldsymbol{r}) \;=\; -\frac{e}{\varepsilon_0\varepsilon_{\mathrm{r}}}\Big[\, zn_0\,\mathrm{e}^{-\beta ze\,\psi_{\mathrm{tot}}(\boldsymbol{r})} \;+\; n_{\mathrm{f}}(\boldsymbol{r})\,\Big]. \tag{19}$$

The standard situation is that the counterions are localized within some region of space, outside of which there is a fixed charge distribution. In this case one solves Eqn. (19) within the inner region, where $n_{\mathrm{f}} \equiv 0$, and incorporates the effects of the outer charges as a boundary condition to the differential equation.

We would finally like to show that Eqn. (3), the rigorous expression for the osmotic pressure of the cell-model in terms of the boundary density, is also valid on the level of Poisson-Boltzmann theory. We therefore have to compute the derivative of the Poisson-Boltzmann free energy with respect to the volume. Since we again assume the outer cell boundary to be hard and neutral, the energy of the system (in particular: $\psi_{\mathrm{f}}(\boldsymbol{r})$) does not explicitly depend on its location R, which hence only enters the boundaries in the volume integration. But changing the volume of the cell could entail a redistribution of the ions, which also may change the free energy. Let us symbolically express this in the following way:

$$\delta F \;=\; \frac{\partial F}{\partial V}\,\delta V \;+\; \int \mathrm{d}^3r\,\frac{\delta F}{\delta n(\boldsymbol{r})}\,\delta n(\boldsymbol{r}). \tag{20}$$

However, since the Poisson-Boltzmann profile renders the functional F stationary, the second contribution vanishes. The pressure is thus given by

$$P \;=\; -\frac{1}{A_R}\frac{\partial F}{\partial R}\bigg|_{\text{PB-profile}}, \tag{21}$$

i. e., the free energy functional is differentiated with respect to the outer cell boundary and evaluated with the Poisson-Boltzmann profile. The quantity A_R is the area of the outer boundary, as introduced in Sec. 2. If one again rewrites $\psi(\boldsymbol{r})$ in terms of $n(\boldsymbol{r})$, the derivative is readily found to be

$$\frac{1}{A_R}\frac{\partial F}{\partial R} = zen(R)\,\psi_{\text{tot}}(R) + k_{\text{B}}T\,n(R)\left[\ln\left(n(R)\lambda_{\text{T}}^3\right) - 1\right] + \mu_0\,n(R). \quad (22)$$

Together with the Poisson-Boltzmann profile from Eqn. (18) and Eqn. (21) this finally yields $P = n(R)k_{\text{B}}T$, as we had set out to show.

The first derivation of this result was given by Marcus [15]. Its intuitive interpretation is as follows: The Poisson-Boltzmann free energy functional describes a system of charged particles in the ideal gas approximation. Since the pressure is constant, we may evaluate it everywhere, *e. g.* at the outer cell boundary, where the electric field vanishes. The latter implies that the rod exerts no force on the ions sitting there, so the only remaining contribution to their pressure is the ideal gas equation of state evaluated at the local density $n(R)$.

It is by no means trivial that Poisson-Boltzmann theory gives the same relation between boundary density and pressure. Rather, it is one of its pleasant features that it retains this exact result. This does of course not mean that Poisson-Boltzmann theory gives the correct osmotic pressure, since its prediction of the boundary density is not correct.

The Poisson-Boltzmann equation has been and remains extremely important as a mean-field approach to charged systems. The above derivation shows how it neglects all correlations (see Eqn. (11)) and that their incorporation will decrease the free energy (see Eqn. (16)). In the lecture notes of Moreira and Netz [16] a different derivation is presented, which shows the Poisson-Boltzmann theory to be the saddle-point approximation of the corresponding field-theoretic action. This latter approach nicely clarifies those observables that have to be small in order for Poisson-Boltzmann theory to be a good approximation — and also what to do if these parameters happen to be large.

4. Concrete example: Poisson-Boltzmann theory for charged rods

In this section we will apply the cell model and its Poisson-Boltzmann solution to the case of linear polyelectrolytes. The main purpose is to demonstrate how the theoretical considerations presented so far can be applied to a realistic situation. We chose the cylindrical geometry since this gives us the opportunity to compare Poisson-Boltzmann theory with Man-

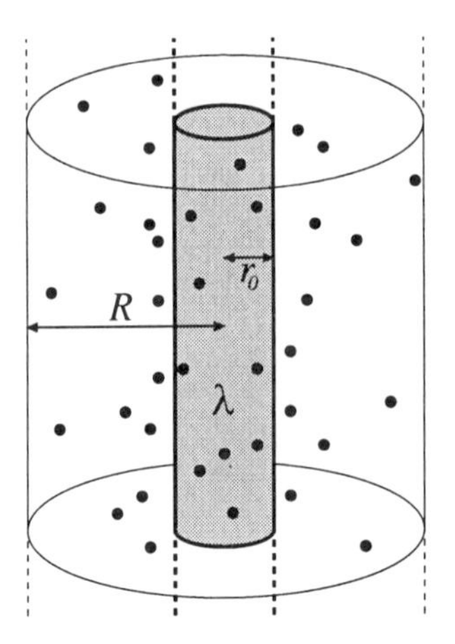

Figure 2. Geometry of the cell model. A cylindrical rod with radius r_0 and line charge density λ is enclosed by a cylindrical cell of radius R. Note that within Poisson-Boltzmann theory the ions are point-like in the sense that no hard core energy term enters the free energy functional F_{PB} from Eqn. (16). However, the theory does not describe individual point-ions but rather an average ionic density.

ning's scenario of counterion condensation. A more comprehensive and very readable introduction, also covering other geometries, is given in Ref. [17].

4.1. SPECIFICATION OF THE CELL MODEL

Consider linear polyelectrolytes of line charge density $\lambda = e\tilde{\lambda} > 0$, radius r_0, and length L that are distributed at a density n_{P} in a solvent characterized by a Bjerrum length ℓ_{B}. We will assume the counterions to be monovalent ($z = -1$), and at first the system does not contain additional salt. Let us enclose the polyelectrolytes by cylindrical cells of radius R and length L. Requiring that the volume of each cell equals the volume per polyelectrolyte in the original solution gives the relation $n_{\text{P}} = 1/\pi R^2 L$, from which we derive the cell radius. If the polyelectrolytes are bent on a large scale, we will require the cell radius R to be small compared to the persistence length of the charged chains and subsequently neglect the bending. We will thus refer to the linear polyelectrolytes simply as "charged rods".

For monovalent counterions each of these rods dissociates $N_{\text{c}} = \tilde{\lambda} L$ of them into the solution, such that their average density is given by $\bar{n}_{\text{c}} = N_{\text{c}} n_{\text{P}} = \tilde{\lambda}/\pi R^2$. If one now neglects end-effects, *i. e.*, if one sets L to infinity, the problem acquires cylindrical symmetry. Observe that $\bar{n}_{\text{c}}$ is independent of L. It is therefore a more convenient measure of the system density, since it is unaffected by the limit $L \to \infty$.

The next step is to replace the individual ion coordinates by a density $n(\boldsymbol{r})$. The cylindrical symmetry will be exploited by assuming that this density also has cylindrical symmetry, even further, that it only depends on the radial coordinate. It is worthwhile pointing out that this statement is not a trivial assumption, for two reasons: First, just because the problem is cylindrically symmetric, the solution need not be.[2] Second, even if the average distribution only depends on the radial coordinate, there may be angular or axial fluctuations that are not taken into account if one right from the start only works with one radial coordinate.

4.2. SOLUTION OF THE POISSON-BOLTZMANN EQUATION

Employing these approximations, the Poisson-Boltzmann equation (19) in the region between r_0 and R can be written as

$$y''(r) + \frac{1}{r}\, y'(r) \;=\; \kappa^2 \mathrm{e}^{y(r)}. \tag{23}$$

Here, $\kappa = \sqrt{4\pi\ell_{\mathrm{B}} n(R)}$ is an inverse length (the Debye screening constant at the outer boundary) and $y(r) = \beta e\psi(r)$ is the dimensionless potential, which is understood to be zero at $r = R$. It will turn out to be convenient to introduce the dimensionless charge parameter

$$\xi \;=\; \tilde{\lambda}\ell_{\mathrm{B}}. \tag{24}$$

It counts the number of charges along a Bjerrum length of rod and is thus a dimensionless way to measure the line charge density λ.

The boundary conditions at $r = r_0$ and $r = R$ follow easily, since by Gauss' theorem we know the value of the electric field there:

$$y'(r_0) \;=\; -\frac{2\xi}{r_0} \qquad \text{and} \qquad y'(R) \;=\; 0. \tag{25}$$

The nonlinear boundary value problem that Eqns. (23) and (25) pose was first solved independently by Fuoss *et. al* [18] and Alfrey *et. al* [19]. It can easily be verified by insertion that the solution is given by

$$y(r) \;=\; -2\,\ln\left\{\frac{r}{R}\sqrt{1+\gamma^{-2}}\,\cos\left(\gamma\,\ln\frac{r}{R_{\mathrm{M}}}\right)\right\}. \tag{26}$$

[2] Broadly speaking: If a physical problem is invariant with respect to some symmetry group $\mathcal{S}$, any solution of the problem is mapped by any element of $\mathcal{S}$ to another solution—the set of solutions is closed under $\mathcal{S}$. However, a particular solution need not be mapped onto itself by *all* elements of $\mathcal{S}$, *i. e.*, it need not possess the full symmetry of the problem. Let us give a simple example: The gravitational field of our sun is spherically symmetric, but the orbit of the earth is not (even if time-averaged).

The dimensionless integration constant γ is related to κ via

$$\kappa^2 R^2 = 2\,(1+\gamma^2). \tag{27}$$

Both γ and $R_{\rm M}$ are found by inserting the general solution (26) into the boundary conditions (25), which yields two coupled transcendental equations:

$$\gamma\,\ln\frac{r_0}{R_{\rm M}} = \arctan\frac{1-\xi}{\gamma} \qquad \text{and} \qquad \gamma\,\ln\frac{R}{R_{\rm M}} = \arctan\frac{1}{\gamma}. \tag{28}$$

Subtracting them eliminates $R_{\rm M}$ and provides a single equation from which γ can be obtained numerically:

$$\gamma\,\ln\frac{R}{r_0} = \arctan\frac{1}{\gamma} + \arctan\frac{\xi-1}{\gamma}. \tag{29}$$

Eqn. (29) only has a real solution for γ if $\xi > \xi_{\rm min} = \ln(R/r_0)/(1+\ln(R/r_0))$. For smaller charge densities γ becomes imaginary, but the solution can be analytically continued by replacing $\gamma \to {\rm i}\gamma$ and using identities like ${\rm i}\gamma\,\tan({\rm i}\gamma) = -\gamma\,\tanh\gamma$. However, in the following we will only be interested in the strongly charged case, in which the charge parameter ξ is larger than $\xi_{\rm min}$.

Let us denote by $\phi(r)$ the fraction of counterions that can be found between r_0 and r. Using $n(r) = n(R)\exp\{y(r)\}$ and Eqns. (26), (27), and (28), we find

$$\phi(r) = \frac{1}{\bar{\lambda}}\int_{r_0}^{r} {\rm d}\bar{r}\;2\pi\bar{r}\,n(\bar{r}) = 1 - \frac{1}{\xi} + \frac{\gamma}{\xi}\tan\Big(\gamma\,\ln\frac{r}{R_{\rm M}}\Big). \tag{30}$$

Observe that $\phi(R_{\rm M}) = 1 - 1/\xi$. Hence, the second integration constant $R_{\rm M}$ is the distance at which the fraction $1 - 1/\xi$ of counterions can be found, which also implies $r_0 \le R_{\rm M} < R$. Due to the importance of this fraction in Manning's theory of counterion condensation (see Sec. (4.3)), $R_{\rm M}$ is sometimes referred to as the "Manning radius".

4.3. MANNING CONDENSATION

The ion distribution around a charged cylindrical rod exhibits a remarkable feature that can be unveiled by the following simple considerations [20]. Assume that the system is infinitely dilute and that there is only one counterion. In the canonical ensemble its radial distribution should be given by ${\rm e}^{-\beta H(r)}/\,{\rm Tr}\,[{\rm e}^{-\beta H(r)}]$ where, up to the kinetic energy and an additive constant, the Hamiltonian is $\beta H(r) = 2\xi\,\ln(r/r_0)$. However, the trace (per unit length) over the coordinate space is

$${\rm Tr}\,[{\rm e}^{-\beta H}] = \int_{r_0}^{\infty} {\rm d}r\;2\pi r\,{\rm e}^{-2\xi\,\ln(r/r_0)} = 2\pi r_0^2\int_1^{\infty}{\rm d}x\;x^{1-2\xi}, \tag{31}$$

which diverges for $\xi < 1$. Hence, the distribution function cannot be normalized. In other words, such rods cannot localize counterions in the limit of infinite dilution, while rods with $\xi > 1$ can. This led Manning to the simple idea that rods with $\xi > 1$ "condense" a fraction of $1-1/\xi$ of all counterions, thereby reducing ("renormalizing") their charge parameter to an effective value of 1, while the rest of the ions remains more or less "free", *i. e.*, not localized [21]. This concept has subsequently been referred to as "Manning condensation" and has led to much insight into the physical chemistry of charged cylindrical macroions.

Similar arguments can be made for the infinite dilution limit in the planar and spherical case. They show that a plane always localizes all its counterions no matter how low its surface charge density is, while a sphere always loses all its counterions no matter how high its surface charge density is.

4.4. LIMITING LAWS OF THE CYLINDRICAL PB-SOLUTION

Although the PB-equation for the cylindrical geometry can be solved analytically, the transcendental equation (29) for the integration constant γ has to be solved numerically. However, since for $\xi > 1$ its right hand side is bounded above by its zeroth and bounded below by its first order Taylor expansion, this gives an allowed interval for γ. All following considerations are restricted to the strongly charged case $\xi > 1$.

$$\pi \geq \gamma \ln \frac{R}{r_0} = \arctan \frac{1}{\gamma} + \arctan \frac{\xi - 1}{\gamma} \geq \pi - \frac{\xi}{\xi - 1}\gamma \qquad (\xi > 1)$$

$$\Rightarrow \frac{\pi}{\ln \frac{R}{r_0}} \geq \gamma \geq \frac{\pi}{\ln \frac{R}{r_0} + \frac{\xi}{\xi - 1}} \tag{32}$$

In the limit $R \to \infty$ the two bounds, and therefore γ, converge to zero. In this limit γ can be approximated by either side of inequality (32), which gives rise to various asymptotic behaviors, known as "limiting laws" for infinite dilution. An immediate first consequence of (32) is that these asymptotic behaviors are reached logarithmically slowly. In the following we will briefly discuss four of these limiting laws.

As we have seen above, the radius $R_{\rm M}$ contains the fraction $1 - 1/\xi$ of ions that are condensed in the sense of Manning. The following limit shows that $R_{\rm M}$ scales asymptotically as the square root of the cell radius:

$$\lim_{R \to \infty} \frac{R_{\rm M}}{\sqrt{R r_0}} = \exp\left\{\frac{\xi - 2}{2\xi - 2}\right\}, \tag{33}$$

It can be shown [22] that a radius that is required to contain any fraction smaller than $1 - 1/\xi$ will remain finite in the limit $R \to \infty$, while a radius containing more than this fraction will diverge asymptotically like R. Roughly speaking, the fraction $1 - 1/\xi$ cannot be diluted away, which is in accordance with the localization argument given in Section 4.3.

Up to a logarithmic[3] correction the electrostatic potential is that of a rod with charge parameter 1:

$$\lim_{R\to\infty} y(r) = y(r_0) - 2\ln\frac{r}{r_0} - 2\ln\Big(1 + (\xi - 1)\ln\frac{r}{r_0}\Big). \tag{34}$$

This is Manning condensation rediscovered on the level of the mean-field potential. Note, however, that the presence of the logarithmic corrections implies that the condensed ions do not sit on top of the charged rod, but rather have a radial distribution. For finite cell radii this distribution is characterized by the length $R_{\rm M}$, which diverges in the dilute limit. Hence, the ions are not particularly closely confined.

The ratio between the boundary density $n(R)$ and the average counterion density $\bar{n}_{\rm c}$ shows the limiting behavior

$$\lim_{R\to\infty}\frac{n(R)}{\bar{n}_{\rm c}} = \lim_{R\to\infty}\frac{1+\gamma^2}{2\xi} = \frac{1}{2\xi}. \tag{35}$$

Since we have seen in Sec. 3 that the boundary density is proportional to the osmotic pressure, the ratio $n(R)/\bar{n}_{\rm c}$ is equal to the ratio $P/P_{\rm id}$ between the actual osmotic pressure and the ideal gas pressure of a fictitious system of non-interacting particles at the same average density. This ratio is called the "osmotic coefficient", and in the dilute limit it converges from above towards $1/2\xi < 1$. The presence of the charged rod hence strongly reduces the osmotic activity of the counterions.

While Eqn. (35) implies that the boundary density goes to zero in the dilute limit, the contact density approaches a finite value:

$$\lim_{R\to\infty} n(r_0) = 2\pi\ell_{\rm B}\tilde{\sigma}^2\Big(1 - \frac{1}{\xi}\Big)^2 = 2\pi\ell_{\rm B}\tilde{\sigma}^2\Big(1 - \frac{1}{2\pi r_0\ell_{\rm B}\tilde{\sigma}}\Big)^2. \tag{36}$$

This as well is a sign that ions must be condensed. We would like to link these two equations to the contact value theorem derived in Section 2, in particular to Eqn. (7). Observe that this equation is *not* satisfied. This is not a bug of the Poisson-Boltzmann approximation but rather a feature of the cylindrical geometry: The contact density is lower than in the corresponding planar case, essentially since only the fraction $1 - 1/\xi$ of condensed ions "contributes" to the ion distribution function. However, in the limit $r_0 \to \infty$ at *fixed* surface charge density $e\tilde{\sigma}$ the charged rod becomes a charged

[3] Since the potential itself is already logarithmic, the logarithmic correction is actually of the form $\ln\ln r$.

plane, $\xi = 2\pi\ell_B\tilde{\sigma}r_0$ diverges, and the correction factor becomes 1, such that the contact value theorem is again satisfied. Note also that ξ can be written as r_0/λ_{GC}, where λ_{GC} is the so called Gouy-Chapman length, the characteristic width of a planar electrical double layer forming at a surface with surface charge density $e\tilde{\sigma}$ [17]. For the contact value theorem to be valid in its planar version it is thus necessary that the characteristic extension of the charged layer is small compared to the radius of curvature of the surface—or, equivalently: $\xi \gg 1$. It is worth pointing out that the solution of the linearized planar Poisson-Boltzmann equation violates the contact value theorem by giving a contact density which is a factor of 2 too large.

5. Additional salt: The Donnan equilibrium

How is the Poisson-Boltzmann equation to be modified, if more than one species of ions is present? First, each ion density $n_i(\boldsymbol{r})$ is assumed to be proportional to the local Boltzmann-factor, thereby generalizing Eqn. (18):

$$n_i(\boldsymbol{r}) = n_{0i}\,\mathrm{e}^{-\beta z_i e\,\psi_{\mathrm{tot}}(\boldsymbol{r})} \qquad \text{with} \qquad n_{0i} = \frac{N_i}{\int_V \mathrm{d}^3r\ \mathrm{e}^{-\beta z_i e\,\psi_{\mathrm{tot}}(\boldsymbol{r})}}. \tag{37}$$

Second, the total charge density $e\sum_i z_i n_i(\boldsymbol{r})$ has to satisfy Poisson's equation. This situation arises if the counterions form a mixture of different valences or if the system contains additional salt. In this section we would like to make a few remarks about the latter case.

The amount of salt can be specified by the number of salt molecules per cell. Since salt molecules are neutral (unlike counterions), their number is not restricted by the constraint of electroneutrality, and different cells may contain different numbers of salt molecules. This variation cannot be taken into account by a description focusing on just one cell and is thus neglected. One assumes instead that the cell contains a number of salt molecules equal to the average salt concentration in the polyelectrolyte solution times the cell volume. In other words: The division of salt between the cells is assumed to be perfectly even.

If the presence of salt is due to the fact that the polyelectrolyte solution is in contact with a salt reservoir, there is a further problem to solve: How is the average concentration of salt molecules in the polyelectrolyte solution related to the concentration of salt molecules in the salt reservoir? This situation is depicted in Fig. 3 and is referred to as a "Donnan equilibrium" [24, 25]. In the following we will address this question on the level of the cell model and Poisson-Boltzmann theory. For simplicity we will only treat the case in which all ions are monovalent.

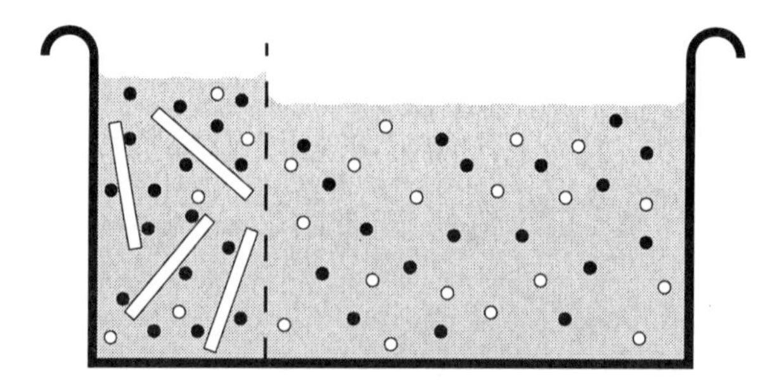

Figure 3. Solution of charged rod-like polyelectrolytes and counterions in equilibrium with a bulk salt reservoir. The membrane is permeable for everything but the macroions. Notice that both compartments are charge neutral.

The compartment containing the macroions will be described by a cell-model, which apart from the central rod and the counterions will contain a certain number of salt molecules, yet to be determined. Since both the counterions and the oppositely charged coions can cross the membrane, they have to be in electrochemical equilibrium. However, before we can write down a condition for that, there is an additional effect that we have to take into account: Since the charged macroions cannot leave their compartment, their counterions also will have to remain there for reasons of electroneutrality. Hence, upon addition of salt there is a tendency for the salt to go into the other compartment, which is less "crowded". However, this implies that in general there must be a discontinuity in the counter- and coion density across the membrane. Such a difference can only be sustained by a corresponding drop in the electrostatic potential across the membrane separating the two compartments. This potential drop is referred to as the "Donnan potential", Φ_{D}, and must be taken into account when balancing the electrochemical potentials.

Having said this, we can now proceed to compute electrochemical potentials on both sides. On the side containing the macroions we will compute the chemical potential at the cell boundary, where the only contribution is the entropy term if we set the potential to zero there. In the bulk salt reservoir we get an entropy term corresponding to the bulk salt density, which we call n_{b}, and a term corresponding to the Donnan potential Φ_{D}. With $n_{\pm}$ being the cation and anion densities at the cell boundary, we obtain

$$\begin{aligned} k_{\mathrm{B}}T \ln n_{\pm} &= k_{\mathrm{B}}T \ln n_{\mathrm{b}} \pm e\,\Phi_{\mathrm{D}} \\ \Rightarrow \qquad n_{\pm} &= n_{\mathrm{b}}\,\mathrm{e}^{\pm\beta e \Phi_{\mathrm{D}}}. \end{aligned} \tag{38}$$

Multiplying cation and anion density gives

$$n_{+}\,n_{-} = n_{\mathrm{b}}^{2}. \tag{39}$$

That is, the bulk salt density is the geometric average of the cation and anion densities at the cell boundary. Dividing the ion densities in Eqn. (38) yields an expression for the Donnan potential:

$$\beta e\,\Phi_{\rm D} \;=\; \frac{1}{2}\ln\frac{n_+}{n_-} \stackrel{(39)}{=} \ln\frac{n_+}{n_{\rm b}}. \tag{40}$$

This also shows that the Donnan potential diverges in the zero salt limit.

For sufficiently dilute solutions the osmotic pressure follows from the van't Hoff equation βP = [solute]. Since the *excess* osmotic pressure is given by the difference between the osmotic pressures at the cell boundary acting from inside and from outside, we find

$$\beta P \;=\; n_+ + n_- - 2n_{\rm b} \stackrel{(39)}{=} \left(\sqrt{n_+} - \sqrt{n_-}\right)^2 \;\geq\; 0. \tag{41}$$

Let $\delta n_+ = n_+ - n_{\rm b}$ denote the difference between the cation density at the outer cell boundary and the cation density in the bulk salt reservoir. Combining Eqns. (38), (40) and (41), we can rewrite the pressure as

$$\begin{aligned}
\frac{\beta P}{n_{\rm b}} \;&=\; \frac{\left(\sqrt{n_+} - \sqrt{n_-}\right)^2}{n_{\rm b}} \;=\; \left({\rm e}^{\beta e\Phi_{\rm D}/2} - {\rm e}^{-\beta e\Phi_{\rm D}/2}\right)^2 \\
&=\; 4\,\sinh^2\left(\frac{1}{2}\ln\frac{n_+}{n_{\rm b}}\right) \;=\; 4\,\sinh^2\left[\frac{1}{2}\ln\left(1+\frac{\delta n_+}{n_{\rm b}}\right)\right] \\
&\stackrel{\delta n_+ \ll n_{\rm b}}{\approx}\; \left(\frac{\delta n_+}{n_{\rm b}}\right)^2.
\end{aligned} \tag{42}$$

If one wishes to determine the osmotic pressure, one has to solve the nonlinear Poisson-Boltzmann equation in the presence of salt, subject to the constraint in Eqn. (39). However, no analytical solution is known for this case. A numerical solution can be obtained in the following way: First "guess" an initial amount of salt to be present in the cell, solve the PB equation[4], compute the bulk salt concentration implied by this amount via Eqn. (39), adjust the salt content, and iterate until self-consistency is achieved.

An approximate treatment of the problem can be obtained from the following two assumptions:

1. If the amount of salt is small, it may not significantly disturb the counterion profile from the salt-free case. Hence, one may hope that the *counterion* concentration n_R at the cell boundary is still given by its value from *salt-free* Poisson-Boltzmann theory.

---

[4] Two simple ways for achieving this are described in Ref. [23]. Although these authors treat the spherical case, their approach works equally well for cylindrical symmetry, since the only important point is that the problem is one-dimensional.

2. At the outer cell boundary the densities of additional cations and anions due to the salt are equal, say $\Delta n_{\rm s}$.

Given these two assumptions, the equilibrium condition (39) requires

$$(n_R + \Delta n_{\rm s})\,\Delta n_{\rm s} \;\approx\; n_{\rm b}^2 \qquad \Rightarrow \qquad 2\Delta n_{\rm s} \;\approx\; \sqrt{n_R^2 + (2n_{\rm b})^2} - n_R. \quad (43)$$

Together with Eqn. (41) this gives the following approximate expression for the excess osmotic pressure:

$$\beta P \;\approx\; n_R + 2\Delta n_{\rm s} - 2n_{\rm b} \;=\; \sqrt{n_R^2 + (2n_{\rm b})^2} - 2n_{\rm b}. \quad (44)$$

The advantage of this expression is that it requires only the knowledge of the counterion density n_R at the outer cell boundary from salt-free Poisson-Boltzmann theory, which is much easier to determine than the solution including the salt explicitly.[5]

The above approximation is good if the counterion concentration is large compared to the salt concentration. In the opposite limit of excess salt concentration Eqn. (44) behaves asymptotically like

$$\frac{\beta P}{n_{\rm b}} \;=\; 2\left[\sqrt{1+\Big(\frac{n_R}{2n_{\rm b}}\Big)^2} - 1\right] \;\approx\; 2\left[1+\frac{1}{2}\Big(\frac{n_R}{2n_{\rm b}}\Big)^2 - 1\right] \;=\; \Big(\frac{n_R}{2n_{\rm b}}\Big)^2. \quad (45)$$

Since n_R is given by $\bar{n}_{\rm c}$ times the osmotic coefficient, which does not strongly vary with density, this equation implies that in the salt dominated case the osmotic pressure varies quadratically with the average counterion concentration. However, this is not born out by a numerical solution of the Poisson-Boltzmann equation with added salt, which shows an exponential behavior (see Fig. 4). The latter can be understood by the following simple argument: For large salt content the counterion and coion density profiles can be expected to merge exponentially with a bulk Debye-Hückel screening constant $\kappa_{\rm D} = \sqrt{8\pi\ell_{\rm B}n_{\rm b}}$. We may thus assume that

$$\delta n_+ \;\propto\; n_{\rm b}\,{\rm e}^{-\kappa_{\rm D}R}. \quad (46)$$

Since the cell radius is related to the average counterion concentration via $\bar{n}_{\rm c} = \tilde{\lambda}/\pi R^2$, we can write together with Eqn. (42)

$$\frac{\beta P}{n_{\rm b}} \;\propto\; \exp\{-2\kappa_{\rm D}R\} \;=\; \exp\Big\{-2\sqrt{8\pi\ell_{\rm B}n_{\rm b}}\sqrt{\tilde{\lambda}/\pi\bar{n}_{\rm c}}\Big\}. \quad (47)$$

Taking the logarithm we finally see

$$\log\left(\frac{\beta P}{n_{\rm b}}\right) \;=\; C_1 - C_2\sqrt{\frac{n_{\rm b}}{\bar{n}_{\rm c}}} \quad (48)$$

[5] It requires only the numerical solution of the transcendental equation (29), not the numerical solution of a nonlinear differential equation.

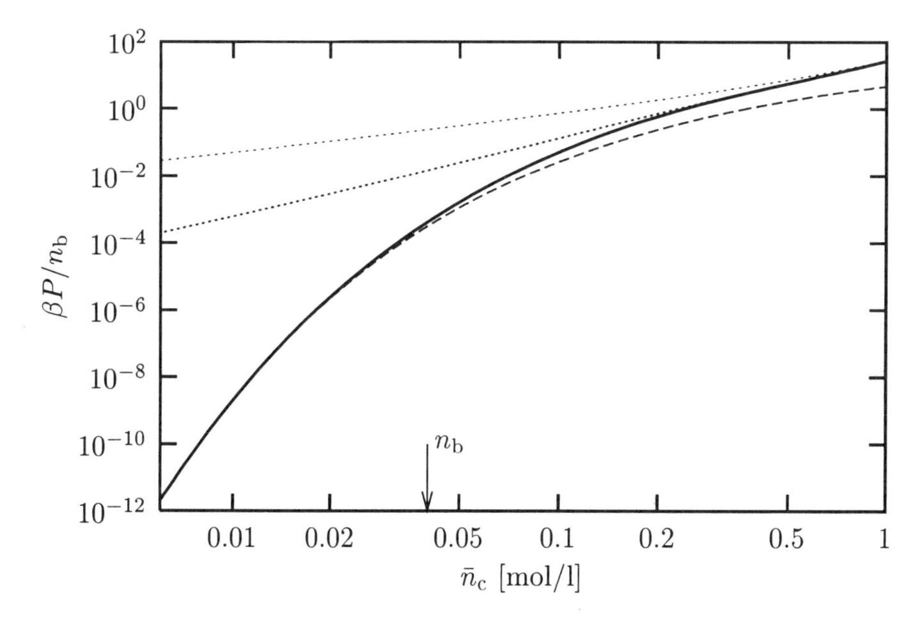

Figure 4. Ionic contribution to the osmotic pressure βP of a DNA solution divided by bulk salt concentration n_b as a function of average counterion concentration $\bar{n}_c$. The bulk salt concentration is $n_b = 40$ mmol/l. The solid line is the prediction of the Poisson-Boltzmann equation (taking into account counterions and salt), the (upper) fine dashed line is from Poisson-Boltzmann theory without salt. The short dashed line is the approximation from Eqn. (44) and the long dashed line is a fit to Eqn. (48) within the range $\bar{n}_c = 6 - 10$ mmol/l.

with some constants C_1 and C_2.

This functional dependence should hold whenever $n_b/\bar{n}_c \gg 1$. However, it also demonstrates that the range of validity of the cell model reaches its limit for high salt concentrations. As increasingly more salt is added to the system the osmotic pressure of the ions vanishes exponentially. However, this does not imply that the total osmotic pressure of the polyelectrolyte solution vanishes, since we have neglected the contribution coming from the macroions [26]. Since the latter generally depends on observables which are largely irrelevant for the cell model (*e.g.*, the degree of polymerization), this model must break down here.

6. Outlook

We presented in our brief introduction two of the most common approximations encountered in the theory of charged macromolecules: The cell-model and Poisson-Boltzmann theory. Where can one go from here?

One of the key deficiencies of Poisson-Boltzmann theory is the neglect of interparticle correlations. We have seen how this arose from the assumption of a product state, which subsequently led to a simple (local) density functional theory. An important theorem originally due to Hohenberg and Kohn states that there actually *exists* a density functional which gives the correct free energy of the full system and which differs from the Poisson-Boltzmann functional by an additional term that takes into account the effects of correlations.[6] Although the theorem does not specify what this functional looks like, it shows that attempts that go beyond Poisson-Boltzmann theory but stay on a density functional level are not futile. Indeed, various local [28] and nonlocal [29] corrections to the Poisson-Boltzmann functional have been suggested in the past.

The Coulomb problem has been treated in a field-theoretic way, *i.e.*, the classical partition function is transformed via a Hubbard-Stratonovich transformation into a functional integral [30]. Poisson-Boltzmann theory is rediscovered as the saddle-point of this field theory, and higher order corrections can in principle be obtained using the large and well-established toolbox of field-theoretic perturbation theory. There also exists the possibility to approximate the functional integral in the limit opposite to Poisson-Boltzmann theory, when correlations dominate the system [31]. All this is thoroughly discussed in the lecture notes by Moreira and Netz [16].

A different route to incorporate correlations is offered by integral equation theories. Their key idea is to first derive exact equations for various correlation functions and then introduce some approximate relations between them, based for instance on perturbation expansions, which lead to integral equations that implicitly give the desired correlation functions. This approach and its relation to Poisson-Boltzmann and Debye-Hückel theory is the topic of the lecture of Kjellander [5].

A further method for dealing with correlations is to simulate the systems on a computer and explicitly keep track of all the ions—or even solvent molecules. This approach has become an increasingly important tool for both describing real systems as well as testing approximate theories. More details can be found in the lectures of Jönsson and Wennerström [2] and Holm and Kremer [32].

Going beyond the cell model and taking the actual shape of polyelectrolytes into account is in general an extremely difficult business. However, a remarkable amount of information can be obtained by using some (or, better yet, a lot of) physical insight and writing the free energy as a sum of a few terms which account for the most relevant physical properties of the system (for instance the chain elasticity, electrostatic self-energy or hydrophobic

[6] A good introduction into this topic can be found in Ref. [27].

interactions) and possibly some variational parameters. Since in doing so all prefactors are neglected (it only matters how one observable "scales" with another) these approaches are known as scaling theories. Joanny gives an introduction and a few famous applications in his lecture [33].

All these approaches reach beyond the cell-model and/or the Poisson-Boltzmann equation. They boldly go where no mean-field theory has gone before. However, we believe that in order to appreciate their efforts it is worthwhile to know where they came from. It was the intention of this chapter to provide some of that knowledge.

Acknowledgments

M. D. would like to thank P. L. Hansen and I. Borukhov for stimulating discussions.

References

1. A. Katchalsky, Pure Appl. Chem. **26**, 327 (1971).
2. B. Jönnson and H. Wennerström, see chapter in this volume.
3. P. Debye and E. Hückel, Phys. Z. **24**, 185 (1923).
4. T. L. Hill, *An Introduction to Statistical Thermodynamics*, Dover Publications, New York (1986). D. A. McQuarrie, *Statistical Mechanics*, Harper Collins, New York (1976).
5. R. Kjellander, see chapter in this volume.
6. F. H. Stillinger Jr. and R. Lovett, J. Chem. Phys. **48**, 3858 (1968); F. H. Stillinger Jr. and R. Lovett, J. Chem. Phys. **49**, 1991 (1968).
7. S. L. Carnie and D. Y. C. Chan, Chem. Phys. Lett. **77**, 437 (1981).
8. D. Henderson and L. Blum, J. Chem. Phys. **69**, 5441 (1978).
9. D. Henderson, L. Blum, and J. L. Lebowitz, J. Electroanal. Chem. **102**, 315 (1979).
10. H. Wennerström, B. Jönsson, and P. Linse, J. Chem. Phys. **76**, 4665 (1982).
11. L. D. Landau and E. M. Lifshitz, *The classical Theory of Fields*, Addison-Wesley, Cambridge (Mass.) (1951).
12. D. C. Grahame, Chem. Rev. **41**, 441 (1947).
13. P. M. Chaikin and T. C. Lubensky, *Principles of condensed matter physics*, Cambridge University Press, Cambridge (1995).
14. The formula (8) goes under many names, for instance "Gibbs-Bogoliubov-inequality" or "Feynman variational principle". Its derivation is remarkably simple: For any two arbitrary states p and p_0 the Gibbs inequality $\operatorname{Tr} p_0 \ln p \leq \operatorname{Tr} p_0 \ln p_0$ holds, since the elementary inequality $\ln x \leq x - 1$ implies

$$\operatorname{Tr} p_0 \ln p - \operatorname{Tr} p_0 \ln p_0 = \operatorname{Tr} p_0 \ln \frac{p}{p_0} \leq \operatorname{Tr} p_0 \Big(\frac{p}{p_0} - 1 \Big) = \operatorname{Tr} p - \operatorname{Tr} p_0 = 0. \quad (49)$$

If one inserts an arbitraty p_0 and the canonical $p = \mathrm{e}^{-\beta H} / \operatorname{Tr} \mathrm{e}^{-\beta H}$, one obtains

$$\begin{aligned} S_0 := -k_{\mathrm{B}} \operatorname{Tr} p_0 \ln p_0 \ &\leq \ -k_{\mathrm{B}} \operatorname{Tr} p_0 \ln p \ = \ -k_{\mathrm{B}} \operatorname{Tr} \left\{ p_0 \left(-\beta H - \ln \operatorname{Tr} \mathrm{e}^{-\beta H} \right) \right\} \\ &= \frac{1}{T} \left(\operatorname{Tr} p_0 H + k_{\mathrm{B}} T \ln \operatorname{Tr} \mathrm{e}^{-\beta H} \right) \ =: \ \frac{1}{T} \left(\langle H \rangle_0 - F \right), \end{aligned} \quad (50)$$

which is formula (8). Incidentally, the Gibbs inequality remains valid if p and p_0 are quantum states — and so does the Gibbs-Bogoliubov-inequality.

15. R. A. Marcus, J. Chem. Phys. **23**, 1057 (1955).
16. A.G. Moreira and R.R. Netz, see chapter in this volume.
17. D. Andelman, in: *Handbook of Biological Physics, I*, ed. by R. Lipowsky and E. Sackmann, Elsevier (1995), Chapter 12.
18. R. M. Fuoss, A. Katchalsky, and S. Lifson, Proc. Natl. Acad. Sci. USA **37**, 579 (1951).
19. T. Alfrey, P. Berg, and H. J. Morawetz, J. Polym. Sci. **7**, 543 (1951).
20. This has been discussec by R. R. Netz in his Les Houches lecture, but is not included in his lecture notes.
21. G. S. Manning, J. Chem. Phys. **51**, 924, 934, 3249 (1969).
22. M. Le Bret and B. H. Zimm, Biopol. **23**, 287 (1984).
23. S. Alexander, P. M. Chaikin, P. Grant, G. J. Morales, P. Pincus, and D. Hone, J. Chem. Phys. **80**, 5776 (1984).
24. F. G. Donnan, Chem. Rev. **1**, 73 (1924).
25. J. Th. G. Overbeek, Prog. Biophys. and Biophys. Chem. **6**, 57 (1956).
26. E. Raspaud, M. da Conceiçao, and F. Livolant, Phys. Rev. Lett. **84**, 2533 (2000).
27. J. P. Hansen and I. R. McDonald, *Theory of Simple Liquids*, 2nd ed., Academic Press, London (1986).
28. S. Nordholm, Chem. Phys. Lett. **105**, 302 (1984); R. Penfold, S. Nordholm, B. Jönsson, and C. E. Woodward, J. Chem. Phys. **92**, 1915 (1990); M. J. Stevens and M. O. Robbins, Europhys. Lett. **12**, 81 (1990); M. C. Barbosa, M. Deserno, and C. Holm, Europhys. Lett. **52**, 80 (2000).
29. R. Groot, J. Chem. Phys. **95**, 9191 (1990); A. Diehl, M. C. Barbosa, M. N. Tamashiro, and Y. Levin, Physica A **274**, 433 (1999); C. N. Patra and A. Yethiraj, J. Phys. Chem. **103**, 6080 (1999).
30. R. R. Netz and H. Orland, Europhys. Lett. **45**, 726 (1999); R. R. Netz and H. Orland, Eur. Phys. J. E **1**, 203 (2000).
31. A. G. Moreira and R. R. Netz, Europhys. Lett. **52**, 705 (2000).
32. C. Holm and K. Kremer, see chapter in this volume.
33. J.-F. Joanny, see chapter in this volume.

DNA CONDENSATION AND COMPLEXATION

Biological Implications

WILLIAM M. GELBART
Department of Chemistry & Biochemistry
University of California, Los Angeles
Los Angeles, CA 90095-1569

1. Introduction

A primary aim of these lectures is to introduce two structural motifs in biology that have provided the motivation for many recent studies of electrostatic effects in solutions of macro-ions.

The first motif involves **condensates of DNA** that are induced by the presence of polyvalent counterions, and corresponds to **the packaging of DNA in viruses**. Here, different portions of double-stranded DNA exert attractive forces on one another when they are sufficiently aligned and close together. These interactions are now known to arise from counterion correlation effects that are neglected in the well-known mean-field theories such as Poisson-Boltzmann and Debye-Hückel. As a consequence, the double helical strands are induced to pack hexagonally at quite high densities, very close to the densities characteristic of crystalline DNA.

The second motif involves **complexes of DNA** with oppositely charged colloidal particles, and corresponds to **nucleosome core particles and the basic fiber of chromatin**. Here a semiflexible chain is induced to adsorb on an oppositely charged sphere or cylinder. The driving force for complexation is an entropic one, associated with release of counterions, which offsets the bending energy price of wrapping the double-helical DNA around the cationic protein aggregate.

In introducing the basic *physics* underlying the above two situations, we also devote considerable time to gaining some feeling for the *biological* systems themselves, even though we will almost always be concerned with their study under highly controlled – i.e., *in vitro*–conditions. In both cases we will attempt to move systematically from the biological context to a

C. Holm et al. (eds.), Electrostatic Effects in Soft Matter and Biophysics, 53–86.
© 2001 *Kluwer Academic Publishers. Printed in the Netherlands.*

well-posed model problem that is amenable to both analytical theory and molecular dynamics and Monte Carlo simulation.

2. Semiflexible Chain Condensation and Viral Infection

2.1. BRIEF INTRODUCTION TO BACTERIAL VIRUSES AND THEIR INFECTION OF CELLS

Viruses that infect bacteria are called bacteriophage – or "phage". They were discovered as early as the 1920's and then investigated vigorously from the 1940's through the 1960's as testing grounds for the evolving ideas and practices of modern molecular biology. [1] Figure 1 shows an

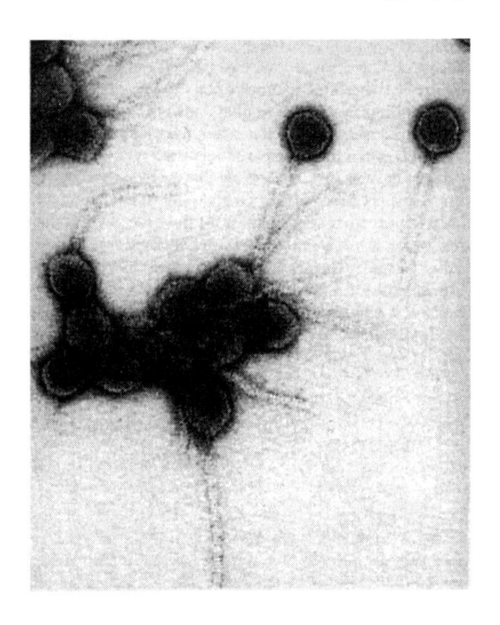

Figure 1.

electron micrograph [2] of bacteriophage lambda (λ), perhaps the best studied example of this incredibly simple biological system (phage). The hexagons are 2D cross-sections of icosahedral protein *capsids*—rigid shells that contain nothing but a single molecule of the viral DNA. Each capsid is attached to a hollow cylindrical "tail", formed from repeated copies of a few other proteins, and the inner diameter of this tail is just big enough for the (double-stranded) DNA to exit the capsid.

For each phage like λ there is a unique membrane protein that acts as viral receptor, binding the tip of the tail to the outer membrane of the bacterial cell. In the electron micrograph [3] of Figure 2 we see a "particle"

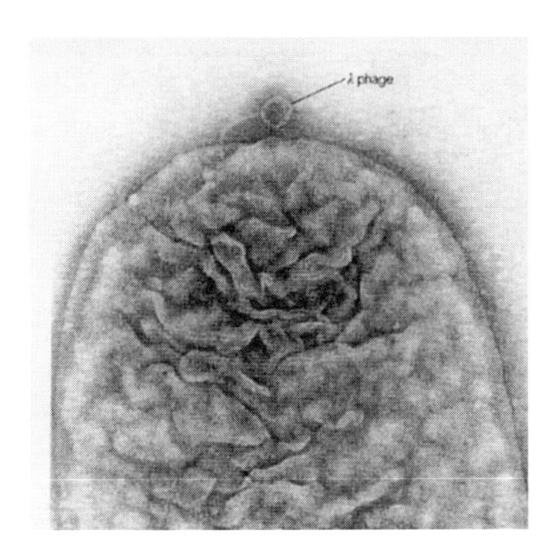

Figure 2.

of λ binding to the receptor protein LamB in the outer membrane of E. coli, arguably the best studied *bacterial* system. Note the size scales in this picture. The capsid diameter is of order 500 Angstroms, or 1/20 of a micron, whereas the bacterial cell has dimensions on the order of several microns. The other set of length scales that we want to focus on now is that involving the viral DNA.

Double-stranded DNA (and henceforth we will be dealing exclusively with double-stranded DNA) has a *persistence length* of about 500 Angstroms. By definition [4] of the persistence length (l_p) of a semiflexible linear polymer, bending it into a radius smaller than this requires more energy than is thermally available ($k_B T$). Equivalently, DNA does not spontaneously bend on smaller length scales than this; it is persistent for distances of 500 Angstroms along its length. But as noted immediately above, the viral capsid size is comparable to this, which means that a source of energy is required to bend the DNA to fit into the capsid, especially because the overall length of the viral DNA is typically *hundreds* of times greater. Consequently, not only is energy required to *bend* the chain but one also must contend with the loss of configurational entropy associated with confining the long chain in the small capsid. Finally, the DNA is highly charged—two phosphate groups per base pair, i.e., per 3.4 Angstroms of double-helical length—and a large amount of work needs to be done in overcoming its strong self-repulsions as the molecule is packaged in the virus at solidlike densities.

A good portion of the ensuing discussion will be concerned with phenomenological descriptions and approximate estimates of these energies involved in packaging viral DNA. Before proceeding to this theory, however, it is important to review more of the biological basics of the viral infection process. Figure 3 shows a cartoon representation of the several

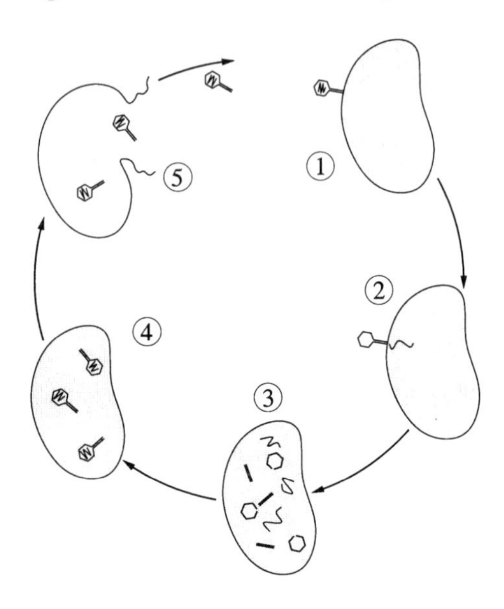

Figure 3.

steps comprising the infection of a bacterial cell by a phage virus. We have already discussed the first step ("1"), in which the tail of the virus binds to its receptor protein in the bacterial membrane; in binding, the tail is opened and DNA is ejected (2) from the capsid into the cell interior. Clearly, what drives (at least the initial stage of) this injection process is the high pressure of DNA in the capsid, and we shall be expressly concerned in the sections below with the nature of this pressure. The viral DNA is then *replicated* and *expressed* by the bacterial cell machinery. More explicitly, many copies of the DNA are made and similarly with many copies of the proteins encoded in its genes. Some of these proteins self-assemble into the icosahedral capsid, while others are organized into the hollow tail (3). Still another is the *motor protein* that does the work of forcing the viral DNA into the preformed capsids; the tail is joined onto the capsid only upon completion of this loading process (4). When a sufficient number (of the order of hundreds) of viral particles have been replicated in this way, still another viral gene product serves to trigger lysis of the cell membrane (5). Each of the released viruses then "attacks" a new cell and the infection proceeds in this way.

We shall be ignoring all of the interesting molecular biology associated with replication and expression of the viral DNA in the bacterial cell.

Instead, we will focus on the first and last steps in the infection process, the ones that are involved most directly in the packaging of DNA in the viral capsid. More explicitly, we discuss the ejection of DNA upon binding and opening of the tail and then the loading of DNA into the pre-formed capsid by the appropriate motor protein. In both cases, the *pressure* in the capsid is the key physical feature, alternately *driving* and *resisting* the ejection and loading processes, respectively. One final piece of experimental background, before proceeding to our discussion of model systems, is the following. Infectious phage can be "synthesized" in the laboratory, outside the bacterial cell and under highly controlled physical conditions. Basically, one purifies from infected cell extract a small number of vital ("minimal") components: the preformed capsids and tails, the viral DNA, and the motor protein that packages the DNA into the capsids. Upon addition of ATP and polyvalent cations to this aqueous solution, fully functional (infectious) phage are formed. Indeed, packaging "kits" of this kind [5] form the basis for a large biotechnology business in which bacteria are made to synthesize commercially interesting proteins that are products of genes inserted into the viral DNA.

2.2. CLASSICAL DNA CONDENSATION EXPERIMENTS: HIGHLY DILUTE BULK SOLUTIONS

It is not coincidental that polyvalent cations are invariably "part of the brew" in the above *in vitro* syntheses of phage, i.e., in the loading of DNA into viral capsids. Normally, the *in vivo* syntheses also involve tri- or tetravalent cations. [6] What, then, is the connection between DNA packaging and the presence of polyvalent cations?

A great deal of concerted theoretical and experimental attention has been devoted to the properties of DNA chains in aqueous solution containing polyvalent counterions. In all instances, the focus is on the effective attractive interactions between chains mediated by the polyvalent counterions. That DNA should end up attracting itself is highly counterintuitive because of the high density of charge associated with the double helix, namely, one fundamental charge per 1.7 Angstroms. As discussed throughout this School, the time-honored role of added salt is to *screen* the Coulomb repulsions between highly charged macro-ions in solution. Indeed, the classical, mean-field (Poisson-Boltzmann, Debye-Hückel), theory can be shown to always give a repulsion—albeit screened—between like-charged colloidal particles, regardless of the valence of any added salts. Polyvalent (and in particular, trivalent and higher) counterions are nevertheless observed to give strong attractive interactions between stiff polyelectrolyte chains when they are close enough and parallel enough. The fundamental statistical mechanical theory underlying this effective attraction began to attract renewed interest just a few years ago in the case of rodlike polyelectrolytes [7] and is the subject of several lectures at this school. We shall focus here on the phenomenological features and consequences of this interaction.

Consider a highly dilute (e.g., nanomolar) solution of λ DNA in simple salt solution–say, 100mM NaCl. The λ genome is about 50,000 base pairs long, corresponding to a length of about 170,000 Angstroms (17 microns). This length exceeds the persistence length by 2-3 orders of magnitude, and the Debye screening length (at 100mM NaCl) is of the order of the "monomer" (base pair) size, about 10 Angstroms. It follows that the size–radius of gyration (R_g) [4]–of this DNA in solution is of the order of microns:

$$R_g \cong (N_p l_p^2)^{1/2} = (L l_p)^{1/2} \approx 10^4 \text{A} \tag{1}$$

Here we have used the fact that the number of persistence lengths, $N_p = L/l_p$, is large compared with unity. Addition of polyvalent cations results [8] in monomolecular "collapse" of each of these coils into highly condensed toroids characterized by the following structural features:

(i) the average radius of the toroid is about 500A (the DNA persistence length!);

(ii) the cross-section of the toroid is approximately circular, with a radius of 150-200A;

(iii) the toroid is formed from circumferentially wound DNA, with local hexagonal packing of the parallel double-strands.

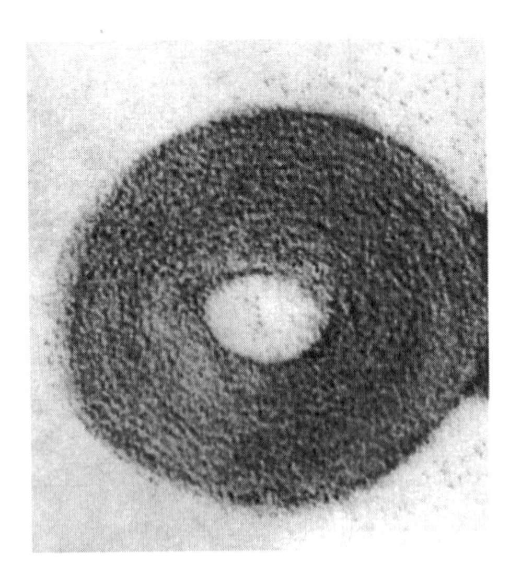

Figure 4.

An electron micrograph (courtesy of J.-L. Sikorav) of such a structure is shown in Figure 4 for the case of dilute solutions of λ DNA condensed by spermidine.

Still more remarkable than such a structure being formed from a single molecule of phage DNA is the fact that similar toroids are observed when much shorter pieces of DNA are condensed by polyvalent cations. [8] Furthermore, these structures arise when very different condensing "agents" are employed, e.g., a wide range of tri- and tetravalent cations, "bad solvent" (methanol, ethanol), or low-molecular-weight polymers that mediate depletion attractions between the strands. [9] Especially interesting about the toroids formed from *short* DNA is the feature that a *limited aggregation* process is involved. In the case of 400 base-pair segments, for example, hundreds of these DNA molecules are involved in the formation of each toroid whose structure is of the kind outlined by (i)-(iii) above. Because of the strong attractive forces mediated by polyvalent cations (or methanol, or PEG, etc.), these molecules organize into "head-to-tail", circumferentially wound, hexagonally packed, toroids containing about the same amount (and hence the same dimensions) of DNA as the single-molecule condensate of lambda phage. This fact immediately raises the question: what is the physics that limits the aggregation of condensing molecules? Equivalently, what is special about a toroid whose cross-sectional diameter accommodates only dozens of double-stranded chains?

A similar phenomenon is observed in the case of a very different biological polyelectrolyte, namely F-actin, a principal structural protein in cells and in muscle tissue. Here the persistence length (of this polymerized version of the anionic protein monomer, G-actin) is almost two orders of magnitude larger than that of DNA, and is often comparable to the overall length of the polymer itself. Consequently, the F-actins behave as essentially "rigid" rods and upon the addition of polyvalent cation these "strands" aggregate into rodlike *bundles.* [10] But as in the case of DNA discussed immediately above, these bundles have finite cross-sections whose diameter corresponds to only dozens of strands. Once, again, then: what limits the aggregation process?

2.3. DIGRESSION ON MICRO-PHASE SEPARATION

Consider for a moment the familiar case of condensation from gas to liquid, or from liquid to solid. In the time-honored theory of these processes, the key idea is that a free energy barrier to growth of condensed phase arises from the increase in surface area of condensate. More explicitly, the bulk energy is being lowered by an amount proportional to the *volume* of the condensed phase that forms, while the surface energy varies with the *area*. For a "drop" of radius R (volume $V = R^3$), then, these two terms scale as

R^3 and R^2 (V and $V^{2/3}$) with coefficients that are negative (arising from the cohesive energy density) and positive (interfacial tension), respectively:

$$E = -eV + \gamma V^{2/3}. \tag{2}$$

Accordingly, once a large enough ("critical") size is achieved–through thermal fluctuations–the barrier is surmounted and the condensate grows without bound, forming a macroscopic phase.

In the case of the aggregation of DNA and actin described above, however, the condensation (aggregation) is obviously limited, suggesting that still another barrier must be operative. Perhaps the most famous example of this kind is that of the charged liquid drop treated over 100 years ago by Lord Rayleigh. [11] He basically asked the question: how much charge can a liquid drop support before it breaks up into two smaller ones? The energy of a drop with radius R is basically a sum of two terms: the Coulomb energy Q^2/R, where Q is the charge of the drop; and the surface energy γR^2, where γ is the surface tension of the liquid. Suppose the drop were to reorganize as two drops, say, each carrying half the total charge and having half the volume of the original drop. Then it is straightforward to show (try it!) that the final energy will be lower if the original charge exceeds a critical value, Q_c, of order $(\gamma R^3)^{1/2}$. Rayleigh actually estimated in this way "what electrification is necessary to render a small rain drop of, say, 1 mm diameter, unstable." He expressed the critical potential Q_c/R in the form γR^2, as above, but included the numerical factors "of order unity" ($(16\pi)^{1/2}$ in this case!) that we have dropped. Using the then best value of the surface tension of water (81cgs), he concluded, "an electrification of about 5000 Daniell cells would cause the division of the drop in question."

This is arguably the simplest physical example of how *long-ranged repulsions* (arising here from the Coulomb interaction) *can limit the size of a condensate* which is stabilized by attractions (here the dispersional forces). A more complicated example, but involving essentially the same physics, is provided by a polyelectrolyte in a bad solvent. "Bad solvent" means nothing, phenomenologically, other than that there is an effective attraction between chain monomers with the result that they want to condense on each other. In the case of neutral polymers this process corresponds to chain collapse in low concentration solutions and to phase separation otherwise. [12] If the chains are charged, *however*, there is a limit to how large the condensed globules of monomers can be–for the same reasons just outlined for the Rayleigh drop. But now, because of the connectivity of the monomers, the chain collapses into a "string of pearls." The precise nature of this process has been worked out in detail on many levels, from scaling theory [13] to computer simulation. [14]

Another important example of size-limited condensation arises in the study of Langmuir monolayers, i.e., 2D systems of amphiphilic molecules at the air-water interface. [15] Upon compressing such a system, a sequence of phase transitions is observed, one of the best studied being the first-order change from "expanded" to "condensed" liquid states. In both these phases the dipolar head groups are known to sit at the interface with a preferred orientation, thereby giving rise to a long-range ($1/r^3$) repulsion characteristic of aligned dipoles. It is this interaction that "frustrates" the condensation being driven by the shorter-ranged, dispersional, attractions (associated mostly with the hydrocarbon chains that point up from the air-water interface). Consequently, the condensed liquid in the two phase region is organized as (hexagonally ordered) circular domains of a characteristic size; at higher concentrations of this phase the condensate is organized as "stripes" whose width is comparable to that of the circular domains. [16] Similar patterns arising from "frustrated" condensation are observed in the study of charged polystyrene spheres [17] and of neutral nanoparticles [18] at the air-water interface. In the former case, the long ranged repulsions are also those of aligned dipoles, arising in this instance from the electrical double layer that forms "under" each charged particle. In the latter case, they are again dipolar in nature but this time associated with the alignment of water molecules at the interface between the aqueous subphase and the hydrophobic nanoparticles (silver or gold crystals passivated by monolayers of alkyl thiols).

2.4. SIZE-LIMITED CONDENSATES OF DNA (AND OTHER STIFF POLYELECTROLYTES)

We have already remarked about the fact that, in ultra-low-concentration aqueous solutions of DNA, addition of polyvalent cations leads to the formation of toroidal aggregates of a limited size. In the case of the significantly stiffer F-actin (with a persistence length of microns rather than hundreds of Angstroms), the condensate consists of bundles with limited-size diameters. We treat first the case of the toroidal condensate, because its unique topology implies some special constraints.

The characteristic DNA toroid has a circular cross-section whose radius we will denote by b. We introduce R to denote the average of the inner ($R_{\rm in}$)- and outer ($R_{\rm out}$)- radii of the toroid (with $R_{\rm out} - R_{\rm in} = 2b$), d for the interaxial spacing of the hexagonally packed, circumferentially wound DNA, and L for the total length of condensed chain (corresponding either to a single DNA molecule, as in the case of lambda phage, or to an aggregate of many shorter double-strands, arranged "head to tail"). The volume of the torus, apart from numerical factors of order unity, can be written as $V \cong Ld^2 \cong b^2R$, and its surface area as $A \cong bR \cong (VR)^{1/2}$. The energy of the toroid is a sum of several terms associated with the "bulk" and "surface" energies arising from the inter-chain attractions and with the bending energy of the chains:

$$E = E_{\text{bulk}} + E_{\text{surface}} + E_{\text{bending}} \cong -eV + ed(VR)^{1/2} + \kappa \frac{1}{R^2} \frac{V}{d^2}. \tag{3}$$

Here e is the cohesive energy density, which when multiplied by the inter-axial spacing d becomes an effective surface tension. κ is the 1D bending modulus of the DNA given by the product of the chain persistence length (l_p) and the thermal energy ($k_B T$). [3] Note that $1/R$ is the average curvature of the toroidally condensed chain, and V/d^2 is its total length. Minimizing the energy in Eq. (3) with respect to R gives $R \cong (\frac{\kappa}{ed^3})^{2/5} V^{1/5}$, [19], which implies the optimum *shape* of the toroid of volume V (with the circular-cross-section radius following from $b \cong (V/R)^{1/2}$). Substituting this result for R into Eq. (3) gives

$$E \cong -eV + [(\kappa e^4 d^2)^{1/5}] V^{3/5}. \tag{4}$$

Note that both the surface and bending energy terms in (3) end up scaling with the same power (3/5) of the volume. Because the first term is dominant at large volume, we see again (compare with the discussion following Eq. (2)) that, once a large enough volume is achieved condensation is expected to proceed without bound as the sub-linear term (barrier) is overwhelmed. The 3/5 power (vs. 2/3 in the case of the spherical condensate appropriate to simple liquids–see Eq. (2)) results simply from the geometry of the torus. Why, then, are finite sized toroids observed upon the addition of condensing agents to dilute DNA solutions?

One answer might lie in our neglect of the topological defects that are impossible to avoid in building up the toroid from DNA strands. More explicitly, it can be shown that there must be on average a new defect for every "circuit" around the toroid. This can be demonstrated by some artful playing with a length of rubber tubing. Start by winding it into a planar, closely packed spiral, i.e., putting it on the table and bending it in a circle back on itself several times around. The "condensed" chain will be parallel along its length to its neighboring chain portions; but–constrained to lie in the plane of the tabletop–it can have only two neighboring chains. For close packing in *three* dimensions, on the other hand, *six* neighboring chains are possible; indeed, this is the preferred organization of stiff chains in condensates that are driven by attractive interactions between parallel strands. Accordingly, to lower the cohesion energy, we need to depart from the planar spiral organization and begin to pull chain up out of the tabletop. But this requires *crossing over* the preceding chain portions, and this process must happen roughly one time each turn as one builds up the 3D toroid with circular cross-section. Each of these crossovers involves, locally, some extra bending of the chain and less optimum cohesion with neighboring chain. The resulting energy arising from these defects has been shown [20] to scale *super*linearly with the volume of condensed chain. Eq. (4) for ideal (defect-free) toroids must be modified to read

$$E = E_{\text{ideal}} + E_{\text{defects}} \cong -eV + []V^{3/5} + \{\}V^{p>1}. \tag{5}$$

It is this last term which provides a *minimum* in $E(V)$, coming at volumes larger than those around the region of the *maximum* (barrier). The position of this minimum corresponds to the preferred/restricted size of the toroidal condensate, and its value depends on the size of the dimensionless ratio of cohesive and bending energies, (ed^4/κ). If this ratio is too small, no condensation occurs, because the bending energy price is too high; if it is too large, the cohesive energy is overwhelming and there is no effective limit to toroid size. For typical values associated with physically relevant interaxial spacing (d), strength of attraction (e), and persistence length (κ), one finds [20] volumes of order those observed in the DNA condensation experiments.

An alternative explanation of the restricted size of aggregates of stiff polyelectrolytes arises from a consideration of the *kinetics* of aggregation/condensation, taking into explicit account the nature–in particular the mutual orientation dependence–of the interchain energies between highly charged, persistent chains. [21] It has long been appreciated that the repulsion between charged rods increases dramatically as one rotates them from a perpendicular to a parallel configuration. [22] In the presence of polyvalent cations, however, there is an attractive component of the interaction which becomes dominant at short interchain spacings and which is optimized by parallel configuration. [21] Figure 5 shows a schematic plot of the angle dependences of the interchain interaction energies for large (see solid curve, corresponding to a separation of 15 σ, where σ is the distance between fixed charges on the chains) and small separations (boldface and dashed curves). In the former case, the interaction is repulsive for all orientations but is smaller for perpendicular rods. This energy is described well by the mean-field theory (Poisson-Boltzmann approximation) because correlation effects involving the condensed counterions are not important at large separations. In the latter case, the interaction is attractive and is optimized for parallel rods; here, it is precisely the corrections to mean-field theory which account for the attractions. Note that ΔE corresponds to an energy barrier that perpendicular rods must overcome as they rotate into the parallel configuration at small separations. Ha and Liu [21] have shown that this barrier height increases as successive rods are brought up to the bundle condensate, scaling in particular as the square root of the number of

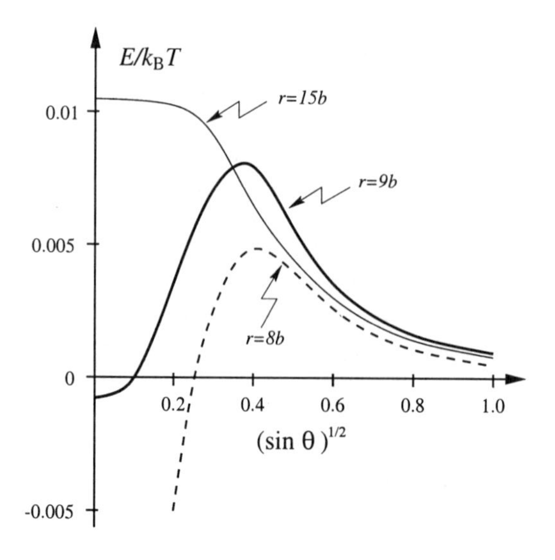

Figure 5.

rods in the bundle. Accordingly, when this energy barrier begins to exceed the thermally available energy (k_BT), further growth of the condensate becomes essentially impossible, even though each added rod would lower the overall energy once it was properly aligned with the others. In this sense, the aggregate size is *kinetically* limited.

2.5. *NON*-CLASSICAL DNA CONDENSATION: UNLIMITED SIZE TOROIDS

Recently toroidal aggregates of DNA, condensed by the presence of polyvalent cations, have been shown to form without any bounds on their size. [23] This is most likely because they have been formed without the need to get "around"/"over" the orientational barrier for successive addition of chains to the condensate. And this is because, as explained below, the DNA that is being added to condensate is *already aligned* with the aggregated chains. By contrast, in the classical dilute-solution studies, each persistence length is–by definition!–randomly oriented respect to the others, so that one is confronted directly by the energy barrier discussed immediately above. The "trick," it turns out, is to exploit Nature's simplest and most effective devices for packaging *and delivering* DNA: the bacterial viruses.

Figure 6 is a cryoTEM picture [23] of a toroidal condensate of DNA whose volume is almost ten times larger than the "classical" toroid shown in Figure 4. It consists of three molecules of the genome of the bacteriophage "T5", corresponding to a total of 366,000 base pairs. These DNA's have been transferred from the phage capsids to the interior of a (neutral,

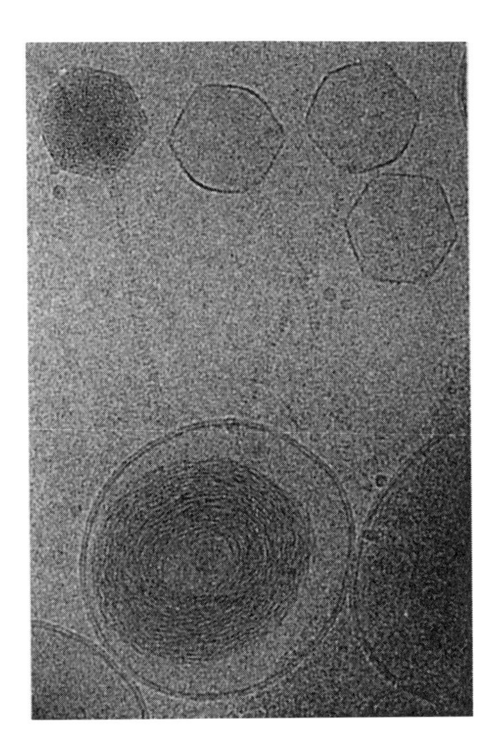

Figure 6.

phosphatidylcholine) lipid vesicle whose bilayer contains several molecules of the T5 receptor protein, FhuA. Upon binding of the tip of the tail of the virus to FhuA, the capsid is opened and the DNA ejected from the capsid into the vesicle interior. The vesicles were prepared in a sodium chloride solution (100mM) that also contained 50mM spermidine chloride, a sufficient concentration of trivalent ion to condense the DNA. A fourth T5 capsid has bound to the vesicle but has not yet ejected its DNA. Note the clear circumferential winding of the DNA's in the toroid, and the scale of its dimensions; no scale bar is explicitly included, because one has instead the capsid diameter–700 Angstroms–as a reference. Figure 7 [23] shows a simpler and more dramatic way to prepare arbitrarily large toroidal condensates. Here one dispenses with the lipid vesicles and directly adds the FhuA protein to a solution containing the purified T5 (as well as sodium [100mM] and spermidine [5mM] chloride salts). Each capsid is opened by a different molecule of FhuA, and yet a large number of neighboring capsids contribute their ejected DNA's to the same toroidal condensate. Figure 7 shows the toroid–$<$ 100 nm in diameter–formed at early times (after about 1 minute), i.e., the capsids are still mostly full; at later times (15 minutes) the toroid is much larger ($>$ 500 nm diameter) and contains more than

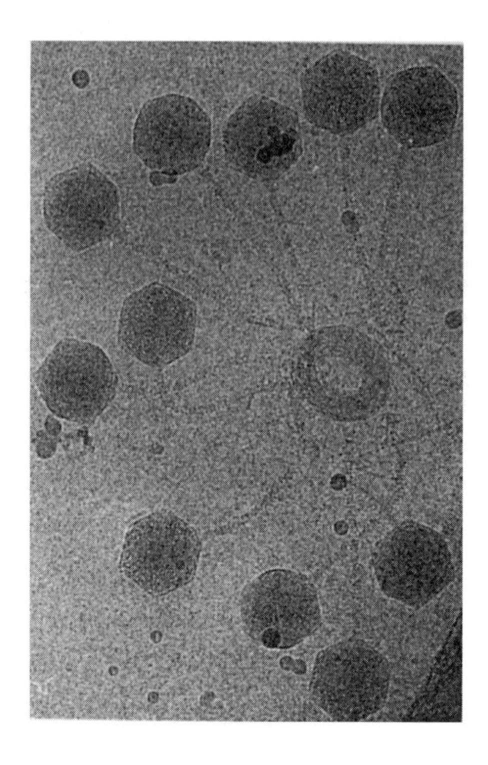

Figure 7.

25 times as much DNA as the toroids observed in the low-concentration, pure DNA studies involving "collapse" and aggregation of molecules already widely dispersed in solution.

How do we understand the fact that there appears to be no limit to the size of the toroids in the case of their preparation from viral ejection? Consider again Figure 7. While most of the DNA from each capsid has not yet been ejected, the fact is that the toroid is already being formed from the DNAs of as many as 10-12 different capsids. These capsids are randomly oriented in solution, and so are, therefore, the DNA's ejected from them. But the (approximately circular) cross-section of the initially formed toroid involves a small number of chains (on the order of dozens), corresponding to a small barrier ($< k_BT$) to orientation, as explained in the analysis above at the end of the preceding section. Since these leading portions of DNA are necessarily aligned, it follows that no *further* barrier develops as the remaining lengths of each DNA are "reeled in" by the ejection/condensation process.

A natural question that arises here is the nature of the force that drives the ejection of DNA from the viral capsids. Since this process occurs *spontaneously*, e.g., without the work of any molecular motor, one has to account for the pressure that is associated with the DNA packaged in the viral particles. (See below for the opposite case of *packaging* DNA in the capsids where such work is performed by "burning"/hydrolyzing ATP.) This point is addressed in the following section, where we treat the problem of confining a semiflexible polyelectrolyte in a small volume.

2.6. FORCES AND PRESSURES ASSOCIATED WITH PACKAGED DNA

Recall that the viral capsid diameter (R_c) is comparable to the persistence length (l_p) of DNA, and that the overall length (L) of the viral genome is hundreds of times greater. It follows that the free energy of the confined chain is much higher than that of the free–unpackaged–one. This free energy difference, *per unit volume*, corresponds to the *pressure* (or *stress*) of the packaged DNA. Its three main contributions come from: the *bending* of the chain (since R_c is of order l_p), the loss of *configurational entropy* (since $L >> l_p, R_c$), and the *inter-chain forces*. Note, further, that the forces between different portions of the chain can be either attractive or repulsive, depending upon whether or not a sufficient concentration of polyvalent cation is present. We are faced, then, with the problem of calculating the thermodynamic properties of such a confined chain, in particular its energy. Estimates of the above several contributions to the energy were made over twenty years ago by Riemer and Bloomfield [24] for the case of a self-attracting chain, i.e., for DNA in the presence of concentrations of di- and trivalent cations typical of those determined for T2 phage under *in vivo* conditions. [25] They decomposed the interchain interactions into their repulsive and attractive components. The former were estimated by their Poisson-Boltzmann counterion-condensation-renormalized form and the latter by invoking the relevant amine phosphate binding energy and polyamine composition. The two components comprise the largest contribution to the packaging free energy but essentially cancel each other, leaving a difference that is comparable with estimates of bending/kinking energies and configuration entropy contributions. The resulting energy densities involved correspond to pressures on the order of tens of atmospheres.

Odijk [26] has analyzed the case of DNA packaging in phage capsids for which the interchain interaction is exclusively repulsive. He considers the particular case of T7, whose DNA packing has been shown to be spoollike by a variety of measurements, including most recently and definitively cryoelectron micrographs of aligned, tailless, capsids [27]. The spool contains a hollow core that arises because the DNA must avoid bending on too small a length scale. Odijk first calculates the bending energy of the viral genome as a function of the radius of the hollow core, using the usual 1D elasticity (quadratic, continuum) theory. Then he does the same for the interchain electrostatic repulsion using the same nonlinear PB approximation mentioned above. Minimizing with respect to the hollow-core radius then gives the equilibrium interchain spacing as a function of total length of packaged DNA, yielding results that compare well with optical diffraction experiments. [27] The key qualitative conclusion is that, because of the small inner-radius of the spool, the bending stress of the DNA is sufficient to balance its electrostatic self-repulsion and form a stable hexagonally packed structure.

One can also investigate the packaged structures of viral DNA and estimate the associated forces and pressures by performing computer simulations. Kindt et al. [28] have done so for a simple model of a semiflexible chain confined in a spherical volume whose dimensions are comparable to the chain's persistence length but small (by orders of magnitude) compared to its overall contour length. Their potential energy is a familiar one from biopolymer simulation. All monomers interact with each other through Lennard-Jones forces; neighboring monomers are connected by tight springs with force constants large enough to essentially constrain the bond lengths to be constant (of order the Lennard-Jones σ). Neighboring bonds interact through a harmonic potential that is quadratic in the angle between them with a force constant that accounts for the chain's persistence length. Finally, a short-range repulsion is introduced between each monomer and the capsid wall, describing the confinement constraint. A Brownian molecular dynamics algorithm [29] is used to simulate the confined chain. More explicitly, at each time step Δt the displacement $\Delta \mathbf{r}_i$ of the ith monomer is given by

$$\Delta \mathbf{r}_i = \mathbf{F}_i \frac{D}{k_B T} \Delta t + \mathbf{R}_i. \tag{6}$$

Here $\mathbf{F}_i$ is the force on the ith monomer from all other monomers, and from the wall, D is a monomer diffusion constant, and $\mathbf{R}_i$ denotes the *random* displacement arising from solvent bath forces chosen from a Gaussian distribution with mean square width equal to $6D\Delta t$. The radius (R_c) of the capsid is chosen to be comparable to chain persistence length (l_p) and so that the chain volume fraction upon loading of the full chain (length L, volume $L\sigma^3$) is comparable to typical experimental values, i.e., just below close packed.

Figure 8 [28] shows equilibrated configurations for loaded volume fractions ($L\sigma^3/R_c^3$) of 0.1 (see 8A) and 0.6 (8B); the Lennard-Jones ε is 0.5kBT, and the persistence length (l_p) and capsid diameter ($2R_c$) are equal to one another. In 8B, two views are shown one from the side and the other down the spool axis. It is found that as more and more chain is loaded the original toroid structure evolves into a spool with nearly empty inner cylindrical core. In the final stages of loading a portion of the chain begins to fill in the

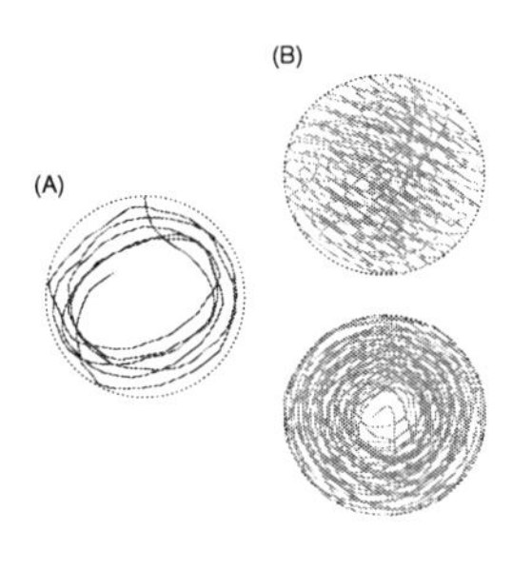

Figure 8.

core with strands parallel to the spool axis instead of continuing to wind further around the spool.

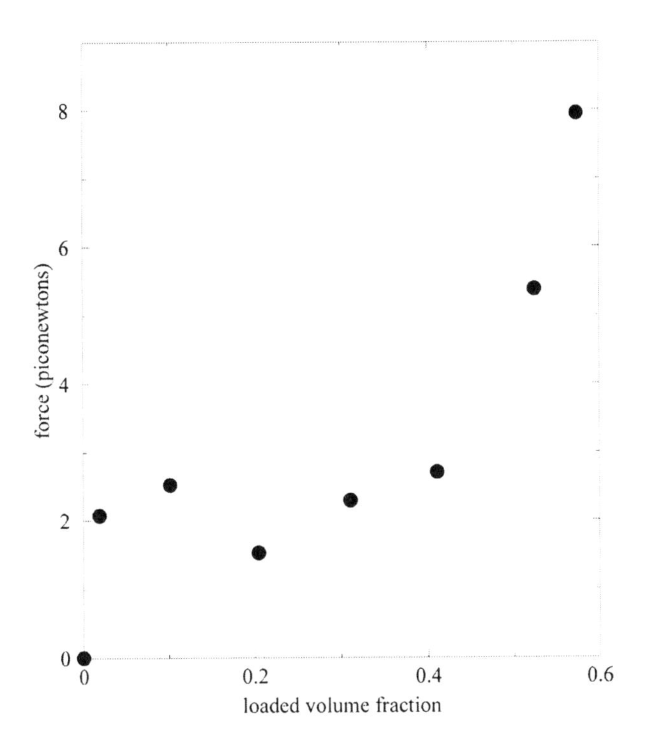

Figure 9.

For each of several loaded chain lengths the force acting radially outward on the last-to-enter monomer has been averaged over equilibrated configurations for that length. Figure 9 [28] is a plot of this force as a function of the loaded volume fraction. Note that the force resisting loading drops significantly at early stages as a condensed toroidal/spool structure begins to form in the capsid, corresponding to the chain being able to benefit from its self-attraction. As loading proceeds further, however, the force climbs steeply, reflecting the increasing curvature and packing strain of the condensate. Indeed, at this stage, neighboring chain portions are up against each other's repulsive walls.

This conclusion has been corroborated and further illustrated [28] by a phenomenological treatment of the same model, i.e., including the same intra- and interchain energetics. The bending energy is treated by the 1D continuum theory while the d (interaxial separation)-dependent DNA-DNA interactions are described by a simple functional that is consistent with osmotic stress measurements on DNA condensed by the addition of polyvalent counterions. [30] In this analytical theory it is possible to follow directly the evolution of packaged DNA from toroidal to spoollike structures upon increase in the loaded volume fraction. The loading force is again found to increase dramatically only at the end of the loading process, where the interaxial spacing begins to drop from the "preferred" value, i.e., the value for which the attractive forces are acting optimally. Beyond this point stress builds up as the elastic energy required to wind around the increasingly narrow inner core of the spool becomes prohibitive and the spool is compressed against the capsid wall, resulting in strong interchain repulsion.

3. Chain/Ball Wrapping and Chromatin Structure

3.1. CONDENSATION VS COMPLEXATION: FACTS OF COMPACTION

Genomic DNA, "naked" in simple salt solution, behaves very much like a flexible coil because its overall contour length is orders of magnitude larger than its persistence length. In the case of a simple bacterial virus, we found in particular–see Eq. (1)–that the radius of gyration (R_g) of the DNA in monovalent salt is $R_g \cong (N_p l_p^2)^{1/2} = (L l_p)^{1/2} \approx 10^4$A. More generally, it follows that the *volume fraction* of "monomer" (nucleotides) in a typical "naked" DNA configuration is of the order of $\phi \cong \frac{Lr^2}{R_g^3} \cong \frac{r^2}{(L l_p^3)^{1/2}}$, where r is the radius of the double helix; using $r \approx 10$A, $l_p \approx 500$A and $L \approx 10^5$A, we find that $\phi \approx 10^{-5}$. Note that close packing, e.g., crystalline densities correspond to $\phi \approx 1$, and it is in this sense that "naked" DNA is highly *ramified* or "dilute". It is clear from electron micrograph pictures like those shown in Figures 4 and 6 that the condition $\phi \approx 1$ obtains in DNA condensed by polyvalent cations in aqueous solution. That is, the double helical strands are hexagonally packed at densities comparable to those characteristic of the crystal.

Now let us consider the situation for DNA in chromatin, i.e., DNA complexed with globular aggregates of oppositely charged proteins (histones). As discussed in some detail in the following section, the basic building block of chromosomal DNA is the *nucleosome*, a complex that consists of about 200 base-pairs length of DNA in association with a 50A-radius "ball" of cationic protein. About 150 base pairs of the DNA are wrapped around the ball with 50 "left over" (25 on each side of the wrapped protein aggregate) to comprise the "linker" chain between nucleosomes. Twenty-five base pairs correspond to a length of about 90A, and hence the "radius" of each nucleosome (R_n) is roughly $50 + 90 = 140$A. One can then make an estimate of the volume fraction of base pairs in a nucleosome: $\phi \cong \frac{\ell r^2}{R_n^3} \approx 10^{-2}$, where ℓ is the wrapped length, $\approx 150\,(3.4\text{A})$. While this is small compared to unity, it is also large compared to 10^{-5}. Accordingly, it is useful to distinguish between the degree of DNA compaction in chromatin vs. in viral capsids; we shall use the term *complexation* to refer to the former ($\phi \approx 10^{-2}$), and *condensation* ($\phi \approx 1$) for the latter. This difference is consistent with the *evolutionary* "purpose" of the compaction in each case. More explicitly, DNA is compacted–but weakly so (hence, "complexed")– in chromatin so that the chromosomes can fit into the cell nucleus *and yet still be accessible for transcription* (gene expression). When DNA is compacted in viral capsids, on the other hand, it is done so more strongly (hence "condensed"), *for storage and protection.* Furthermore, as we have seen in the preceding section, this strong compaction leads to sufficient pressurization of the DNA for ultimate ejection upon infection of a bacterial cell by the virus.

3.2. CHROMATIN/NUCLEOSOME STRUCTURE

Figure 10 shows the high-resolution (2.8A) structure of the nucleosome "core particle" (NCP) as determined recently by X-ray diffraction. [31] "Core particle" describes nucleosomes whose linker chains have been digested by a nuclease (an enzyme that cuts up nucleic acids), leaving only the wrapped length–146 base pairs (bp)–of DNA. This length corresponds to almost two superhelical turns of the DNA (see white in Figure 10) around the histone octamer. In fact, the high-resolution crystal structures are obtained by complexing a *specific* 146 bp sequence of nucleotides that binds the histones especially strongly. The globular octamer of histones is well approximated by a short, squat, right-circular cylinder with radius

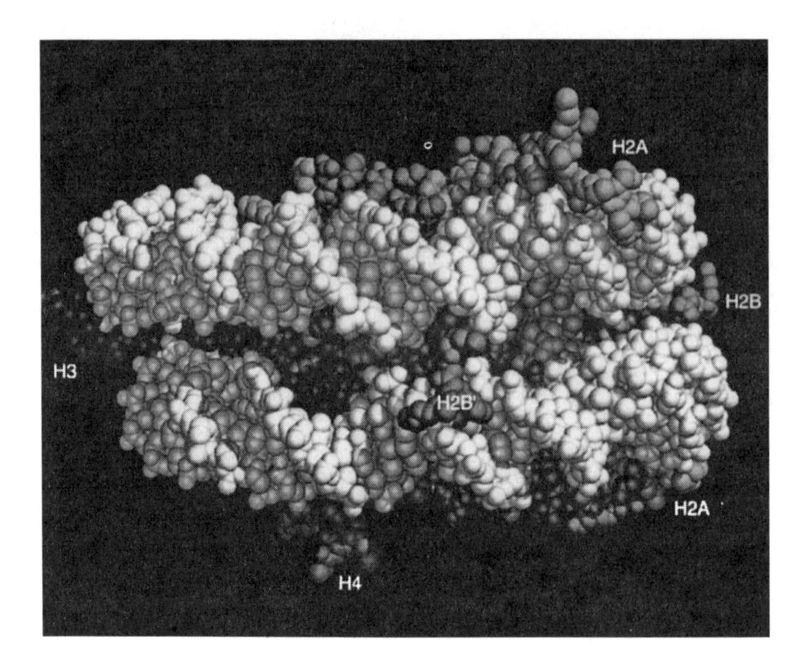

Figure 10.

50A and height 70A; in all the theoretical work discussed in the rest of this chapter this aggregate is simply treated as a sphere. The positive charge on this ball, due predominantly to the lysine and arginine amino acid residues of the histones, is more than compensated for by the negative charge of the phosphates on the DNA backbone strands; the NCP behaves as an *anionic* particle in electrophoresis experiments. This "overcompensation" is commonly referred to as *overcharging*, and will be discussed at length in the final section below.

The structural motif of chromatin–nucleosomes linked as wrapped "beads on a necklace"–is clear in the electron micrograph shown in Figure 11. [32] Immediately preceding Kornberg's nucleosomal model [33], which has been confirmed and refined many times since, the Olinses reported [34] electron micrographs (similar to Figure 11) of purified chromatin extracts whose structural properties were presciently deduced. They visualized chains of "ν bodies" that were estimated to involve a 400A length of DNA, suggesting (at 3.4A/bp) about 120 bp per nucleosome and quite close to the 146bp established by much subsequent work. Furthermore, under their conditions of 200mM monovalent salt concentration, the string of beads did not appear to be organized into any higher order structure. This low ionic strength motif–the "necklace of beads"–is often referred to as the *10nm fiber* because of the 10nm diameter of the individual nucleosomes.

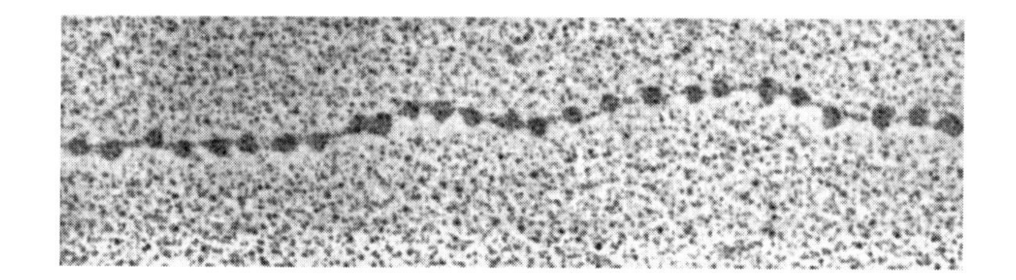

Figure 11.

3.3. SALT EFFECTS

Because of the high linear charge density of the linker DNA and the over-compensation of the charge of the histone octamer, one expects significant effects of added salts on the structure of both the chromatin fiber itself and the individual nucleosome particles. Figure 12 shows a "structural

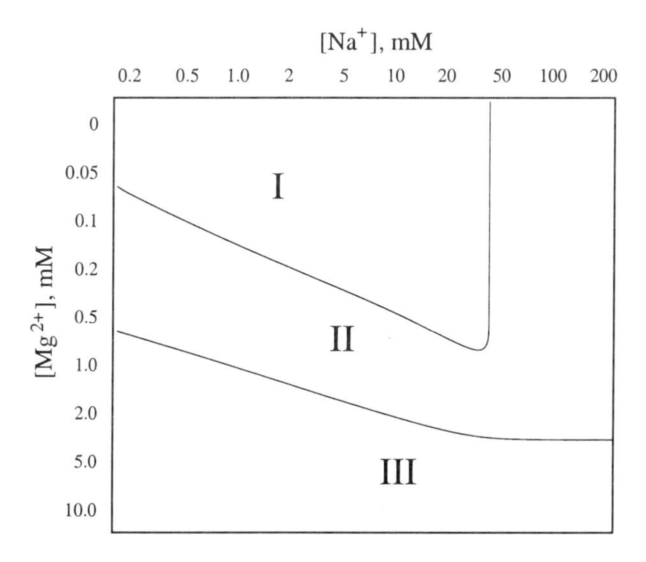

Figure 12.

phase diagram" of chromatin, adapted from the work of Widom [35], as a function of mono- and divalent cations, with structures determined from a combination of electron microscopy and X-ray diffraction experiments. [35] All pairs of Na^+ and Mg^{2+} concentrations in region I correspond to the basic 10nm fiber described immediately above. Region II refers to a set of structures that involve a somewhat condensed configuration of the beads on a necklace. Considerable controversy remains to this day concerning the details of this higher order structure, much data and argument supporting alternately a solenoidal, superhelical structure or a nonplanar zigzag configuration with internal linker. [36] In both cases the apparent diameter of this folded structure is approximately 30nm, and these higher ionic strength configurations are referred to accordingly as the *30nm fiber*. Finally, at significantly higher concentrations still, of mono- and/or divalent salts, the 30nm fibers are observed to be packed tightly together so that contrast between individual DNA chains is lost (see region III).

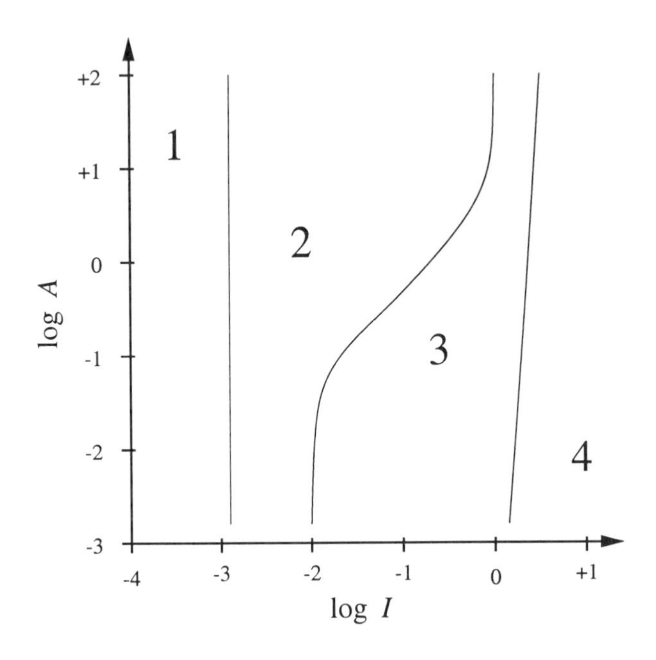

Figure 13.

The effects of salt on *individual nucleosome core particles*(NCP's) is shown in Figure 13, as a function of DNA concentration (measured by the 260 nm absorbance A) and ionic strength (I), adapted from the work of Yager, McMurray and van Holde. [37] At sufficiently *low* salt (NaCl) concentration the NCP's appear (region 1) as "expanded;" i.e., the DNA is partially unwrapped from the histone octamer. At sufficiently *high* ionic strengths there is no longer any trace of complexation between the DNA and histones (region 4), whereas at somewhat lower values the nucleosomes and octamers are only partially dissociated (region 3). Finally, at intermediate salt concentrations (region 2), one has NCP's with their characteristic fully wrapped structure.

In the final section below we address theoretically both the overcharging of the nucleosomes and the effects of salt on their unwrapping.

3.4. CHAIN/BALL WRAPPING

3.4.1. *Flexible Chains with No Counterions or Salt*

The simplest chain/ball complex is one involving a perfectly flexible chain, with no compensating counterion charges or added salt. While this is clearly an unphysical model for charged colloid and polyelectrolyte in aqueous solution, it provides nevertheless a good starting point for exposing some of the basic electrostatic aspects underlying more realistic complexes.

Let the chain consist of N monomers each with diameter σ and each carrying a fundamental charge that is negative, say; let the ball have a diameter D, and a total charge of Ze. $N > Z$ provides the possibility of overcharging, i.e., that the lowest-energy configuration of the complex will involve a number of adsorbed chain monomers exceeding Z. Let $N - \alpha$ be the number of adsorbed monomers, implying a length $\alpha\sigma$ of chain that does *not* wrap the ball. $Z_{\text{eff}} = Z - (N - \alpha)$ is the effective charge associated with the complex; $Z < 0$ corresponds to overcharging, since the bare charge of the ball is positive. Assuming that the adsorbed monomer charges are smeared out uniformly over the surface of the sphere, Mateescu et al. [38] have obtained analytical expressions for each of the following contributions to the total energy of the chain/ball system: self-energy of the adsorbed chain, attraction between Ze and the adsorbed anionic charge, interaction between the "complex" [charge Z_{eff}] and "tails" (unwrapped chain portions), and energy of the tails, both "self" and "inter" (tail-tail repulsion). Each of these terms is an explicit function of α, and minimization with respect to α gives directly the zero-temperature configuration and energy. In this way it is determined that the effective charge of the complex depends sensitively on the diameter of the ball. At large D values one finds strong overcharging, i.e., $Z_{\text{eff}} \approx -(N - Z)$, corresponding to all of the chain being wrapped and hence bringing to the ball (+Ze) *all* of its charge (-Ne). Around a critical value D^*, however, the effective charge of the complex drops quickly to zero, suggesting nothing more than neutralization of the ball charge by the chain. Monte Carlo simulations at finite temperature, for the same simple model, [38,39] confirm this dependence of overcharging on size of the ball.

3.4.2. *Effects of Chain Persistence and Added Salt*

Now suppose we allow for some rigidity of the wrapping chain, and for the presence of added salt. As a consequence there is a new, elastic, contribution to the energy, and all electrostatic interactions are exponentially screened. Let $l_B = e^2/\varepsilon k_B T$ denote the usual Bjerrum length (with ε the relevant dielectric constant), i.e., the distance at which the Coulomb interaction between two fundamental charges is equal to the thermal energy $k_B T$. κ, the inverse screening length, is given by $(8\pi l_B c_B)^{1/2}$, with c_B the concentration of added slat. The ball diameter is D, as before; the persistence and total length of the chain are l_p and L, respectively; and the number of charges per unit length of chain is denoted by τ. Finally, the chain configuration is described by the space curve $\mathbf{r}(s)$, where $\mathbf{r}$ is referenced to an origin at the center of the ball, and s is the arc length along the curve. It follows that the total energy can be written in the simple form

$$\frac{E}{k_BT} = \frac{1}{2}\ell_p \int ds\ddot{r}^2(s) - \frac{\ell_B Z\tau}{1+\kappa(D/2)} \int_0^L ds \frac{e^{-\kappa(|\mathbf{r}(s)|-D/2)}}{|\mathbf{r}(s)|} + \ell_B\tau^2 \int_0^L ds \int_s^L ds' \frac{e^{-\kappa(|\mathbf{r}(s)-\mathbf{r}(s')|)}}{|\mathbf{r}(s)-\mathbf{r}(s')|}. \quad (7)$$

The first term corresponds to the continuum elastic energy of the chain, locally proportional–through the 1D bending modulus, l_pk_BT–to the square of the curvature. The second and third terms are the (screened) electrostatic interaction energies between the chain and the ball, and amongst the monomers of the chain, respectively. A hard-core interaction between the chain monomers and the ball is included implicitly.

Kunze and Netz [40] have solved the above model numerically for a range of physical constants appropriate to the case of histone octamer and a nucleosomal length of DNA chain in simple salt solution. In the spirit of the *linearized* Poisson-Boltzmann theory, i.e., the Debye-Hückel approximation, the charge density per unit length of the chain is described by its bare value rather than its counterion-condensation-renormalized one, hence $\tau = 2/3.4$A. Their results are shown in Figure 14, for $Z = 40$,

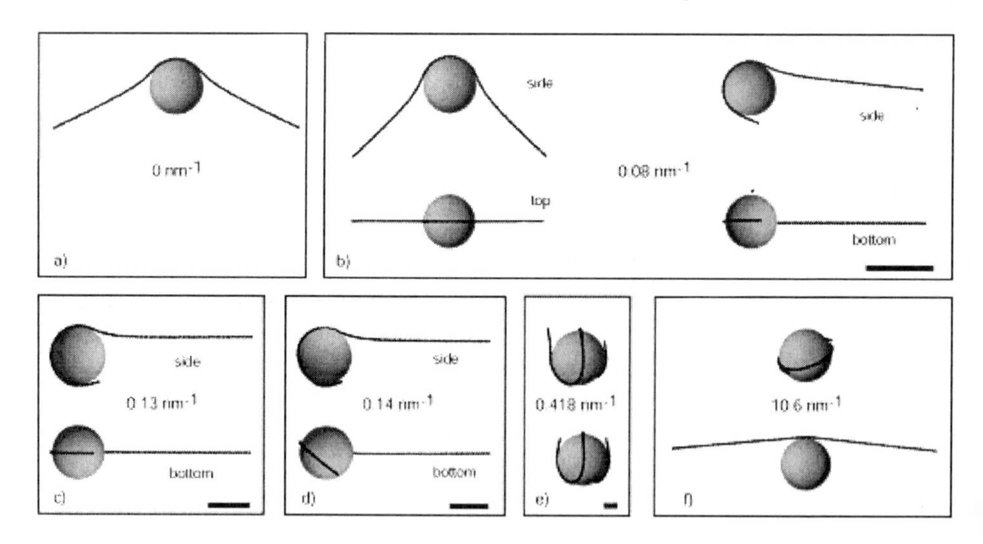

Figure 14.

$l_B = 7$A, $l_p = 30$nm, $D = 10$nm, $L = 50$nm (146 base pairs), and for a large range of screening lengths (κ^{-1}) varying from one Angstrom (high salt) to infinity (no salt). The screening length is indicated by the boldface scale bar in the lower right hand corner of each box, except for the first and last, where it is ∞ and 1 Å, respectively. It is pleasing that this simple model accounts qualitatively for the effects of ionic strength on nucleosome structure as discussed in the preceding section. More explicitly, under low salt conditions, where the screening length is large compared to ball size and chain length, the complex is significantly "expanded" (see region "1" in Figure 13); even though the chain binds to the oppositely charged ball, the repulsions between the monomers prevent wrapping of the chain. At intermediate ionic strengths the chain is able to wrap (see regions "2" and "3"), corresponding to the chain-ball attractions no longer being overwhelmed by the Coulomb interactions amongst monomers. Finally, at high salt concentrations (region "4" of Figure 13), *all* electrostatic interactions are screened out and the bending energy is dominant; the chain unwraps and becomes essentially straight along its full length ($L \approx \ell_p$). Kunze and Netz have worked out many interesting details of the wrapped states, including symmetry changes with respect to the ball equator, and transitions between equal and discrepant lengths of unwrapped chain portions; they have also calculated phase diagrams as a function of ball charge and ionic strength.

3.4.3. *Overcharging and Counterion Release: "Sandwich Model"*

Here we begin to take into account the explicit presence of counterions and the consequences of their "condensation" on highly charged surfaces. This requires, at the very least, dealing directly with the full *non-linear* Poisson-Boltzmann theory. For this purpose it is useful to consider the case of planar geometry, e.g., chains adsorbing on oppositely charged planar surfaces. A key point will be the correspondence between the maximum attraction, minimum-energy, situation and the maximum release of condensed counterions.

Figure 15 shows [41] schematically the complex formed from aligned rods intercalating into a lamellar phase of oppositely charged bilayer. This is the structure that has been discovered and extensively investigated for complexes of DNA and mixtures of cationic and neutral lipids [41]; a critical review of this work is given elsewhere in this volume. [42] Suppose one keeps fixed the composition of lipid, i.e., the fraction of them that is charged. Then, by changing the ratio of cationic lipid to phosphate, $\rho = N_+/N_-$, we can vary the d spacing between rods; the h spacing between bilayers is found to remain essentially constant, at about 28A, corresponding to the double-helix diameter augmented by a hydration shell. Note that $\rho = 1$ defines the "isoelectric point", where the lipid/DNA complex is charge-

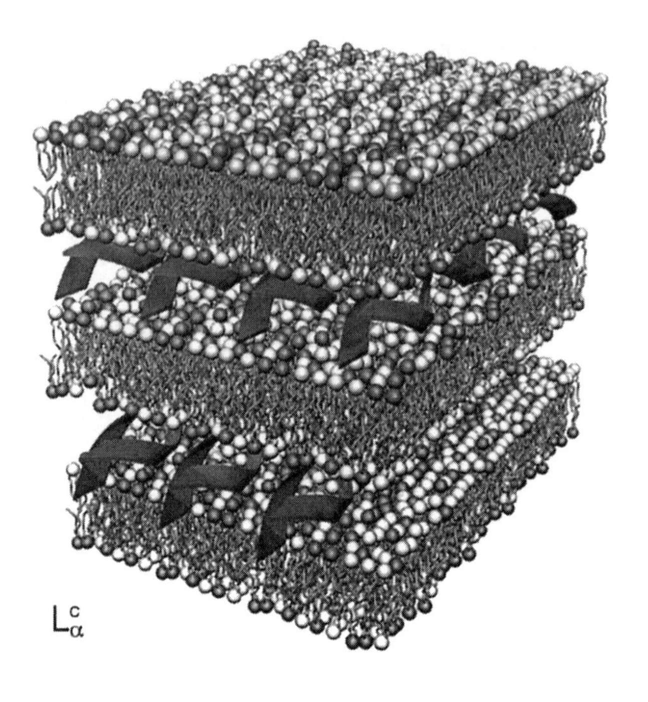

Figure 15.

neutral overall. One of the remarkable experimental facts about this system is that there exists a rather large range of $\rho < 1$ and $\rho > 1$ for which a single phase of lipid/DNA complex is observed. More explicitly, for fixed lipid composition–say 50/50 (and hence fixed charge densities σ_- and σ_+ on the DNA and bilayer surfaces, respectively), one finds [41]: excess DNA in equilibrium with complex at $\rho < 0.75$; excess lipid (vesicles) in equilibrium with complex at $\rho > 1.25$; and complex *with no excess phase* for $0.75 < \rho < 1.25$. In this latter, intermediate, regime, the complex of DNA intercalated in the lamellar phase of cationic/neutral lipid bilayers is first "overcharged" (excess DNA, $0.75 < \rho < 1.0$) and then "undercharged" (excess lipid, $1.0 < \rho < 1.25$). Alternatively put, for $0.75 < \rho < 1.0$ ($1.0 < \rho < 1.25$) the DNA associates with the cationic bilayers so as to *over (under)* compensate for their charge.

The above behavior can be treated by a solution to the nonlinear Poisson-Boltzmann equation applied to the pure DNA, pure bilayer, and DNA/ bilayer-sandwich situations. This has been done numerically by Harries et al., [43] who demonstrate that one can account directly in this way for the excess-DNA/ single-phase-complex/excess lipid progression described above. The d spacing in the single-phase regime is found to increase (linearly with ρ from 30 to 55A as the lipid/DNA ratio is increased from 0.75

to 1.25, corresponding to the observed 25% over- and undercharging of the complex. Furthermore, this behavior can be rationalized in terms of a simple "box model" for the distribution of excess charge in the complex. More explicitly, at the isoelectric point, where the DNA rods are spaced so as to bring just enough charge to balance that of the lipid bilayers, the spacing between rods (d) is related to that between bilayers (h) by $d = (\sigma_-/\sigma_+)h \equiv d^*$ ($\leftrightarrow d^{\text{isoelectric}}$). Imagine now that we are *away* from the isoelectric point, on say the $\rho > 1$ side, where there is excess cationic charge; here $d > d^*$ and the excess charge density can be expressed as $\sigma_+^{\text{excess}} = \sigma_+ \frac{d-d^*}{d}$. Suppose that we smear out this excess surface charge on the lipid bilayer surface and calculate the self-energy of the resulting electrical double layers and the interaction between two such opposing surfaces in the complex. The former energy (per unit area and in units of k_BT) is simply $(\sigma_+^{\text{excess}}/e)\ln(l_D/l_{GC})$, where l_D is the inverse screening length ($\leftrightarrow \kappa^{-1}$) and l_{GC} is the electrical double-layer thickness ("Gouy-Chapman length"); the latter energy is $1/l_B h$. Note that $\ln(l_D/l_{GC})$ is the free energy change per ion due to the difference in counterion concentrations in the condensed (double-layer) and bulk-solution conditions; $(\sigma_+^{\text{excess}}/e)$ is the number of ions that must be associated with the complex to ensure electroneutrality. Similar expressions are derived for the overcharging ($\rho < 1$) case where the excess charge is negative, and smeared out on the DNA surfaces. This approximate way of calculating the free energy of the complex works well to account for the phase behavior and degree of over- and undercharging outlined above.

The "charge smearing/box model" also provides a simple physical interpretation of why and how the complex can deviate so significantly from its isoelectric (charge-neutral) point. The isoelectric point is special because no counterions need to be confined in the complex; charge neutrality is assured by the "fixed" charges in the surfaces of the DNA and bilayers. Away from $\rho = 1$, on the other hand, there is an entropic cost for each extra counterion needed (see the $\ln(l_D/l_{GC})$ term from above), as well as an interaction between electrical double layers (see $1/l_B h$). Nevertheless, one does observe a significant range of deviation of ρ from unity. For example, as lipid mixture is added beyond $\rho = 1$ the extra lipid bilayer adds to the complex (rather than remaining in an excess bilayer phase), even though this involves repulsion between the charged bilayers in the complex. This happens as long as the excess charge density σ_+^{excess} remains small enough compared to σ_+, so that the self-energy of the bilayer is significantly lower in the complex than in the excess phase. At some point, namely as ρ nears 1.25, this energy lowering is no longer sufficient to offset the interaction (see $1/l_B h$) between bilayers in the complex; at this point all additionally added lipid goes into an excess phase.

Finally, it is interesting to remark on how the isoelectric point can be correlated with the maximum extent of counterion release. Solving the nonlinear PB equation for the complex allows one to calculate directly the distributions–and hence the *number*–of counterions associated with both the positive (lipid) and negative (DNA) surfaces. In this way Wagner et al. [44] have determined numerically the fraction, f, of counterions that is released upon complex formation. They find that f is essentially 1 at the isoelectric point; it decreases to significantly lower values on either side of $\rho = 1$. They have also *measured* the fraction of released counterions and found it to agree quantitatively with the nonlinear PB predictions.

3.4.4. *Chain/Ball Model with Counterions and Added Salt*

We now apply the ideas from the above discussion to a treatment of the original problem of a colloidal ball being wrapped by a polyelectrolyte of opposite charge. Consider the following physical situation. A semiflexible chain of radius r, length L, persistence length l_p, and a charge density of e per length b, is complexed with a ball of radius R ($\leftrightarrow D/2$) and total charge +Ze. By "complexed," we mean specifically that some length l of the chain is wrapped around the ball. The complex and its unwrapped portion of chain sit in an aqueous salt solution characterized by a Bjerrum length l_B and a Debye screening length κ^{-1}. What is the free energy F associated with this complex? In answering this question we shall decompose F into four contributions, each free energy term being "measured" in units of k_BT and relative to the situation where each charge (i.e., the fundamental charge e) is set equal to zero. More explicitly, we write:[48,49]

$$F = F_{\text{free chain}} + F_{\text{complex}} + F_{\text{free chain/complex}} + F_{\text{elastic}}. \tag{8}$$

The first term describes the free energy of a charged cylinder in salt solution. In the limit of b small compared to l_B, i.e., of large "Manning parameter" $\xi \equiv (l_B/b)$, this free energy term is dominated by the entropy associated with confinement of the "condensed" counterions, and the corresponding free energy can be estimated by exploiting the well-known results obtained from solutions to the Poisson-Boltzmann equation applied to a highly charged *planar* surface in salt solution. In this latter case, the relevant length scale describing the counterion distribution is the Gouy-Chapman length, l_{GC}. This is basically the thickness of the electrical double layer, and is related to the number density of surface charge σ by the well-known relation $l_{GC} = 1/2\pi l_B \sigma$. We can apply this "picture" to the case of a charged cylinder, iff the radius r of the cylinder is large compared to l_{GC}. [45] Then σ is given by $1/2\pi rb$, and hence l_{GC} by rb/l_B. In treating the free energy cost of confining the condensed counterions within l_{GC}, we estimate this quantity by its ideal value $\ln(c_{\text{cond}}/c_{\text{bulk}})$, on a per-particle (counterion) basis. (Note that this is the same as the $\ln(l_D/l_{GC})$ term discussed in 3.4.3 above.) Here c_{cond} and c_{bulk} refer to the concentrations of counterions within l_{GC} and in the surrounding salt solution, respectively. From the definition of the Debye screening length, c_{bulk} is given by $\kappa^2/8\pi l_B$. For simplifying our estimate of c_{cond}, on the other hand, we assume that *all* the counterions are condensed, e.g., one per area $2\pi rb$, consistent with the large ξ limit. It then follows that $c_{\text{cond}} = l_B/2\pi r^2 b^2$ and hence that

$$F_{\text{chain}} \cong \Omega \frac{L-l}{b}, \qquad \Omega = 2\ln(\xi\kappa^{-1}/r). \tag{9}$$

Here we have used the fact that Ω refers to a free energy per condensed counterion, whereas there are $(L-l)/b$ of these associated with the total length of unwrapped chain.

The second term is somewhat more complicated, because there are two regimes of sphere charge that must be considered. In the analysis immediately above we have used the fact that counterions condense on a rod as soon as the charged density of the cylinder exceeds a certain value, e/l_B. Furthermore, we treated a chain whose charge lay well above this limit and therefore whose counterions were virtually completely condensed, so that the effective charge density was essentially e/l_B, no matter how small b might be relative to l_B. In the case of a spherical macro-ion the situation is more subtle, because a nonzero concentration of macro-ions is necessary for them to be able to "hold on to" their counterions, i.e., in order for counterion condensation to occur. As argued by Oosawa, [46] and by Alexander et al., [47] however, the concentration of macro-ions enters only through a logarithmic term. More explicitly, the counterions begin to condense on a sphere as soon as the sphere charge Z exceeds the value

$$Z_{\text{max}} \cong \frac{R}{l_B}\ln(1/\varphi) \cong \frac{R}{l_B}, \tag{10}$$

where φ is the volume fraction of spheres in solution.

Now, how does a typical value of Z compare with this "critical", or "maximum", effective charge? For the case of nucleosomes that interest us here, the "bare" charge Z associated with the histone octamer is on the order of 250, while R is roughly 50 Angstroms (as compared with a Bjerrum length of 7), thereby implying $Z >> Z_{\text{max}}$. But for the general case of a wrapped-sphere complex, the relevant "bare" charge for discussing counterion condensation on the *complex* is

$$Z(l) \equiv Z - \frac{l}{b}, \tag{11}$$

since the length l of wrapped chain brings an opposing charge of l/b to the sphere. We consider, then, two cases: $Z(l) < Z_{\text{max}}$ and $Z(l) >> Z_{\text{max}}$. In the former case, the effective charge of the complex is $Z(l)$ and the corresponding free energy is simply the familiar charging energy $[Z(l)]^2 l_B/2R$. In the latter case, the free energy is dominated by the entropy contribution associated with the confinement of the condensed counterions. Via arguments similar to those detailed above for the highly charged chain (but with $\sigma = Z/4\pi R^2$ now), we find that this free energy cost–per counterion–is $\overline{\Omega} = 2\ln(Z(l)l_B\kappa^{-1}/R^2)$. Since in the high-charge limit all counterions are assumed to be condensed, i.e., there are $Z(l)$ of them, it follows that the free energy of the wrapped sphere complex can be written as

$$F_{\text{complex}} = \begin{array}{ll} [Z(l)]^2 \dfrac{l_B}{2R}, & Z(l) < Z_{\max} \\ |Z(l)|\overline{\Omega}, & |Z(l)| >> Z_{\max} \end{array} \tag{12}$$

The third term in Eq. (8) is the electrostatic interaction between the wrapped-chain complex and the remainder of the chain. Assuming that the screening length κ^{-1} is large compared to R, this free energy can be approximated by the simple form

$$F_{\text{free chain/complex}} \approx Z^*(l)\ln(\kappa R), \tag{13}$$

where $Z^*(l)$ is the effective charge of the complex, i.e., equal to $Z(l)$ and $Z_{\max}$ for $Z(l)$ smaller than or larger than $Z_{\max}$, respectively.

The final term in Eq. (8) is simply the elastic energy arising from the fact that the wrapped length l of chain is bent with a radius of curvature R which is small compared to the persistence length l_p:

$$F_{\text{elastic}} = \frac{l_p}{R^2} l. \tag{14}$$

Here we have again used the fact that the bending modulus of a semi-flexible chain is proportional to its persistence through a factor of k_BT.

Returning to the total energy, given by Eq. (8) with (9-12), we need to minimize with respect to wrapped length l to find the equilibrium state of the complex. In the case of sufficiently weak over- and under- charging, i.e., for $|Z(l)| < Z_{\max}$, we find [48], neglecting logarithmic corrections, that

$$l = Zb + \frac{R}{l_B/b}\{\ln\frac{l_D}{l_{GC}} - \frac{l_p b}{R^2}\}. \tag{15}$$

Note that: Zb is the *isoelectric* length, i.e., the wrapped length that neutralizes the complex. For physically relevant values of all the constants in (15), the second (negative) term in curly brackets is small compared to the first. This implies that the wrapped length *exceeds* the isoelectric value–the complex is *overcharged* (since the first term in brackets is positive with the Debye length much larger than the double-layer thickness). Obviously, when the chain is sufficiently rigid for the second term to dominate the first, the complex will be *under*charged. Finally, it is important to remark that all of the above analyses implicitly assume that the wrapped chain charge can be treated as smoothed out on the ball surface, an approximation that breaks down in the case of weaker wrapping. In this latter case, counterion correlation effects need to be explicitly taken into account, [50] and in this instance the overcharging is no longer dominated by counterion release effects.

References

1. Cairns, J., Stent, G.S. and Watson, J.D., eds. (1992) *Phage and the Origins of Molecular Biology*, Cold Spring Harbor Laboratory Press, Plainview.
2. See Figure 10-27, p. 288 in Griffiths, A. J. F., Miller, J.H , Suzuki, Lewontin, R. C., and Gelbart, W. M. (1993) *An Introduction to Genetic Analysis*, Freeman, New York.
3. See Figure 10-19 (a), p. 283 in reference 2.
4. Khokhlov, A. N. and Grosberg, Y. A. (1994) *Statistical Physics of Macromolecules*, AIP Press, New York.
5. Lambda DNA Packaging System (trademark: Packagene), information found at http://www.promega.com.
6. Hafner, E.W., Tabor, C.W. and Tabor, H. (1979) Mutants of E. Coli That Do Not Contain Putrescine or Spermidine, *J. Biol. Chem.* **254**, 12419-12426.
7. See, for example, Gronbech-Jensen, N., Mashl, R.J., Bruinsma, R.F. and Gelbart, W.M. (1997) Counterion-Inducted Attraction between Rigid Polyelectrolytes, *Phys. Rev. Lett.* **78**. 2477-2480; Ha, B.-Y. and Liu, A.J. (1997) Counterion-Mediated Attraction between Two Like-Charged Rods, *Phys. Rev. Lett.* **79**, 1289-1292 and references to earlier works contained therein.
8. Bloomfield, V.A. (1991) Condensation of DNA by Multivalent Cations: Considerations of Mechanism, *Biopolymers* **31**, 1471-1481.
9. Lerman, L.S. (1995) Chromosomal Analogues: Long-range Order in Ψ-Condensed DNA, *Cold Spring Harbor Symposia of Quantitative Biology* **38**, 59-73 and Ubbink, J. and Odijk, T. (1995) Polymer- and Salt-Induced (PSI) Toroids of Hexagonal DNA, *Biophys. J.* **68**, 54-61.
10. Tang, J.X., Ito, T., Tao, T., Traub, P. and Janmey, P. (1997) Opposite Effects of Electrostatics and Steric Exclusion on Bundle Formation by F-Actin and other Filamentous Polyelectrolytes, *Biochem.* **36**, 12600-12607 and references cited therein.
11. Rayleigh, Lord J.W.S. (1882) On the Equilibrium of Liquid Conducting Masses Charged with Electricity, *Phil. Mag.* **14**, 184-186.
12. Post, C.B. and Zimm, B.H. (1982) Theory of DNA Condensation: Collapse vs. Aggregation, *Biopolymers* **21**, 2123-2137.
13. See, for example, Dobrynin, A.V., Rubinstein, M. and Obukhov, S.P. (1996) Cascade of Transitions of Polyelectrolytes in Poor Solvents, *Macromolecules* **29**, 2974-79 and references contained therein.

14. Micka, U., Holm, C. and Kremer, K. (1999) Strongly Charged, Flexible Polyelectrolytes in Poor Solvent: Molecular Dynamics Simulations, *Langmuir* **15**, 4033-4044.
15. Andelman, D., Brochard, F., Knobler, C.M. and Rondelez, F. (1994) Chapter 12 in Gelbart, W.M. (ed.) *Microemulsions and Monolayers*, Springer-Verlag, New York.
16. Seul, M. and Andelman, D. (1995) Shapes and Patterns: The Phenomenology of Modulated Phases, *Science* **267**, 476-483 and references contained therein.
17. Ruiz-Garcia, J., Gamez-Corrales, R. and Ivlev, B. (1997) Foam and Cluster Structure Formation by Latex Particles at the Air/Water Interface, *Physica A* **236**, 97-104.
18. Sear, R. P., Chung, S.-W., Markovich, G., Gelbart, W.M. and Heath, J.R. (1999) Spontaneous Patterning of Quantum Dots at the Air-Water Interface, *Phys. Rev. E* **59**, R6255-R6258.
19. Grosberg, A. Yu and Zhestov, A.V. (1986) On the Compact Form of Linear Duplex DNA, *J. Biomol. Struct. Dynam.* **3**, 859-872.
20. Park, S.Y., Harries, D. and Gelbart, W.M. (1998) Topological Defects and the Optimum Size of DNA Condensates, *Biophys. J.* **75**, 714-720.
21. Ha, B.-Y. and Liu, A.J. (1999) Kinetics of Bundle Growth in DNA Condensation, *Europhys. Lett.* **46**, 624-630.
22. Stroobants, A., Lekkerkerker, H.M.W. and Odijk, T. (1986) Effect of Electrostatic Interaction on the Liquid-Crystal Phase-Transition in Solutions of Rodlike Polyelectrolytes, *Macromolecules* **19**, 2232-2238.
23. Lambert, O., Letellier, L., Gelbart, W.M. and Rigaud, J.-L. (2000) Delivery by Phage as a Strategy for Encapsulating Toroidal Condensates of Arbitrary Size into Liposomes, *Proc. Nat. Acad. Sci.* (USA) **97**, 7248-7253.
24. Riemer, S.C. and Bloomfield, V.A. (1978) Bacteriophage Heads: Some Considerations on Energetics, *Biopolymers* **17**, 785-794.
25. Ames, B. and Dubin, D.T. (1960) The Role of Polyamines in the Neutralization of Bacteriophage DNA, *J. Biol. Chem.* **235**, 769-775.
26. Odijk, T. (1998) Hexagonally Packed DNA within Bacteriophage T7 Stabilized by Curvature Stress, *Biophys. J.* **75**, 1223-1227.
27. Cerritelli, M.E., Cheng, N., Rosenberg, A.H., McPherson, C.E., Booy, F.P. and Steven, A.C. (1997) Encapsidated Conformation of Bacteriophage T7 DNA, *Cell* **91**, 271-280.
28. Kindt, J., Tzlil, S., Ben-Shaul, A. and Gelbart, W.M. (preprint) DNA Packaging and Ejection Forces in Bacteriophage.
29. Ermak, D. L. and McCammon, J. A. (1978), Brownian Dynamics with Hydrodynamic Interactions, *J. Chem. Phys.* **69**, 1352-1360.
30. Rau, D. C. and Parsegian, V. A. (1992) Direct Measurement of the Intermolecular Forces between Counterion- Condensed DNA Double Helices. Evidence for Long Range Attractive Hydration Forces, *Biophys. J.* **61**, 246-259.
31. Luger, K., Mader, A.W., Richmond, R.K., Sargent, D.F. and Richmond, T.J. (1997) Crystal Structure of the Nucleosome Core Particle at 2.8 Å Resolution, *Nature* **389**, 251-260.
32. This figure is a copy of Figure 8-9 (B), attributed to Victoria Foe, page 343, in: Alberts, B., Bray, D., Lewis, J., Raff, M., Roberts, K. and Watson, J.D. (1994) *Molecular Biology of the Cell*, 3rd edition, Garland, New York.
33. Kornberg, R.D. (1974) Chromatin Structure: A Repeating Unit of Histones and DNA, *Science* **184**, 868-71.

34. Olins, A.L. and Olins, D.E. (1974) Spheroid Chromatin Units (ν Bodies), *Science* **183**, 330-332.
35. Widom, J. (1986) Physicochemical Studies of the Folding of the 100A Nucleosome Filament into the 300A Filament, *J. Mol. Biol.* **190**, 411-424.
36. See, for example, the systematic and critical review by Widom, J. (1998) Structure, Dynamics, and Function of Chromatin in Vitro, *Ann. Rev. Biophys. Biomol. Struct.* **27**, 285-327 and reference contained therein.
37. Yager, T.D., McMurray, C.T. and van Holde, K.E. (1989) Salt-Induced Release of DNA from Nucleosome Core Particles, *Biochemistry* **28**, 2271-81.
38. Mateescu, E.M., Jeppesen, C. and Pincus, P. (1999) Overcharging of a Spherical Macroion by an Oppositely Charged Polyelectrolyte, *Europhys. Lett.* **46**, 493-498.
39. Nguyen, T.T. and Shklovskii, B.I. (preprint) Overcharging of a Macroion by an Oppositely Charged Polyelectrolyte.
40. Kunze, K.-K. and Netz, R. (2000) Salt-Induced DNA-Histone Complexation, *Phys. Rev. Lett.* **85**, 4389-4392.
41. Radler, J.O., Koltover, I., Salditt, T., and Safinya, C.R. (1997) Structure of DNA-Cationic Liposome Complexes: DNA Intercalation in Multilamellar Membranes in Distinct Interhelical Packing Regimes, *Science* **275**, 810-814.
42. See chapter by J. O. Radler in this book.
43. Harries, D., May, S., Gelbart, W.M. and Ben-Shaul, A. (1998) Structure, Stability, and Thermodynamics of Lamellar DNA-Lipid Complexes, *Biophys. J.* **75**, 159-173.
44. Wagner, K., Harries, D., May, S., Kahl, V., Radler, J.O. and Ben-Shaul, A. (2000) Direct Evidence for Counterion Release upon Cationic Lipid-DNA Condensation, *Langmuir* **16**, 303-306.
45. Rouzina, I. and Bloomfield, V.A.(1996) Competitive Electrostatic Binding of Charged Ligands to Polyelectrolytes: Planar and Cylindrical Geometries, *J. Phys. Chem.* **100**, 4292-4304.
46. Oosawa, F. (1971) *Polyelectrolytes*, Marcel Dekker, New York.
47. Alexander, S., Chaikin, P.M., Grant, P., Morales, G.J., Pincus, P. and Hone, D. (1984) Charge Renormalization, Osmotic Pressure, and Bulk Modulus of Colloidal Crystals: Theory, *J. Chem. Phys.* **80**, 5776-5781.
48. Park, S.Y., Bruinsma, R.F. and Gelbart, W.M. (1999) Spontaneous Overcharging of Macro-ion Complexes, *Europhys. Lett.* **46**, 454-460.
49. Schiessel, H., Bruinsma R. and Gelbart W.M., *J. Chem. Phys.*, in press
50. Nguyen, T.T. and Shklovskii, B.I. (2001) Complexation of a Polyelectrolyte with Oppositely Charged Spherical Macroions: Giant Inversion of Charge, *J. Chem. Phys.* **114**, 5905-5916 and Nguyen, T.T., Grosberg, A. Yu and Shklovskii, B.I. (2000) Screening of a Charged particle by Multivalent Counterions in Salty Water: Strong charge Inversion, *J. Chem. Phys.* **113**, 1110-1125.

INTERACTIONS IN COLLOIDAL SUSPENSIONS

Electrostatics, Hydrodynamics and their Interplay

DAVID G. GRIER and SVEN H. BEHRENS
Dept. of Physics, The James Franck Institute,
and The Institute for Biophysical Dynamics
The University of Chicago, Chicago, IL 60637, USA

Abstract. These lecture notes address some recent advances in our understanding of macroionic interactions inspired in part by the evolution of new techniques for studying macroions' dynamics.

Introduction

Charged colloidal particles suspended in water interact through hard core repulsions, van der Waals attractions, Coulomb interactions, and hydrodynamic coupling. The particles' influence on the surrounding medium modifies these interactions, for instance leading to screening of Coulomb interactions by atomic-scale simple ions. It also can give rise to entirely new effects such as entropically driven depletion interactions in heterogeneous suspensions. Bounding surfaces can modify all of these interactions, particularly if they carry their own charges. Competition and cooperation among these influences give rise to a bewildering variety of many-body phenomena including colloidal crystallization (of various kinds), gel formation (of various kinds), glass formation, and a host of collective responses to external forces and fields. Confidence in our ability to explain and control such cooperative behavior is undermined by the all-too-frequent discovery of new mysteries regarding colloidal interactions.

This contribution focuses on recent experimental investigations of interactions in dilute suspensions of monodisperse charge-stabilized spheres whose behavior should be governed principally by screened Coulomb and hydrodynamic interactions. Both classes of interactions have come under renewed scrutiny because of an accumulation of anomalous observations. For example, spheres carrying the same sign charge appear to attract each other under circumstances for which existing theories predict repulsion [1]. Such

C. Holm et al. (eds.), Electrostatic Effects in Soft Matter and Biophysics, 87–116.
© 2001 *Kluwer Academic Publishers. Printed in the Netherlands.*

anomalies may highlight previously overlooked many-body contributions to colloidal dynamics.

1. Experimental Techniques

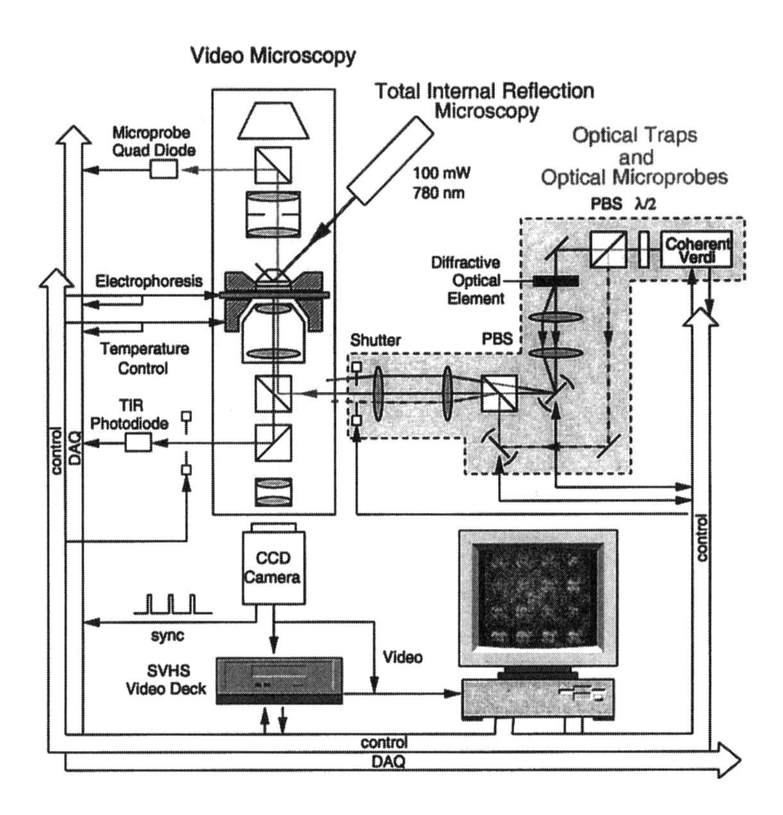

Figure 1. Experimental apparatus for measuring colloidal interactions and dynamics. Based on a conventional optical microscope, this system combines video imaging with optical tweezer manipulation, total internal reflection (TIR) particle tracking, and single-particle light scattering. The entire system, including environmental control and monitoring for the sample, is computer controlled.

1.1. DIGITAL VIDEO MICROSCOPY

Colloidal spheres larger than a couple hundred nanometers are easily observed with a good quality light microscope. Their images can be captured by a video camera, digitized, and analyzed with a computer to measure their positions precisely in each video frame. Identifying each particle's centroid with its scattering pattern's center of brightness yields resolutions exceeding 20 nm in the plane [2]. Consecutive snapshots of N particles' positions can be linked [2] into trajectories, $\vec{r}_i(t)$, to yield the in-plane distribution function

$$\rho(\vec{r},t) = \sum_{i=1}^{N} \delta(\vec{r} - \vec{r}_i(t)). \tag{1}$$

Our goal is to use these measured trajectories to shed new light on the interactions responsible for colloidal dynamics.

1.2. OPTICAL TRAPPING

Imaging colloidal particles' motions offers valuable insights into processes such as structural phase transitions, aggregation and gelation, and complex fluids' response to external forces. Complementary information becomes available with the ability to manipulate individual colloidal particles. Optical tweezers introduced by Ashkin, Dziedzic, Bjorkholm and Chu [3] provide this capability with an unparalleled combination of precision, sensitivity and reproducibility.

An optical tweezer uses forces exerted by gradients in light intensity to capture small particles at the focal point of a tightly focused laser beam. A high degree of convergence is necessary to prevent the particles from being dispersed by radiation pressure. Consequently, most optical tweezers are built around optical microscopes, as shown in Fig. 1, taking advantage of the objective lens' high numerical aperture and optimized optics to achieve a diffraction-limited focus. A collimated laser beam entering the objective's back aperture comes to a focus and forms a trap in the microscope's focal plane. Typically, a few hundred microwatts of visible light suffices to trap a micron-scale dielectric particle. Multiple beams entering the objective's back aperture form multiple optical traps, with the desired configuration of beams being created with beam splitters [4], holograms [5, 6], spatial light modulators [7] or by timesharing a single beam with high-speed deflectors [8, 9].

The combination of optical tweezer manipulation and digital video microscopy makes possible precise measurements of colloidal interactions and dynamics [2, 4]. In the following sections, we apply these methods to test long-accepted and still-evolving theories for colloidal interactions.

2. Electrostatic Interactions

The Coulomb interaction between charged colloidal particles dispersed in a polar solvent is moderated and mediated by a diffuse cloud of atomic-scale ions. These simple ions are much smaller and less highly charged than the macroionic colloid. The resulting disparity in dynamical time scales inspires the notion of an effective macroionic interaction averaged over the simple ions' degrees of freedom. How to formulate this effective interaction has inspired debate for half a century.

Fueling this controversy, experimental observations have raised the surprising possibility that like-charged colloidal spheres sometimes attract each other. If we view the spheres in isolation, their attraction seems counterintuitive. Recalling instead that the overall suspension is electroneutral suggests that unexpected features in the spheres' effective pair potential must reflect unanticipated dynamics in the surrounding medium. The phenomena discussed in the following sections are noteworthy because they appear to contradict long-accepted predictions of mean field theory. Such discrepancies raise concern about our understanding of such related problems as protein folding and colloidal stability.

2.1. THE DLVO THEORY

The partition function for N simple ions of charge q_i arrayed at positions $\vec{r}_i$ in the electrostatic potential $\phi(\vec{r})$ is

$$\mathcal{Q} = \mathcal{Q}_0 \int_\Omega d\vec{r}_1 \cdots d\vec{r}_N \, \exp\left[-\beta V\left(\{\vec{r}_i\}\right)\right], \tag{2}$$

where

$$V\left(\{\vec{r}_i\}\right) = \frac{1}{\epsilon} \sum_{i=1}^{N} q_i \phi(\vec{r}_i) \tag{3}$$

is the total Coulomb energy, the prefactor $\mathcal{Q}_0$ results from integrals over momenta, and $\beta^{-1} = k_B T$ is the thermal energy at temperature T. All charged species in the system, including the fixed macroions, contribute to $\phi(\vec{r})$. The macroions also exclude simple ions from their interiors, so their volumes are excluded from volume of integration Ω. Equation (3) implicitly adopts the primitive model, approximating the solvent's influence through its dielectric constant, ϵ. We will reconsider the solvent's role in Sec. 3 when we examine hydrodynamic and electro-hydrodynamic coupling.

The partition function can be expressed as a functional integral over all possible simple-ion distributions

$$\mathcal{Q} = \mathcal{Q}_0 \int' Dn \, \exp\left(-\beta f[n]\right), \tag{4}$$

where $n(\vec{r})$ is one particular simple-ion distribution with action

$$\beta f[n] \approx \beta V[n] + \int_\Omega n \ln n \, d\Omega. \tag{5}$$

The prime on the integral in Eq. (4) indicates that the simple ions' number is conserved: $\int_\Omega n \, d\Omega = N$.

Equation (5) differs from the exact action by terms accounting for higher-order correlations among simple ions. Dropping these terms, as we have in Eq. (5), yields a tractable but thermodynamically inconsistent theory [10]. Minimizing $f[n]$ to implement the mean field approximation yields the Poisson-Boltzmann equation

$$\nabla^2 \phi = -\frac{4\pi}{\epsilon} \sum_\alpha n_\alpha q_\alpha \exp\left(-\beta q_\alpha \phi\right), \tag{6}$$

where n_α is the concentration of simple ions of type α far from charged surfaces.

By considering only one ionic distribution, the mean field approximation also neglects fluctuations in the simple ions' distributions. Even this simplified formulation has no analytic solution except for the simplest geometries. Derjaguin, Landau [11], Verwey and Overbeek [12] (DLVO) invoked the Debye-Hückel approximation to linearize the Poisson-Boltzmann equation. Solving for the potential outside a sphere of radius a carrying charge $-Ze$ yields [13]

$$\phi(r) = -\frac{Ze}{\epsilon} \frac{\exp(\kappa a)}{1 + \kappa a} \frac{\exp(-\kappa r)}{r}. \tag{7}$$

The monotonic decay of correlations within the simple ion distribution is described by the Debye-Hückel screening length, κ^{-1}, given by

$$\kappa^2 = \frac{4\pi}{\epsilon k_B T} \sum_\alpha n_\alpha {q_\alpha}^2. \tag{8}$$

We will consider only monovalent simple ions with $q_\alpha = \pm e$.

Although the Debye-Hückel approximation cannot be valid near the surface of a highly charged sphere, nonlinear effects should be confined to a thin surface layer. Viewed at longer length scales, nonlinear screening should only renormalize Z and κ [14, 15, 16, 17, 18].

In this approximation, we obtain the effective pair potential by integrating Eq. (7) over the surface of a second sphere separated from the first by a center-to-center distance r. This integration is facilitated by assuming the second sphere's presence does not disrupt the first sphere's ion cloud. The resulting superposition approximation yields a screened Coulomb repulsion for the effective inter-sphere interaction,

$$U(r) = \frac{Z^2 e^2}{\epsilon} \left[\frac{\exp(\kappa a)}{1 + \kappa a}\right]^2 \frac{\exp(-\kappa r)}{r}. \tag{9}$$

The full DLVO potential includes a term accounting for dispersion interactions. These are negligibly weak for well-separated spheres [19, 20], however, and are omitted from $U(r)$.

2.2. ISOLATED COLLOIDAL PAIRS

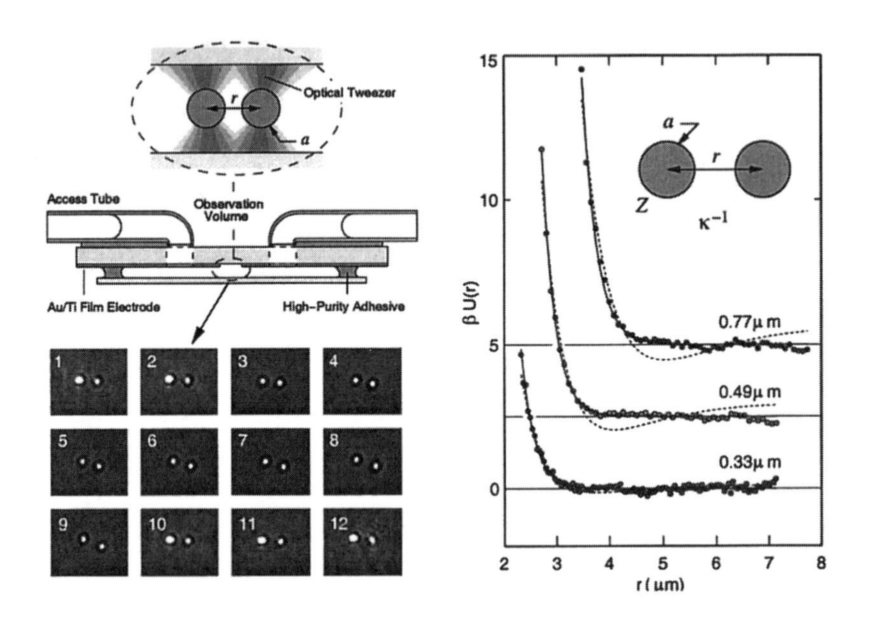

Figure 2. Left: Colloidal interactions are measured in a hermetically sealed glass sample container, shown in cross-section. Two spheres are selected with optical tweezers and alternately trapped and released, their motions being recorded at 1/30 sec intervals as shown in the sequence of images. Spheres appear brighter when trapped because of light backscattered from the optical tweezers. Right: Interaction potentials [21] for pairs of polystyrene sulfate spheres in deionized water at $T = 25°C$. Curves are labeled by the spheres' radii. Solid lines are nonlinear least squares fits to Eq. (9). Dashed lines are fits to the Sogami-Ise theory, Eq. (13).

Since its development, the DLVO theory has profoundly influenced the study of macroionic systems. Testing its predictions directly through measurements on pairs of spheres has become possible with the recent development of experimental techniques capable of resolving colloids' delicate interactions without disturbing them. These fall into three categories: (1) measurements based on the equilibrium structure of low density suspensions [22, 23, 24], (2) measurements based on the non-equilibrium trajectories [4, 25, 26, 27] of spheres positioned and released by optical tweezers [28] and (3) measurements based on the dynamics of optically trapped spheres [20, 29].

Methods (1) and (2) take advantage of the Boltzmann relationship

$$\lim_{\bar{\rho}\to 0} g(r) = \exp\left[-\beta U(r)\right] \tag{10}$$

between the spheres' pair potential and their equilibrium pair correlation function, which, for an ergodic sample and $r \neq 0$ can be evaluated as

$$g(r) = \left\langle \frac{1}{\bar{\rho}^2 \Omega} \int_\Omega \rho(\vec{x} - \vec{r}, t)\rho(\vec{x}, t)\, d\vec{x} \right\rangle \tag{11}$$

where $\bar{\rho}$ is the mean concentration of spheres and the angle brackets indicate an average over both time and angles. The two approaches differ in how they measure $g(r)$, but agree in their principal result: isolated pairs of spheres, far from walls and unconfined by their neighbors, repel each other as predicted by Eq. (9).

Non-equilibrium optical tweezer measurements, such as the examples in Fig. 2 use a pair of optical traps to position two spheres at reproducible separations in a microscope's focal plane. Extinguishing the traps eliminates any perturbing influence of the intense optical field and frees the particles to move under random thermal forces and their mutual interaction. Their motions are captured in individual video fields at $\tau = 1/30$ sec intervals and digitized. Each pair of consecutive images, such as the examples in Fig. 2, provides one discrete sampling of the probability $P(x, \tau | x', 0)$ that two spheres initially separated by x' will have moved to x a time τ later. Repeatedly trapping and releasing the spheres over a range of initial separations enables us to sample $P(x, \tau | x', 0)$ uniformly. This probability density is the kernel of the master equation [30] describing how the non-equilibrium pair distribution function, $\varrho(x, t)$, evolves in time:

$$\varrho(x, t + \tau) = \int_0^\infty P(x, \tau | x', 0)\, \varrho(x', t)\, dx'. \tag{12}$$

The domain of integration is conveniently limited in practice by the core repulsion which suppresses $\varrho(x', t)$ at small x', and by the finite range of the interaction which renders $P(x, \tau | x', 0)$ diagonal and $\varrho(x', t)$ independent of x' for large x'. Consequently, Eq. (12) may be discretized and solved as an eigenvalue problem for the equilibrium distribution $\lim_{t\to\infty} \varrho(x, t) \equiv \lim_{\bar{\rho}\to 0} g(x)$, whose logarithm is proportional to the pair potential through Eq. (10).

Data from Ref. [25] for three sizes of anionic polystyrene sulfate spheres dispersed in deionized water are reproduced in Fig. 2. Solid curves passing through the data points result from nonlinear least squares fits to Eq. (9) for the spheres' effective charges and the electrolyte's screening length. As expected [14, 15, 16, 17, 18], the effective charges, ranging from $Z = 6000$

for $a = 0.327$ μm up to 22,800 for $a = 0.765$ μm, are one or two orders of magnitude smaller than the spheres' titratable charges [2, 21]. The screening length of $\kappa^{-1} = 280 \pm 10$ nm is comparable to the spheres' diameters and corresponds to a total ionic strength around 10^{-6} M, a reasonable value for deionized water at $T = 25°$C. These numbers will be useful for comparison with results in Sec. 5. The apparent success of the screened-Coulomb functional form does not validate the Debye-Hückel approximation, however, since the exact theory including all ion correlations has the same leading-order behavior [31].

Comparable results were obtained by Vondermassen *et al.* [22] from measurements on optical cross-sections of dilute bulk suspensions at low ionic strength. Sugimoto *et al.* [20] studied pairs of spheres at higher ionic strength trapped in optical tweezers and were able to measure the van der Waals contribution. In all cases, the measured pair potentials agree at least semi-quantitatively with predictions of the DLVO theory.

The observed pair repulsions at low ionic strength pose a challenge to theories predicting long-ranged pair-wise attractions. For example, Sogami and Ise [32, 33] proposed that the colloidal pair potential can develop an attractive tail in the grand canonical ensemble when the simple ions' number is allowed to vary:

$$U_{SI}(r) = \frac{Z^2 e^2}{\epsilon} \left[1 + \kappa a \coth \kappa a - \frac{\kappa r}{2}\right] \frac{\exp(-\kappa r)}{r}. \tag{13}$$

This controversial theory has been quoted widely as a possible explanation for anomalous colloidal phenomena. However, Eq. (13) fails to describe the long-range pair repulsions in Fig. 2, and so cannot be expected to describe many-sphere behavior [21] through superposition.

2.3. METASTABLE SUPERHEATED CRYSTALS

Suspensions of pairwise repulsive monodisperse spheres are believed to exist in three equilibrium phases: fluid, face-centered cubic (FCC) crystal or body-centered cubic (BCC) crystal. Experiments on charge-stabilized colloidal suspensions have revealed other states and transitions, however, including (controversial [34]) equilibrium liquid-vapor coexistence [35, 36], reentrant solid-liquid transitions [37], stable voids [36], and metastable superheated crystals [27, 38]. These are most easily explained if the colloidal pair potential includes an attractive component [32, 33, 35, 36, 37, 39, 40] or if charge-stabilized suspensions develop many-body cohesions [41, 42].

Metastable crystals such as the example in Fig. 3 are made by forcing the spheres in a low density suspension against glass walls through an as-yet-unexplained electro-hydrodynamic instability [38]. An oscillating

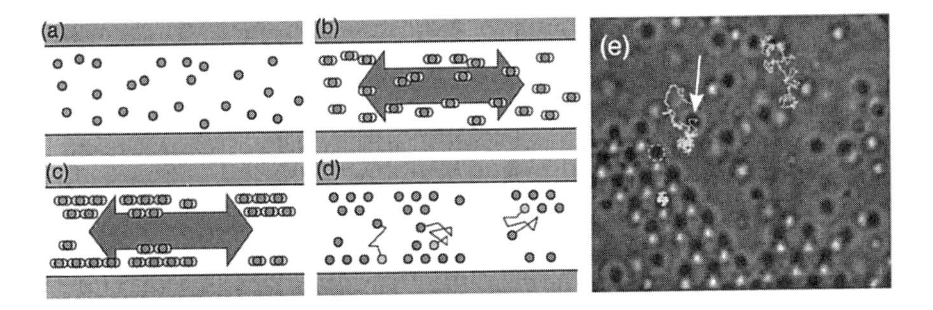

Figure 3. Electro-hydrodynamic crystallization: (a) Initial fluid. (b) Oscillating electric field drives spheres in the plane and also drives them transversely to the walls (c) where they crystallize. (d) Removing the field allows the now-superheated crystals to melt. (e) Video microscope image [27] of a metastable FCC colloidal crystallite coexisting with a low density fluid. Superimposed trajectories of selected spheres distinguish those localized in the crystal from others diffusing in the fluid. Some spheres, such as the one indicated with the arrow, temporarily bind to the crystal.

electric field drives the spheres back and forth electrophoretically while simultaneously creating a surface-driven ion flux through electro-osmosis. Under a limited set of conditions, the interplay of hydrodynamic and electric forces drives spheres out of the bulk of the suspension and toward the walls. The volume fraction near the walls increases until several epitaxial layers of close-packed crystal form. The particular example in Fig. 3 consist of polystyrene sulfate spheres of radius $a = 0.326 \pm 0.003$ μm (Catalog #5065A, Duke Scientific, Palo Alto, CA) compressed by a 10 V peak-to-peak 60 Hz signal applied across a 1 mm gap in a cell 90 μm thick and 1 cm wide. Created in deionized water, such compression-generated crystals can have lattice constants extending to more than 3 μm. The crystals should melt in a matter of seconds once the compressing field is turned off, at a rate limited by diffusion. Indeed, this is what happens [38] for suspensions whose ionic strength is greater than 10^{-6} M. Crystallites in more strongly deionized suspensions, however, can persist for as long as an hour [27, 38], their facets and interfacial dynamics attesting to a large stabilizing latent heat [27]. Such metastability cannot arise from the superposition of pair-wise repulsions [27].

To identify how many-body correlations could induce many-body attractions, Van Roij, Dijkstra and Hansen (vRDH) expanded a charge-stabilized suspensions' free energy functional to lowest non-trivial order in the simple ions' mutual interactions [41, 42]. The resulting free energy includes not only a superposition of DLVO-like repulsions but also a many-body cohesion. Failing to account for the simple ions' exclusion from the spheres' volume in this expansion yields the Sogami-Ise result.

vRDH use their free energy functional to map out the macroions' phase diagram and, under conditions similar to those reported in Refs. [27] and [38], predict crystal-fluid coexistence with an exceptionally large volume fraction contrast. Comparisons with other experiments suggest comparable explanations for void formation and other anomalous bulk phenomena. Despite its apparent success, the vRDH theory is controversial, in part because of concerns regarding the convergence of the expansions involved [43], and in part because it does not directly explain anomalous attractions measured for confined pairs of spheres.

2.4. CONFINED PAIR ATTRACTIONS

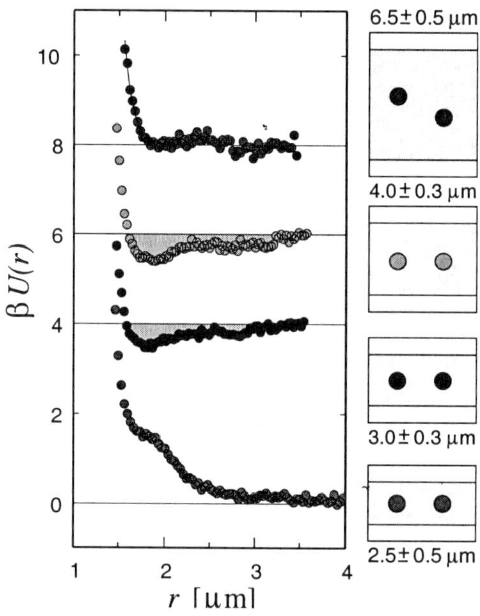

Figure 4. Confinement-induced attraction between pairs of colloidal spheres. Pair potential for spheres of radius 0.327 μm confined between parallel glass walls. For large separations, spheres are free to move in all three dimensions and their pair potential resembles that for unconfined spheres. Once the spheres are confined to the midplane by electrostatic interactions with the walls, the pair potential has a minimum, indicating long-ranged attractions. At very small wall spacings, spheres also experience an unscreened repulsion mediated by the glass walls.

The same spheres which repel each other in isolation and form metastable crystals when compressed also can develop a strong and long-ranged pair attraction when confined by one or two glass walls. Such attractions were first observed by Kepler and Fraden [23] who measured $g(r)$ by tracking particles in dilute monolayers sandwiched between parallel glass plates, and supplemented Eq. (10) with molecular dynamics simulations to obtain $U(r)$. Carbajal-Tinoco, Castro-Román and Arauz-Lara reproduced this result, further applying liquid structure theory to demonstrate that many-body correlations could not account for the apparent pair attraction [24]. This interpretation was supported by Rao and Rajagopalan's analysis of the experimental data with a predictor-corrector algorithm [44].

Optical tweezer measurements [25] such as the middle two examples in Fig. 4 confirm the walls' role by demonstrating their influence on two otherwise isolated spheres. Spheres between widely separated walls (Fig. 4 top) move freely in three dimensions; their pair potential is described accurately by the DLVO theory, in this case with $Z = 6,000$ and $\kappa^{-1} = 0.280\ \mu$m.

The sample container's glass walls also carry a negative surface charge due to the dissociation of silanol groups. As they move together, they tend to confine the negatively charged spheres to their midplane. This confinement appears to induce a long-ranged attraction which, together with the screened electrostatic repulsion, yields a minimum in the pair potential roughly 0.5 $k_B T$ deep at a separation of about 3 diameters.

Neu [45] and Sader and Chan [46] (NSC) recently proved that these observations cannot be explained by mean-field theory. The NSC proof holds for constant potential boundary conditions, at least some variants of constant charge boundary conditions, and for confining pores of arbitrary cross-section. Trizac and Raimbault extended the proof to include steric effects based on the the small ions' finite size [47]. These proofs appear to contradict [43] recent perturbative many-body calculations [48] which find wall-induced attractions for constant charge boundary conditions.

As a further complication, still closer confinement appears to induce a long-ranged *repulsion* in the pair potential, as in the bottom trace of Fig. 4. Stillinger [49] and Hurd [50] analyzed the related problem of charged colloid interacting near an air-water interface. The resulting unscreened dipole repulsion appears to be too weak [51] to explain the measurement, however.

2.5. DISCUSSION

No consensus has yet emerged regarding the origin of confinement-induced attractions in charge-stabilized colloidal suspensions. Even so, the existing body of experimental evidence constrains possible explanations.

Isolated pairs of like-charged spheres appear to repel each other without exception. And yet evidence for many-body attractions abounds when the number density of the same spheres is increased. On this basis, we provisionally rule out theories which predict pairwise attractions for isolated spheres.

This leaves open the question of how neighboring spheres or bounding surfaces induce long-ranged attractions.

Mean-field explanations appear to be excluded by the NSC and Trizac proofs. Fluctuations in the simple-ion distributions are not captured by mean-field treatments and thus offer a fruitful line of inquiry. Existing calculations suggest that such fluctuations do indeed lead to attractions, but that these are doubly-screened and thus short-ranged [52, 53]. Surface charge fluctuations due to ion adsorption similarly are found to induce short-ranged attraction [54]. Attractions such as those in Fig. 4, on the other hand, are longer-ranged than the singly-screened pair repulsion. If ionic fluctuations are responsible, the mechanism must involve modes of fluctuation not yet considered [43].

The primitive model on which the mean-field theory is based treats the supporting fluid as a continuum. While the solvent's discrete structure induces attractions at molecular length scales [55], it is unlikely to influence colloidal interactions at a range of several micrometers.

Very recently, Squires and Brenner [56] reported a previously overlooked role for the solvent in mediating confined colloidal interactions. The interplay of electrostatic and hydrodynamic interactions between charged spheres near a bounding charged wall can explain at least some anomalous observations, according to their theory. Exploring this suggestion invites an examination of colloidal hydrodynamics in confined geometries.

3. Hydrodynamic Interactions

Particles immersed in a fluid excite long-ranged flows as they move, and similarly move in response to fluid motion. By generating and reacting to a fluid's local velocity, colloidal particles experience hydrodynamic interactions with each other and with the walls of their container. Despite its long-recognized ubiquity, such hydrodynamic coupling is incompletely understood. This section addresses the hydrodynamically coupled dynamics of well-separated particles moving slowly through a viscous fluid, a comparatively simple limiting case of considerable practical importance. In particular, it applies the method of singularities [57], also known as stokeslet analysis, to the hydrodynamics of small colloidal systems and compares the results to recent measurements. The favorable outcome suggests that stokeslet analysis offers a tractable, general, and scalable formulation for many-body colloidal hydrodynamics.

3.1. STOKESLET ANALYSIS

The local velocity $\vec{u}$ in a slowly flowing incompressible viscous fluid is described by the Stokes equation

$$\eta \nabla^2 \vec{u} = \vec{\nabla} p. \tag{14}$$

where η is the fluid's viscosity, and p is the local pressure. In the absence of sources or sinks,

$$\vec{\nabla} \cdot \vec{u} = 0 \tag{15}$$

completes the flow's description. Eq. (14) is a good approximation for flows at low Reynolds numbers for which viscous damping dominates inertial effects – typical conditions for small collections of diffusing spheres.

Flows vanish on solid surfaces thus setting boundary conditions for solutions to Eqs. (14) and (15). Once these boundary conditions are satisfied, we can calculate the viscous drag on a particle by integrating the pressure tensor

$$\Pi_{\alpha\beta} = -p\,\delta_{\alpha\beta} + \eta\,(\nabla_\alpha u_\beta + \nabla_\beta u_\alpha) \tag{16}$$

over its surface:

$$F_\alpha = \int_S \Pi_{\alpha\beta}\, dS_\beta. \tag{17}$$

3.1.1. *A sphere in an unbounded fluid*

Stokes used this approach in 1851 to obtain the force needed to translate a sphere of radius a through an otherwise quiescent fluid of viscosity η at a constant velocity $\vec{v} = v\hat{z}$. The flow past the sphere has the form

$$\frac{u_\alpha(\vec{r})}{v} = \frac{3}{4}\,a\left(\frac{\delta_{\alpha z}}{r} + \frac{z r_\alpha}{r^3}\right) + \frac{1}{4}\,a^3\left(\frac{\delta_{\alpha z}}{r^3} - \frac{3 z r_\alpha}{r^5}\right), \tag{18}$$

where $\vec{r}$ is the distance from the sphere's center and z the displacement along the direction of motion. The flow's contribution to the pressure is

$$p(\vec{r}) = \frac{3}{2}\,\eta a \frac{\vec{r}\cdot\vec{v}}{r^3}. \tag{19}$$

Substituting these into Eqs. (16) and (17) yields

$$\vec{F} = \gamma_0\, \vec{v}, \tag{20}$$

for the drag, where the sphere's drag coefficient is

$$\gamma_0 = 6\pi\eta a. \tag{21}$$

The same drag coefficient parameterizes the force needed to hold the sphere stationary in a uniform fluid flow $\vec{u}$. Conversely, a constant force $\vec{F}$ applied to the sphere causes it to attain a steady-state velocity

$$\vec{v} = b_0 \vec{F}, \tag{22}$$

where $b_0 = 1/\gamma_0$ is the sphere's mobility.

Similar calculations for more complicated systems can be acutely difficult; few have analytic solutions. For this reason, a great many highly specialized approximation schemes have been devised for hydrodynamic problems. Progress for many-body systems such as colloidal suspensions has been steady, but slow.

3.1.2. *Faxén's Law*

The foundation for the analysis which follows is provided by Faxén's first law [58], introduced in 1922, which relates a sphere's velocity $\vec{v}$ to the forces and flows it experiences:

$$\vec{v} = \vec{u}(\vec{r}) + b_0 \vec{F} + \frac{a^2}{6} \nabla^2 \vec{u}(\vec{r}). \tag{23}$$

If no external forces act on the sphere, and if pressure gradients are small over its diameter, then the last two terms on the right-hand side of Eq. (23) may be ignored. In this approximation, the sphere is simply advected by the external flow.

3.1.3. *The mobility tensor*

This observation immediately suggests a simplified description of colloidal hydrodynamic coupling. Consider a sphere labeled j, located at $\vec{r}_j$ within a suspension. A force $\vec{f}(\vec{r}_j)$ applied to this sphere excites a flow $\vec{u}(\vec{r}) = \mathsf{H}(\vec{r} - \vec{r}_j)\, \vec{f}(\vec{r}_j)$ at position $\vec{r}$. This linear relationship is guaranteed by the Stokes equation's linearity. H is known as the Oseen tensor and is obtained from Eqs. (18), (20) and (21). At large enough distances, $\vec{u}$ appears sufficiently uniform that a sphere at $\vec{r}_i$ is simply advected according to Faxén's law:

$$v_\alpha(\vec{r}_i) = u_\alpha(\vec{r}_i) \tag{24}$$

$$= \mathsf{b}_{i\alpha,j\beta}\, f_\beta(\vec{r}_j), \tag{25}$$

where $\mathsf{b}_{i\alpha,j\beta}$ is a component of a mobility tensor describing sphere i's motion in the α direction due to a force applied to sphere j in the β direction. More to the point, $\mathsf{b}_{i\alpha,j\beta} = \mathsf{H}_{\alpha\beta}(\vec{r}_i - \vec{r}_j)$. The diagonal elements, $\mathsf{b}_{i\alpha,i\alpha} = b_0$, describe the sphere's own response to an external force.

Any number of sources may contribute to the flow past sphere i, with each contributing linearly in the Stokes approximation. The mobility tensor therefore may be factored into a self-mobility and a mobility, b^e, due to all external sources,

$$\mathsf{b}_{i\alpha,j\beta} = b_0\,\delta_{ij}\,\delta_{\alpha\beta} + \mathsf{b}^e_{i\alpha,j\beta}. \tag{26}$$

The results in this section all follow from this equation, once the necessary flow fields have been calculated. In particular, we will consider contributions to a sphere's mobility from (1) neighboring spheres and (2) bounding planar surfaces.

3.1.4. *The stokeslet approximation*

The flow field generated by a moving sphere (Eq. (18)) is essentially independent of the sphere's radius when viewed at sufficiently large distances. Consequently, we may approximate the flow field around a sphere driven through an unbounded fluid by the flow due to a unit force on a point within the fluid. The corresponding Oseen tensor is the Green's function for the system:

$$\mathsf{G}^S_{\alpha\beta}(\vec{r}) = \frac{1}{8\pi\eta}\left(\frac{\delta_{\alpha\beta}}{r} + \frac{r_\alpha r_\beta}{r^3}\right). \tag{27}$$

This Green's function plays an important role in many hydrodynamic problems and is known as a *stokeslet*. Comparison with Eq. (18) shows that G^S is indeed a good approximation for H when $r \gg a$.

3.2. PAIR DIFFUSION

As a first application of the stokeslet approximation to colloidal hydrodynamics, we calculate two colloidal spheres' hydrodynamic coupling in an unbounded fluid. Each sphere moves in the flow field G^S due to its neighbor. The corresponding mobility tensor is

$$\mathsf{b}_{i\alpha,j\beta} = b_0\,\delta_{ij}\delta_{\alpha\beta} + (1-\delta_{ij})\,\mathsf{G}^S_{\alpha\beta}(\vec{r}_i - \vec{r}_j) \tag{28}$$

with $i, j = 1, 2$. This result describes the spheres' hydrodynamic interactions at separations large compared with their diameters. It is not in the most convenient form for comparison with optical tweezer measurements, however. For those purposes, we relate spheres' mobility to their more easily measured diffusivity.

3.2.1. *Mobility and diffusivity*

A sphere suspended in a fluid follows a random thermally-driven trajectory $\vec{r}(t)$ satisfying

$$\langle |r_\alpha(t) - r_\alpha(0)|^2\rangle = 2D_\alpha t, \tag{29}$$

where the angle brackets indicate an ensemble average, and D_α is the sphere's diffusion coefficient in the α direction. In 1905, Einstein demonstrated that a sphere's diffusivity at temperature T is simply related to its mobility through the now-familiar Stokes-Einstein relation

$$D_0 = k_B T \, b_0. \tag{30}$$

This was the first statement of the more general fluctuation-dissipation theorem which plays an important role in statistical mechanics. More generally, components of the N-particle diffusivity tensor

$$\mathsf{D} = k_B T \, \mathsf{b}. \tag{31}$$

parameterize generalized diffusion relations [59]

$$\langle \Delta r_{i\alpha}(\tau) \Delta r_{j\beta}(\tau) \rangle = 2 \, D_{i\alpha, j\beta} \, \tau. \tag{32}$$

describing how particle i's motion in the α direction influences particle j's in the β direction. Off-diagonal terms in D thus encode spheres' hydrodynamic interactions.

In general, D can be diagonalized by projection onto a system of coordinates ψ consisting of linear combinations of the $\vec{r}_i$. These normal modes evolve independently over time; they offer natural experimental probes of the particles' dynamics.

3.2.2. *Normal mode diffusivity*

In the particular case of two identical spheres, diagonalizing the diffusivity tensor yields diffusion coefficients for collective (C) motion of the center-of-mass coordinate $\vec{R} = \vec{r}_1 + \vec{r}_2$ and relative (R) motion $\vec{r} = \vec{r}_1 - \vec{r}_2$, with one set of normal modes directed perpendicular ($\perp$) to the initial separation, and the other parallel ($\|$):

$$D_\perp^{C,R}(r) = \frac{D_0}{2} \left[1 \pm \frac{3}{2} \frac{a}{r} + \mathcal{O}\left(\frac{a^3}{r^3}\right) \right] \tag{33}$$

$$D_\|^{C,R}(r) = \frac{D_0}{2} \left[1 \pm \frac{3}{4} \frac{a}{r} + \mathcal{O}\left(\frac{a^3}{r^3}\right) \right]. \tag{34}$$

Positive corrections apply to collective modes and negative to relative. The collective diffusion coefficients $D_\perp^C$ and $D_\|^C$ are enhanced by hydrodynamic coupling because fluid displaced by one sphere entrains the other. Relative diffusion is suppressed by the need to transport fluid into and out of the space between the spheres.

Batchelor [60] obtained a series solution of the Stokes equation for this system in 1976, including additional terms up to $\mathcal{O}\left(a^6/r^6\right)$. This observation highlights both the weakness and the strength of stokeslet analysis.

The stokeslet G^S depends only on a sphere's position and not on its radius. Consequently, Eq. (28) contains information only to linear order in a. On the other hand, the same approach can be applied readily to other configurations, and the associated diffusivity tensor diagonalized to obtain the independent dynamical modes.

Deriving Eqs. (33) and (34) required enough approximations that the reader might be concerned about the results' accuracy. Figure 5 compares their predictions with measured diffusion coefficients for silica spheres of radius $a = 0.495 \pm 0.025$ μm (Lot 21024, Duke Scientific, Palo Alto, CA) suspended in water. These measurements were obtained with optical tweezers using techniques similar to those in Sec. 2 except that electrostatic interactions were minimized with added salt.

Two spheres were captured with optical tweezers, raised to $h = 25$ μm above the nearest surface and released, their motions being tracked through digital video microscopy for 5/30 sec thereafter. Repeatedly positioning and releasing the spheres over a range of initial separation yielded statistically large samples of the spheres' hydrodynamically coupled motions in the plane. These were binned in r and analyzed according to Eq. (32) to extract the plotted diffusivities. In the absence of hydrodynamic coupling, and thus at large separations, the spheres' pair diffusivities should be $2D_0$, with the free self-diffusion coefficient expected to be $D_0 = 0.550 \pm 0.028$ $\mu\text{m}^2/\text{sec}$ at the experimental temperature $T = 29.00 \pm 0.05°$C. This limiting behavior appears as a dashed line in Fig. 5 and agrees with the experimental data at large separations. Eqs. (33) and (34) yield the dashed curves in Fig. 5 and agree well with the measured diffusivities, with no adjustable parameters.

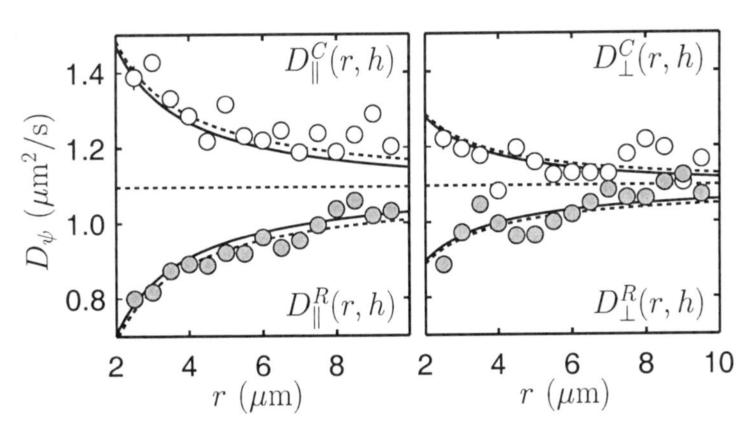

Figure 5. Measured pair diffusion for 1 μm-diameter silica spheres in water at $T = 29°$C. Dashed curves result from stokeslet analysis (Eqs. (33) and (34)). Solid curves include corrections for proximity to a wall (Sec. 3.4).

Crocker [61] previously reported a larger data set for two polystyrene microspheres somewhat closer to the nearest bounding wall whose dynamics agreed less well with Eqs. (33) and (34). Crocker surmised that the spheres' hydrodynamic coupling to the nearer wall reduced their diffusivities, and that this affected collective diffusion more than relative. However, a quantitative analysis of a wall's influence was not available and the correction was treated semi-empirically. Stokeslet analysis provides the necessary accounting for confinement's influence.

3.3. HYDRODYNAMIC COUPLING TO A WALL

The flow field around a moving sphere spans an unbounded system. A wall's no-flow boundary condition modifies this flow, breaking its symmetry and increasing the drag on the sphere. Faxén introduced the method of reflections in 1927 to address this question and obtained the diffusivity for a sphere at height h above a wall moving parallel ($\|$) to the surface:

$$D_{\|}(h) = D_0 \left[1 - \frac{9}{16}\frac{a}{h} + \frac{1}{8}\frac{a^3}{h^3} - \frac{45}{256}\frac{a^4}{h^4} - \frac{1}{16}\frac{a^5}{h^5} + \mathcal{O}\left(\frac{a^6}{h^6}\right) \right]. \tag{35}$$

The method of reflections is challenging even for this configuration and essentially unworkable for more complex systems.

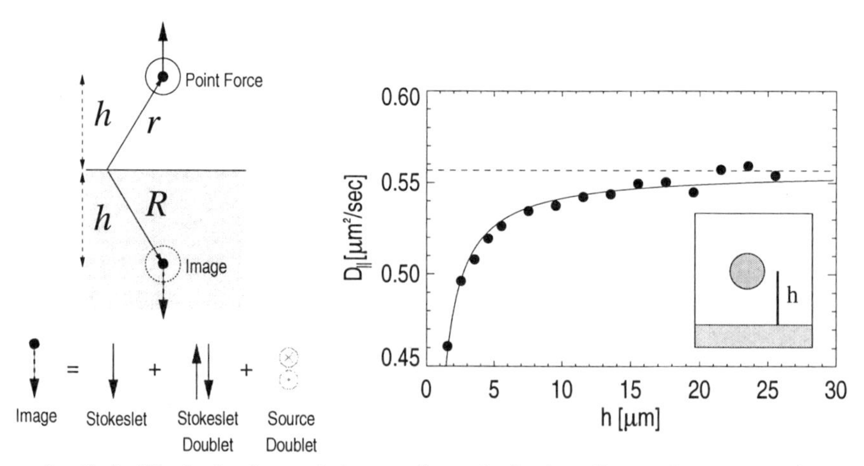

Figure 6. Left: The hydrodynamic image of a stokeslet in a flat surface is a combination of an oppositely directed stokeslet, a stokes doublet and a source doublet. Right: The in-plane diffusivity of a 1 μm-diameter silica sphere is suppressed by proximity to a surface. The solid curve shows the stokeslet prediction (Eq. (39)) while the essentially indistinguishable dashed curve is Faxén's prediction (Eq. (35)).

As early as 1906, Lorentz reported the Green's function for flow near a flat surface. Seventy years later, Blake [62] recognized that Lorentz's result could be reinterpreted by analogy to electrostatics. He suggested that the flow due to a stokeslet could be canceled on a bounding surface by conceptually placing its hydrodynamic image on the opposite side. Solutions to the Stokes equations being unique, the resulting flow must be Lorentz's Green's function for bounded flow.

The electrostatic image needed to cancel a charge distribution's field on a surface is just an appropriately scaled mirror image of the initial source. In hydrodynamics, the image of a stokeslet is not simply another stokeslet, but rather a more complicated construction including sources which Blake dubbed a stokeslet doublet (D) and a source doublet (SD). This combination is depicted schematically in Fig. 6. The flow due to the entire image system is described by the Green's function [62]

$$\mathsf{G}^W_{\alpha\beta}(\vec{r}-\vec{R}) = -\mathsf{G}^S_{\alpha\beta}(\vec{r}-\vec{R}) + 2h^2\,\mathsf{G}^D_{\alpha\beta}(\vec{r}-\vec{R}) - 2h\,\mathsf{G}^{SD}_{\alpha\beta}(\vec{r}-\vec{R}), \quad (36)$$

where $\vec{R} = \vec{r} - 2h\hat{z}$ is the position of the image and

$$\mathsf{G}^D_{\alpha\beta}(\vec{x}) = \frac{1-2\delta_{\beta z}}{8\pi\eta}\frac{\partial}{\partial x_\beta}\left(\frac{x_\alpha}{x^3}\right) \quad \text{and} \quad (37)$$

$$\mathsf{G}^{SD}_{\alpha\beta}(\vec{x}) = (1-2\delta_{\beta z})\frac{\partial}{\partial x_\beta}\,\mathsf{G}^S_{\alpha\beta}(\vec{x}) \quad (38)$$

are Green's functions for a source doublet and a stokeslet doublet, respectively.

Applying Faxén's first law and identifying $\mathsf{b}^e = \mathsf{G}^W$ leads to

$$D^S_{\parallel}(h) = D_0\left[1 - \frac{9}{16}\frac{a}{h} + \mathcal{O}\left(\frac{a^3}{h^3}\right)\right], \quad (39)$$

where the S superscript distinguishes this result from Eq. (35). As before, we are left wondering about the higher-order terms in Faxén's more specialized analysis.

Figure 6 shows typical data obtained with optical tweezers and digital video microscopy for a silica sphere's diffusion above a wall. The particular sphere for this data set was one of the pair studied in the previous section. The sphere's height h above the wall was repeatedly reset by the optical tweezer at $\tau = 83$ msec intervals. During this period of free motion, it could diffuse out of plane only $\Delta z = \sqrt{2D_0\tau} = 0.3\ \mu$m, on average. Advancing the microscope's focus in $1.0 \pm 0.3\ \mu$m steps allowed us to sample the dynamics' dependence on h. The measured height-dependent diffusivity agrees well with both Eqs. (35) and (39).

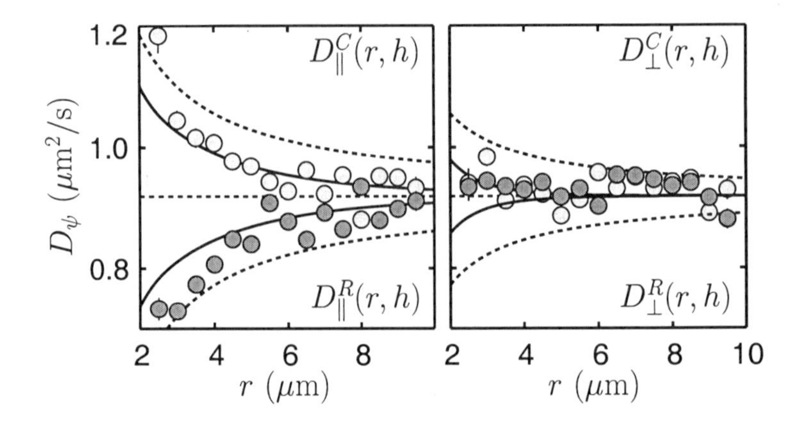

Figure 7. Measured in-plane pair diffusion coefficients for two 1 μm-diameter silica spheres in water at $T = 29°$C at height $h = 1.6$ μm above a glass wall. Dashed curves result from linear superposition of drag coefficients, while solid curves result from stokeslet analysis.

3.4. PAIR DIFFUSION NEAR A WALL

In his 1927 treatise, Oseen suggested that Faxén's results might be applied to more complicated systems even if his methods could not [63]. Oseen pointed out that the drag on a sphere moving in direction α near a wall can be factored into the sphere's drag in an unbounded system and an additional contribution $\gamma_\alpha^W(h)$ due to the wall:

$$\gamma_\alpha(h) = \gamma_0 + \gamma_\alpha^W(h). \tag{40}$$

He proposed that the drag coefficient for motion in a given direction might be approximated by linearly superposing individual contributions $\gamma_{\alpha i}$ from all bounding surfaces and neighboring particles:

$$\gamma_\alpha(\vec{r}) \approx \gamma_0 + \sum_{i=1}^{N} \gamma_{\alpha i}(\vec{r} - \vec{r}_i). \tag{41}$$

Oseen emphasized that this cannot be rigorously correct because it violates boundary conditions on all surfaces. Even so, if the surfaces are well-separated, the errors may be acceptably small. Given this hope, Oseen's linear superposition approximation has been widely adopted.

The data in Fig. 7 show measured in-plane pair diffusion coefficients for 1 μm-diameter spheres positioned by optical tweezers at $h = 1.55 \pm 0.66$ μm above a glass wall. Naively adding the drag coefficients [58] due to sphere-sphere and sphere-wall interactions yields $D_\psi^{-1}(r,h) = D_\psi^{-1}(r) + [D_\parallel^{-1}(h) - D_0^{-1}]/2$, whose predictions appear as dashed curves in Fig. 7 and agree poorly with measured diffusivities.

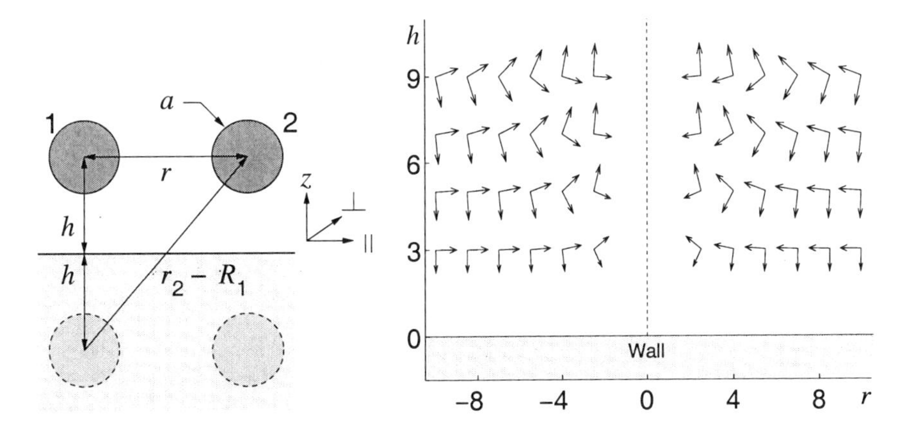

Figure 8. Left: Hydrodynamic interactions for two spheres near a planar boundary. Right: Normal modes of motion obtained from stokeslet analysis of this system [64].

A more complete treatment not only resolves these quantitative discrepancies but also reveals an additional influence of the bounding surface: the highly symmetric and experimentally accessible modes parallel to the wall are no longer independent. As shown in Fig. 8, each sphere interacts with its own image, its neighbor, and its neighbor's image. These influences contribute $\mathsf{b}^e_{i\alpha,j\beta} = (1 - \delta_{ij})\, \mathsf{G}^S_{\alpha\beta}(\vec{r}_i - \vec{r}_j) + \mathsf{G}^W_{\alpha\beta}(\vec{r}_i - \vec{R}_j)$ to the mobility of sphere i in the α direction. Eigenvectors of the corresponding diffusivity tensor appear in Fig. 8. The independent modes of motion are rotated with respect to the bounding wall by an amount which depends strongly on both r and h. Even though the experimentally measured in-plane motions are not independent, they still satisfy Eq. (32) with pair-diffusion coefficients $D^{C,R}_\alpha(r,h) = D_{1\alpha,1\alpha}(r,h) \pm D_{1\alpha,2\alpha}(r,h)$, where the positive sign corresponds to collective motion, the negative to relative motion, and α indicates directions either perpendicular or parallel to the line connecting the spheres' centers. Explicitly, we obtain [64]

$$\frac{D^{C,R}_\perp(r,h)}{2D_0} = 1 - \frac{9}{16}\frac{a}{h} \pm \frac{3}{4}\frac{a}{r}\left[1 - \frac{1 + \frac{3}{2}\xi}{(1+\xi)^{3/2}}\right] \quad \text{and} \tag{42}$$

$$\frac{D^{C,R}_\parallel(r,h)}{2D_0} = 1 - \frac{9}{16}\frac{a}{h} \pm \frac{3}{2}\frac{a}{r}\left[1 - \frac{1 + \xi + \frac{3}{4}\xi^2}{(1+\xi)^{5/2}}\right] \tag{43}$$

up to $\mathcal{O}(a^3/r^3)$ and $\mathcal{O}(a^3/h^3)$, where $\xi = 4h^2/r^2$. These results appear as solid curves in Figs. 6 and 7. Not only does the stokeslet approximation perform better than linear superposition for spheres close to a wall, it performs equally well at all separations we have examined [64].

4. Electro-hydrodynamic Coupling

The hydrodynamic measurements described in the previous section were carried out under conditions for which electrostatic interactions could be ignored. By contrast, hydrodynamic interactions may well have influenced the interaction measurements described in Sec. 2. Squires and Brenner [56] pointed out that a charged wall repels a nearby charged sphere with a force proportional to the wall's surface charge density, σ_W:

$$F_z^{(1)}(h) = 4\pi \frac{Ze\sigma_W}{\epsilon} \frac{\exp(\kappa a)}{1+\kappa a} \exp(-\kappa h). \tag{44}$$

Now consider two spheres released from optical tweezers at height h above the wall with an initial center-to-center separation r. The force, $F_{2z}^{(1)}(h)$, on sphere 2 establishes a flow which advects sphere 1 according to Faxén's first law [58]. The advection velocity's in-plane component,

$$v_{1x}(r,h) = b_{1x,2z}^{(1)}(r,h)\, F_{2z}^{(1)}(h), \tag{45}$$

draws the sphere toward its neighbor, a like-charge attraction of electro-hydrodynamic origin. The relevant component,

$$b_{1x,2z}^{(1)}(r,h) \approx \frac{3}{32\pi\eta h} \frac{\xi^2}{(1+\xi^2)^{\frac{5}{2}}}, \tag{46}$$

of the mobility tensor arises from the sphere's hydrodynamic interaction with its neighbor's image in the wall. The corresponding component of the force on sphere 1,

$$f_{1x}(r,h) = b_{1x,2z}^{(1)}(r,h)\, F_{2z}^{(1)}(h) / \left(b_{1x,1x}^{(1)}(h) - b_{1x,2x}^{(1)}(r,h) \right), \tag{47}$$

is purely kinematic rather than thermodynamic [56]. Nevertheless, it could be mistaken for an equilibrium attraction in optical tweezer measurements such as those described in Sec. 2.

Using the wall-corrected in-plane mobility $b_{1x,1x}^{(1)}(h) \approx b_0[1 - 9a/(16h)]$ (Sec. 3.3), Squires and Brenner [56] found that Eqs. (9) and (47) together account for Larsen and Grier's optical tweezer measurement of like-charge attractions near a single charged wall [27]. The polystyrene sulfate spheres in those measurements had a radius of $a = 0.326$ μm and an effective charge number of $Z = 7300$ established by interaction measurements far from walls [27]. When the same spheres were released at $h = 2.5$ μm above the nearest glass wall, their measured interaction developed a minimum 0.5 k_BT deep at a separation of $r \approx 3$ μm. The Squires-Brenner theory reproduces these features for $\sigma_W = 2200\, e$ μm^{-2}, a value consistent with silanol dissociation in pure water.

Squires and Brenner's success at explaining Larsen and Grier's optical tweezer measurements casts the problem of like-charge colloidal attractions in a new light. Similar electro-hydrodynamic coupling might explain optical tweezer measurements of attractions measured between two parallel walls [25] if the spheres were not released precisely at the midplane. However, electro-hydrodynamic coupling cannot have influenced the two independent equilibrium measurements of confined colloids' interactions [23, 24] which revealed attractions comparable to those measured with optical tweezers [25, 27]. If charged colloids' electrostatic interactions are purely repulsive, as suggested by mean-field theory and assumed by Squires and Brenner, attractions measured with optical tweezers might be ascribed to hydrodynamics while those observed by equilibrium structural measurements would have to be in error.

The sequence of measurements in Fig. 4 suggests instead that the success of the Squires-Brenner theory for Larsen and Grier's measurement near one wall does not rule out an *equilibrium* origin for confinement-induced like-charge attractions under slightly different conditions. This interaction measurement was performed a bit further than $h = 3$ μm away from two charged walls [2] and shows no sign of a long-range attraction, kinematic or otherwise. If charged walls *do* induce equilibrium attractions between nearby spheres, Larsen and Grier's measurements may have been performed too far from the wall for these to have been evident. In this case, the electro-hydrodynamic mechanism may explain the data in Ref. [27] while leaving all of the other anomalous observations unresolved.

There would be little room for alternative interpretations if the available equilibrium interaction measurements on confined monolayers [23, 24] were known to be free of artifacts. However, such measurements are subject to subtle sources of error, some of which could mimic long-ranged attractions [44]. To explore this possibility, we measured the equilibrium pair interactions in a monolayer of silica spheres near a single charged planar surface.

5. Electrostatics Near One Wall

We studied suspensions of silica spheres with diameter $\sigma = 2a = 1.58$ μm (Catalog # 8150, Duke Scientific, Palo Alto, CA) sedimented in deionized water into monolayers above the bottom wall of a 200 μm-thick glass sample cell. The balance of electrostatic and gravitational forces maintained the spheres in a two-dimensional layer at $h = 0.9 \pm 0.1$ μm. As in previous studies [23, 24], we extracted the sphere's equilibrium pair potential from their pair correlation function $g(r)$. Unfortunately, $g(r)$ cannot be related directly to $U(r)$ using Eq. (10) because we cannot be sure *a priori* that the

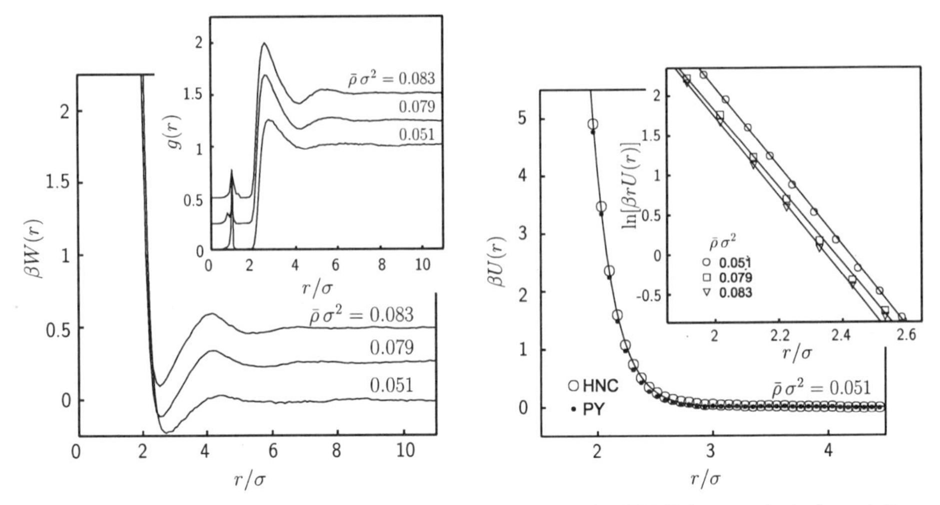

Figure 10. The pair interaction energy calculated in the HNC (open circles) and Percus-Yevick (full dots) approximations together with a fit of the HNC result to Eq. (9). Inset: Logarithmic representation of HNC results and best fits for data obtained at different areal densities.

monolayer is sufficiently dilute. Finite concentration introduces many-body correlations into $g(r)$ and, in general,

$$W(r) = -k_B T \ln g(r) \tag{48}$$

is the density-dependent potential of mean force, rather than the pair potential. To make matters worse, no exact relationship is known between $W(r)$ and $U(r)$, although approximations are available provided that errors in $g(r)$ are sufficiently small. Achieving the necessary accuracy requires avoiding several subtle sources of error.

The average number of spheres separated from a given sphere by $r \pm \delta/2$ is $2\pi r \delta \, \bar{\rho} \, g(r)$, where $\bar{\rho}$ is the two-dimensional particle concentration. Given $A\bar{\rho}$ spheres in field of view A, the total number of such r-pairs should yield the pair correlation function through $N(r) = N_0(r) \, g(r)$, with $N_0(r) = \pi \delta A \, \bar{\rho}^2 \, r$. In a limited field of view, however, spheres near the edge have fewer neighbors to sample at range r than those near the middle. Applying periodic boundary conditions would introduce spurious correlations, while limiting the calculation to a central region of the field of view would drastically degrade statistics. Instead, we count the $\vec{r}$-pairs in a recorded image, normalize by the number of particles further than $\vec{r}$ from any edge, and sum over angles to calculate $g(r)$.

The limited field of view contains very few particles when $\bar{\rho}$ is small, so that $N_0(\sigma)$ typically is not much larger than 1 for a resolution $\delta \approx 0.1\sigma$. Limiting relative errors in $g(r)$ to ε would require $N_0(r = \sigma) > 1/\varepsilon$. Measuring $g(r)$ to this accuracy therefore requires averaging $M > 1/(\pi\delta A\bar{\rho}^2\sigma\varepsilon)$ uncorrelated snapshots of the suspension's structure. The time required for correlations to decay is roughly ten percent of the interval $\tau = (4D\bar{\rho})^{-1}$ a particle needs to diffuse the mean inter-particle separation. Thus the time $T = M\tau$ required to adequately sample $g(r)$ is $T > 0.1/(4\pi\delta AD\sigma\varepsilon\,\bar{\rho}^3)$.

On this basis, we suspect that some previously reported equilibrium measurements [23] of like-charge attractions may have been undersampled by as much as a factor of 20 in time. Excess correlations observed under these conditions might be interpreted, therefore, as transients, rather than as evidence for equilibrium attractions. This raises concern because the attractions reported in these studies are comparable to the quoted energy resolutions and thus could be particularly sensitive to incomplete averaging.

Because of difficulties in controlling concentration, temperature, and ionic strength constant over long periods, statistical accuracy strongly favors larger particle concentrations. On the other hand, many-body contributions obscure pair interactions for larger concentrations. The areal densities between $\bar{\rho}\sigma^2 = 0.05$ and 0.1 chosen for the present study represent a compromise between statistical and interpretive accuracy. Adequate sampling for our lowest density monolayer required half an hour of sampling.

Even if $g(r)$ is calculated appropriately, spurious correlations could be introduced by lateral variations in the glass wall's interaction with the spheres. Preferred regions would collect spheres, mimicking an inter-sphere attraction. We compared two-dimensional histograms of recorded particle positions with analogous histograms for uniformly distributed random data sets with the same number of particles. Differences in the first two moments of these histograms vanish with increasing delay time between consecutive snapshots, suggesting that each particle's position becomes uncorrelated over time as expected. Thus the substrate potential appears to be featureless on the length- and timescales of our experiment, to within our resolution.

Our experimental results for $g(r)$ and $W(r)$ at different concentrations appear in Fig. 9. The curves indicate a repulsive core interaction causing particle depletion extending to about 2σ. Beyond this, they reveal a preferred nearest-neighbor separation between two and three diameters, as well as the onset of the oscillatory correlations typical of a structured fluid. The depth of the minimum in the potential of mean force depends on $\bar{\rho}$ and so reflects at least some many-body contributions.

Provided that $g(r)$ is free from experimental artifacts, reliable approximations for $U(r)$ can be obtained from the Ornstein-Zernike integral equation with appropriate closure relations [65]. Good results for "soft" potentials typically are achieved with the hypernetted chain (HNC) approximation, whereas the Percus-Yevick (PY) approximation is known to be a better choice for hard spheres. The pair interaction potential can be evaluated numerically as

$$\beta U(r) = -\ln[g(r)] + \begin{cases} \bar{\rho}\mathcal{I}(r) & \text{(HNC)} \\ \ln[1+\bar{\rho}\mathcal{I}(r)] & \text{(PY)} \end{cases} \tag{49}$$

where $\mathcal{I}(r)$ is the convolution integral

$$\mathcal{I}(r) = \int [g(r') - 1 - \bar{\rho}\mathcal{I}(r')]\,[g(|\mathbf{r}' - \mathbf{r}|) - 1]\, d^2r' \tag{50}$$

to be solved iteratively with $\mathcal{I}(r) = 0$ as initial value [66]. Evaluating $\mathcal{I}(r)$ directly rather than with numerical Fourier transforms avoids introducing misleading features into $U(r)$ due to experimental noise.

Figure 10 shows the pair potential obtained for $\bar{\rho}\sigma^2 = 0.051$ together with a fit of the HNC approximation to Eq. (9) for $Z = 5000 \pm 500$, $\kappa = 0.32 \pm 0.05$ μm, and an arbitrary additive offset. Having obtained comparable results in both the HNC and PY approximations lends confidence in the accuracy of $U(r)$. Corresponding pair potentials extracted from data at $\bar{\rho}\sigma^2 = 0.079$ and 0.083 are essentially indistinguishable on this scale, and the extracted coupling constants are consistent with those obtained at lower areal density. The absence of minima in the pair potential confirms that oscillations in the potential of mean force (Fig. 9) resulted from crowding, while the underlying interaction is purely repulsive. The similarity of the results at different areal densities suggests that the interaction responsible for the monolayer's liquid structure indeed can be resolved into an additive pair potential. Values extracted for the charge number and screening length are consistent with surface charging due to silanol dissociation in a solution with an ionic strength around 10^{-6} M.

Monte Carlo simulations of particles interacting with comparable pair potentials confirm our energy resolution to be about 0.1 k_BT. This uncertainty is too large to resolve van der Waals attraction or image charge repulsions [49, 51]. Long-range attractions of the previously reported strength [2, 23, 24, 27], however, would have been resolved.

6. Conclusions

Does geometric confinement induce equilibrium attractions between like-charged colloidal spheres? At first glance, the observation of purely repulsive interactions in Sec. 5 suggests that attractions measured with optical tweezers in Sec. 2.4 should be interpreted as a kinematic effect along the lines proposed by the Squires-Brenner theory in Sec. 4. However, the silica spheres discussed in Sec. 5 carry a much lower effective charge than the polystyrene spheres for which confinement-induced attractions have been observed. If a putative wall-induced attraction depends at least linearly on the spheres' surface charge density, then it might not have been observable at the energy resolution of our equilibrium interaction measurements. Furthermore, the silica spheres' interactions were measured near one wall, rather than between two. The degree of confinement may make a quantitative, or even qualitative difference to the nature of the effective pair interaction.

Are reports of equilibrium like-charge attractions to be trusted? Deliberately undersampling the data used to compute $g(r)$ in Sec. 5 does indeed lead to the appearance of long-ranged attractions, but only in rare instances, and not systematically. While we cannot yet rule out experimental error as the origin of the attractions observed in Refs. [23] and [24], the possibility appears remote. Further measurements are in progress to isolate the role of confinement, surface charge, and screening length on the nature of colloidal electrostatic interactions in equilibrium.

For the time being, we are left with the tentative conclusion that geometric confinement can induce like-charge attractions in charge-stabilized colloid. Such an attraction appears to be outside the realm of local density theories.

If like-charge colloidal attractions are an equilibrium phenomenon, it cannot result from electro-hydrodynamic coupling. Even so, the Squires-Brenner theory opens a new chapter in our understanding of electrokinetic effects in charge stabilized suspensions. Certainly, it sheds new light on the compression mechanism exploited in Sec. 2.3 to create metastable superheated crystals. Future advances along these lines will inspire further exploration into the boundless variety of complex, beautiful and useful properties engendered by the interplay of electrostatic and hydrodynamic interactions in colloidal suspensions.

Acknowledgements

This work was supported by the National Science Foundation, by a Fellowship from the David and Lucile Packard Foundation, and by the Swiss National Science Foundation. We are grateful for enlightening conversations with Eric Dufresne, Vladimir Lobaskin, Roland Kjellander, Roland Netz, Stuart Rice and Binhua Lin.

References

1. D. G. Grier, J. Phys.: Condens. Matt. **12**, A85 (2000).
2. J. C. Crocker and D. G. Grier, J. Colloid Int. Sci. **179**, 298 (1996).
3. A. Ashkin, J. M. Dziedzic, J. E. Bjorkholm, and S. Chu, Opt. Lett. **11**, 288 (1986).
4. J. C. Crocker and D. G. Grier, Phys. Rev. Lett. **73**, 352 (1994).
5. E. R. Dufresne and D. G. Grier, Rev. Sci. Instr. **69**, 1974 (1998).
6. E. R. Dufresne *et al.*, Rev. Sci. Instr. **72**, 1810 (2001).
7. P. C. Mogensen and J. Gluckstad, Opt. Comm. **175**, 75 (2000).
8. K. Sasaki *et al.*, Opt. Lett. **16**, 1463 (1991).
9. C. Mio, T. Gong, A. Terray, and D. W. M. Marr, Rev. Sci. Instr. **71**, 2196 (2000).
10. L. Onsager, Chem. Rev. **13**, 73 (1933).
11. B. V. Derjaguin and L. Landau, Acta Physicochimica (URSS) **14**, 633 (1941).
12. E. J. Verwey and J. T. G. Overbeek, *Theory of the Stability of Lyophobic Colloids* (Elsevier, Amsterdam, 1948).
13. G. M. Bell, S. Levine, and L. N. McCartney, J. Colloid Int. Sci. **33**, 335 (1970).
14. S. Alexander *et al.*, J. Chem. Phys. **80**, 5776 (1984).
15. H. Löwen, P. A. Madden, and J.-P. Hansen, Phys. Rev. Lett. **68**, 1081 (1992).
16. H. Löwen, T. Palberg, and R. Simon, Phys. Rev. Lett. **70**, 1557 (1993).
17. T. Gisler *et al.*, J. Chem. Phys. **101**, 9924 (1994).
18. L. Belloni, Colloids Surfaces **A 140**, 227 (1998).
19. B. A. Pailthorpe and W. B. Russel, J. Colloid Int. Sci. **89**, 563 (1982).
20. T. Sugimoto *et al.*, Langmuir **13**, 5528 (1997).
21. D. G. Grier and J. C. Crocker, Phys. Rev. E **61**, 980 (2000).
22. K. Vondermassen, J. Bongers, A. Mueller, and H. Versmold, Langmuir **10**, 1351 (1994).
23. G. M. Kepler and S. Fraden, Phys. Rev. Lett. **73**, 356 (1994).
24. M. D. Carbajal-Tinoco, F. Castro-Román, and J. L. Arauz-Lara, Phys. Rev. E **53**, 3745 (1996).
25. J. C. Crocker and D. G. Grier, Phys. Rev. Lett. **77**, 1897 (1996).
26. A. D. Dinsmore, A. G. Yodh, and D. J. Pine, Nature **383**, 239 (1996).
27. A. E. Larsen and D. G. Grier, Nature **385**, 230 (1997).
28. D. G. Grier, Cur. Opin. Colloid Int. Sci. **2**, 264 (1997).
29. J. C. Crocker, J. A. Matteo, A. D. Dinsmore, and A. G. Yodh, Phys. Rev. Lett. **82**, 4352 (1999).
30. H. Risken, *The Fokker-Planck Equation, Springer series in synergetics*, 2 ed. (Springer-Verlag, Berlin, 1989).
31. R. Kjellander and D. J. Mitchell, Mol. Phys. **91**, 173 (1997).
32. I. Sogami, Phys. Lett. **96A**, 199 (1983).
33. I. Sogami and N. Ise, J. Chem. Phys. **81**, 6320 (1984).
34. T. Palberg and M. Würth, Phys. Rev. Lett. **72**, 786 (1994).
35. B. V. R. Tata, M. Rajalakshmi, and A. K. Arora, Phys. Rev. Lett. **69**, 3778 (1992).
36. H. Yoshida, N. Ise, and T. Hashimoto, J. Chem. Phys. **103**, 10146 (1995).
37. J. Yamanaka *et al.*, Phys. Rev. Lett. **80**, 5806 (1998).
38. A. E. Larsen and D. G. Grier, Phys. Rev. Lett. **76**, 3862 (1996).
39. R. Rajagopalan and K. S. Rao, Phys. Rev. E **55**, 4423 (1997).
40. K. Ito, K. Sumaru, and N. Ise, Phys. Rev. B **46**, 3105 (1992).
41. R. van Roij and J.-P. Hansen, Phys. Rev. Lett. **79**, 3082 (1997).
42. R. van Roij, M. Dijkstra, and J.-P. Hansen, Phys. Rev. E **59**, 2010 (1999).
43. E. Trizac, Phys. Rev. E **62**, 1 (2000).

44. K. S. Rao and R. Rajagopalan, Phys. Rev. E **57**, 3227 (1998).
45. J. C. Neu, Phys. Rev. Lett. **82**, 1072 (1999).
46. J. E. Sader and D. Y. C. Chan, J. Colloid Int. Sci. **213**, 268 (1999).
47. E. Trizac and J.-L. Raimbault, Phys. Rev. E **60**, 6530 (1999).
48. D. Goulding and J.-P. Hansen, Mol. Phys. **95**, 649 (1999).
49. F. H. Stillinger, J. Chem. Phys. **35**, 1584 (1961).
50. A. J. Hurd, J. Phys. A **18**, L1055 (1985).
51. R. R. Netz and H. Orland, Eur. Phys. J. E **1**, 203 (2000).
52. R. Podgornik and V. A. Parsegian, Biophys. J. **74**, A177 (1998).
53. Y. Levin, Phys. A **265**, 432 (1999).
54. O. Spalla and L. Belloni, Phys. Rev. Lett. **74**, 2515 (1995).
55. J. Israelachvili, *Intermolecular and Surface Forces*, 2 ed. (Academic Press, London, 1992).
56. T. Squires and M. P. Brenner, Phys. Rev. Lett. **85**, 4976 (2000).
57. C. Pozrikidis, *Boundary Integral and Singularity Methods for Linearized Viscous Flow* (Cambridge University Press, New York, 1992).
58. J. Happel and H. Brenner, *Low Reynolds Number Hydrodynamics* (Kluwer, Dordrecht, 1991).
59. D. L. Ermak and J. A. McCammon, J. Chem. Phys. **69**, 1352 (1978).
60. G. K. Batchelor, J. Fluid Mech. **74**, 1 (1976).
61. J. Crocker, J. Chem. Phys. **106**, 2837 (1997).
62. J. R. Blake, Prog. Colloid Polymer Sci. **70**, 303 (1971).
63. C. Oseen, *Neuere Methoden und Ergebnisse in der Hydrodynamik* (Akademische Verlagsgesellschaft, Leipzig, 1927).
64. E. R. Dufresne, T. M. Squires, M. P. Brenner, and D. G. Grier, Phys. Rev. Lett. **85**, 3317 (2000).
65. J.-P. Hansen and I. R. McDonald, *Theory of Simple Liquids*, 2nd ed. (Academic Press, London, 1986).
66. E. M. Chan, J. Phys. C **10**, 3477 (1977).

Polyelectrolyte
POOR Solvent
Good solvent

COMPUTER SIMULATIONS OF CHARGED SYSTEMS

CHRISTIAN HOLM, KURT KREMER
Max-Planck-Institut für Polymerforschung
55128 Mainz, Germany

1. Introduction

Polyelectrolytes represent a broad and interesting class of materials [1] that enjoy an increasing attention in the scientific community. For example, in technical applications polyelectrolytes are widely used as viscosity modifiers, precipitation agents, superabsorbers, or leak protectors. In biochemistry and molecular biology they are of interest because virtually all proteins, as well as DNA, are polyelectrolytes.

In contrast to the theory of neutral polymer systems, which is well developed, the theory of polyelectrolytes faces several difficulties. Simple scaling theories, which have been proven so successfully in neutral polymer theory, have to deal with additional length scales set by the long range Coulomb interaction[2]. Furthermore there is a delicate interplay between the electrostatic interaction of the distribution of the counterions and the conformational degrees of freedom, which in turn are governed by a host of short range interactions, which renders the problem difficult. There are only two limiting cases that are easy to solve. These are the case of high excess salt, effectively screening out the electrostatic interaction (which in turn allows one to treat it as a perturbation), or the case of an overwhelming dominance of the Coulomb force, which results in a strongly elongated chain. Unfortunately it is often just the intermediate case, which proves to be the most interesting regime in terms of experiment, application and theory.

Computational simulations provide some unique ways to elucidate the properties of charged systems. We first give a more general introduction to the relevant simulation methods, and focus then on some recently obtained results.

C. Holm et al. (eds.), Electrostatic Effects in Soft Matter and Biophysics, 117–148.
© 2001 *Kluwer Academic Publishers. Printed in the Netherlands.*

2. Simulations techniques

2.1. SOME BASICS ON SIMULATIONS

There exist a number of nice reviews and books [3, 4, 5] which deal extensively with various aspects of computer simulations of complex, not necessarily charged, systems. Thus, the present introduction can also be viewed as a guide to the literature.

There are two basic concepts, which are used in computer simulations of complex systems. The conceptionally most direct approach is the molecular dynamics (MD) method. One numerically solves Newton's equation of motion for a collection of particles, which interact via a suitable interaction potential $U(\vec{r}_i)$, where $\vec{r}_i$ are the positions of the particles. Through the equation of motion a natural time scale is built in, however, this might not be the physically realistic time scale (e. g. if the solvent is replaced by a dielectric background). Running such a simulation samples phase space for the considered system deterministically. Though this sounds very simple, there are many technical and conceptual complications, which one encounters on the way. The second approach, the Monte Carlo (MC) method, samples phase space stochastically. Monte Carlo is intrinsically stable but has no natural time scale built in. This can be reinterpreted, however, by an adjustment of suitable "time scales" for some MC procedures [4]. The MD and MC approaches are the basic simulation methods for exploring the statistical properties of complex fluids. At present, many applications employ variants thereof, or even hybrid methods, where combinations of both are used. Before going into detail we ask when is either kind of method more?

At first sight it is tempting to perform a computer simulation of a polyelectrolyte solution where all details of the chemical structure of the monomers are included. For instance, the chain diffusion constant D could be measured by monitoring the mean square displacement of the monomers of the chains. This, however, is tempting only at the very first glance. Even for the fastest computers one would need an exceeding amount of computer time. As for all disordered, complex, macromolecular materials, (charged) polymers are characterized by a hierarchy of different length and time scales, and these length and especially the time scales span an extremely wide range [6]. On the atomistic level the properties are dominated by the local oscillations of bond angles and lengths.[1] The typical time constant of about 10^{-13} sec results in a simulation time step of 10^{-15} sec. On the

[1] For reactions or to study exited states, the electronic structure is treated explicitely. Such methods (Carr-Parrinello simulations, quantum chemistry etc.) are beyond the scope of the present paper.

semi-macroscopic level the behavior is dominated by the overall relaxations of conformation of the objects or even larger units (domains etc). The times, depending on chain length, temperature and density, can easily reach seconds. To cover that many decades in time within a conventional computer simulation is certainly impossible at present. On the other hand, it is important to relate the chemical structure of a system to its macroscopic properties.

2.2. MOLECULAR DYNAMICS

MD simulations date back to the early fifties. For a rather complete overview about simulations in condensed matter we refer to [3]. Consider a cubic box of Volume $V = L^3$, containing N identical particles. In order to avoid surface effects and (as much as possible) finite size effects, one typically uses periodic boundary conditions. The particle number density is given by $\rho = N/L^3$. The first simulations employed hard spheres of radius R_o, leading to a volume fraction $\rho_v = 4/3\pi R_o^3 \rho$. Though still used extensively for some studies on the glass transition of colloidal systems we focus here on soft potentials.

A key thermodynamic quantity, the temperature, is imposed via the equipartition theorem $\frac{m}{2}\langle \vec{\dot{r}}_i^2 \rangle = \frac{3}{2}kT$, m being the particle masses. Note that in hard sphere systems temperature only defines a time scale, but is otherwise irrelevant. One can find many different soft potentials in the literature. However, most widely used is the Lennard-Jones potential $U^{LJ}(r_{ij})$, derived originally for interactions of noble gases (Ar, Kr ...), r_{ij} being the distance between particle i and j. In its simplest form for two identical particles it reads

$$U^{LJ}(r_{ij}) = 4\epsilon \left[(\frac{\sigma}{r_{ij}})^{12} - (\frac{\sigma}{r_{ij}})^{6} \right] \tag{1}$$

Usually, a cutoff r_c is introduced for the range of the interaction. This typically varies between 2.5 σ (classical LJ interaction with an attractive well of depth ϵ, used for the poor solvent chains in Sec. 4.2) and $2^{1/6}\sigma$ (the potential is cut off at the minimum, leading to a pure repulsive interaction, as is typically done for good solvent chains). For chain molecules a bonding interaction for $r < R_0$

$$U_{\mathrm{FENE}}(r) = -\frac{1}{2}kR_0^2 \ln(1 - \frac{r^2}{R_0^2}) \tag{2}$$

is added, which keeps the bond length below a maximum of R_0. The spring constant k varies between 5 and 30 ϵ/σ^2. Electrostatics is included via the

Coulomb interaction

$$U_c(r_{ij}) = \frac{l_B k_B T}{r_{ij}} q_i q_j \tag{3}$$

Here $l_B := \frac{e_0^2}{4\pi\varepsilon_0\varepsilon_r k_B T}$, where e_0 is the elementary unit charge, k_B is the Boltzmann constant, T denotes temperature, ε_0 and ε_r are the vacuum and relative dielectric permeability of the solvent, respectively, and $q_{i,j}$ are charges measured in units of e_0. Two monovalent charges separated by the Bjerrum length l_B have an interaction energy equal to $k_B T$. The Bjerrum length thus is a measure of the interaction strength. It is equal to 7.14 Å for water at room temperature. The computational aspects of the long range potential will be discussed shortly in Sec.2.4.

The unit of energy is ϵ, of length σ and of mass m. This defines the "LJ-units" for temperature $[T] = \epsilon/(k_B T)$, time $[t] = \sqrt{\sigma^2 m/\epsilon}$ and number density $[\rho] = \sigma^{-3}$. In most practical programs σ, m, ϵ are used as the basic units and set to one. The straight forward simulation technique is to integrate Newton's equations of motion for the particles:

$$m_i \ddot{\vec{r}}_i = -\vec{\nabla} \sum_{j,j\neq i} U(r_{ij}) \tag{4}$$

Since energy in such a simulation is conserved we have a microcanonical ensemble. Presently other thermodynamic ensembles are commonly used for practical applications (NPT: isobaric-isothermal, NVT: isothermal (canonic) ...). Because we often employ a stochastic MD method, known as the Langevin thermostat[7], we will briefly describe its main ingredients. Instead of integrating Newton's equations of motion, one solves a set of Langevin equations

$$m_i \ddot{\vec{r}}_i = -\vec{\nabla} \sum_{j,j\neq i} U(r_{ij}) - \Gamma \dot{\vec{r}}_i + \vec{\xi}_i(t) \tag{5}$$

with $\vec{\xi}_i(t)$ being a δ-correlated Gaussian noise source with its first and second moments given by

$$\langle \vec{\xi}_i(t) \rangle = 0 \qquad \text{and} \qquad \langle \vec{\xi}_i(t) \cdot \vec{\xi}_j(t') \rangle = 6\, k_B T\, \Gamma \delta_{ij} \delta(t - t'). \tag{6}$$

The friction term $-\Gamma \dot{\vec{r}}_i$ and the noise $\vec{\xi}_i(t)$ is thought of as imitating the presence of a surrounding viscous medium responsible for a drag force and random collisions, respectively. The second moment of $\vec{\xi}_i(t)$ is adjusted via an Einstein relation in order to reach the canonical state in the limit $t \to \infty$. The dynamics generated by the Langevin equation can alternatively be written as a general Fokker-Planck process. This permits a transparent proof of two important facts: *(i)* the stationary state of the process is the

Boltzmann distribution and *(ii)* the system will evolve and converge to the Boltzmann distribution [8]. Since for small times the stochastic part is more important than the deterministic one ($\sqrt{t} \gg t$ for small t) it is actually not necessary to use Gaussian random variables in the simulation [9]. It suffices to use equidistributed random variables with first and second moment being identical to the Gaussian deviate, as required by the equipartition theorem.

A simple but very efficient and stable integration scheme is the Verlet algorithm (more complicated methods, which do not have time inversion symmetry, do not, in general, perform significantly better). With a simulation time step δt , where $\delta t << 2\pi/\omega_{max}$ and ω_{max} is the typical highest frequency of the system (for crystals the Einstein frequency), we have in one dimension

$$\begin{aligned} r_i(t+\delta t) &= r_i(t) + \delta t v_i(t) + \frac{\delta t^2}{2} a_i(t) + \frac{\delta t^3}{6} \dot{a}_i(t) + \mathcal{O}(\delta t^4) \\ r_i(t-\delta t) &= r_i(t) - \delta t v_i(t) + \frac{\delta t^2}{2} a_i(t) - \frac{\delta t^3}{6} \dot{a}_i(t) + \mathcal{O}(\delta t^4), \end{aligned}$$

where $v_i(t) = \dot{r}_i(t)$ and $a_i(t) = \dot{v}_i(t)$. An addition of the two lines yields

$$r_i(t+\delta t) = 2r_i(t) - r_i(t-\delta t) + \delta t^2 a_i(t) + \mathcal{O}(\delta t^4) \tag{7}$$

Therefore the position calculations have an algorithmic error of $\mathcal{O}(\delta t^4)$. Subtraction of the lines yields

$$v_i(t) = \frac{1}{2\delta t}[r_i(t+\delta t) - r_i(t-\delta t)] + \mathcal{O}(\delta t^3) \tag{8}$$

leading to errors of $\mathcal{O}(\delta t^3)$. There are many variants of this basic integrating scheme used throughout the literature [5]. One can follow the realistic time evolution of a system, as long as the forces/potentials are realistic for the modeled system and as long as classical mechanics is sufficient. In a purely deterministic simulation the accumulation of small errors can cause significant deviations from the real trajectory. If the system is ergodic, which requires mixing of normal modes (recall the well-known Fermi-Pasta-Ulam problem, where one asks how anharmonic a potential has to be in order to equilibrate a one dimensional chain of particles[10]) one can determine ensemble averages from time averages

$$< A >= \frac{1}{M} \sum_{i=1}^{M} A(t_i) \tag{9}$$

of any physical quantity A of interest. This describes the most elementary Ansatz for a microcanonical simulation [5]. Here all extensive thermodynamic variables of the system, namely N, V, E are kept constant. Sometimes this is also called NVE ensemble. As mentioned before, most applications employ other ensembles such as the canonic (NVT), the isobaric-isothermal (NPT) or even the grand canonical (μ, P, T) ensemble, μ being the chemical potential. As a general rule, in cases such as two phase coexistence or calculations of transport properties, one should choose an ensemble with many intensive variables kept constant as possible. For charged systems, however, it is rather difficult to perform efficient simulations in the (N, P, T) or (μ, P, T) ensembles. Therefore the most common ensemble is the NVT since it is easy to use the deterministic equations with additional stochastic terms to constrain the temperature (Eq. 5) or variants thereof [5].

2.3. MONTE CARLO METHOD

The classical Monte Carlo approach goes to the other extreme, namely to purely stochastic sampling. Starting from a particular configuration, randomly a particle (or a number of particles) is selected and displaced by a random jump. For hard sphere systems the move is accepted if the new configuration complies with the excluded volume; if not the old configuration is retained. The approach is also called simple sampling. This cycle is repeated over and over. Once every particle on average has a chance to move, one Monte Carlo step is completed. This is the most basic Monte Carlo simulation (see e.g.[3, 11]). Since there is no energy involved, it trivially fulfills detailed balance

$$W(\{x\} \to \{y\})P_{eq}(\{x\}) = W(\{y\} \to \{x\})P_{eq}(\{y\}) \tag{10}$$

where $W(\{x\} \to \{y\})$ is the probability to jump from state $\{x\}$ to state $\{y\}$ and $P_{eq}(\{x\})$ the equilibrium probability of state $\{x\}$. All states have exactly the same probability. Detailed balance is a sufficient condition for a MC simulation to relax into thermal equilibrium, though this may take a very long time. Special cases of algorithms without detailed balance will not be discussed here.

It is useful to compare the basic aspects of MC simulation to the examples discussed above for molecular dynamics simulations. The Hamiltonian depends for simplicity only on the positions of all particles $\{r_i\}$ and is denoted by $H(\{r_i\}$. The expectation value of any observable A is given by

$$< A >= \sum_{\{r_i\}} A(\{r_i\})P_{eq}(\{r_i\}) \tag{11}$$

with

$$\begin{aligned} P_{eq}(\{r_i\}) &= \exp(-H(\{r_i\})/k_BT)/Z \\ Z &= \sum_{\{r_i\}} \exp(-H/k_BT) \end{aligned} \tag{12}$$

An exact way would be to sample all possible states, which in all but the most trivial cases is impossible. Thus we sample phase space stochastically. Taking a particle at random, calculating its energy, one moves it and calculates the new energy. With $P(\{x\})$ being the Boltzmann-probability of the original state and $P(\{y\})$ being that of the new state, detailed balance is obeyed if

$$\frac{W(\{x\} \to \{y\})}{W(\{y\} \to \{x\})} = \exp\{-(H(\{x\}) - H(\{y\})/k_B T)\} \tag{13}$$

Under this condition the algorithm is ergodic and the system relaxes into e-quilibrium. The Metropolis method is the most frequently used prescription one to accept or reject a move:

$$W(\{x\} \to \{y\}) = \Gamma \begin{cases} \exp(H(\{x\}) - H(\{y\})/k_B T) & , \ \triangle H > 0 \\ 1 & , \ \triangle H < 0. \end{cases} \tag{14}$$

Since only the ratio of the W's is relevant, Γ is an arbitrary constant between zero and one, usually $\Gamma = 1$. A random number ζ, equally distributed between 0 and 1, is used to decide upon the acceptance of a move. If $\zeta < W(\{x\} \to \{y\})$ the move is accepted, otherwise rejected. (For $\Gamma = 1$ any move, which lowers the energy is accepted.) This is the basic MC procedure used for sampling phase space in statistical physics.

In many cases, however, one also would like to gain information on the dynamics of a system understudy. How can a MC simulation, with no intrinsic time scale, be used to obtain information on the dynamics? In the method described above the system evolves from one state to another by a "move". Through these stochastic moves the configurations of particles change with "time". This is a dynamic MC method based on a Markov process, where subsequent configurations $\{x\} \to \{x^i\} \to \{x^{ii}\} \to \ldots$ are generated with a transition probability $W(\{x^i\} \to \{x^{ii}\})$. To a large extent the choice of the move is arbitrary. In order to interprete this motion in terms of Langevin dynamics of the system , the moves have to be **local**. (They have to be confined at each move to a small number of particles in a spacially small region!) Then one can interpret it as a local elementary unit of motion. The prefactor Γ can actually be viewed as an attempt rate $\Gamma = \tau_o^{-1}$ for the moves and introduces a timescale. This "changes" the purely statistical transition probability into a transition probability per unit time[4, 11]. To compare simulated (overdamped) dynamics with an experiment, it essentially requires determination of τ_o (e. g. via diffusion constants). It is obvious, that this simulation does **not** include any hydrodynamic effects since there is no momentum involved. There are very interesting, more advanced methods like DPD (dissipative particle dynamics) and Lattice-Boltzmann methods, currently under development in order to include this efficiently [12]). Using the interpretation of a MC step as a time step, ensemble averages can be written as time averages:

$$< A >= \frac{1}{M - M_o} \sum_{i=M_o+1}^{M} A(\{x^i\}) \cong \frac{1}{t - t_o} \int_{t_o}^{t} dt' A(t') \ . \tag{15}$$

We view one attempted move per system particle as one time-step. The first configurations in a simulation are usually not yet equilibrium configurations. One first has to "relax" the system into equilibrium, meaning the data for the first M_o steps are omitted.

In this interpretation the dynamic Monte Carlo procedure is nothing but a numerical realization of a Markov process described by a Markovian master equation

$$\begin{aligned} \frac{d}{dt} P(\{x\}, t) &= - \sum_{\{x^i\}} W(\{x\} \to \{x^i\}) P(\{x\}, t) \\ &\quad + \sum_{\{x^i\}} W(\{x^i\} \to \{x\}) P(\{x^i\}, t) \end{aligned} \tag{16}$$

with $P(\{x\}, t)$ the time dependent probability of state $\{x\}$. The condition of detailed balance is sufficient that $P_{eq}(\{x\})$ is the steady-state solution of the master equation. If all states are mutually accessible $P(\{x\}, t)$ must relax towards $P_{eq}(\{x\})$ as $t \to \infty$ irrespective of the starting state. Note however that the choice of a "good" starting state can save enormous amounts of CPU time.

So far, the two extreme cases for classical, particle based computer simulations were discussed, microcanonical MD and canonical MC. There are many approaches "in between" which are used depending on the problem under consideration. The techniques range from pure MD, where Newton's equations of motion are solved ($\ddot{x} = -\nabla U$), MD coupled to a heat bath and added friction ("Langevin MD", "Noisy MD"), ($\ddot{x} = -\nabla U - \zeta \dot{x} + f(t)$), ζ friction, f (t) random force), Brownian Dynamics (BD) ($\dot{x} = -\nabla U +$ random displacement), force biased MC (attempted moves are selected from the very beginning according to local forces), to plain MC as described above.

For the application to polymers one should keep in mind, that the conformational entropy of the chains add additional complications, which make proper equilibration especially difficult or time consuming. Thus, wherever possible, methods should be used, which are faster than the slow intrinsic dynamics of the chains [4].

One of the biggest problems for the simulations of charged systems is the long range nature of the Coulomb interactions. In principle, each charge interacts with all others, leading to a computational effort of $\mathcal{O}(N^2)$ already within the central simulation box. For many physical investigations one wants to simulate bulk properties and therefore introduces periodic boundary conditions to avoid boundary effects. The standard method to compute the merely conditionally convergent Coulomb sum

$$E = \frac{1}{2} \sum_{\vec{n}}{}' \sum_{ij} \frac{q_i q_j}{|\vec{r}_{ij} + \vec{n}L|}, \tag{17}$$

where the prime denotes that for $\vec{n} = \vec{0}$ the term $i = j$ has to be omitted, is the traditional Ewald summation [13]. The basic idea is to split the original sum via a simple transformation into two exponentially convergent parts, where the first one, ϕ_r, is short ranged and evaluated in real space, the other one, ϕ_k, is long ranged and can be analytically Fourier transformed and evaluated in Fourier space:

$$\frac{1}{r} = \frac{1 - f(r)}{r} + \frac{f(r)}{r} \simeq \phi_r(r_c, \alpha) + \phi_k(k_c, \alpha) \tag{18}$$

Traditionally, one uses for f the error function $\mathrm{erf}(\alpha r) := 2\pi^{1/2} \int_0^{\alpha r} \exp -t^2$ dt, though other choices are possible and sometimes more advantageous[14, 15, 16]. For any choice of the Ewald parameter α and no truncation in the sums the formula yields the exact result. In practice one wants to cut off the infinite sum at some finite values r_c and k_c to obtain E to a user controlled accuracy, which is possible by using error estimates [17]. The aformentionend procedure results in the well known Ewald formula for the energy of the box

$$E = E^{(r)} + E^{(k)} + E^{(s)} + E^{(d)}, \tag{19}$$

where the contributions from left to right are the real space, Fourier space, self , and dipole-correction energy terms. These are given by

$$E^{(r)} = \frac{1}{2}\sum_{i,j}\sum_{\mathbf{m}\in\mathbb{Z}^3}{}' q_i q_j \frac{\mathrm{erfc}(\alpha|\mathbf{r}_{ij}+\mathbf{m}L|)}{|\mathbf{r}_{ij}+\mathbf{m}L|} \tag{20}$$

$$E^{(k)} = \frac{1}{2}\frac{1}{L^3}\sum_{\mathbf{k}\neq 0}\frac{4\pi}{k^2}e^{-k^2/4\alpha^2}|\tilde{\rho}(\mathbf{k})|^2 \tag{21}$$

$$E^{(s)} = -\frac{\alpha}{\sqrt{\pi}}\sum_i q_i^2 \tag{22}$$

$$E^{(d)} = \frac{2\pi}{(1+2\epsilon')L^3}\left(\sum_i q_i \mathbf{r}_i\right)^2, \tag{23}$$

and the Fourier transformed charge density $\tilde{\rho}(\mathbf{k})$ is defined as

$$\tilde{\rho}(\mathbf{k}) = \int_{V_b} \mathrm{d}^3r\, \rho(\mathbf{r}) e^{-i\,\mathbf{k}\cdot\mathbf{r}} = \sum_{j=1}^{N} q_j\, e^{-i\,\mathbf{k}\cdot\mathbf{r}_j}. \tag{24}$$

The dipole term depends not on α, hence is independent of the splitting function. It reflects the way Eq. (17) is summed up, here in a spherical way towards infinity[18]. It also includes a correction for the dielectric constant ϵ' outside the summed up sphere volume. For metallic boundary conditions, $\epsilon' = \infty$, the dipole term vanishes. In principle the thermodynamic properties should be independent of the choice of boundary conditions[19]. The Ewald sum has complexity $\mathcal{O}(N^{3/2})$ in its optimal implementation [20], and therefore is not suitable for the study of large systems ($N > \mathcal{O}(1000)$). Implementing a fast Fourier transformation (FFT) for the Fourier part results in the so-called particle-mesh-Ewald formulations, which improve the efficiency to $\mathcal{O}(N \log N)$ [21, 22, 23, 24], and which can also efficiently be parallelized[25, 26]. The most versatile and accurate method of all mesh-methods is the oldest one, the P3M algorithm[21, 27], for which also precise error estimates exist[28]. Another way of computing Eq.(17) is via a convergence factor

$$E = \lim_{\beta\to 0}\frac{1}{2}\sum_{\vec{n}}{}'\sum_{ij}\frac{q_i q_j \exp\left(-\beta|\vec{r}_{ij}+\vec{n}L|\right)}{|\vec{r}_{ij}+\vec{n}L|}. \tag{25}$$

This approach is used in the Lekner [29] and Sperb [30] methods to efficiently sum up the 3D Coulomb sum. Although the method in its original versions has $\mathcal{O}(N^2)$ complexity, Sperb et al. have developed a factorization approach which yields an $\mathcal{O}(N \log N)$ algorithm [31] as well .

Other advanced methods of $\mathcal{O}(N \log N)$ are tree algorithms[32], which are the first order approximation of even better, so-called fast multipole methods [33]. These can reach a linear complexity, but at the expense of a heavy computational overhead which makes these methods advantageous only for a very large number of charges ($N \approx 100\,000$) [34].

For thin polyelectrolytes films or membrane interactions one is also interested in summations where only 2 dimensions are periodically replicated and the third one is of finite thickness h ($2D + h$ geometry). For this geometry Ewald based formulas are only slowly convergent, have mostly $\mathcal{O}(N^2)$ scalings and no "a priori" error estimates exist [35]. Recently Arnold [36, 37] developed a method which is based on convergent factors, whose errors are well controlled, and which uses a factorization approach resulting in an $\mathcal{O}(N^{5/3})$ scaling (MMM2D). In two dimensions the convergence factor based methods and the Ewald sum methods yield exactly the same results, there is no dipolar correction term needed[36]. This is in contrast to the 3D methods[18]. However, an even better scaling can be achieved, if one returns to the 3D Ewald formula, and allows for a large empty space between the unwanted replicas in the third dimension [38]. However, so far the method has been only checked on a trial and error basis. We recently improved this situation by computing an analytic error term which accounts for the contributions of the unwanted replicas. By simply subtracting this term from the 3D sum, which is a linear operation in N, one can in principle come as close as desired to the real $2D + h$ sum, by allowing for just some arbitrary small amount of empty space between the layers[39]. Using then again the P3M method we obtain an $N \log N$ scaling with well controlled errors also for the $2D + h$ geometry, which up to now seems to be the optimal choice.

To simulate the structure of water (or other dipolar solvents), one also needs to treat the dipolar interactions in a similar fashion. Also here the Ewald method is applicable[20], and error estimates exist[40].

3. Mean-field models: Debye-Hückel chains

While we deal here mainly with systems where the charges are explicitly taken into account, historically and even up to now many studies consider the ions solely in a mean field approximation. In the first step all non bonded charges are considered as a smeared continuous charge density. Such a situation is described by the Poisson-Boltzmann (PB) equation. While this in most cases is not exactly solvable, many studies employ the Debye-Hückel approximation, which is the solution of the linearized PB equation[41, 42, 43, 44]. The resulting potential between charges is the screened Coulomb potential

$$V_{DH} = \frac{l_B}{e_0} k_B T \frac{\exp(-\kappa r)}{r}, \tag{26}$$

with $1/\kappa$ being the Debye screening length. The polymer can now easily be modeled as a random walk of N monomers from which a fraction f is monovalently charged. If one in addition introduces the stiffness along the backbone of the chain, e. g. by a cosine-potential, the total Hamiltonian reads

$$\frac{H}{k_B T} = -A \sum_{i=1}^{N-1} (\vec{b}_i \cdot \vec{b}_{i+1}) + \sum_{i=2}^{N} \sum_{j=1}^{i-1} \theta(q_i q_j) l_B \frac{\exp(-\kappa r_{ij})}{r_{ij} k_B T}. \qquad (27)$$

The positive amplitude A defines the strength of the angular potential and the Heavyside θ-function is one if the monomers i and j are charged, and zero otherwise. The symbol $\vec{b}_i$ is the bond vector between monomer number i and $i+1$. For the present simulations the bond length $|\vec{b}|$ and the Bjerrum length l_B are fixed to 1σ, and thus set the basic length scale. One bond mimics several neutral monomers as usual for coarse grained simulation models. The longest chains considered contained up to 2049 repeat units with a charge fraction of $f = \frac{1}{16}$. Mapping this onto a PS-NaPSS copolymer for a more flexible case and keeping in mind that for these polymers roughly three repeat units fit into one Bjerrum length one arrives at a molecular weight of more than 600 000 g/mol. Thus, for these kind of questions computer simulations are quite capable of covering the experimentally interesting regime. This allows us to systematically vary not only the chain length and the screening length but also the chain bending stiffness. This is only of limited experimental relevance since very long isolated chains in dilute solutions cannot be experimentally analyzed so far. However one of the central questions in the theory of polyelectrolytes is whether the characteristic electrostatic length is a linear or quadratic function of the Debye length (i.e. proportional to the square of charge density or to the charge density itself). Analytic results mainly predict a κ^2 asymptotic dependency of the persistence length[2]. However, an estimate of the required chain length to reach the asymptotic regime for flexible weakly charged polyelectrolytes shows that the chains are so long that all experimentally realizable concentrations are in the semi-dilute regime. Thus, this is a typical question of interest which is of no direct importance for experiments. Nevertheless, it can add significantly to our understanding which in turn can also influence the interpretation of experiments in the semi-dilute regime. Therefore, it is worthwhile to undertake some efforts in computer simulation to investigate this question. There are a number of attempts to do this, which so far do not lead to a clear-cut answer. A typical result from such a simulation is given in Fig. 1.

It shows the dependence of the electrostatic persistence length on the screening length. The two dotted lines indicate the variational (Var) and the Odijk-Skolnick-Fixman (OSF) results ($L_e \propto \kappa^{-1}$, or κ^{-2} respectively). As

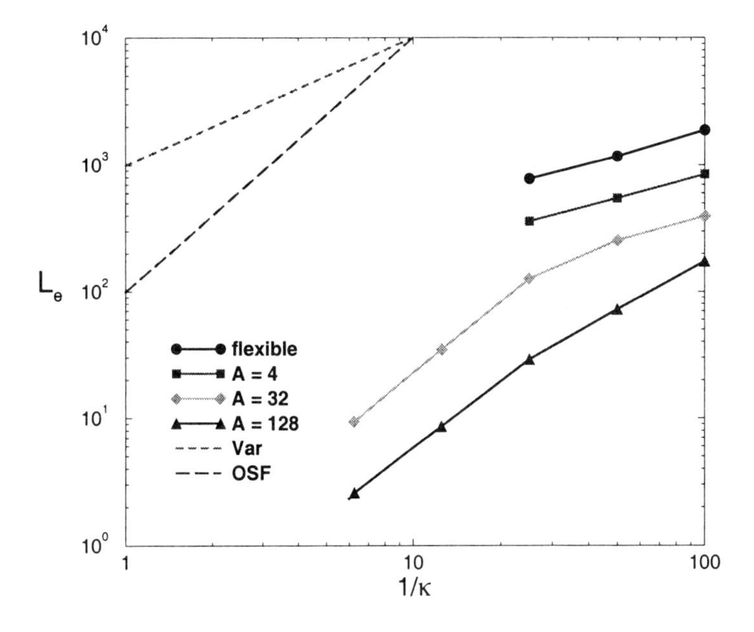

Figure 1. Electrostatic persistence length as a function of $1/\kappa$ for several A values.

one can see for the regime covered there seems to be a continuous transition from a very weak dependency on κ towards the asymptotically expected κ^{-2} behavior for the semi-flexible polyelectrolyte. In the limit of very stiff chains the OSF result becomes exact and the data has to agree with that. Even though the simulated chains lie very well within the experimentally relevant regime, there is, at least for typical experimental flexibilities, no clear sign of the crossover into the asymptotic OSF regime. This allows for two important conclusions. First of all, the chains are certainly not long enough to display asymptotic behavior. Secondly, typical experimental chains are also not long enough to display the asymptotic behavior as predicted by mean-field theories. Unlike neutral polymers, polyelectrolytes normally are not in an asymptotic limit where the predicted scaling laws can be cleanly observed. Another consequence is revealed in a closer analysis of the simulation data. For semiflexible polyelectrolyte chains there is no unique persistence length anymore, as all theoretical pictures based on the worm like chain model assume. Over short distances the intrinsic stiffness dominates and gives a clear signal in the bond direction correlation function. Only over large distances along the backbone of the chain does the electrostatic effect show up and introduces a second characteristic length scale into the system. This can lead to the special situation that the electrostatic contribution to the persistence length for stiff chains is smaller than for flexible chains. The reason simply lies in the fact that for the stiff chains the charges along the backbone are already much further apart because of the intrinsic stiffness as opposed to flexible chains. These larger distances then experience a much weaker electrostatic repulsion due to the screening of the electrostatic interaction resulting in a weaker effect on the persistence length. For details we refer to Ref. [45].

Altogether, the main conclusion from these simulations of simplified mean-field like models is on the one hand that polyelectrolytes have to be extremely long to be in the asymptotic regime, and that one has to be very careful in deriving general statements from either simulations, which typically are not asymptotic just as experiments, and analytic theory, which is usually in an asymptotic limit.

Another conclusion is that the concept of persistence length for polyelectrolytes certainly is not well defined in terms of the properties of the chains. As soon as intrinsic stiffness is included there is no longer a unique length scale that describes the internal structure of the chain. On the other hand this is the basis of all classical models employed in analytic theory.

4. Solutions of flexible polyelectrolytes

4.1. GOOD SOLVENT CHAINS WITH EXPLICIT COUNTERIONS

The first investigation of totally flexible many chain polyelectrolyte systems in good solvent with explicit monovalent counterions was performed several years ago[46]. The simulations were carried out mostly with systems of 8 or 16 chains with $N_m = 16$, 32 and 64. Instead of the P3M algorithm a spherical approximation in a truncated octahedral simulation box was used which, for values smaller than $N_{total} \approx 500$ is faster than the PME method. More details of the whole study can be found in [46]. All beads and counterions interacted with the truncated Lennard-Jones potential plus the full Coulomb interaction.

In this work experimental values of the osmotic pressure and the maximum position in the interchain structure facture were successfully reproduced. One of the important findings was that the chains essentially are never rodlike. Counterion-chain correlations can dramatically shrink the polyelectrolyte chain. The end-to-end distance shortens significantly as the density increases from dilute towards the overlap density. The chain structure is highly anisotropic in the very dilute limit, and the scaling with respect to N_m is asymmetric; but as the overlap density is approached, the structural anisotropy dissipates and the scaling becomes approximately symmetric. On long length scales the chain structure continuously changes from very elongated to neutral-like coils. Yet, on short length scales, the

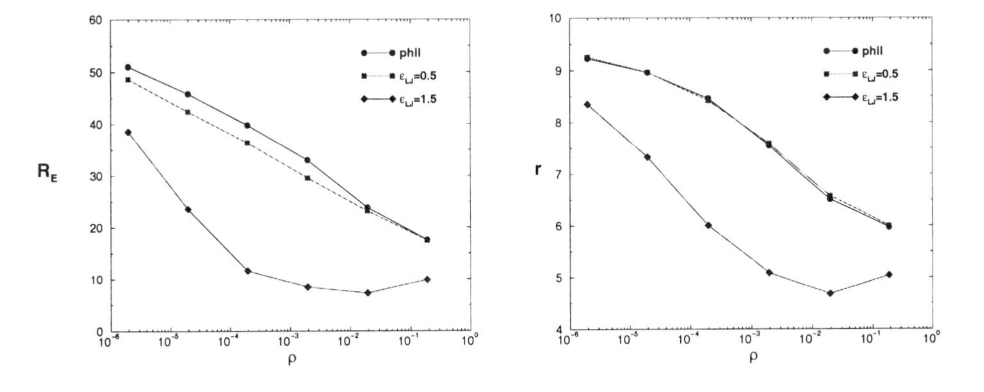

Figure 2. $\mathrm{R_E}$ (left) and r (right) versus density ρ for hydrophilic (phil), weak hydrophobic ($\epsilon_{LJ} = 0.5$), and strongly hydrophobic ($\epsilon_{LJ} = 1.5$) chains

chain structure is density independent and elongated more than neutral chains.

It was found that in the dilute limit the scaling for the extension perpendicular to the chain was $R_\perp \propto N^{0.65-0.70}$, and for the extension parallel $R_{||} \propto N^{0.90-1.00}$. Near the density, where the rodlike chains in disordered solution overlap, $\rho \sim N^{-2}, R_\perp$ grows at the expense of $R_{||}$ until at the overlap density $\rho^\star$, the effective exponent is about 0.82. The transition regime ranges from $\rho \sim N^{-2}$ to about $\rho \sim N^{-1.4}$ where the coils start to overlap and one eventually reaches $\nu = 1/2$ in the semidilute regime. The exponents reported should not be taken as asymptotic ($N \to \infty$), however they should be relevant for many experimental situations.

4.2. POOR SOLVENT CHAINS WITH EXPLICIT COUNTERIONS

Many polyelectrolytes possess a carbon based backbone for which water is a poor solvent. Therefore, in aqueous solution, there is a competition between the solvent quality, the Coulombic repulsion, and the entropic degrees of freedom. The conformation in these systems can under certain conditions assume pearl-necklace like structures[47], which have also been seen on the Debye-Hückel level [48]. These also exist for strongly charged polyelectrolytes at finite densities in the presence of counterions[49, 26]. The simulations in Ref.[49] used up to 16 chains of length $N_m = 94$, with a charge fraction of $f = 1/3$, and monovalent counterions. The hydrophobic interaction strength was tuned by means of the Lennard-Jones parameter ϵ. There we showed that the polymer density ρ can be used as a very simple parameter to separate different conformation regimes. This can already

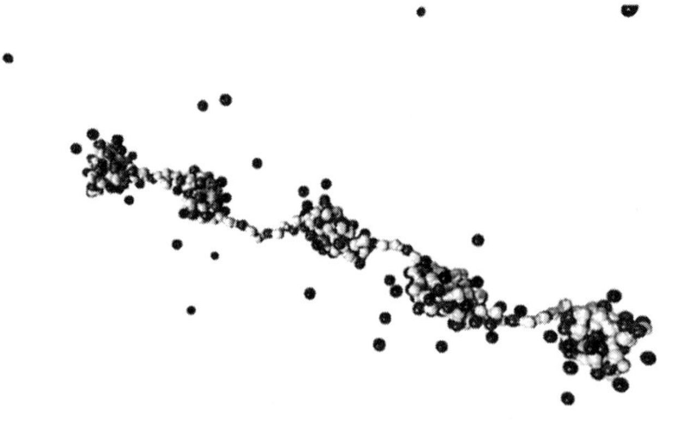

Figure 3. Typical polyelectrolyte conformation for a density $\rho = 2 \cdot 10^{-4}\sigma^{-3}$, showing 5 pearls. The chain had 382 monomers with a charge fraction $f = 1/3$, $l_B = 1.5$, and $\epsilon = 1.75$.

be seen in the plots of the end-to-end distance R_e and $r = \frac{R_E^2}{R_G^2}$ versus ρ in Fig. 2. At very high densities the electrostatic interaction is highly screened, so that the hydrophobic interaction wins, and the chains collapse to dense globules. If one slightly decreases the density, the chains can even contract further, because there are no more steric hinderences from the other chains or counterions, and the screening is smaller. The collapsed globules, however, have still a net charge, and repel each other, so that this phase resembles a charged stabilized colloid or microgel phase. With decreasing density the electrostatic interaction will dominate over the hydrophobic one. The chains will tend to elongate, assuming pearl-necklace conformations, like in Fig. 3, as they have been predicted for weakly charged polyelectrolytes in Ref. [47]. The more the chain stretches, the smaller become the locally compact regions. Note that in contrast to the analytical theories[50, 51], the pearls are stable, even though there are counterions localized near and/or inside the pearls.

Experimentally there are some hints for the existence of pearl-necklace chains[52, 53]. One of the obstacles to observing them in scattering experiments could be related to the strong fluctuations of the pearl number. Even in equilibrium we have found coexistence of several pearl states[26]. In Fig. 4 we see the time evolution of one single chain composed of 382 monomers with a charge fraction $f = 1/3$, $l_B = 1.5$, and $\epsilon = 1.75$ in a many chain system at density $\rho = 1.48 \times 10^{-5}$. One observes jumps between a five and four pearl configuration. Also the position of the pearls move quite vividly.

The different length scales appearing in a chain can be analyzed by

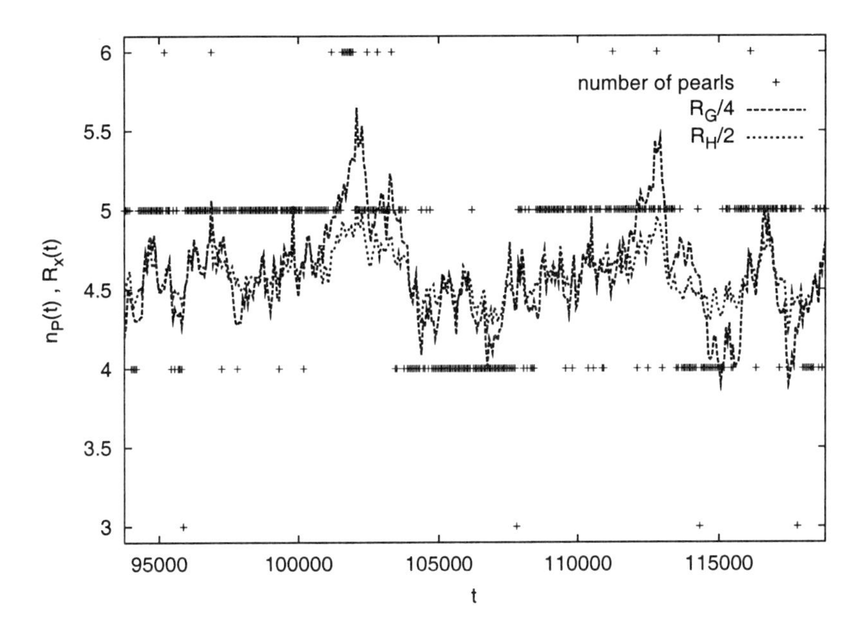

Figure 4. Time development of the radius of gyration R_G, the hydrodynamic radius R_H and the number of observed pearls for the same system as in Fig. 3.

looking at the spherically averaged form factor $S_1(q)$ of the chain. The maximum seen at $q \approx 6$ comes from the bond length. In the range $1 < q < 2$ we observe a sharp decrease in S_1, which comes from the scattering from the pearls, because it shows the typical Porod scattering of $S_1(q) \simeq q^{-4}$. The kink at $q \approx 1.66$ appears at the position expected from the pearl size, but is broadly smeared out due to large size fluctuations. The shoulder which can be seen at $q \approx 0.5$ does not come from the intra-pearl scattering but is due to the scattering of neighboring pearls along the chain (inter-pearl contribution), which have a mean distance of $\langle r_{PP} \rangle = 13.3.\sigma$. It is also smeared out due to large distribution of inter-pearl distances. We conclude that the signatures of the pearl-necklaces are weak already for monodisperse samples. A possible improvement could be achieved for chains of very large molecular weights and only few pearl numbers, which could lead to stable and large signatures. Many more interesting results on poor solvent polyelectrolytes can be found in Ref.[26] and will be published soon.

4.3. COUNTERION DISTRIBUTION AROUND FINITE POLYELECTROLYTES

We recently completed a study of the spatial distribution of the counterions around strongly charged, flexible polyelectrolytes[54] in good and poor sol-

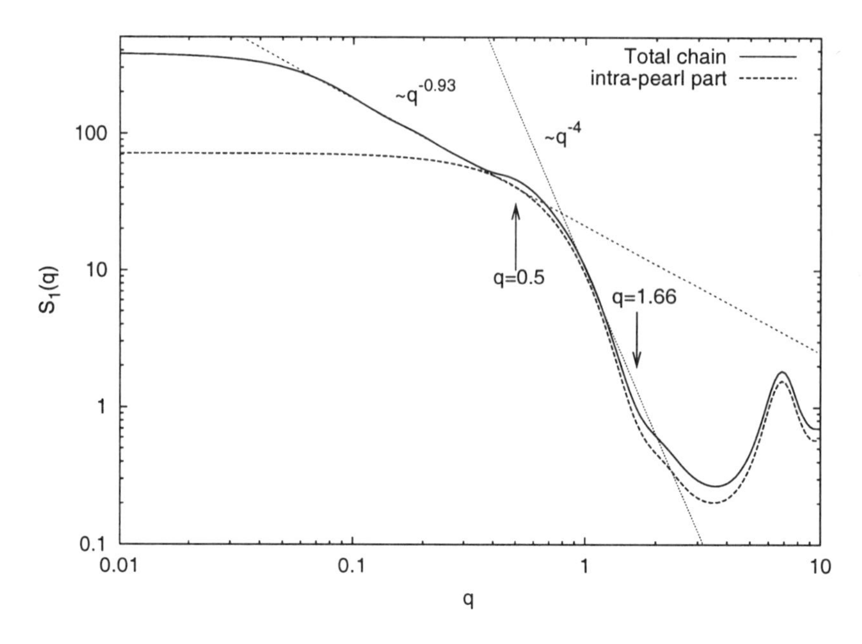

Figure 5. Spherically averaged form-factor $S_1(q)$ for the same system as in Fig. 4. The solid line denotes the single chain form-factor, the dashed line shows only the contributions of monomers within the same pearl.

vent. There we demonstrated that by partially neutralizing the quenched charged distribution on the chain backbone the inhomogeneous distribution of counterions leads to the same qualitative effects that are observed in weakly charged polyelectrolytes with an annealed charge distribution[2]. This is due to the presence of the mobile partially neutralizing counterions, which results in an annealed backbone charge distribution. The common underlying physical mechanism for the end-effect is the differences in the electrostatic field of the chain along its backbone. The strength of the end-effect depends on parameters like chain length, charge fraction and ionic strength, and those dependencies were found in agreement with the scaling predictions. We found a saturation of the end-effect for long chains, when the chain extension, namely R_e, is at least twice as large as the Debye screening length. A simple Debye-length criterion appeared to be sufficient to explain the penetration depth of the end-effect. However, looking at the amplitude dependency on density and ionic strength of the solution, we found that both parameters, the number of annealing ions and the ionic strength of the solution, influence the end-effect and that the first one dominated. The amplitude of the end-effect was shown to depend strongly on the charge parameter $\xi := l_B/b$, where b is the distance of the bare charges on the backbone of the chain. The definition of such an end-effect via close mobile counterions can not be made for an effective charge $\xi << 1$, because under dilute conditions there are almost no counterions close to the chain.

Even though the chain conformation is very different in the poor solvent case the end-effect was found to be qualitatively the same, namely the counterions are more likely to be found at the middle of the chain than at the ends. We could also clearly see the necklace structure by looking at the effective charge along the contour length. However, the string length of our simulated pearl-necklaces was too short to show any charge difference between the pearls and the strings, as has been predicted in Ref. [55].

We also obtained a fairly good agreement of the simulated ion distribution with the PB solution of the cell model of an infinitely extended charged rod[41]. This supports the idea that the description of polyelectrolytes as rodlike objects in mean-field theory is valid in the dilute regime. Further improvements could probably be achieved along the lines of Ref.[56], where a combination of a cylindrical and spherical cell model is used to describe the solution properties of polyelectrolytes.

4.4. RODLIKE POLYELECTROLYTES

Stiff linear polyelectrolytes can be approximated by charged cylinders. This is a relevant special case, applying to quite a few biologically important polyelectrolytes with a large persistence length, like DNA, actin filaments or microtubules. Within PB theory [57] and on the level of a cell model the cylindrical geometry can be treated exactly in the salt-free case [41, 58, 59, 60, 61, 62, 63], providing for instance new insights into the phenomenon of the Manning condensation [64, 65]. For low line charges, the agreement between PB theory and the simulations of the full interaction system is rather nice. However, for highly charged rods or/and multivalent counterions PB theory fails quantitatively (underestimated condensation) and qualitatively (overcharging, charge oscillations and attractive interactions); see, e.g. Ref.[63, 66, 67].

Recently the osmotic coefficient of a synthetic stiff polyelectrolyte, a poly(para-phenylene), was measured in a salt-free environment[68, 69]. We have compared this data to predictions of PB theory, and a local density functional theory which includes a correlation correction of the basis of a recently proposed Debye-Hückel-Hole-cavity theory (DHHC) [70], and simulational results within the cell model. We find that correlation effects enhance condensation and lower the osmotic pressure, yet are not fully able to explain the discrepancy with the experimental data. Here the approach of working within the "primitive model" breaks down. In our opinion, specific interactions between the counterions, the macroion, and the solvent particles are needed to explain the discrepancy. Other theoretical approaches beyond the cell model which try to incorporate finite-size effects and interactions of the macroion itself will in general lead to a higher osmotic coefficient which is in contrast to the experimental data[71].

Attractive interactions have been observed[72, 73, 74, 75] and predicted between like-charged macromolecules. However, there are nice rigorous results which prove that these effects cannot be described by mean-field theories[76, 77, 78]. Especially in the community of biological inspired physics[44, 79, 80, 81], these interactions are thought to be important for the clarification of the mechanism behind DNA compactification in viral heads[82], the chromatin structure[83], and novel methods for gene delivery[84], to name just the most prominent examples. There are numerous simulations which show similar attractions on a distance of few counterion diameters [66, 85, 86, 87, 88, 89, 90].

The mechanism which is driving the observed attractions for rod-like systems has been speculated to be correlations between the counterion layers around the macroion. However, until now, no unique theoretical picture has emerged that can clarify the detailed mechanism behind the attractions. There is the low temperature Wigner crystal theory, initiated by Refs. [91, 92, 81], which postulates an ordered ground state of the counterions. Then there are theories which are based on Van der Waals type correlated fluctuations [60, 93, 94, 95], that are in principle hight T theories. There are also theories which are fluctuation based, but are valid at low T[43, 96, 97]. Integral equation [42, 98, 99] theories on various approximation levels have been demonstrating the existence of these attractions for a long time, but from these theories it is difficult to extract the detailed mechanism behind the observed correlations. Here also simulations can be helpful, because they have in principal access to all correlations[100]. More details of our results in rod-like geometries can be found in Refs.[63, 66, 67, 70, 71, 101, 102].

5. The energetic path to understand overcharging

There has been a recent interest in the study of systems which are strongly coupled by Coulomb interactions. These systems show a variety of, at first sight, surprising behaviors, which cannot be accounted for by the mean-field PB theory. For example, there are attractions between like charged objects and a charge reversal of macroions occurs when viewed from some distance. This means that there are more ions of the opposite charge within a certain radius around the macroion then necessary to charge neutralize it. This overcompensation is called "overcharging".

In this section we want to demonstrate that there are situations for charged colloidal objects in which one can understand the phenomenon of overcharging by very simple energetic arguments. By overcharging, in general, we mean that the bare charge of the macroion is overcompensated at some distance by oppositely charged "microions". To achieve this in nature we have to add salt to the system. For the sake of simplicity, however, we will consider non-neutral systems, because they can on a very simple basis explain why colloids prefer to be overcharged.

5.1. THE MODEL

Our model is solely based on electrostatic energy considerations, meaning that we only look at the ground state of a system of charges. We consider a colloid of radius a with a central charge Z. In the ground state the counterions of this colloid are located on the surface, because there they are closest to the central charge. On the other hand they want to be in such a configuration that they minimize their mutual repulsion. For two, three, and four counterions these configurations correspond to a line, an equilateral triangle, and a tetrahedron, respectively, regardless of the central charge magnitude. The problem of the minimal energy configuration of electrons disposed on the surface of a sphere dates back to Thomson [103], and is actually unsolved for large N. The reason is, that there are many metastable states which differ only minimally in energy, and their number seems to grow exponentially with N. Also chemists developed the valence-shell electron-pair repulsion (VSEPR) theory [104] which uses similar arguments to predict the molecular geometry in covalent compounds, also known as the Gillespie rule.

A simple illustration of energetically driven overcharging is depicted in Fig. 6. The central charge is $+2e$, and the neutral system has two counterions of valence 1. If we add successively more counterions of the same valence, and put them on the surface such that their mutual repulsion is minimized, we can compute the total electrostatic energy according to

$$E(n) = k_B T (l_B/a) \left[-n Z_m + f(\theta_i) \right], \tag{28}$$

where $f(\theta_i)$ is the repulsive energy part which is only a function of the ground state configuration. We surprisingly find, that actually the minimal energy is obtained when *four* counterions are present, hence we overcharged the colloid by two counterions, or by 100 %! That is, the excess counterions gain more energy by assuming a energetically favorable configuration around the macroion than by escaping to infinity, the simple reason behind overcharging. In our example, the minimum is reached when four counterions are present. The colloid radius and the Bjerrum length enter as prefactors and change only the energy difference between neighboring states.

The spatial correlations of the counterions are fundamental to obtain overcharging. Indeed, if we apply the same procedure and smear Z counte-

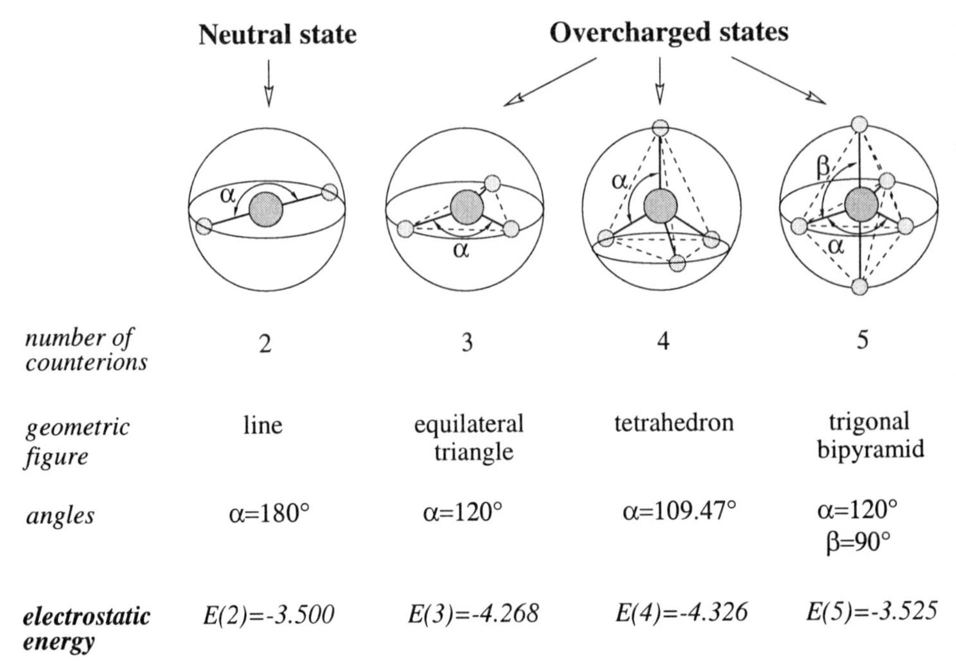

Figure 6. Ground state configurations for two, three, four and five electrons. The corresponding geometrical figure repulsion and their typical angles are given. The electrostatic energy (in units of $k_B T l_B / a$) is given for a central charge of $+2e$.

rions onto the surface of the colloid of radius a, we obtain for the energy

$$E = l_B \left[\frac{1}{2} \frac{Z^2}{a} - \frac{Z_m Z}{a} \right]. \tag{29}$$

The minimum is reached for $Z = Z_m$, hence no overcharging can occur.

The important message to be learned is that, from an energetic point of view, a colloid *always* tends to be overcharged by discrete charges. Other important geometries like infinite rods or infinitely extended plates cannot be treated in such a simple fashion because they are not finite in all directions. One needs therefore enough screening charges in the environment to limit the range of the interactions in the infinite directions, there is a need for a *minimal* amount of salt present to allow for overcharging[102], which is not the case for a colloid.

Obviously, for a large number of counterions the direct computation of the electrostatic energy by using the exact equation (28) becomes unfeasible. Therefore we resort to simulations for highly charged spheres.

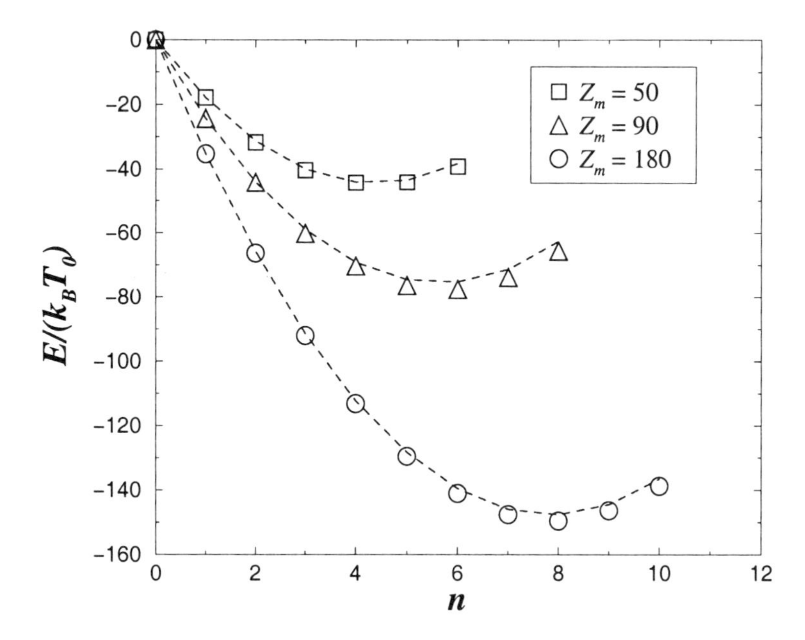

Figure 7. Electrostatic energy (in units of $k_B T_0$) for *ground state* configurations of a single charged macroion as a function of the number of *overcharging* counterions n for three different bare charges Z_m. The neutral case was chosen as the potential energy origin, and the curves were produced using the theory of Eq. (33), compare text.

5.2. ONE COLLOID

The electrostatic energy as a function of the number of overcharging counterions n is displayed in Fig. 7. We note that the maximal (critical) acceptance of n (4, 6 and 8) increases with the macroionic charge Z_m (50, 90 and 180 respectively). Furthermore for fixed n, the gain in energy is always increasing with Z_m. Also, for a given macroionic charge, the gain in energy between two successive overcharged states is decreasing with n.

In the ground state the counterions are highly ordered. Rouzina and Bloomfield [91] first stressed the special importance of these crystalline arrays for interactions of multivalent ions with DNA strands, and later Shklovskii [81] showed that the Wigner crystal (WC) theory can be applied to determine the interactions in strongly correlated systems. In two recent short contributions [106, 107] we showed that the overcharging curves obtained by simulations of the ground state, like Fig. 7, can be simply explained by assuming that the energy ε per counterion on the surface of a macroion scales as $\sqrt{c}$, where c denotes the counterion concentration $c = N/A$, N is the *total* number of counterions on the surface and A the total macroion area. This can be justified by a simple argument, where each ion interacts in first approximation only with the oppositely charged background of its Wigner-Seitz (WS) cell, which can be approximated by a disk of radius h, yielding the same WS cell area.

For fixed macroion area we can write the energy per counterion as

$$\varepsilon^{(h)}(N) = -\frac{\bar{\alpha}^{(h)}\ell}{\sqrt{A}}\sqrt{N} = -\bar{\alpha}^{(h)}\ell\sqrt{c}, \tag{30}$$

where $\ell = l_B Z_c^2$ and the simple hole theory gives $\bar{\alpha}^{(h)} = 2\sqrt{\pi} \approx 3.54$[108].

For an infinite plane, where the counterions form an exact triangular lattice, one obtains the same *functional* form as in Eq. (30), but the prefactor $\bar{\alpha}^{(h)}$ gets replaced by the numerical value $\bar{\alpha}^{WC} = 1.96$[109].

Not knowing the precise value of $\bar{\alpha}$ we can still use the simple scaling behavior with c to set up an equation to quantify the energy gain ΔE_1 by adding the first overcharging counterion to the colloid. To keep the OCP neutral we imagine adding a homogeneous surface charge density of opposite charge ($\frac{-Z_c e}{A}$) to the colloid[81]. This ensures that the background still neutralizes the incoming overcharging counterion and we can apply Eq. (30). To cancel our surface charge addition we add another homogeneous surface charge density of opposite sign $\frac{Z_c e}{A}$. This surface charge does not interact with the now neutral OCP, but adds a self-energy term of magnitude $\frac{1}{2}\frac{\ell}{a}$, so that the total energy difference for the first overcharging counterion reads as

$$\Delta E_1 = (N_c + 1)\varepsilon(N_c + 1) - N_c\varepsilon(N_c) + \frac{\ell}{2a}. \tag{31}$$

By using Eq. (30) this can be rewritten as[110]

$$\Delta E_1 = -\frac{\bar{\alpha}\ell}{\sqrt{A}}\left[(N_c + 1)^{3/2} - N_c^{3/2}\right] + \frac{\ell}{2a}. \tag{32}$$

Completely analogously one derives for the energy gain ΔE_n for n overcharging counterions

$$\Delta E_n = -\frac{\bar{\alpha}\ell}{\sqrt{A}}\left[(N_c + n)^{3/2} - N_c^{3/2}\right] + \frac{\ell}{a}\frac{n^2}{2}. \tag{33}$$

Using Eq. (33), where we determined the unknown $\bar{\alpha}$ from the simulation data for ΔE_1, we obtain a curve that matches the simulation data almost perfectly (Fig. 7). The second term in Equation (33) also shows why the overcharging curves of Fig. 7 are shaped parabolically upwards for larger values of n. If one successively removes each of n counterions from a neutral colloid, one can derive in a similar fashion the ionization energy cost

$$\Delta E_n^{ion} = -\frac{\bar{\alpha}\ell}{\sqrt{A}}\left[(N_c - n)^{3/2} - N_c^{3/2}\right] + \frac{\ell}{a}\frac{n^2}{2}. \tag{34}$$

Using the measured value of $\bar{\alpha}$ we can simply determine the maximally obtainable number n_{max} of overcharging counterions by finding the stationary point of Eq. (33) with respect to n:

$$n_{max} = \frac{9\bar{\alpha}^2}{32\pi} + \frac{3\bar{\alpha}}{4\sqrt{\pi}}\sqrt{N_c}\left[1 + \frac{9\bar{\alpha}^2}{64\pi N_c}\right]^{1/2}. \qquad (35)$$

The value of n_{max} depends only on the number of counterions N_c and $\bar{\alpha}$. For large N_c Eq. (35) reduces to $n_{max} \approx \frac{3\bar{\alpha}}{4\sqrt{\pi}}\sqrt{N_c}$ which was derived in Ref. [105] as the low temperature limit of a a neutral system in the presence of salt. What we have shown is that the overcharging in this limit has a pure electrostatic origin, namely it originates from the energetically favorable arrangement of the ions around a central charge. We also showed in Ref. [110] that $\bar{\alpha}$ reaches the perfect WC value of 1.96 if the colloid radius a gets very large at fixed c, or when c becomes large at fixed a.

If instead of a central charge scheme one uses discrete charge centers distributed randomly over the colloidal surface we find counterion structures which are quite far away from the WC array, especially when the counterions are pinned to their counter charges. This depends on the interaction energy at contact, which depends of course on l and distance of closest approach. However, we still find overcharging, although reduced in value, of the form given by Eq. 33 [108, 111]

5.2.1. *Macroion-counterion interaction profile at* $T = 0K$

The interaction profile between a completely neutralized macroion and one excess counterion is obtained by displacing adiabatically the excess counterion from infinity towards the macroion. From far away the counterion sees only a neutral object and has no measurable interaction, whereas upon approach to the macroion the WC hole gets created in the counterion layer, and we observe a distance dependant attraction towards the macroion. We investigated cases of $Z_m = 2 \ldots 288$. All curves can be nicely fitted with an exponential fit of the form

$$E_1(r) = \Delta E_1 e^{-\tau(r-a)}, \qquad (36)$$

where ΔE_1 is the measured value for the first overcharging counterion, and τ is the only fit parameter. In all our results for τ versus $\sqrt{N_c}$ we observe a linear dependence for a wide range of values for N_c, $\tau \propto \sqrt{N_c}$, which again can be explained by applying the WC hole picture [110].

5.3. TWO COLLOIDS

Now we apply what we have learned about a single colloid to two equal-sized, fixed charged spheres of bare charge Q_A and Q_B separated by a center-center separation R and surrounded by their neutralizing counterions, which give concentrations c_A and c_B, respectively.

All these ions making up the system are immersed in a cubic box of length $L = 80\sigma$, and the two macroions are held fixed and disposed symmetrically along the axis passing through the centers of opposite faces. This leads to a colloid volume fraction $f_m = 2 \cdot \frac{4}{3}\pi(a/L)^3 \approx 8.4 \times 10^{-3}$. For *finite* colloidal volume fraction f_m and temperature, we know from the study carried out above that in the strong Coulomb coupling regime all counterions are located in a spherical "monolayer" in contact with the macroion. Here, we investigate the mechanism of *strong, long range* attraction stemming from *monopole* contributions; that is, one colloid is overcharged and the other one undercharged.

5.3.1. *Observation of metastable ionized states*

For the charge symmetrical situation we have $c_A = c_B$. When we brought this system to room temperature T_0 and generated initially the counterions randomly inside the box we observed in some cases that one of the colloids remained undercharged, and the other one was overcharged, and these configurations turned out to be extremely long lived in the course of our MD simulations(more than 10^8 MD time steps). However it is clear that such a state is "metastable" because by symmetry arguments it cannot be the lowest energy state. The observed barrier is the result of the WC attraction, because close to the macroion surface the energy is reduced. For very distant macroions the barrier height for the first overcharged state has to equal ΔE_1 from Eq. 33. The barrier profile at $T = 0$ can also be extremely well approximated by an application of Eqs. (33) and (34), plus taking into account the distant dependent monopole contribution[106]. This leads to a barrier height which scales as $\sqrt{c}$ for large separations. For smaller separations one has to take into account also the effect of strong mutual polarization of both macroions, which leads effectively to a sharing of their proximal counterion layer into a superlattice. This can be taken into account by a higher effective counterion density close to the surface, leading to an almost linear scaling of the barrier height with c [106, 110].

5.3.2. *Asymmetrically charged colloids*

The most interesting phenomenon, however, appears when the two colloids have different counterion concentrations, here $c_A > c_B$, since then **stable ionized states** can appear. The physical reason is that a counterion can gain more energy by overcharging the colloid with c_A then it loses by ionizing colloid B. A straight forward application of the procedure outlined for the barrier calculation [107, 110] yields a simple criterion (more specifically a sufficient condition), valid for large macroionic separations, for the charge asymmetry $\sqrt{N_A} - \sqrt{N_B}$ to produce an ionized ground state of two unlike charged colloids with the same size:

$$\left(\sqrt{N_A} - \sqrt{N_B}\right) > \frac{4\sqrt{\pi}}{3\tilde{\alpha}^A} \approx 1.2. \tag{37}$$

5.3.3. *Finite temperature analysis*

We have also demonstrated that the ground state phenomena survive for finite temperatures, i.e. an ionized state can also exist at room temperature T_0. The left part of Figure 8 shows the time evolution of the electrostatic energy of a system $Z_A = 180$ with $Z_B = 30$, $R/a = 2.4$ and a colloidal volume fraction of $7 \cdot 10^{-3}$, where the starting configuration is the neutral state ($DI = 0$). One clearly observes two jumps in energy, $\Delta E_1 = -19.5$ and $\Delta E_2 = -17.4$, which corresponds each to a counterion transfer from colloid B to colloid A. These values are consistent with the ones obtained for the ground state, which are -20.1 and -16.3 respectively. Note that this ionized state ($DI = 2$) is more stable than the neutral but is expected to be metastable, since it was shown previously that the most stable ground state corresponds to $DI = 5$. The other stable ionized states for higher DI are not accessible with reasonable computer time because of the high energy barrier made up of the correlational term and the monopole term which increase with DI. In the right part of Fig. 8 we display a typical snapshot of the ionized state ($DI = 2$) of this system at room temperature.

Obviously, these results are not expected by the DLVO theory even in the asymmetric case (see e. g. [112]). Previous simulations of asymmetric (charge and size) spherical macroions [113] were also unable to predict such a phenomenon since the Coulomb coupling was weak (water, monovalent counterions). Note that the appearance of (meta-)stable ionized states can alter the effective interactions between charged colloids in solution. The monopole attraction will lead to attraction between like charged colloids, flocculation, and related phenomena.

At this stage, we would like to stress again, that the appearance of a stable ionized ground state is due merely to correlation. An analogous consideration with smeared out counterion distributions along the lines of Eq. (29) will again always lead to two colloids exactly neutralized by their counterions [114]. Our energetic arguments are quite different from the situation encountered at finite temperatures, because in this case even a PB description would lead to an asymmetric counterion distribution. However,

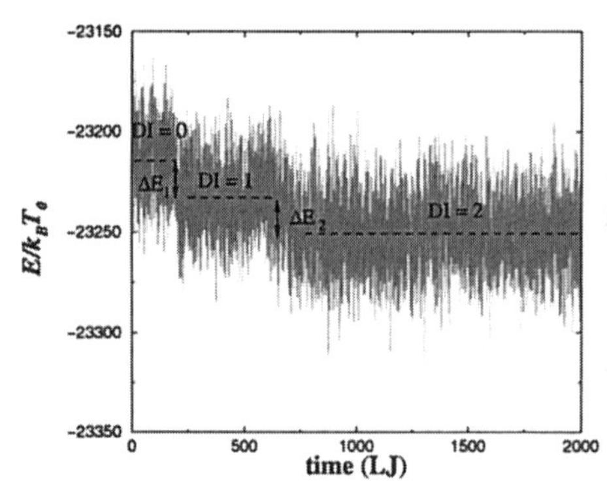

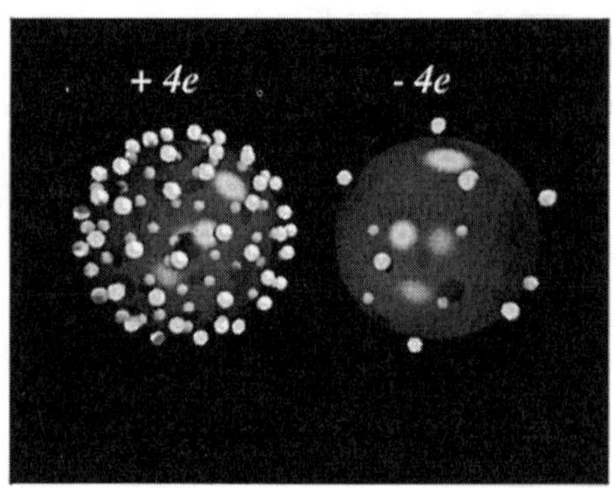

Figure 8. Relaxation, at room temperature $T_0 = 298K$, of an initial unstable neutral state towards ionized state. Plotted is the total electrostatic energy versus time (LJ units), for $Z_B = 30$ and $R/a = 2.4$. Dashed lines lines represent the mean energy for each *DI* state. Each jump in energy corresponds to a counterion transfer from the macroion *B* to macroion *A* leading to an ionized state that is lower in energy than the neutral one. The right figure is a snapshot of the final ionized state, with net charges *+4e* and *-4e* as indicated.

in the latter case this happens due to purely entropic reasons, namely in the limit of high temperatures, the counterions want to be evenly distributed in space, leading to an effective charge asymmetry.

Note also, that there can exist parameter regions, such as high molar electrolytes, where the overcharging of a single macroion is due to mainly entropic effects [98, 115, 116], whose exact mechanism is currently under investigation[117].

6. Acknowledgments

We gratefully acknowledge collaborations at various stages with A. Arnold, J. DeJoannis, M. Deserno, H.J. Limbach, R. Messina, U. Micka, and Z. Wang, financial support by the DFG, the BMBF nano-center and LEA, and a large computer time grant No. hkf06 from NIC Jülich.

References

1. M. Hara, *Polyelectrolytes: Science and Technology* (Marcel Dekker, New York, 1993); K. S. Schmitz,*Macroions in solution and colloidal suspension* (VCH Publishers, New York, 1 edition, 1993); H. Dautzenberg, W. Jaeger, J. Kötz, B. Philipp, Ch. Seidel, D. Stscherbina, *Polyelectrolytes*, (Hanser Publishers, Munich, 1994); S. Förster, M. Schmidt, Adv. Poly. Sci. **120**, Springer Verlag Berlin, Heidelberg (1995); J.-L. Barrat, J.-F. Joanny, Adv. Chem. Phys. **94**, 1 (1995).
2. J.-F. Joanny, Chapter in this volume.
3. K. Binder, in *Monte Carlo and Molecular Dynamics of Condensed Matter Systems, Como Conference Proceeding*, edited by K. Binder and G. Ciccotti (Società Italiana

di Fisica, Bologna, 1996); Baumgärtner and K. Binder, *Applications of the Monte Carlo Method in Statistical Physics* (Springer, Berlin, 1984); S. G. Whittington, *Numerical Methods for Polymeric Systems, The IMA Volumes in Mathematics and its Applications* (Springer, New York, 1998); L. Monnerie and U. W. Suter, *Atomistic Modelling of Physical Properties*, Vol. 116 of *Advances in Polymer Science* (Springer, Heidelberg, 1994); K. Binder, in *Computational Modeling of Polymers*, edited by J. Bicerano (Springer, Berlin, Heidelberg, New York, 1992); *Computer Simulation of Polymers*, edited by R. J. Roe (Prentice Hall, Englewood Cliffs, NJ, 1991); *Elastomeric Polymer Networks*, edited by H. E. Mark and B. Erman (Prentice Hall, Englewood Cliffs, 1992).

4. K. Kremer and K. Binder, Comp. Phys. Reports **7**, 259 (1988); K. Kremer, in *Monte Carlo and Molecular Dynamics of Condensed Matter Systems*, Como Conf. Proceedings 1995, edited by K. Binder and G. Ciccotti (Societa Italiana di Fisica, Bologna, 1996), p. 671.
5. M. P. Allen and D. J. Tildesley, *Computer Simulations of Liquids*, 2nd ed. (Oxford Univ. Press, London, 1989); D. Frenkel and B. Smit, *Understanding Molecular Simulation: From Basic Algorithms to Applications* (Academic Press, San Diego, CA, 1996).
6. J. Baschnagel *et al.*, in *Bridging the Gap Between Atomistic and Coarse Grained Model of Polymers: Status and Perspectives, Advances in Polymer Science* (Springer, Berlin, 1998).
7. G. S. Grest and K. Kremer, Phys. Rev. A, **33**, 3628 (1986).
8. H. Risken, *The Fokker-Planck Equation* (Springer, Berlin, second edition, 1989); B. Dünweg, J. Chem. Phys., **99**, 6977 (1993).
9. B. Dünweg and W. Paul, Int. J. Mod. Phys. C, **2**, 817 (1991).
10. E. Fermi, J.R. Pasta, and S. Ulam, in *Collected Works of Enrico Fermi* **2**, 978 (University of Chicago Press, Chicago, 1965).
11. *Monte Carlo Methods in Statistical Physics* edited by K. Binder (Springer Verlag, Berlin, Heidelberg, New York, 1979); *Applications of the Monte Carlo Method in Statistical Physics*, edited by K. Binder (Springer Verlag, Heidelberg, New York, 1984); *Monte Carlo Methods in Condensed Matter Physics*, edited by K. Binder (Springer Verlag, Berlin, Heidelberg, New York, 1992).
12. P. Español, P. Warren, Europhys. Lett. **30**, 191 (1995); A. J. C. Ladd, J. Fluid Mech. **271**, 285 (1994); J. Fluid Mech. **271**, 311 (1994); Phys. Rev. Lett. **76**, 1392 (1996); P. Ahlrichs and B. Dünweg, J. Chem. Phys. **111**, 8225 (1999); P. J. Hoogerbrugge and J. M. V. A. Kroelman, Europhys. Lett. **19**, 155 (1992); S. Chen and G. D. Doolen, Annu. Rev. Fluid Mech. **30**, 329 (1998).
13. P. Ewald, Ann. Phys. **64**, 253 (1921).
14. D. M. Heyes, J. Chem. Phys. **74**, 1924 (1981).
15. H. J. C. Berendsen, in *Computer Simulation of Biomolecular Systems*, edited by W. F. van Gunsteren, P. K. Weiner, and A. J. Wilkinson (ESCOM, The Netherlands, 1993), Vol. 2, pp. 161–81.
16. P. H. Hünenberger, J. Chem. Phys. **113**, 10464 (2000).
17. J. Kolafa and J. W. Perram, Molecular Simulation **9**, 351 (1992).
18. S. W. de Leeuw, J. W. Perram, and E. R. Smith, Proc. R. Soc. Lond. A **373**, 27 (1980).
19. S. W. de Leeuw, J. W. Perram, and E. R. Smith, Proc. R. Soc. Lond. A **373**, 57 (1980).
20. J. Perram, H. G. Petersen, and S. de Leeuw, Mol. Phys. **65**, 875 (1988).

21. R. W. Hockney and J. W. Eastwood, *Computer Simulation Using Particles* (IOP, London, 1988).
22. D. Y. T. Darden and L. Pedersen, J. Chem. Phys. **98**, 10089 (1993).
23. U. Essmann *et al.*, J. Chem. Phys. **103**, 8577 (1995).
24. H. G. Petersen, J. Chem. Phys. **103**, 3668 (1995).
25. E. L. Pollock and J. Glosli, Comp. Phys. Comm. **95**, 93 (1996).
26. H. J. Limbach, Ph.D. thesis, Universität, Mainz, Germany, 2001.
27. M. Deserno and C. Holm, J. Chem. Phys. **109**, 7678 (1998).
28. M. Deserno and C. Holm, J. Chem. Phys. **109**, 7694 (1998).
29. J. Lekner, Physica A **176**, 485 (1991).
30. R. Sperb, Molecular Simulation **20**, 179 (1998); ibid. **22**, 199 (1999).
31. R. Sperb and R. Strebel, ETH Research Report No 2000-02.
32. J. Barnes and P. Hut, Nature **324**, 446 (1986).
33. L.Greengard and V. Rhoklin, J. Comp. Phys. **73**, 325 (1987).
34. K. Esselink, Comp. Phys. Comm. **87**, 375 (1995).
35. A. H. Widmann and D. B. Adolf, Comp. Phys. Comm. **107**, 167 (1997).
36. A. Arnold, Diploma thesis, Johannes Gutenberg-Universität, 2001.
37. A. Arnold and C. Holm, to be published (2001).
38. I.-C. Yeh and M. L. Berkowitz, J. Chem. Phys. **111**, 3155 (1999).
39. A. Arnold, J. Dejoannis, and C. Holm, to be published (2001).
40. Z. W. Wang and C. Holm, J. Chem. Phys., in press (2001).
41. M. Deserno and C. Holm, Chapter in this volume.
42. R. Kjellander, Chapter in this volume.
43. A. G. Moreira and R. R. Netz, Chapter in this volume.
44. R. Podgornik, Chapter in this volume.
45. U. Micka, K. Kremer, Europhys. Lett. **38**, 279 (1997).
46. M. J. Stevens and K. Kremer, J. Chem. Phys. **103**, 1669 (1995); M. Stevens, K. Kremer, Phys. Rev. Lett. **71**, 2228 (1993); M. Stevens, K. Kremer, Macromolecules **26**, 4717 (1993)).
47. A. V. Dobrynin, M. Rubinstein, and S. P. Obukhov, Macromolecules **29**, 2974 (1996).
48. A. V. Lyulin and B. Dünweg and O. V. Borisov and A. A. Darinskii, Macromolecules **32**, 3264 (1999); P. Chodanowski, and S. Stoll, J. Chem. Phys **111**, 6069 (1999).
49. U. Micka, C. Holm, and K. Kremer, Langmuir **15**, 4033 (1999); U. Micka and K. Kremer, Europhys. Lett.,**49**, 189 (2000).
50. H. Schiessel and P. Pincus, Macromolecules **31**, 7953 (1998); H. Schiessel, Macromolecules **32**, 5673 (1999).
51. A. V. Dobrynin, and M. Rubinstein, Macromolecules **32**, 915 (1999).
52. M. Rawiso, Chapter in this volume.
53. C. E. Williams, Chapter in this volume.
54. H. J. Limbach and C. Holm, J. Chem. Phys. **114**, 9674 (2001).
55. M. Castelnovo, P. Sens, and J.-F. Joanny, Eur. Phys. J. E **1**, 115 (2000).
56. A. Deshkovski, S. Obukhov, and M. Rubinstein, Phys. Rev. Lett. **86**, 2341 (2001).
57. G. L. Gouy, J. de Phys. **9**, 457 (1910).
58. T. Alfrey, P. W. Berg, and H. J. Morawetz, J. Polym. Sci. **7**, 543 (1951).
59. A. Katchalsky, Pure Appl. Chem. **26**, 327 (1971).
60. B. Jönsson and H. Wennerström, Chapter in this volume.
61. M. L. Bret and B. Zimm, Biopolymers **23**, 271 (1984).
62. M. L. Bret and B. Zimm, Biopolymers **23**, 287 (1984).

63. M. Deserno, C. Holm, and S. May, Macromolecules **33**, 199 (2000).
64. G. Manning, J. Chem. Phys. **51**, 924 (1969).
65. F. Oosawa, *Polyelectrolytes* (Marcel Dekker, New York, 1971).
66. M. Deserno, Ph.D. thesis, Universität Mainz, 2000.
67. M. Deserno, C. Holm, and K. Kremer, in *Physical Chemistry of Polyelectrolytes*, Vol. 99 of *Surfactant science series*, edited by T. Radeva (Marcel Decker, New York, 2001), Chap. 2, pp. 59–110.
68. B. Guilleaume *et al.*, J. Phys. Cond. Mat. **12**, A245 (2000).
69. J. Blaul, M. Wittemann, M. Ballauff, and M. Rehahn, J. Phys. Chem. B **104**, 7077 (2000).
70. M. C. Barbosa, M. Deserno, and C. Holm, Europhys. Lett. **52**, 80 (2000).
71. M. Deserno *et al.*, Eur. Phys. J. E **5**, 97 (2001).
72. R. Podgornik, D. Rau, and A. Parsegian, Biophys. J. **66**, 962 (1994).
73. J. X. Tang, S. Wong, P. T. Tran, and P. Janmey, Ber. Bunsenges. Phys. Chem. **100**, 796 (1996).
74. V. A. Bloomfield, Current Opin. Struct. Biol. **6**, 334 (1996).
75. A. P. Lyubartsev, J. X. Tang, P. A. Janmey, and L. Nordenskiöld, Phys. Rev. Lett. **81**, 5465 (1998).
76. J. Neu, Phys. Rev. Lett. **82**, 1072 (1999).
77. J. Sader and D. Y. Chan, J. Colloid Interface Sci. **213**, 268 (1999).
78. E. Trizac and J.-L. Raimbault, Phys. Rev. E (2000).
79. W. M. Gelbart, Chapter in this volume.
80. A. R. Khokhlov, K. Zeldovich, and E. Y. Kramarenko, Chapter in this volume.
81. T. T. Nguyen, A. Y. Grosberg, and B. I. Shklovskii, see Chapter in this volume.
82. O. Lambert, L. Letellier, W. Gelbart, and J. Rigaud, Proceedings of the National Academy of Sciences (USA) **97**, 7248 (2000).
83. K. E. van Holde, *Chromatin* (Springer, New York, 1989).
84. A. V. Kabanov and V. A. Kabanov, Bioconjugate Chem. **6**, 7 (1995).
85. L. Guldbrand, B. Jönsson, H. Wennerström, and P. Linse, J. Chem. Phys. **80**, 2221 (1984).
86. L. G. Nilsson, L. Guldbrand, and Nordenskiöld, Mol. Phys. **72**, 177 (1991).
87. A. P. Lyubartsev and L. Nordenskiöld, J. Phys. Chem. **101**, 4335 (1997).
88. N. Grønbech-Jensen, R. J. Mashl, R. F. Bruinsma, and W. M. Gelbart, Phys. Rev. Lett. **78**, 2477 (1997).
89. M. J. Stevens, Phys. Rev. Lett. **82**, 101 (1999).
90. E. Allahyarov and H. Löwen, Phys. Rev. E **62**, 5542 (2000).
91. I. Rouzina and V. Bloomfield, Journal of Phys. Chem. **100**, 9977 (1996).
92. B. I. Shklovskii, Phys. Rev. Lett. **82**, 3268 (1999).
93. F. Oosawa, Biopolymers **6**, 1633 (1968).
94. O. Spalla and L. Belloni, Phys. Rev. Lett. **74**, 2515 (1995).
95. B.-Y. Ha and A. J. Liu, Phys. Rev. Lett. **79**, 1289 (1997).
96. A. W. C. Lau, D. Levine, and P. Pincus, Phys. Rev. Lett. **84**, 4116 (2000).
97. A. W. C. Lau, P. Pincus, D. Levine, and H. A. Fertig, cond-mat/0006264.
98. E. Gonzales-Tovar, M. Lozada-Cassou, and D. Henderson, J. Chem. Phys. **83**, 361 (1985).
99. R. Kjellander and S. Marcelja, Chem. Phys. Lett. **112**, 49 (1984).
100. M. Deserno and C. Holm, to be published.
101. M. Deserno and C. Holm, submitted (2001).
102. M. Deserno, F. Jiménez-Ángeles, C. Holm, and M. Lozada-Cassou, cond-mat/0104002, and Journal Phys. Chem. B, in press.

103. J. J. Thomson, Philos. Mag. **7**, 237 (1904); A. Perz-Garrido and M. Moore, Phys. Rev. B **60**, 15628 (1999).
104. D. W. Oxtoby, H. P. Gillis, and N. H. Nachtrieb, in *Principles of Modern Chemistry* (Saunders College Publishing, Philadelphia, 1999), Chap. 3, p. 80.
105. B. I. Shklovskii, Phys. Rev. E **60**, 5802 (1999).
106. R. Messina, C. Holm, and K. Kremer, Phys. Rev. Lett. **85**, 872 (2000).
107. R. Messina, C. Holm, and K. Kremer, Europhys. Lett. **51**, 461 (2000).
108. R. Messina, C. Holm, and K. Kremer, Euro. Phys. J. E. **4**, 363 (2001).
109. L. Bonsall and A. A. Maradudin, Phys. Rev. B **15**, 1959 (1977).
110. R. Messina, C. Holm, and K. Kremer, Phys. Rev. E **64**, 021405 (2001).
111. R. Messina, cond-mat/0104076, submitted.
112. B. D'Aguanno and R. Klein, Phys. Rev. A **46**, 7652 (1992).
113. E. Allahyarov, H. Löwen, and S. Trigger, Phys. Rev. E **57**, 5818 (1998).
114. H. Schiessel, private communication.
115. H. Greberg and R. Kjellander, J. Chem. Phys **108**, 2940 (1998).
116. M. Lozada-Cassou and F. Jiménez-Ángeles, eprint physics/0105043.
117. R. Messina, E. González-Tovar, M. Lozada-Cassou, and C. Holm, to be published.

SCALING DESCRIPTION OF CHARGED POLYMERS

JEAN-FRANÇOIS JOANNY
Institut Charles Sadron, 6 rue Boussingault, 67083 Strasbourg Cedex, France.

1. Introduction

The series of 3 lectures that I gave in Les Houches was meant to be an introduction to scaling methods applied to polyelectrolytes they were divided into three main sections: a general introduction on the conformation of polyelectrolyte chains in (mostly dilute) solutions, a discussion of the behavior of polyelectrolytes close to surfaces and a presentation of more recent results on problems where charge fluctuations are important such as polyampholytes or polyelectrolyte complexes. In these lecture notes, I will skip completely the section on interfaces which is covered by a recent review [1] and I will emphasize the general introduction on the scaling description of polyelectrolyte chains [2]; the last section will give a brief summary on the properties of polyampholytes and polyelectrolyte complexes.

Polyelectrolytes are polymers that carry ionic charges along the chemical backbone. They can be obtained in two ways. Quenched (or strong) polyelectrolytes are obtained by a copolymerization of neutral and ionic monomers. On each chain, the fraction of charged monomers is fixed by chemistry as well as the position of the ionic groups along the chain. The nature of the distribution of the charged monomers along the chains and the charge fluctuations from chain to chain are also characteristics of the synthesis process. The properties of quenched polyelectrolytes do not seem to depend strongly on these parameters as long as the charges are rather uniformly distributed: they are not very different for regularly distributed charges or randomly distributed charges; they are of course very different for a diblock copolymer with a charged and a neutral block. In the following, we will not discuss these structural parameters and consider that the polymer has a uniform linear charge density along the chain η or a constant fraction of charged monomers $f = \eta a$ where a is the size of a monomer. Annealed (or weak) polyelectrolytes are polyacids or polybases where the

C. Holm et al. (eds.), Electrostatic Effects in Soft Matter and Biophysics, 149–170.
© 2001 *Kluwer Academic Publishers. Printed in the Netherlands.*

charges result from the chemical equilibrium associated to the dissociation of the acid or the base. The charge of the chain is then monitored by the pH of the solution which is up to trivial factors the chemical potential of the charges. The main difference with annealed polyelectrolytes is that the charges are not fixed along the chain but are mobile (due to recombination and dissociation events) and that the number of charges on a chain is not fixed but fluctuates (at constant chemical potential).

The properties of polyelectrolytes are more complex than those of organic neutral polymers in solution because the electrostatic interactions are long ranged. In polymer problems, entropic effects play a major role and it is thus interesting to measure the interaction between two elementary charges q at a distance r in units of kT and to write it as $v(r) = kT\frac{\ell_B}{r}$ where $\ell_B = \frac{q^2}{4\pi\varepsilon kT}$ is the so-called Bjerrum length, ε being the dielectric constant of water. The direct interaction between charges (the potential of mean force) is however screened by the existence of free ions in the water solution. In most of the following, we will use the linear Debye-Hückel approximation $v(r) = kT\frac{\ell_B}{r}\exp -\kappa r$ where the screening length $1/\kappa$ is related to the monovalent salt concentration n by $\kappa^2 = 8\pi n\ell_B$. We thus implicitly assume here that all the small ions are point-like and we ignore the specificity of the counterions (which is at the origin of the differences between solutions containing Na and Li counterions that are observed experimentally; but these differences are in many cases not too large); we also assume that the Debye-Hückel theory of simple electrolytes can be used, i.e. that the concentration of small ions is not too large. Another general important feature of polyelectrolyte solutions is that, even in the absence of added salt, they also contain the counterions that neutralize the polymer. Some of the properties of polyelectrolyte solutions are as we will see below entirely governed by the counterions and only very weakly depend on the polymer.

In these notes, we will put an emphasis on single chain properties and discuss first weakly charged polyelectrolytes in section 2 and strongly charged polyelectrolytes in section 3. The effects of screening by added salt are considered in section 4 and the concentration effects in section 5. Section 6 gives a brief review of the properties associated with charge fluctuations for polyampholytes and polyelectrolyte complexes and the last section gives some concluding remarks and discusses some possible issues.

2. Weakly charged polyelectrolytes in dilute solutions

In this section, we discuss flexible weakly charged polyelectrolytes with a fraction of charged monomers $f \ll 1$. We consider both Gaussian polymer chains with only electrostatic interactions that correspond to polymers for which water is a θ solvent (this is approximately the case for polyacrylic acid) and polymers in a poor solvent (this is the case of most polyelectrolytes where the organic backbone is highly insoluble in water). The last two parts of this section are devoted to annealed polyelectrolytes and to the stretching of polyelectrolyte chains and its application to charged gel elasticity.

2.1. FLORY THEORY AND ELECTROSTATIC BLOBS

For a neutral polymer in a θ solvent comprising N monomers, the root mean square end to end distance is given by the Gaussian statistics $R_0 = N^{1/2}a$. Electrostatic interactions are repulsive and tend to stretch the chain; the end to end distance of a polyelectrolyte R is then larger than R_0. The electrostatic energy of a chain of radius R can be estimated up to a numerical prefactor of order one by assuming that the charges are uniformly distributed in a sphere of radius R: $F_{el} \approx kT\ell_B \frac{(fN)^2}{R}$. If the fraction f of charged monomers is small, electrostatics is only a perturbation and the chain remains Gaussian $R = R_0$. This remains true as long as the entropy of the Gaussian chain remains dominant i.e. if the electrostatic free energy is smaller than kT. This condition can be written as $N < g_{el} = (f^2\ell_B/a)^{-2/3}$

Polyelectrolyte chains with a number of monomers larger than g_{el} are thus extended by electrostatic interactions. The end to end distance can be found by a Flory type of argument. The total free energy of the chain is the sum of the electrostatic free energy and of the entropic elasticity of the Gaussian chain $F_{ent} = \frac{3kTR^2}{R_0^2}$. The minimization of the total free energy gives the equilibrium size[3]

$$R \approx Na(f^2\ell_B/a)^{1/3} \tag{1}$$

The chain is strongly elongated since the size scales linearly with the number of monomers, note however that because the charge fraction f is smaller than one the chain is far from being fully extended. A geometrical interpretation of this result can be given in terms of electrostatic blobs[3]. We first divide the chain into subsections each containing g_{el} monomers that we call electrostatic blobs. Inside each blob, the electrostatic interactions are of order kT and the blobs can be considered as Gaussian; they therefore have a size $\xi_{el} \approx a(f^2\ell_B/a)^{-1/3}$. The blobs on the other side strongly interact and minimize their energy by making a linear array. The polyelectrolyte chain has then a cigar-like structure shown on figure (1)The chain radius is imposed by the geometry $R \approx (N/g_{el})\xi_{el}$ which gives back the result of equation (1).

The blob picture calls for several remarks. First, a polyelectrolyte chain must not be considered as a frozen cigar of blobs; there are thermal fluctua-

Figure 1. Electrostatic blob model for a weakly charged polyelectrolyte chain.

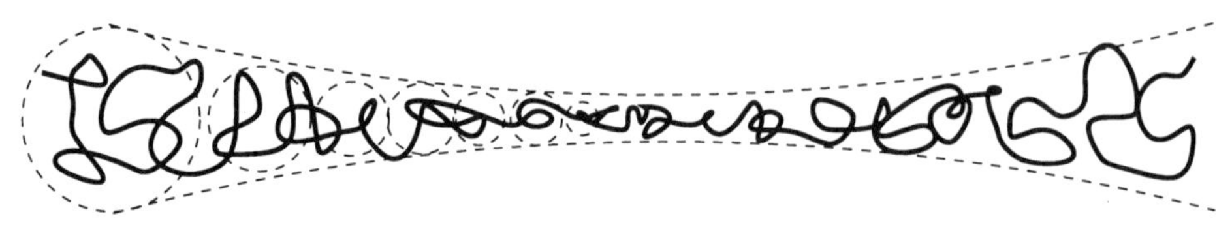

Figure 2. Trumpet-like shape of a weakly charged polyelectrolyte

tions both along the chain and in the transverse direction. In the transverse direction, the fluctuations are almost random and the transverse size is roughly given in this blob picture by the Gaussian statistics $R_\perp \approx N^{1/2}a$. Second, the blob construction ignores the fact that the electrostatic interactions are long ranged and that monomers inside each blob interact with all the monomers of the chain. This must be corrected for by calculating the electrostatic energy for a cylinder of length R and radius ξ_{el} and not for a sphere; it adds an extra factor $\log^{1/3}(N/g_{el})$ to the radius given by equation (1); this factor is always of order one and thus does not change the qualitative result.

A more quantitative calculation can be made by using a variational approach and the Gibbs-Bogoliubov inequality[4]. The simplest trial energy is that of a Gaussian chain with a constant force applied at its end point. This gives a radius in agreement with equation (1) including the logarithmic factor but with precise numerical prefactors that seem to compare quantitatively with numerical simulations. More refined variational energies have also been used.

One last limitation of the blob model is that it implicitly assumes that the tension is constant along the chain, this is not true since the electrostatic potential is higher at the center of the chain than close to the end points. One thus expects a higher tension at the center and thus a smaller blob size[5]. A polyelectrolyte chain has thus a trumpetlike shape as shown on figure (2). The trumpet effect can be calculated explicitly in the strong stretching approximation where the chain conformation is characterized only by the average position z of monomer s. The equilibrium conformation then results from an equilibrium between the chain tension $\frac{3kT}{a^2}\frac{\partial z}{\partial s}$ and the

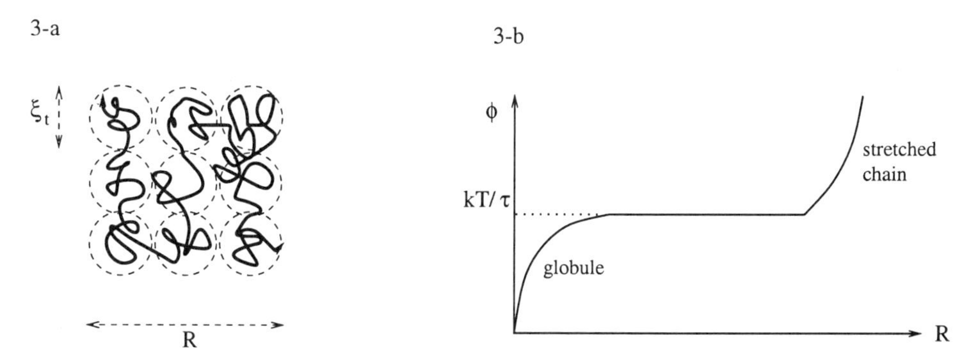

Figure 3. Collapsed polymer chain: figure 3-a: thermal blobmodel; figure 3-b: elongation R under an external applied force ϕ.

local electrostatic force $\frac{3kT}{a^2}\frac{\partial^2 z}{\partial s^2} = -qE$. The local electric field can be calculated from the conformation $z(s)$ using Coulomb law. The trumpet effect is weak and the size of the blobs (inversely proportional to the local tension) differs only by a logarithmic factor between the endpoints and the center of the chain.

2.2. POLYELECTROLYTES IN A POOR SOLVENT: THE PEARL-NECKLACE MODEL

Most polyelectrolytes are organic polymers and their backbone is in a poor solvent in water. We characterize the solvent quality by a negative virial coefficient(or excluded volume) between monomers $v = -\tau a^3$. A neutral polymer chain in a poor solvent forms a collapsed globule that can be viewed as a close-packed assembly of Gaussian thermal blobs of size $\xi_t \approx a/\tau$[7, 8]. The radius of the globule is given by $R \approx (\frac{N}{\tau})^{1/3} a$ as sketched on figure (3). It is important to note that a collapsed globule behaves as a small oil drop in water and that its free energy is dominated by its surface energy. The relevant surface tension is $\gamma = kT/\xi_t^2$. If a constant force, φ, is applied at the chain ends of a collapsed globule, the globule only weakly deforms at small values of φ but it undergoes a discontinuous transition for a force $\varphi_c \approx kT/\xi_t$ between a collapsed structure which is weakly deformed and a stretched structure which is a linear array of thermal blobs. If the applied force is equal to φ_c and only for this value, the collapsed and the extended states have the same free energies and can coexist at equilibrium[9].

A very weakly charged polyelectrolyte in a poor solvent is in a collapsed globule conformation. Its properties are then very similar to that of a charged oil droplet in water[10]. When its charge increases, a charged liquid drop undergoes a Rayleigh instability and breaks into two smaller droplets. Upon further increase of the charge each of these two droplets can undergo

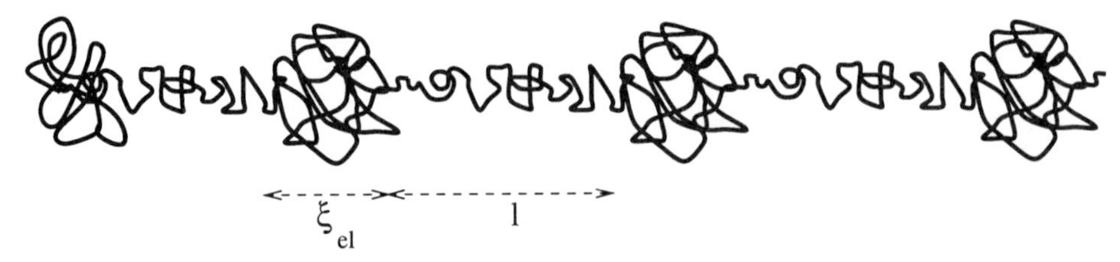

Figure 4. Pearl-necklace of a polyelectrolyte chain in a poor solvent

itself a Rayleigh instability and so forth. The instability occurs when the electrostatic energy $F_{el} = kT\ell_B\frac{(fN)^2}{R}$ is of the order of the surface tension energy $F_s \approx \gamma R^2$ i.e., using the values of collapsed globules, if the droplet has a size equal to the electrostatic blob size ξ_{el}. For a given charge fraction f any larger drop is unstable. The number of monomers in the drop at the Rayleigh instability is $g_R = \frac{\tau a}{f^2\ell_B}$.

When a liquid drop undergoes the Rayleigh instability, the two daughter droplets separate apart. For a polyelectrolyte in a poor solvent the drops cannot separate apart due to the chain connectivity[11]. This leads to a pearl-necklace structure where the drops (the pearls) are connected by elongated strands as displayed on figure (4). In this structure, most of the polymer is in the pearls and there is an equilibrium between the elongated strands and the pearls as for a collapsed chain under tension. This imposes that the strands are elongated arrays of length l of blobs of size ξ_t where the tension has the critical value φ_c. The tension is due to the electrostatic interaction between the pearls $\varphi_c = kT\ell_B\frac{(fg_R)^2}{l^2}$ and the length of the strands is $l \approx (\frac{\tau a}{\ell_B})^{1/2}\ \frac{a}{f}$. The pearl-necklace geometry gives then the chain end to end distance

$$R \approx Nf(\frac{a\ell_B}{\tau})^{1/2} \tag{2}$$

Note that as the charge fraction is increased at constant solvent quality, the pearl-necklace structure develops via a discrete series of discontinuous transitions between a structure with k pearls and a structure with $k+1$ pearls when $f \approx (\frac{k\tau a}{N\ell_B})^{1/2}$. The pearls become irrelevant when the total mass in the strands is larger than in the pearls i.e. when $f \approx (\frac{\tau^3 a}{\ell_B})^{1/2}$.

2.3. ANNEALED POLYELECTROLYTES

In annealed polyelectrolytes, the chemical potential of the charges μ is fixed by the pH of the solution. It imposes the average charge on the chain[12]. We first discuss here the conformation of annealed polyelectrolytes in a θ solvent and then briefly annealed polyelectrolytes in a poor solvent and titration experiments.

A weakly charged Gaussian polyelectrolyte is best described within the strong stretching approximation outlined in the previous paragraph. The chemical potential of the charges is

$$\mu = kT \log \frac{f}{1-f} + q\psi(z) \tag{3}$$

where $\psi(z)$ is the local electrostatic potential at position z along the chain. Within the strong stretching approximation, the average potential is written as $q\psi(z) = \int_0^N ds' \frac{\ell_B f(s')}{|z - z(s')|}$ where $f(s')$ is the actual charge fraction at monomer s'. If we assume that the electrostatic potential is constant along the chain, the charge fraction is constant and increases monotonically as the pH increases for a polyacid. When the (small) variation of the potential which is larger at the chain center is taken into account, we find a larger charge close to the end points than close to the chain center[5]

$$f(z)/ < f >= 1 - (f\ell_B/a)^{1/3}\{\log[1 - (2z/R)^2] + 2(1 - \log 2)\}$$

where the origin for z has been taken at the chain midpoint. Again the effect is not very large but it has been observed in some simulations[6].

In a poor solvent, polyelectrolytes have a pearl-necklace structure. Even if we neglect this end effect, the electrostatic potential is higher on the pearls than on the strands. The charge on the pearls is thus lower and the number of monomers in a pearl is slightly larger for a same average charge. This is a weak effect if the solvent is not too poor. If the solvent is too poor (if $\tau > (a/\ell_B)^{3/5} N^{-1/5}$), this "condensation" of the charges makes the pearl-necklace structure unstable and the polyelectrolyte can either have a globular conformation or an extended conformation with a first-order discontinuous transition between these two states. This manifests itself as a plateau in the titration curve (a plot of μ as a function of f) as observed experimentally for many annealed polyelectrolytes.

2.4. POLYELECTROLYTE STRETCHING AND POLYELECTROLYTE GELS

Recent developments of experimental techniques such as atomic force microscopy have allowed direct measurements of the elasticity of polyelectrolyte chains. In a θ solvent, a polyelectrolyte acts as a Gaussian spring (even when it is stretched by electrostatic interactions) with a spring constant $\frac{3kT}{R_0^2}$[13]. In a poor solvent, the elasticity of the pearl-necklace structure can be studied quantitatively by minimizing its free energy in the presence of an external force φ[14]. The size and composition of the pearls does not change upon stretching. The application of a force essentially provokes a dissolution or an unwinding of the pearls and an increase of the distance l between pearls. The force balance that determines l must now include the external force and must be rewritten as $\varphi_c = kT\ell_B \frac{(fg_R)^2}{l^2} + \varphi$. This leads to a highly nonlinear elasticity $R = R(\varphi = 0)\frac{1}{[1-\varphi/\varphi_c]^{1/2}}$.

This result remains valid as long as most of the polymer mass is in the pearls i.e. as long as $R < N\tau a$. If the size is larger, the attractive interactions no longer play any role and the Gaussian elasticity is recovered. The derivation of the non linear elasticity also assumes that the number of pearls is large enough that it can be considered as a continuous variables. If there are only a few pearls one expects a force plateau with a discontinuous stretching each time a pearl is unwound. This discontinuous stretching has been studied in details in references [14, 15].

An important application of the stretching of polyelectrolyte chains is the swelling of polyelectrolyte gels. In a gel, the polyelectrolyte chains are crosslinked to make a macroscopic network. In the presence of water charged gels swell so much that they are called superabsorbents. At equilibrium a gel is in equilibrium with pure solvent and its osmotic pressure (which is the chemical potential of the solvent) vanishes. For simplicity, we assume here that the different meshes of the gel are not entangled and do not overlap and that if the chains between crosslink have a size R, the monomer concentration is $c \sim N/R^3$; this is known as the "c* theorem"[7]. The gel pressure is the sum of an attractive elastic pressure and of the swelling pressure which has two contributions, the counterion contribution and the electrostatic contribution. It was realized a long time ago by Katchalsky that the counterion contribution always dominates the electrostatic contribution[16]. Considering the counterions as an ideal gas which is a good assumption for weakly charged gels, the counterion osmotic pressure is $\Pi_c = kTfc$. In a θ solvent, the elastic pressure is due to the Gaussian elasticity of the chains and is equal to $\Pi_s = -kT\frac{c}{N}\frac{R^2}{R_0^2}$. In a poor solvent, the elastic pressure is dominated by the stretching of the strands, the stretching energy is roughly kT per thermal blob $\Pi_s = -kT\frac{c}{N}\frac{N}{g_R}\frac{l}{\xi_t}$. This gives the concentration in the gel in a θ solvent $c \approx \frac{1}{N^2 f^{3/2} a^3}$ and in a poor solvent $c \approx \frac{\tau^3}{N^2 f^3}$; the crossover between these two behaviors occurring when $\tau \sim f^{1/2}$[17]. It is important to notice that both in a θ solvent and in a poor solvent, the concentration is very low (or the swelling very large) and independent of the strength of the electrostatic interactions (ℓ_B). The only effect of the electrostatic interactions is to confine the counterions inside the gel to maintain the macroscopic electroneutrality. If one calculate the size of the elastic chains in a θ solvent ($R \approx Nf^{1/2}a$) and in a poor solvent ($R \approx Nfa/\tau$) one can check that it is larger that the size of the corresponding free chain because of the swelling of the gel is due to the counterion pressure.

3. Strongly charged polyelectrolytes and counterion condensation

So far, we have only considered weakly charged polyelectrolytes and neglected the role of the counterions that we have either ignored or considered as an ideal gas. The attractive electrostatic interactions between the polymers and the counterions induce correlations that can be strong if the polymer is highly charged[18]. These correlations remain small and can be ignored (as we did for weakly charged polyelectrolytes) if the electrostatic potential seen by a free ion in contact with the polymer chain ψ is small $q\psi/kT < 1$. In this section, we briefly discuss strongly charged polyelectrolytes and counterion condensation; we refer to the other chapters of this book for a more detailed account of these effects.

3.1. NON LINEAR ELECTROSTATICS AND ION CONDENSATION

Strongly charged polyelectrolytes are very strongly extended and may be modeled as rods to study counterion condensation. At the scaling level, this is consistent with equation (1), where the blob size becomes of the order of the monomer size when $f \approx 1$. Note that numerical simulations have shown that strongly charged flexible polyelectrolytes have a size that scales linearly with their mass but that their conformation is not quite rodlike and they often show bent conformations. For simplicity we will discuss only infinitely long charged rods.

At the mean field level, the electrostatic potential and the distribution of counterions around a rod carrying a charge per unit length η, are given by the Poisson-Boltzmann equation

$$\nabla^2\psi = -4\pi\ell_B n_0 \exp\psi \tag{4}$$

where $kT\psi/q$ is the electrostatic potential and n_0 a constant. This equation has been first solved by Fuoss et al.[19] using a cell model where each rodlike polymer is embedded in a cylindrical cell of radius b (perpendicular to the rod) that is electrically neutral. This model actually shows a condensation "transition" corresponding to a change in the mathematical nature of the solution of the Poisson-Boltzmann equation. We will only discuss here the limit of very dilute solutions ($b \to \infty$) which has been rederived in an appendix of reference [20]. If the charge per unit length of the rod is small ($\eta\ell_B < 1$) the counterions are free; the counterion density vanishes in the vicinity of the rod and the electrostatic potential around the rod is identical to that of a uniformly charged rod in the absence of the counterions. If $\eta\ell_B \geq 1$, some of the counterions are bound in the sense that the counterion concentration in the vicinity of the rod is finite. The total number of counterions per unit length is $\lambda = \eta - 1/\ell_B$. If one wants to replace the rod with its condensed counterions by an effective charged rod, the charge of this effective rod is $\widetilde{\eta} = 1/\ell_B$. Indeed, at large distances from the rod, the asymptotic solution of the Poisson-Boltzmann equation gives an electrostatic potential identical to that of a rod of charge $\widetilde{\eta}$ with no counterions. It is important to notice that condensation occurs at a lower value of the charge for multivalent counterions: for counterions of valency z, the condensation threshold is given by $\eta\ell_B = 1/z$.

A very physical description of counterion condensation has been given by Oosawa [21]. His model is a two-states model for the counterions. The free counterions form an ideal gas in equilibrium with a one-dimensional gas of condensed counterions localized on the rod (for point-like ions). The condensation is driven by the attractive electrostatic interaction between the rod and the condensed counterions. Although extremely simplified, this model gives results in good agreement with the full Poisson-Boltzmann theory. It can be easily extended to include for example a finite ionic strength of the solution.

For many purposes, a strongly charged polyelectrolyte can be roughly treated as a weakly charged polyelectrolyte with an effective charge $\widetilde{\eta}$. One of the distinctive properties is that the condensed counterion cloud is highly polarizable due to the possible motion of the counterions along the polymer. A highly charged polyelectrolyte looks thus like an annealed polyelectrolyte. One of the limitations of the Poisson-Boltzmann theory is that the gas of condensed counterions is rather dense and cannot be treated as an ideal gas. Theories going beyond the Poisson-Boltzmann approximation are thus needed to get a quantitative description.

The condensation of counterions on a polyelectrolyte in a poor solvent in a pearl-necklace conformation has also been studied. The linear charge density of the necklace is higher on the pearls and counterion condensation occurs first on the pearls. At the scaling level, one finds that if the charge is large enough that the counterions condense on the pearls, the necklace structure collapses to a spherical globule.

3.2. BEYOND THE POISSON-BOLTZMANN APPROXIMATION, ATTRACTIVE INTERACTIONS

Interesting new effects are observed by going beyond the Poisson-Boltzmann approximation. At "high temperatures" one can consider the thermal fluctuations in the density of condensed counterions along the rod and treat them perturbatively. This was first done by Oosawa [21] who calculated the interaction between two parallel rods due to these fluctuations. This interaction results from a coupling of the density fluctuations on the two rods due to the electrostatic interactions and is very similar to a Van der Waals interaction; it is attractive and proportional to $-\ell_B^2/d$ where d is the distance between the rods. The interaction between rods at a finite angle has been recently calculated by Ha and Liu [22] who have shown that the interaction is attractive if the angle between the rods is small but that it can be repulsive if the angle is close to $\pi/2$.

The other extreme approximation is a low temperature approximation where the condensed counterions are treated as a Wigner crystal. This clearly leads at short distances to an attractive interaction that competes with the direct repulsive electrostatic interaction and can dominate as shown for example in the case of DNA by numerical simulations [23]. A more refined description has been proposed where the counterion cloud is treated by using the equation of state obtained for the so-called one component plasma [24]. This is discussed in details in the lecture notes of Shklovskii.

4. Screening and persistence length

As discussed in the introduction, the range of the electrostatic interaction can be modulated by adjusting the salt concentration (the ionic strength) of the solution. At high ionic strength, the idea due to Odijk and independently to Skolnick and Fixman is that the only effect of electrostatic interactions is to modify the rigidity of the polymer chains and can be characterized by an electrostatic persistence length. The persistence length is defined by considering a polymer chain as a flexible wire and by expanding its free energy in powers of its curvature. If the interactions are short range and if the scale over which the curvature varies is larger than the range of the interaction, the expansion is local and can thus be written as

$$\mathcal{F} = kT\frac{l_p}{2}\int_0^L ds\ \rho^2(s) \tag{5}$$

where s is the curvilinear abscissa along the chain and L the total contour length; the local curvature is $\rho(s)$. This expansion has no linear term by symmetry and defines the persistence length l_p.

4.1. ELECTROSTATIC PERSISTENCE LENGTH OF RIGID POLYELECTROLYTES

A rigid chain has a finite persistence length l_0 in the absence of electrostatic interaction. The total persistence length is then the sum of the bare persistence length and of the electrostatic contribution l_{el}. If the electrostatic interactions are screened over the Debye length κ^{-1}, pieces of chain of size κ^{-1} are independent and the electrostatic contribution to the persistence length can be estimated by bending on a circle a chain segment of length κ^{-1}. By assuming that all the lengths are proportional to the Debye length, the cost in electrostatic energy is estimated as $\mathcal{F} \sim kT\ell_B\eta^2\kappa^{-1}$; the corresponding bending energy from equation (5) is $kTl_{el}/\kappa^{-1}$the identification between these two results gives the electrostatic persistence length $l_{el} \sim \eta^2\ell_B/\kappa^2$. A more detailed summation of the pairwise interactions for a bent wire with an interaction energy per unit length $v(r)$ between monomers at a distance r gives the persistence length $l_{el} = \frac{1}{16\pi}\int d\mathbf{r}\ v(r)$. Inserting in this equation the Debye Hückel interaction, we find the result of Odijk or Skolnick and Fixman[25, 26]

$$l_p = l_0 + \frac{\ell_B\eta^2}{4\kappa^2} \tag{6}$$

It is important to notice that the electrostatic contribution to the persistence length is not proportional to the Debye screening length and that it can be larger. This is due to the fact that pieces of chains of size κ^{-1}indeed behave as rods but that two consecutive such rods strongly interact and their orientations are highly correlated.

For a highly charged polymer above the condensation threshold, a naive argument is to replace the charge per unit length by the effective charge per unit length $\tilde{\eta} = 1/\ell_B$ and the electrostatic persistence length is $l_{el} = \frac{1}{4\kappa^2\ell_B}$. More precise results have been obtained by solutions of the Poisson-Boltzmann equation in a curved geometry either numerically [28]or more recently analytically [29].

4.2. LENGTHSCALE DEPENDENCE OF THE PERSISTENCE LENGTH

The purely mechanical argument of Odijk ignores thermal fluctuations and the polymer behavior at very short length scale. A more refined theory has been constructed in reference [27]. The orientation of the chain is fixed at a given point and the average squared angle made by the chain with this direction $\theta^2(s)$ is calculated as a function of the contour length s. The calculation is made by assuming that the angle remains small and that the electrostatic energy can be expanded at Gaussian order. The local persistence length is then identified as the slope of $\theta^2(s)$ as a function of s. A critical value of the contour length appears in the calculation $s_c \approx (\frac{l_0}{\tau^2\ell_B})^{1/2}$. The persistence length depends therefore on the length scale considered. At short contour length $s < s_c$, electrostatic interactions are irrelevant and the persistence length is l_0. At large length scales $s > \kappa^{-1} > s_c$ the persistence length is given by equation (6).

The validity of this approach is insured if the angle remains small, this requires that $s_c < l_0$ or that $\tau^2 \ell_B l_0 > 1$. This gives the range of validity of the rigid chain theory where the polymer chain remains rigid at all length scales with a persistence length depending on length scale. If this condition is not satisfied, the bending angle becomes of order one at distances where the electrostatic interactions are not yet relevant and the chain coils to form Gaussian blobs. Indeed if a local persistence length is included in the blob description, the condition for the chain to remain flexible is that the blob size be larger than the persistence length. This leads to the opposite condition $\tau^2 \ell_B l_0 < 1$. This condition thus clearly distinguishes between rigid and flexible polyelectrolyte chains.

4.3. PERSISTENCE LENGTH OF FLEXIBLE CHAINS

Flexible chains are locally Gaussian and can be described by the electrostatic blob model as long as the Debye-Hückel screening length is larger than the electrostatic blob size. It is then tempting to assume that the effect of the electrostatic interactions on the conformation of the chain of blobs can be described in terms of a persistence length. This is not however obvious since the chain of blobs has no intrinsic rigidity and it does not have a finite bare persistence length; there could be couplings between stretching and bending.

Nevertheless the existence of a well-defined persistence length is generally accepted. There exist however big controversies on the value of this persistence length. The most natural assumption first used by Khokhlov and Kachaturian [30]is to coarse-grain the monomers to blobs and to calculate the persistence length using the formula obtained for rigid chains. One finds then $l_p = \frac{1}{\kappa^2 \xi_{el}}$. There is no direct derivation of this persistence length from a microscopic description of the chain except for a variational calculation of Netz and Orland[31] based on a Gaussian approximation of the free energy of the chain in Rouse modes.

Several other variational calculations have been performed and start from an effective Hamiltonian which is that of a semi-flexible chain with a well-defined persistence length. They all lead to a persistence length of the order of the screening length κ^{-1}[27]. The variational techniques are however based on not well-controlled approximations and the error made is unknown.

Simulations on flexible polyelectrolyte chains are so far not able to discriminate between these two theories and most of them lead to results in disagreement with both scaling approaches [32]. This may be well due to the fact that the parameters chosen do not lead to any well-defined asymptotic behavior and that cross-over effects remain important. The comparison with experiments does not seem conclusive either. Several experiments lead to a persistence length proportional to the screening length but the determination of the persistence length from the actually measured quantity is often based on very approximate theoretical models.

5. Semidilute solutions

In a very dilute solution, polyelectrolyte chains behave as individual objects and their conformation is the same as that of isolated polyelectrolytes. The chains are however highly stretched and highly charged and they start to interact and overlap at very low concentrations. In most experimental situations, a polyelectrolyte solution is thus in the semi-dilute range. We first present here the mean field theory based on the random phase approximation which is valid at large enough concentration and then give scaling arguments.

5.1. DENSE SOLUTIONS, RANDOM PHASE APPROXIMATION

If the polyelectrolyte chains are Gaussian, the structure factor matrix of a polyelectrolyte solution can be calculated using the random phase approximation introduced for polymers by deGennes [33, 34]. We only discuss here the monomer-monomer structure factor $S(q)$ in a polyelectrolyte solution containing added salt. Within the random phase approximation, it is given by

$$\frac{1}{S(q)} = \frac{1}{S_0(q)} + v + w^2 c + \frac{4\pi \ell_B f^2}{\kappa^2 + q^2} \tag{7}$$

$S_0(q) = \frac{12c}{q^2 a^2}$ is the structure factor of an isolated Gaussian chain in the intermediate range and the non electrostatic interactions have been expanded in a virial expansion: the second virial coefficient is the excluded volume v and w^2 is the third virial coefficient. The Debye screening length used here includes both the counterions and the monovalent added salt at a density n, $\kappa^2 = 4\pi\ell_B(2n + fc)$. This result is very similar to that obtained by Edwards for the structure factor of a semi-dilute solution of neutral polymers but the screened electrostatic interaction given by the last term must be added to the second virial coefficient. The main feature of equation (7) is that the structure factor has a peak at a finite wave vector q_* if the ionic strength is not too high. This polyelectrolyte peak is considered as the signature of electrostatic interactions in scattering experiments on polyelectrolytes. Within the random phase approximation, the peak position is given by $(q_*^2 + \kappa^2)^2 = q_0^4$ where $q_0^4 = \frac{48\pi\ell_B f^2 c}{a^2}$. When the ionic strength is increased the peak shifts to lower wave vectors and disappears when $\kappa = q_0$.

The random phase approximation implicitly makes a mean field assumption and the chain statistics is the Gaussian statistics. A criterion for validity is that the wavelength at the peak q_0^{-1} be smaller than the electrostatic blob size or $ca^3 > (f^2\ell_B/a)^{2/3}$. Experimentally the peak position in scattering experiments scales with concentration with an exponent larger than predicted by the random phase approximation $q_* \sim c^{1/2}$; at high concentration the exponent decreases and gets closer to the RPA value $q_* \sim c^{1/4}$.

When the solvent gets poorer, the excluded volume v becomes negative, and the structure factor value at the peak increases and eventually diverges. This instability is the signature of a mesophase formation transition. In a poorer solvent the stable structure of the solution is predicted to be a periodic array of dense and dilute polyelectrolyte regions with a period close to q_*. The equilibrium symmetry of the mesophases and their period have been studied in great detail by the Russian group; however, a clear experimental evidence of these mesophases is still lacking.

5.2. ORDERING IN SEMIDILUTE POLYELECTROLYTE SOLUTIONS

The essential experimental result on the structure of polyelectrolyte solutions is that the structure factor has a peak even if the solution is rather dilute. Below the overlap concentration the peak position is given by the distance between chains $q_* \sim (N/c)^{1/3}$. The most naive interpretation of such a peak is that there is a translational order in the solution. Indeed, just below the overlap concentration the interaction between neighboring chains is very large and if one considers the polyelectrolyte chains as colloidal objects, they should form a colloidal Wigner crystal. There is however no clear evidence for the existence of a Wigner crystal in a dilute polyelectrolyte solution. The situation is even less clear when the chains overlap; the crystal could melt because of screening of the electrostatic interactions both by the small ions and by the polymer itself (this last effect is in my opinion poorly understood).

Another type of ordering that could occur in polyelectrolyte solutions is nematic ordering. The chains have at least locally a rodlike behavior and could have an Onsager transition to a nematic phase as other semiflexible polymers. For flexible polymers with a small bare persistence length l_0, no such transition seems to be observed experimentally.

Although this point certainly needs further theoretical study, we will assume in the next section that semi-dilute polyelectrolyte solutions are isotropic disordered solutions. The peak in the scattering curve is then associated to a liquid-like order.

5.3. SCALING DESCRIPTION OF SEMIDILUTE SOLUTIONS

We first discuss semidilute solutions in a θ solvent and in the absence of added salt. The basic idea of the scaling approach is to assume that there is only one characteristic length scale and to derive it from a scaling argument. Actually, there are several length scales in a semidilute polyelectrolyte solution and we therefore implicitly assume that they are all equal (or that they differ only by a numerical prefactor): the mesh size of the temporary network formed by the overlapping chains ξ, the screening length of the electrostatic interactions, the wavelength corresponding to the peak of the structure factor, the total persistence length of the polyelectrolyte chains. The persistence length deserves a detailed analysis and is not fully understood, a discussion can be found in reference[2]. The scaling argument is then very similar to that made for neutral polymers in a good solvent [3]. The overlap concentration is given by $c^* \sim N/R^3$ where R is the radius of an isolated polyelectrolyte chain and is given by equation (1). The mesh size is then written as $\xi \sim R(c/c^*)^\alpha$ where α is an unknown exponent. The exponent is determined by imposing that ξ is independent of molecular weight. The result is

$$\xi \simeq 2\pi/q_* \sim c^{-1/2} f^{-1/3} \tag{8}$$

this result is in good agreement with many scattering experiments.

The osmotic pressure of a semi-dilute polyelectrolyte solution is not dominated by the polymer but by the small ions; in a first approximation, it is the ideal gas pressure of the counterion gas $\Pi = kTfc$. The polyelectrolyte chains contribute only a small correction to this pressure. Scaling laws can also be derived in the presence of added salt. In order to use the scaling laws derived for a semi-dilute solution of semi-flexible polymers, one needs a model for the persistence length. For the two models discussed above for the persistence length, this was done by Khokhlov and Kachaturian[30] and deGennes et al.[3].

A similar scaling approach can also be applied to polyelectrolytes in a poor solvent [35]. The structure of the semi-dilute solution is sketched on figure (5). Starting from the size of the pearl-necklace conformation, the scaling argument gives a mesh size

$$\xi \simeq 2\pi/q_* \sim c^{-1/2} f^{-1/2} \tau^{1/4} \tag{9}$$

This result is valid as long as there is more than on pearl per blob i.e. when the mesh size is larger than the strand length $l \approx (\frac{\tau a}{\ell_B})^{1/2} \frac{a}{f}$ at small enough concentration $c < f/\tau^{1/2}$. Above this concentration the pearls

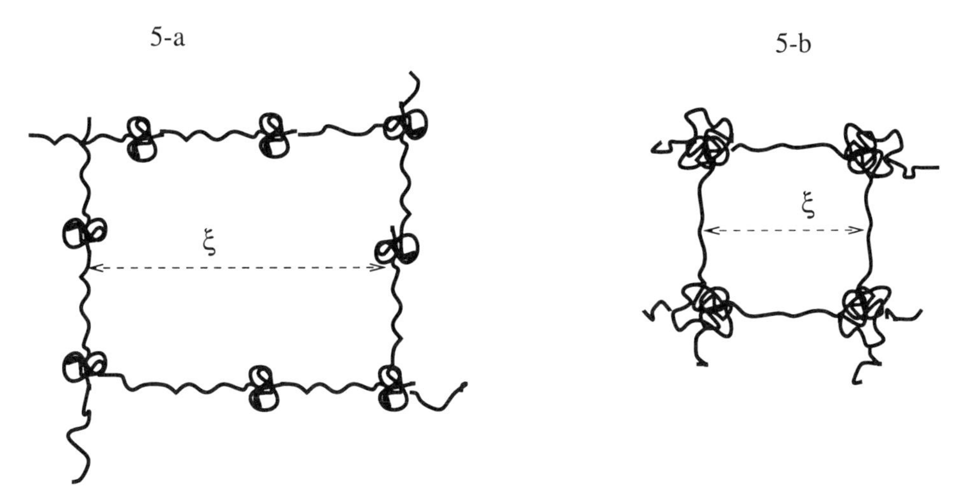

Figure 5. Semi-dilute polyelectrolyte solution in a poor solvent close to the overlap concentration (figure 5-a) and at high concentration (figure 5-b).

remain pinned at the crosslinking points of the pseudo-network as shown on figure (5) and $\xi \sim (g_R/c)^{1/3} \sim (\tau/f^2c)^{1/3}$. At higher concentration, the pearls become close-packed and the properties of the solution are well described by the random phase approximation (section 5.1).

6. Polyampholytes and polyelectrolyte complexes

6.1. POLYAMPHOLYTES

Polyampholytes are polymers carrying on the same chain monomers with charges of both signs. As for polyelectrolytes, the charges can be quenched or annealed. We consider here only annealed polyampholytes comprising a fraction f_+ of positive charges and a fraction f_- of negative charges. The total fraction of charged monomers is $f = f_+ + f_-$ and the fraction of neutral monomers is $1 - f$. If the number of positive and negative charges are not equal, a polyampholyte can be considered as a polyelectrolyte with a charge fraction $\delta f = f_+ - f_-$. The main effect for polyampholytes is the attraction between positive and negative charges that drives a collapse of the chains. The polyampholyte effect competes with the polyelectrolyte effect that tends to stretch the chains. The behavior of polyampholytes is thus rather similar to the behavior of polyelectrolytes in a poor solvent described above.

For quenched polyampholytes, the number and the positions of the charges on each chain is fixed but they fluctuate from chain to chain. If the distribution is random the distribution of the charge fraction is Gaussian $p(\delta f) \sim \exp -(\frac{\delta f^2 N}{f})$. A simple non Gaussian distribution is a Markov process where the probability to have a charge of one sign only depends on the previous charge. The distribution is then characterized by the non trivial eigenvalue of the transfer matrix λ. If we number the charges on the chain and impose a positive charge at the origin, the average value of the nth charge is λ^n. For an alternating distribution $\lambda = -1$, for a random distribution $\lambda = 0$, and for a blocky distribution, $\lambda = 1$.

A neutral polyampholyte chain collapses as a polymer chain in a poor solvent. The attractive free energy can be calculated by assuming that the density inside the chain has a constant value c and by using a generalized version of the Debye-Hückel theory of simple electrolytes[36]. The free energy per unit volume is

$$F_{DH} = -\frac{kT}{12\pi}\left(\frac{1+\lambda}{1-\lambda}\right)^{3/2}\kappa^3 \tag{10}$$

where the Debye screening length is calculated with the charge density on the chain $\kappa^2 = 4\pi\ell_B fc$. For a random polyampholyte, this is the same result as the polarization energy of a simple electrolyte which means that at this order of approximation, the connectivity of the chains does not play any role. For alternating polyampholytes, the polarization free energy vanishes in equation (10), it must be calculated more precisely and is proportional to κ^4. This means that the attractions are much stronger in a polyampholyte where the charges are randomly distributed than in a chain where they are alternating. This is consistent with certain experiments that show a precipitation for randomly distributed charges but no precipitation for alternating charges.

If we first assume that the chain is Gaussian with randomly distributed charges, the attractions are non relevant if the total Debye-Hückel free energy is smaller than kT i.e. if the number of monomers is small enough $N < g_a = b/\ell_B f^2$. If the number of monomers is larger the chain collapses and as for a polymer in a poor solvent, its structure can be pictured as a close-packed assembly of Gaussian blobs containing each g_a monomers and of size $\xi_a = a g_a^{1/2}$[37]. The radius of the chain is then $R \simeq (\frac{Na}{f\ell_B})^{1/3} a$.

When the charge does not vanish, the conformation is very similar to that of a polyelectrolyte in a poor solvent[38]. If $(\frac{f}{N})^{1/2} < \delta f < \frac{f^{3/2}\ell_B}{a}$, the polymer is in a pearl-necklace conformation with a size $R \simeq \frac{N\delta f}{f^{1/2}} a$. At larger values of the overall charge, the polyampholyte effect is irrelevant and the chain behaves as a polyelectrolyte in a θ solvent with a radius given by equation (1): $R \simeq Na(\frac{\delta f^2 \ell_B}{a})^{1/3}$. Note that if the charge distribution is random, the Gaussian average is dominated by chains in the pearl-necklace conformation (containing a few pearls) and the average size is equal to the Gaussian radius R_0[39].

In a solution of finite concentration the attraction between chains leads to a precipitation of neutral polyampholytes. Complexes between polyampholytes of opposite charges form in the dilute phase and a subtle fractionation of the charge distribution has been predicted in reference [40].

6.2. POLYELECTROLYTE COMPLEXES

A behavior very similar to that of polyampholytes is expected when two polyelectrolytes of opposite charges are mixed to form a polyelectrolyte complex. For simplicity, we only consider here the symmetric case where the polyelectrolytes have the same number of monomers and the same charge fraction. At large enough concentration, the solution phase separates into a dense complex phase containing the two polymers and a very dilute solution.

In the dense phase, the polarization energy can be calculated from the Debye-Hückel theory[41]. It depends on two length scales, the Debye-Hückel screening length κ^{-1} associated with the screening of the electrostatic interactions by the added salt and the length q_0^{-1} introduced in section (5.1), $q_0^4 = \frac{96\pi \ell_B f^2 c}{a^2}$, which is associated to the screening by the polymer. The polarization free energy reads

$$F = -kT\kappa^3 \left\{ (1 - s^{-1})(1 + 2s^{-1})^{1/2} \right\} \tag{11}$$

where $s = \kappa^2/q_0^2$.

When the solution separates into two phases, the dilute phase is extremely dilute and its osmotic pressure is that of a pure salt solution. In the dense phase, the attractive pressure due to the polarization effect, must therefore balance the repulsive pressure due to excluded volume. In a θ-solvent, the osmotic pressure is due to the three body interaction $\Pi_{ev} = kTw^2 \frac{(2c)^3}{3}$. If the salt density is low, the screening by the polymer is dominant and the polarization pressure is $\Pi_{pol} \sim -kTq_0^3$. The polymer concentration in the complex scales then as $c \sim w^{-8/9}(\frac{f^2 \ell_B}{a^2})^{1/3}$. If the ionic strength is large ($n > f^{4/3}/\ell_B a^8$), the polarization pressure can be written as $\Pi_{pol} \sim -kT\xi_e^{-3}$ where the relevant correlation length is given by κ/q_0^2. The complex concentration is then $c \sim w^{-4/3}\frac{f^2}{a^2 n}$.

A very important application of polyelectrolytes that has been developed over the recent years is the formation of polyelectrolyte multilayers. Starting form a charged solid surface, say positively charged, the multilayer builds up by consecutive adsorption of a negative polymer and a positive polymer (separated by a rinsing of the solution). Up to 50 layers have been piled up this way. Recently, we have proposed a model of polyelectrolyte multilayers where each layer is attached to the previous one by polyelectrolyte complexation.

7. Concluding remarks

In these lecture notes, we have discussed some theoretical aspects of the properties of polyelectrolytes in solution emphasizing the scaling aspects. The conformational properties of polyelectrolytes in a dilute solution are now rather well understood and the theoretical description is in good agreement with both simulations and experiments. One experimental difficulty is that the overlap concentration is often very low and that real solutions are very seldom in the dilute range; experiments in the dilute regime are difficult to perform. One point however that remains unclear is the question of the electrostatic persistence length. Recent simulations do not give a clearcut answer on this point either and further work is needed. Another complication is that it is difficult in the experiments to sort out the purely electrostatic effects that we have discussed from the effects of the short range hydrophobic interactions. A final point that has not been tested quantitatively is the pearl-necklace structure. There are now several experiments that are not in contradiction with this description but a more quantitative comparison with a model system would be necessary.

All these difficulties with the theoretical approaches of dilute solutions make difficult a quantitative description of the structure of semidilute solutions. A scaling description is easily constructed but it is rather difficult to assess its range of validity.

We have only considered here structural properties of polyelectrolyte solutions. Experimentally the variation of the reduced viscosity that shows a peak as a function of concentration and the so-called Fuoss law are considered as the signature of electrostatic interactions in polymer solution. Scaling laws in rather good agreement with experiments have been constructed but a detailed theory is still missing. Another class of dynamic effects that occur almost systematically in polyelectrolyte solutions is the existence of slow modes in quasi-elastic light scattering experiments. There has been huge debates on whether this is an intrinsic property of polyelectrolyte solutions or an experimental artifact. Finally there is a whole range of problems associated to transport in an external electric field and electrophoresis. There has been a large amount of recent experimental and theoretical work stimulated by DNA gel electrophoresis and some very subtle effects are now well understood.

Acknowledgements

I am grateful to M.Rawiso and A.Johner (I.C.S. Strasbourg) for a critical reading of the manuscript.

References

1. Andelman D., Joanny J.F., Comptes Rendus Acad.Sci IV, **1**, 1053 (2000).
2. Barrat J.L., Joanny J.F., Adv.Chem.Phys. **XCIV**, 1 (1996).
3. DeGennes P.G., Pincus P., Velasco R.M., Brochard F., J.Phys. 37, 1461 (1967).
4. Barrat J.L., Boyer D., J.Phys.II, **3**, 343 (1993).
5. Castelnovo M., Sens P., Joanny J.F., Eur.Phys.J.E **1**, 115 (2000).
6. Berghold G., Van der Schoot P., Seidel C., J.Chem.Phys. **107**, 8083 (1997); Limbach H.J., Holm C., J. Chem. Phys. **114**, 9674-9682 (2001).
7. DeGennes P.G. *Scaling concepts in polymer physics*, Cornell University Press (Ithaca, 1985).
8. Grosberg A., Khokhlov A., *Statistical physics of macromolecules*, AIP press (New York, 1994).
9. Halperin A., Zhulina E., Europhys.Lett. **15**, 417 (1991).
10. Rayleigh L., Phil.Mag., **14**, 184 (1882).
11. Dobrynin A., Rubinstein M., Obukhov S., Macromolecules **29**, 2974 (1996).
12. Raphaël E., Joanny J.F., Europhys.Lett., **13**, 6323 (1990).
13. Châtellier X., Senden T., DiMeglio J.M., Joanny J.F., Europhys.Lett. 41, 303 (1998).
14. Johner A., Vilgis T., Joanny J.F., Eur.Phys.J.E **2**, 289 (2000).
15. Tamashiro M., Schiessel H., Macromolecules **33**, 5263 (2000).
16. Barrat J.L., Joanny J.F., Pincus P., J.Phys.II **2**, 1531 (1992).
17. Vilgis T., Johner A., Joanny J.F., Eur.Phys.J.E **3**, 283 (2000).
18. Manning G., J.Chem.Phys. **51**, 954 (1969).
19. Fuoss R., Katchalsky A., Lifson S., Proc.Natl.Acad.Sci. **37**, 579 (1951).
20. Netz R., Joanny J.F., Macromolecules **31**, 5123 (1998).
21. Oosawa F., *Polyelectrolytes*, Dekker (New York, 1971).
22. Ha B., Liu A. Phys.Rev.Lett. **79**, 1289 (1997).
23. Groenbech-Jensen N., Mashl R., Bruinsma R., Gelbart W., Phys.Rev.Lett. **78**, 2477 (1997).
24. Shklovskii B., Phys.Rev.Lett. **82**, 3268 (1999); see also his lecture notes in this book and the references therein.
25. Odijk T., J.Polym.Sci. **15**, 477 (1977).
26. Skolnick J., Fixman M., Macromolecules **10**, 944 (1977).
27. Barrat J.L., Joanny J.F., Europhys.Lett., **24**, 333 (1983).
28. LeBret M., J.Chem.Phys. **76**, 6243 (1982); Fixman J.Chem.Phys. **76**, 6346 (1982).
29. Obukhov S. Private communication.
30. Khokhlov A., Kachaturian K., Polymer **23**, 1793 (1982).
31. Netz R., Orland H., Eur.Phys.J.B 8, 81 (1999).
32. The results are summarized in the lecture notes of K.Kremer in this volume.
33. Borue V., Erukhimovich I., Macromolecules **21**, 3240 (1988).
34. Joanny J.F., Leibler L., J.Phys. **51**, 545 (1990).
35. Dobrynin A., Rubinstein M., Macromolecules **32**, 915 (1999).
36. Wittmer J., Johner A., Joanny J.F., Europhys.Lett. **24**, 263 (1993).
37. Higgs P., Joanny J.F., J.Chem.Phys. **94**, 1543 (1991).
38. Kantor Y., Kardar M., Europhys. Lett., **27**, 643 (1994).
39. Yamakov V., Milchev A., Limbach H.J., Dünweg B., Everaers R., Phys.Rev.Lett. 85, 4305 (2000).
40. Everaers R., Johner A., Joanny J.F., Macromolecules **30**, 8478 (1997).
41. Castelnovo M., Joanny J.F., Langmuir **16**, 7524 (2000).

P + Ca2+
apCa
ap aCa
μ0 + RT ln
= 0

WHEN ION-ION CORRELATIONS ARE IMPORTANT IN CHARGED COLLOIDAL SYSTEMS

BO JÖNSSON AND HåKAN WENNERSTRÖM
Theoretical Chemistry and Physical Chemistry 1,
Chemical Center, Lund University, POB 124, SE-22100 Lund, Sweden

Introduction

Colloidal particles, biopolymers and membranes all carry charges in an aqueous environment. The molecular source of these charges can be covalently bound ionic groups like phosphates, sulfates, carboxylates, quartenary ammoniums or protonated amines. The carboxylates and amines can titrate in response to pH changes, while the other groups remain charged except at extreme conditions. A particle, a self assembled aggregate or a polymer can also acquire a charge by adsorption of a small charged molecule like an amphiphile. The interactions between charged mesoscopic objects is strongly influenced by the net charge and the electrostatic interactions provide one of the basic organizing principles in both colloidal sols and in living cells. These electrostatic interactions can be both attractive, leading to association, and repulsive resulting in dispersion.

The basic description of electrostatic interactions between colloidal particles was worked out during the 1940-ties independently by Derjaguin and Landau in the Sovjet Union [1] and by Verwey and Overbeek in the Netherlands [2]. Both groups based their description of the electrostatic effects on the Poisson-Boltzmann (PB) equation, as in the Gouy-Chapman theory of a single charged surface [3, 4], and the Debye-Hückel theory of electrolyte solutions. Combined with a description of van der Waals interactions, the resulting DLVO theory has played an immense role for our understanding and description of interactions in colloidal systems.

As all theories the DLVO approach has its limitations coming from both the model and approximations. The theory is based on a continuum description of two media separated by a sharp interface. All real interfaces

C. Holm et al. (eds.), Electrostatic Effects in Soft Matter and Biophysics, 171–204.
© 2001 *Kluwer Academic Publishers. Printed in the Netherlands.*

have a finite width and the DLVO theory can be expected to work properly only at separations that exceed this width. For smooth liquid-solid interfaces, *e.g.* mica-water, this is not a severe limitation, while for surfaces with adsorbed polymers the DLVO contribution to the interaction can become irrelevant. Similarly, one can at interfaces often have lateral correlations, or inhomogeneities, causing deviations from the continuum description for perpendicular separations of the order of the lateral correlation length. Another, more subtle, source of a breakdown of the DLVO description is the mean field approximation inherent in the Poisson-Boltzmann equation. The present review is focused on this effect. We start by discussing the arguments leading to the description of an aqueous charged colloidal system in terms of the PB equation. We then discuss important successful applications of the PB equation. This is followed by a conceptual discussion of ion-ion correlation effects, which are illustrated by simple models. Then the effects of the geometrical shape of the colloidal entities as well as that of internal degrees of freedom in counterions are treated. We conclude by giving a number of examples of experimental manifestations of ion-ion correlation effects.

The Poisson-Boltzmann equation

The PB equation relates the mean electrostatic potential $\Phi(\mathbf{r})$ to the mean ion distribution $c_i(\mathbf{r})$ of ionic species i,

$$\epsilon_r \epsilon_0 \nabla \cdot \nabla \Phi(\mathbf{r}) = -\rho(\mathbf{r}) = -e \sum_i z_i c_{i0} \exp(-z_i e \Phi / k_B T) \tag{1}$$

Here $\epsilon_r \epsilon_0$ is the permittivity of the medium, ρ the charge density and c_{io} is the ion concentration at some reference point, where the electrostatic potential is set to zero (a shift Φ to $\Phi + C$ only implies a redefinition of $c_{i0} \rightarrow c_{i0} \exp(-z_i e C / k_B T)$). In order to find a unique solution we must also specify a number of boundary conditions. Physically the non-uniform ion distribution is caused by some spatially fixed source charges, which typically are placed at a surface. Mathematically these source charges either generate a fixed charge distribution or they can be seen to vary in magnitude in such a way that the potential at the surface remains constant. Another physical condition is that the system is electro-neutral with no external fields. By applying Gauss' law these conditions are sufficient to give a unique solution specifying the mean ion distribution and the mean electric field.

The PB equation is approximate and it can not be derived from a specified physical model and first principles. The development of the PB description from an established physical picture of the system can be seen as occurring in two major steps. Consider a macroscopic system that contains

N_1 water molecules as solvent, N_i small charged molecules (atoms) with valency $z_i = \pm1, \pm2, \pm3$ and N_M macroions with a charge $Z_M e$, where typically $| Z_M | >> 1$. The electroneutrality condition does not apply strictly for a macroscopic system (thunderstorms do occur) but only numerically, that is

$$\frac{\sum_i z_i N_i + Z_M N_M}{\sum_i N_i + N_M} \approx 0 \tag{2}$$

All the species interact and in general one should include many-body terms in the interaction potential

$$V_{tot} = \sum_{\alpha,\beta} \sum_{i=1}^{N_\alpha} \sum_{j=1}^{N_\beta} V_{\alpha\beta}(\mathbf{r}_i^\alpha, \mathbf{r}_j^\beta) + \sum_{\alpha,\beta,\gamma} \sum_{i,j,k} V_{\alpha\beta\gamma}(\mathbf{r}_i^\alpha, \mathbf{r}_j^\beta, \mathbf{r}_k^\gamma) + ... \tag{3}$$

where $V_{\alpha\beta}$ and $V_{\alpha\beta\gamma}$ are the two-body and three-body interaction terms, respectively. Most often it is an impractical task to evaluate the partition function

$$Z(N_1, N_i, N_M) = \int_V ... \int_V d\mathbf{r}^{N_1} d\mathbf{r}^{N_i} d\mathbf{r}^{N_M} \exp(-V_{tot}/k_B T) \tag{4}$$

but with present day computers it is becoming possible to obtain estimates for models where the total number of particles is of the order of 10^5, provided that one neglects most many particle contributions in the interaction potential, and assumes $V_{\alpha\beta}$ to be short ranged. However, for analytically resolvable models one must introduce even more drastic approximations. A standard procedure is to formally integrate out the solvent degrees of freedom in eq.(4)

$$\int d\mathbf{r}^{N_1} \exp(-V_{tot}/k_B T) \Rightarrow \exp[-V_{eff}(\mathbf{r}^{N_i}, \mathbf{r}^{N_M})/k_B T] \tag{5}$$

The criterion on the effective potential V_{eff} is that it should generate the same distribution functions in all orders for species i and M as obtained from the original potential, V_{tot}. To satisfy this criterion at a range of temperatures, V_{eff} has to be temperature dependent. It is important to realize, when comparing with actual experimental data, that V_{eff} has the character of a free energy in contrast to V_{tot}, which is a pure energy.

To perform the reduction of eq.(5) exactly, is essentially equivalent to solving the original problem. Instead one uses physical arguments based on pair interactions in simplified models to find V_{eff}. For ionic species the simplest choice is to use the asymptotic ion-ion interaction in a dielectric medium,

$$V_{ij}(\mathbf{r}_i, \mathbf{r}_j) = \frac{z_i z_j e^2}{4\pi\epsilon_0\epsilon_r r_{ij}} \tag{6}$$

where $\mathbf{r}_i$ is the position of ion i and $r_{ij} = \mid \mathbf{r}_i - \mathbf{r}_j \mid$, and supplement this equation with an approximation of the short range interactions in the form of hard spheres

$$V_{ij}(\mathbf{r}_i, \mathbf{r}_j) = \infty \text{ if } r_{ij} < \frac{\sigma_i + \sigma_j}{2} \tag{7}$$

where σ_i and σ_j are the hard sphere diameters of species i and j, respectively. This constitutes the essence of the so-called primitive model of electrolyte solutions. One way to elaborate on this model is to use effective ion-ion pair potentials obtained from computer simulations [5]. The primitive model is simple enough that it lends itself to extensive Monte Carlo simulations studies of the properties of the model system.

To arrive at the PB equation there still remains a crucial step. We first select one or more macroions, which are fixed in space. If we can impose some translational symmetry we can let $N_M \to \infty$. For example, imagine a set of parallel hexagonally arranged cylinders or cubically placed spherical macroions. If translational symmetry does not apply, we are in practice limited to a small ($N_M \simeq 10$) number of charged aggregates [6]. To be more explicit, if we want to study a collection of highly charged macromolecules and their neutralizing counterions as a two component liquid in a computer simulation, we are in practice limited to a small number of rather small aggregates. In order to avoid the complexities of the statistical mechanical description of the small ions, i, it is a common approach to reduce the N-particle description to low order correlation functions. One transparent way to arrive at the PB equation is by starting from the first equation in the Born-Green-Yvon (BGY) hierarchy

$$-k_B T \nabla \ln c_\alpha(\mathbf{r}) = \nabla V_\alpha(\mathbf{r}) + \sum_\beta \int d\mathbf{r}' \nabla V_{\alpha\beta}(\mathbf{r}, \mathbf{r}') \frac{c_{\alpha\beta}(\mathbf{r}, \mathbf{r}')}{c_\alpha(\mathbf{r})} \tag{8}$$

relating the gradient of the one particle density, c_α, to an integral involving also the two particle density $c_{\alpha\beta}$. The hierarchy can be decoupled by neglecting the small ion-small ion pair correlation so that

$$c_{\alpha\beta}(\mathbf{r}, \mathbf{r}') \simeq c_\alpha(\mathbf{r}) c_\beta(\mathbf{r}') \tag{9}$$

where α and β refer to small ions [7]. This approximation provides a well-defined link between the primitive model and the PB equation. Such links are never unique and there exist several other ways of connecting the primitive model and the PB equation [2, 8, 9].

In this formalism $c_i(\mathbf{r})$ corresponds to the pair correlation function $g_{Mi}(\mathbf{r}, \mathbf{r}')$. The PB equation is non-linear in Φ and it is sometimes stated that the equation is in conflict with the principle of linear superposition of electrostatics. This is erroneous and the statement has its source in a confusion between the mathematical property of linear superposition of solutions to linear differential equations and the principle of superposition in electrostatics. The latter principle is derived from the Poisson equation using the former and the Poisson-Boltzmann equation is certainly compatible with the Poisson equation. An inconsistency does appear, however, when the PB equation is used for a system where the central ion M is in in fact identical to some small ion i. This is the case when in the mean field approximation, eq.(9), we set $c_{\alpha\alpha}(\mathbf{r},\mathbf{r}') = c_\alpha(\mathbf{r})c_\alpha(\mathbf{r}')$, but later *derive* that $c_{\alpha\alpha}(\mathbf{r},\mathbf{r}') \neq c_\alpha(\mathbf{r})c_\alpha(\mathbf{r}')$. The consequences of this inconsistency are eliminated, when the PB equation is linearized as in the Debye-Hückel theory of electrolyte solutions [9].

FREE ENERGY AND FORCES FROM THE PB EQUATION

The PB equation can be analytically solved for a number of physically relevant systems and numerical solutions are available in even more cases. As a consequence the PB equation is often used as the primary theory for electrostatic effects. One can distinguish between three types of properties that are commonly discussed in applications of the PB theory. These are:

- the total electrostatic free energy of a single system
- the ion distributions
- the interaction between two subsystems

The ion distributions follow directly from the solution of the equation. In general, there is a significant step from distribution functions to free energies, but it is one of the virtues of the PB equation that the free energy, A_{el}, is readily obtained from

$$\begin{aligned} E_{el} &= \frac{1}{2}\int dV\rho\Phi = \frac{\epsilon_0\epsilon_r}{2}\int dV(\nabla\Phi)^2 \\ S_{el} &= k_B\sum_i[\int dV c_i(\ln c_i - 1) - <c_i> (\ln <c_i> -1)V] \\ A_{el} &= E_{el} - TS_{el} \end{aligned} \tag{10}$$

where V is the system volume. In fact, based on these intuitively transparent equations, one can show that the ion distribution minimizing the free energy in eq.(10), is the PB distribution. Thus, this is yet another way of "deriving" the PB equation.

Given the free energy expression and the PB equation, it is a mathematical challenge to find the appropriate free energy derivatives like chemical potentials, forces etc. Even in the cases where there is no analytical solution to the PB equation, the derivatives can often be expressed in simple terms without having to explicitly evaluate the free energy. A further virtue of the free energy expression in eq.(10) is that one can explicitly separate the contributions from the direct ion-ion interactions and from the entropy of the inhomogeneous ion distribution. This makes the conceptual interpretation of results much clearer. Alternative ways of evaluating the free energy based on charge or temperature integration lead to somewhat less transparent expressions, although they are mathematically equivalent.

A technical point of some importance is that the connection between eq.(10) and the PB equation, as the corresponding Euler-Lagrange equation minimizing the free energy, presupposed incompressibility of the medium. This approximation introduces only minor errors except under extreme conditions, *e.g.* under supercritical conditions for the medium. Incompressibility has the consequence that the Gibbs and Helmholtz free energies only differ by a known constant and are thus for practical purposes equivalent.

Above we have discussed the PB equation as a description of a aqueous system and that is were we intend to apply it. The PB equation is also relevant for plasmas. In this context there is no solvent, but by imposing constant volume we get the same formal description. Thermodynamically we have in this case a non-trivial difference between the Gibbs and Helmholtz free energies. The pV term for the plasma is with the incompressible dielectric medium replaced by the osmotic pressure term $\Pi V_{solv} = N_1 \mu_{solv}$. Quite a substantial confusion has arisen from the mixing of these two internally consistent approaches [10, 11, 12, 13, 14, 15]. In our view it is the appropriate procedure to use the description that is thermodynamically consistent with the system under study. Thus, it is likely to lead astray, if one discusses the thermodynamics of an aqueous liquid system as if it was a plasma. The theory for electrostatic colloidal forces due to Sogami and Ise [10] suffers from this confusion.

The PB equation has been instrumental in establishing a qualitative physical picture of charged particles in a solvent of high dielectric permittivity. Outside a charged surface there is a counterion cloud, whose thickness is determined by the Debye screening length $\kappa^{-1} = \sqrt{\epsilon_0 \epsilon_r k_B T} / \sqrt{\sum c_i z_i^2 e^2}$, or at low salt concentration by the Gouy-Chapman length, see below. Coions are similarly repelled and the source charge together with the diffuse neutralizing charge form a so-called electrical double layer. Starting from an electrically neutral surface with titrable groups we can see that the double layer is built up from a compromise between the entropy of the counterions, which acts to distribute the counterions uniformly throughout the solution, and the electrostatic attraction between fixed charges on the surface and the counterions. Generating a charged surface from a neutral one implies an increase in energy, which is balanced by an increasing entropy. Thus, the diffuse double layer will expand to a point where the entropy increase by further expansion is perfectly balanced by an increase in energy. Conversely, for an initially neutral surface with a chemical affinity for an ion in solution a charge will build up until the short range attraction is balanced by the electrostatic energy and the entropy of the non-uniform distributions of co- and counterions.

When two charged particles come in proximity, their diffuse double layers will start to overlap. This leads to a disturbance of the entropy-energy balance of the single double layer. Since these terms are optimally balanced, the first order changes of the two terms cancel and what remains is a second order entropy term, whose magnitude is given by the ion concentration in the electrically neutral plane between the particles. The expression for the force between two parallel similarly charged surfaces a distance h apart is

$$\frac{F_{plane}}{area} = k_B T \sum_i c_i(\text{mid-plane}) \approx 64 k_B T \Gamma_0^2 \exp(-\kappa h) \sum_i c_i(\text{bulk}) \quad (11)$$

where the parameter Γ_0 is related to the surface potential [16, 17]. The second equality is only valid for large separations. To obtain the force for homogeneously curved surfaces the Derjaguin approximation [18] offers a significant simplification and the asymptotic expression for two uniformly charged spheres of radius R is

$$F_{sphere}(h) \approx \frac{\pi R}{\kappa} 64 k_B T \Gamma_0^2 \exp(-\kappa h) \sum_i c_i(bulk) \quad (12)$$

The force expressions were extensively tested indirectly during a range of years following the presentation of the DLVO theory. However, direct force measurements with an accurate distance resolution, first introduced by Israelachvili and Adams [19], gave a more stringent test of the theory. It was found that the particular case of mica in an aqueous solution is in quantitative agreement with theory except at very short range. In particular, the asymptotic exponential tail with a decay length equal to the inverse Debye length was verified [20, 21, 22, 23]. A change in salt gives the expected change in decay length. One can note that Γ_0 has a maximum value of unity giving an upper limit to the force. In the experiments one found that the measured force was slightly lower than this upper limit. Independent measurements of the streaming potential gave a surface potential of the same magnitude. In this particular case, the PB equation appears to predict the force with good accuracy, a fact that sometimes seems to be forgotten in subsequent experimental studies of the validity of the PB equation. There is a number of experimental variables that depend on the local distribution of ions outside charged surfaces, *e.g.* reaction rates involving charged reactant(s). In catalysis in charged micellar systems there is a very characteristic dependence of the reaction rate on surfactant concentration,

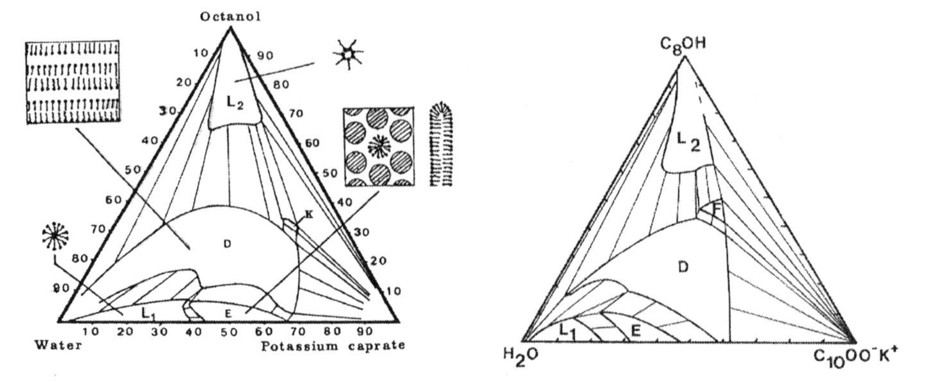

Figure 1. a) The experimental phase diagram for the ternary system water-octanol--K-caprate. The extension of the different phases, micellar, hexagonal, lamellar etc are shown together with schematic pictures of the corresponding aggregates present. b) The same as in a), but calculated on the basis of the PB equation.

which can readily be explained on the basis of the PB equation [17]. The same is true for spectroscopic properties involving charged species [24]. The understanding of diffusion paths of charged reactants outside enzymes has been helped by solving the (linearized) PB equation [25, 26].

An important application of the PB equation is in the calculation of the free energy of self assembly of a charged aggregate like a micelle or a lipid bilayer. It has been demonstrated, that the PB equation with good accuracy predicts both how the free energy of formation changes with salt content and with the concentration of aggregates [27]. It even captures the rather delicate free energy balance between different aggregate shapes [28] and can be used to rationalize the complex phase behavior of a ternary systems [29]. Figure 1 shows a comparison between experimentally determined phase diagram and the one calculated from the Poisson-Boltzmann equation.

The titration of charged groups at interfaces and on polyelectrolytes is influenced by long range electrostatic interactions and this can be modeled with reasonable success using the PB equation [30, 31], although there appears to be substantial problems once one tries to incorporate the effects of a dielectric inhomogeneity [32, 33, 34].

A more delicate test is when one decomposes the free energy into an enthalpic and an entropic part. In eq.(10) we give the contribution to the free energy from the charge-charge interactions, but as pointed out previously, this has the character of a free energy, since it involves an average over solvent degrees of freedom. It is a manifestation of the unique physical properties of water that the dielectric permittivity varies more strongly than $1/T$ in the normal temperature range. The consequence is that an

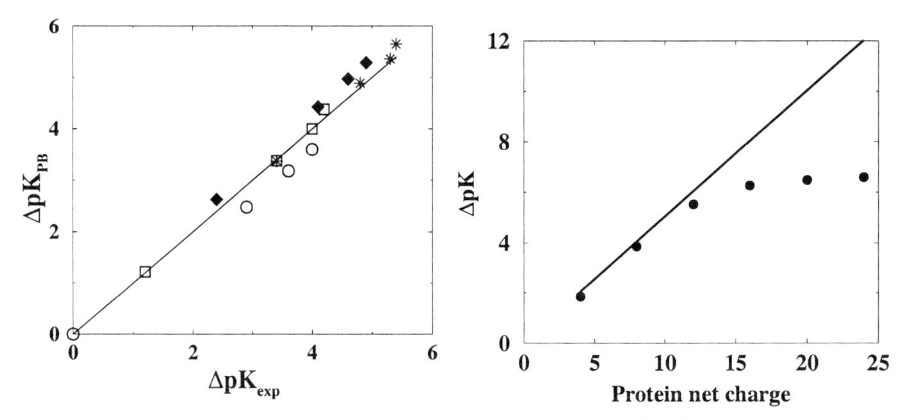

Figure 2. a) A comparison of experimental and theoretical(PB) binding constant shifts for a calcium binding protein. The electrostatic interactions have been modified by adding salt in the range 2-150 mM and by mutating (neutralizing) charge residues in the protein [40]. The symbols represent different mutations; native protein (circles), single mutation (squares), double mutation (diamonds) and triple mutation (stars). The shifts are calculated relative to the native protein at 2 mM salt concentration. b) ΔpK shifts as a function of protein net charge - comparison of DH (line) and MC simulations (symbols). The protein radius is 14 Å and the shift refers to a change in salt concentration from 1 to 500 mM. The protein concentration is $20 \mu M$ and the binding process involves two divalent ions.

electrostatic free energy in the PB approximation is dominantly entropic and the enthalpy is small and with a sign opposite to that of the free energy. That is, the *energy* of interaction between two positively charged particles is attractive! Also in this case the PB equation gives qualitatively correct predictions both when tested against the temperature dependence of phase equilibria and against direct calorimetric titration measurements [31, 35, 36].

The PB equation and its linearized version have also been used to investigate electrostatic interactions in protein binding of small ligands [37, 38, 32, 39]. Most proteins carry a relatively low net charge compared to many colloidal particles and both the PB and the Debye-Hückel(DH) equations give, in general, good agreement with experiments, see Fig.2a. If the protein charge is increased, by for example high or low pH, then at least the DH theory starts to fail. This can be seen in Fig.2b, where the binding of two divalent ions to a hypothetical protein is studied.

The Poisson-Boltzmann equation has turned out to be very useful for a range of applications and it provides a natural first step in an analysis of the effects of electrostatic interactions in an aqueous medium. Due to the usefulness of the approach it becomes important to also analyze when and why the approximation fails.

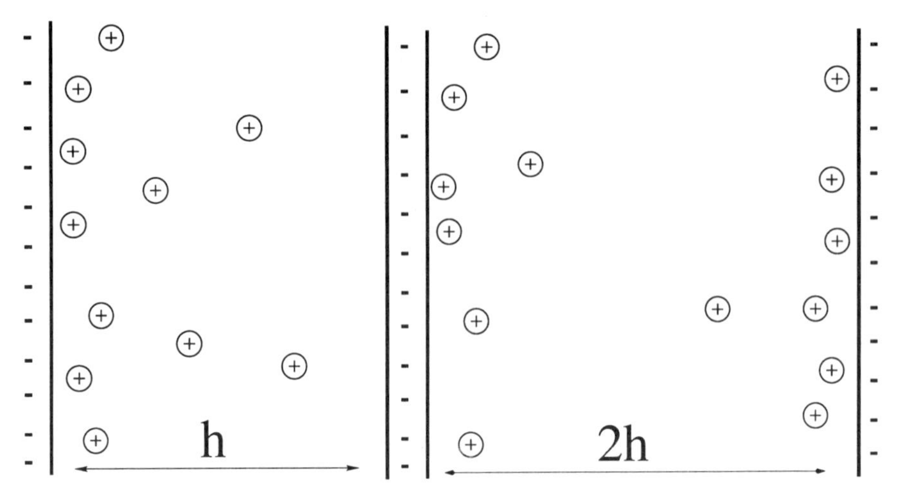

Figure 3. a) Schematic representation of a single charge wall and its counterions. b) Same as in a), but with two charge surfaces. Additional salt pairs have been left out for clarity.

Ion-Ion Correlations

Above we saw that one can arrive at the PB equation by truncating the BGY hierarchy replacing the ion-ion pair correlation function by a product of one particle densities. To get an illustration of the implications of such an approximation let us compare two situations in a planar geometry. In the first case we have one charged wall with surface charge density σ and one neutral surface separated a distance h. Between the walls we have counterions and possibly some electrolyte. In the second case we have two charged surfaces, but now separated by $2h$ and counterions plus electrolyte in between, see Fig.3. Physically the situations are clearly different. If we apply the standard boundary conditions of the PB equation, one finds that ion distributions in the first case are identical to the ion distributions of the two halves in the second case and that the forces between the walls are identical. In an exact treatment of the model this will not be true. In the first case the ion distribution is somewhat changed, but the force is still given by eq.(11) and it is repulsive. In the second case there are interactions between ions on either half of the mid-plane and we can write two expressions for the osmotic pressure [41],

$$
\begin{aligned}
p_{osm} &= k_B T \sum_i c_i(\text{wall}) - \frac{\sigma^2}{2\epsilon_r \epsilon_0} \\
p_{osm} &= k_B T \sum_i [c_i(\text{mid-plane}) + p_i^{corr} + p_i^{hc}]
\end{aligned} \tag{13}
$$

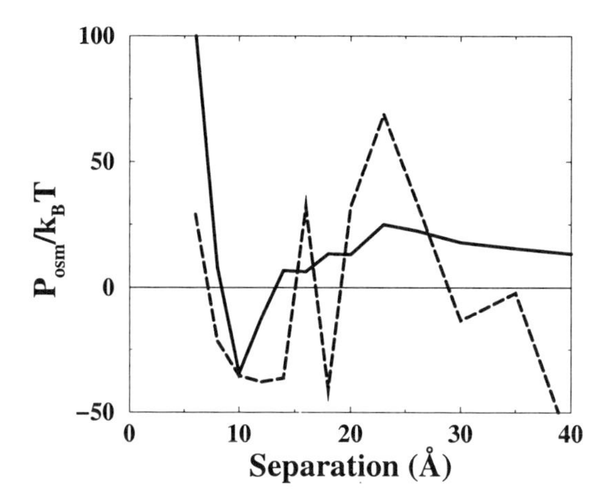

Figure 4. The osmotic pressure as a function of surface separation calculated from an MC simulation according to eq.(13). The dashed line is obtained from the concentration at the charged wall and the solid line is obtained from the second formulation using the mid-plane concentration.

These relations are exact within the primitive model. The term p_i^{corr} comes from the fact that the ions on either side of the mid-plane correlate and it will give an attractive contribution to the pressure. In the mean field description this interaction disappears due to electroneutrality. In other words, there are no correlations across the mid-plane. We will see that these interactions can become strong enough to overcome the entropic repulsive term and lead to a contraction of the system. A further difference from the PB description is that in the exact treatment the size of the ions plays a role, *i.e.* we get the hard core term p_i^{hc}, which in principle could be included in the correlation term, but it is sometimes informative to calculate the two terms, *i.e.* the pressure due to electrostatic and hard core correlations, separately. In many cases these approximations, the neglect of correlations and ion size, will compensate each other and the numerical validity of the PB results is extended [42].

An important technical point with eq.(13) is that the second formulation gives a significantly better numerical estimate of the pressure in an MC simulation. This is particularly important for strongly coupled double layers, where the counterion distribution varies rapidly close to the charged surfaces. Trying to use the first formulation for such systems leads to the calculation of the difference between two large and approximately equal numbers, one of which is suffering from small relative but large absolute fluctuations. Fig.4 demonstrates the virtue of using the mid-plane formulation. The accuracy of both relations in eq.(13) can be improved by using the Widom technique [43, 44, 45] for calculating the ion concentration at a particular position in the double layer. The pressure contribution due to the hard core interaction can be calculated from radial distribution functions for particles at the mid-plane [46], but the Widom ansatz can also be used [44].

In order to study the role of ion-ion correlations it is a virtue to go to a simple yet realistic model where such effects can appear. The case of two parallel planar similarly charged walls separated by a medium containing only counterions is in the PB description characterized by a single dimensionless parameter [47].

$$K = \frac{-\sigma zeh}{2k_BT\epsilon_0\epsilon_r} = \frac{h}{\lambda_{GC}} \tag{14}$$

where we have introduced the Gouy-Chapman length, $\lambda_{GC} = e/2\pi z\sigma l_B$ and the Bjerrum length, $l_B = e^2/4\pi\epsilon_0\epsilon_r k_BT$. In the simplest form of the primitive model only two parameters determine the exact solution of the statistical mechanical model provided one lets the counterion radius go to zero [48]. (In the presence of an electrolyte this would not be possible, since then there is an infinite attraction between coion and counterion.)

$$K_1 = \frac{h}{\lambda_{GC}} \qquad \text{and} \qquad K_2 = \frac{z^4 l_B^2}{h\lambda_{GC}} \tag{15}$$

As long as the ion radius is small enough, *i.e.* the Bjerrum length is much larger than the ion radius, the effect of the counterion radius is small in the counterion only case. Excluding electrolyte ions sometimes appears unphysical, but there is in fact a number of systems where the counterion concentration is much larger than the concentration of neutral salt. Lyotropic liquid crystals formed by ionic amphiphiles is one example where this condition is met [49].

We should expect ion-ion correlations to be important when the Coulomb interaction between the ions is substantial relative k_BT, but as the dimensionless parameter K_2 shows, this effect also increases with increasing surface charge density and decreasing separation between the surfaces. The effect of these changes is to decrease the mean separation between counterions, thus increasing the electrostatic coupling. The most important factor, however, is the ion valency, which also happens to be easily varied under experimental conditions.

It turns out that for aqueous systems at ambient temperatures the ion-ion correlation effects become important for systems with di- or multivalent counterions. This was demonstrated by Monte Carlo simulations by Guldbrand *et al.* [48] and soon after confirmed by solutions of the inhomogeneous hypernetted chain equation (HNC) by Kjellander and Marčelja [50, 51] and

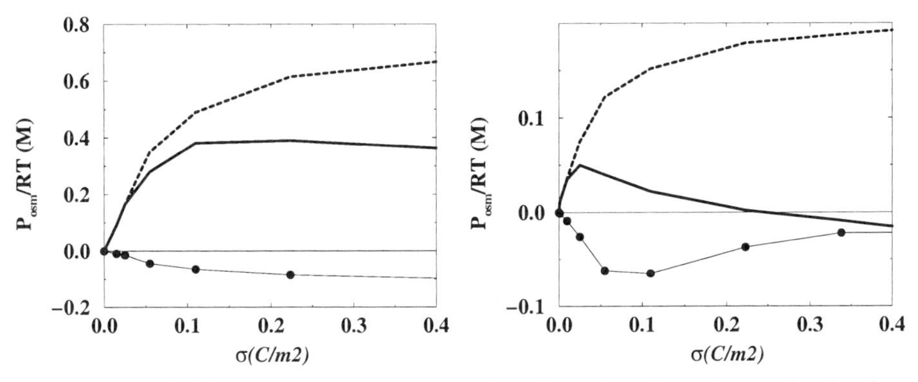

Figure 5. a) The osmotic pressure as a function of surface charge density for two planar double layers with neutralizing monovalent counterions. The solid line is from MC simulations and the dashed line from the PB equation. The thin line with symbols show the attractive contribution to the total pressure. b) The same as a) but with divalent counterions. Note the different scales in a) and b)!

by additional MC simulations by Valleau *et al.* [46] and also by Bratko and Vlachy [52]. The modified PB equation, due to Outhwaite and Bhuiyan [53, 54], has also provided further insight into the approximations in the PB equation [55]. In Fig.5 we show results from these early studies of ion-ion correlation effects. There are regimes where there is a quantitative failure of the PB approximation and there are also a realistic regime where the approximation fails qualitatively and predicts a repulsive rather than an attractive force. Even for monovalent ions substantial corrections can occur in particular with highly charged surfaces at close separations. A typical example would be a concentrated emulsion [56].

One should note that the error in terms of ion distributions are rather modest and also the total electrostatic free energy is reasonable at large separations, although showing a qualitatively incorrect distance dependence.

TWO SIMPLE MODELS FOR ION-ION CORRELATIONS

In previous sections we have emphasized results from various MC simulations when studying ion correlations. The HNC integral equation has been shown to agree with computer simulations [57, 45, 58]. Both methods rely on extensive numerical calculations and it can sometimes be difficult to extract a satisfactory understanding from the methods. Density functionals are becoming more common in many branches of physical sciences and has also been used for investigations of electrical double layers [59, 60, 61, 62, 63]. It seems to us that the latter approach belong to a different category and although very useful on many occasions, it is not of the same rigor as MC

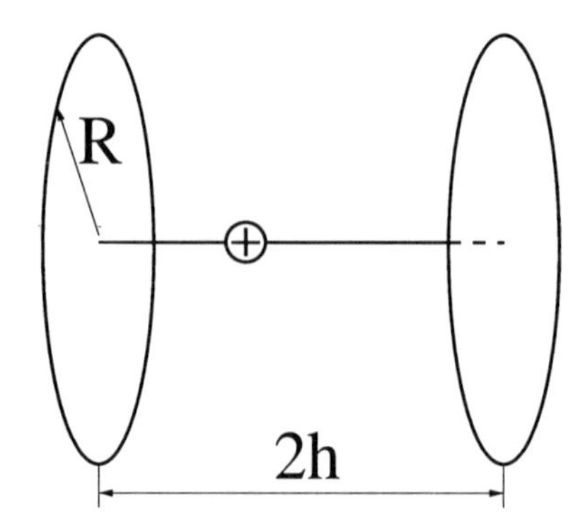

Figure 6. Schematic picture of an ultimately simplified electric double layer consisting of two circular parallel sheets carrying a negative charge of ze uniformly spread out over the two circles. The neutralizing counterion of charge ze is allowed to move on the symmetry line between the circular plates.

or HNC computations.

In order to give an alternative illustration to ion-ion correlation effects, we suggest a simplification of the usual planar double layer invoking just one counterion! Consider two parallel circular surfaces, with radius R, symmetrically arranged as in Fig.6. The two surfaces are uniformly charged and the only counterion of charge ze present, is constrained to the symmetry axis. The whole system is electro-neutral, which means that

$$2\pi R^2\sigma = ze \tag{16}$$

The two surfaces are separated a distance $2h$, thus the ion-surface interaction, V_{iw}, is

$$\frac{V_{iw}}{k_BT} = -\frac{ze\sigma}{4\pi\epsilon_0\epsilon_r k_BT}[\int_0^R \frac{2\pi r dr}{\sqrt{r^2+(x-h)^2}} + \int_0^R \frac{2\pi r dr}{\sqrt{r^2+(x+h)^2}}] \tag{17}$$

Here we have made use of the symmetry leading to simple integrals. Let us introduce the two dimensionless variables

$$\gamma_1 = \frac{z^2 l_B h}{R^2} \text{ and } \gamma_2 = \frac{R}{h} \tag{18}$$

and also scale the coordinates, $\hat{x} = x/h$. γ_1 and γ_2 can be related to the dimension less variable of eq.(15). The configurational integral then takes a very simple form

$$Z = h\exp(-\frac{V_{ww}}{k_BT})\int_{-1}^{1} d\hat{x}\exp[\gamma_1(\sqrt{\gamma_2^2+(\hat{x}-1)^2}+\sqrt{\gamma_2^2+(\hat{x}+1)^2}-2)]$$

Where V_{ww} is the surface-surface interaction, which can easily be computed by numerical integration. The free energy can be calculated from eq.(19)

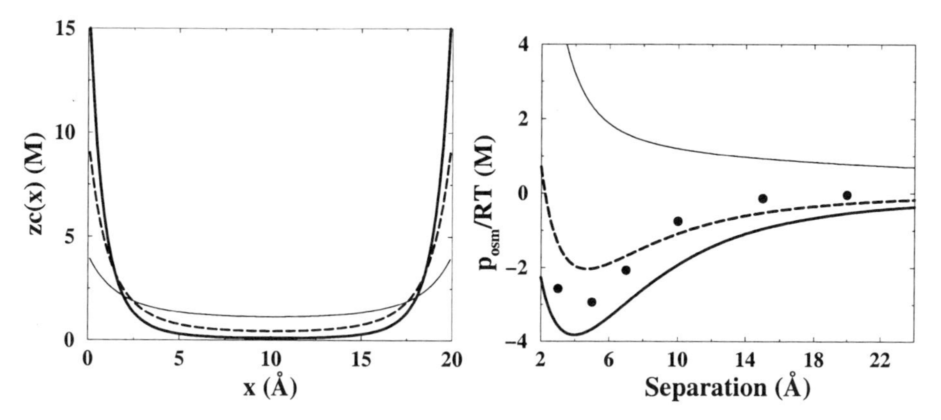

Figure 7. a) The counterion distribution between the two walls as calculated from the simple model system, eq.(19). The Bjerrum length is 7.14 Å, the surface charge density is 0.01 e/Å^2 and the counterion valency has been varied; thin solid line=monovalent, dashed line=divalent and thick solid line=trivalent. b) The pressure between the two walls for the simple model system with parameters and symbols as in a). The filled circles represent MC results for two infinite walls with trivalent counterions and with the same l_B and σ as the simple model.

and together with that all physical quantities of interest. The counterion distribution is shown in Fig.7a for different counterion valency. The qualitative behavior is as could be expected for an electric double layer with an accumulation of counterion density close to the oppositely charged surfaces.

The force as a function of separation is repulsive and monotonically decaying for weakly coupled systems, *i.e.* for low values of γ_1, but with increasing coupling strength the force becomes non-monotonic and even attractive as we by now should expect from a strongly correlated system - Fig.7b. How well this simple model captures the physics in the system can be judged from the comparison also presented in Fig.7b. The model is not quantitatively correct, but it illustrates how the force turns from being monotonically repulsive to showing an attractive regime when the ion valency increases. The change in behavior occurs for approximately the expected parameter values.

The model can be simplified one step further by replacing the two circular surfaces by point charges appropriately placed. Thus, if the counterion is allowed to move along the x-axis between $\pm h$, then the two "surface" charges should be placed at $\pm(h+d)$. The distance d is conveniently chosen so that the electrostatic potential created by the nearest surface in position $\pm h$ is unchanged. This condition gives the following relation for d,

$$d = \sqrt{\frac{ze}{8\pi\sigma}} \tag{19}$$

The ion-"surface" and "surface"-"surface" interactions are easily obtained

$$\frac{V_i w + V_w w}{k_B T} = \frac{z^2 l_B}{2}[-\frac{1}{(d+h)+x} - \frac{1}{(d+h)-x} + \frac{1}{4(d+h)}] \quad (20)$$

The final expression for the configurational integral can be given a very simple appearance by introducing the dimensionless variable

$$\gamma_3 = \frac{z^2 l_B}{2(d+h)} \quad (21)$$

complemented with the scaled coordinate $\hat{x} = x/(d+h)$.

$$Z = (d+h)\int_{-\hat{h}}^{\hat{h}} d\hat{x} \exp[\gamma_3 \frac{7+\hat{x}^2}{4(1-\hat{x}^2)}] \quad (22)$$

The integral is trivial to solve and the corresponding pressure curves show the same qualitative behavior as is obtained from the configurational integral eq.(19). The long range behavior of the pressure curves is not correct, since by only including one counterion we implicitly assume an "infinite" correlation of ions across the double layer. At short surface separations this is an acceptable approximation, but it fails when $h \to \infty$.

EFFECTS OF ELECTROLYTE, GEOMETRY, ION SIZE, TEMPERATURE ETC.

So far we have discussed a simple case for ion-ion correlation effects on double layer forces. Even though the model is actually applicable to some systems, its virtue is that it reveals a mechanism that applies under more general conditions. In many applications involving planar surfaces, like soap films or concentrated emulsions, electrolyte is present. In this case the chemical potential of the electrolyte is an additional parameter and it is also necessary to specify ion sizes. When the concentration or, more correctly, the volume fraction of neutralizing counterions close to the charged surfaces increases, one should expect major effects due to the excluded volume and this can lead to a layering of counterions [64, 65]. The excluded volume effect will of course depend on the bulk salt concentration and for sufficiently high bulk salinity a "depletion mechanism" will also give an attractive contribution to the net osmotic pressure [59, 66]. At more modest ion radii the excluded volume still provides a correction term typically resulting in increased repulsion [46, 42]. From the analysis of the counterions only case, we expect the mean field description to apply at sufficiently large separations and this should hold also in the presence of electrolyte showing that asymptotically one expects the ordinary repulsive double layer interaction.

For short separations we can estimate the amount of electrolyte present between the charged surfaces and for cases where the counterions only situation shows a substantial correction to the PB force the amount of electrolyte in the gap is very small. If we express the net pressure between two surfaces

$$p_{osm}^{net} = p_{osm} - p_{osm}^{bulk}, \tag{23}$$

where p_{osm} is the pressure in the double layer, then the effect of added bulk electrolyte is mainly to change p_{osm}^{bulk} leaving p_{osm} essentially unchanged. This is true for short surface separations, *i.e.* $h \leq \kappa^{-1}$ [67]. It requires a detailed study to find the transition from the long range repulsive regime, where the mean field approximation is applicable, to the short range regime with strong ion-ion correlation effects.

The deviation from the mean field description due to ion-ion correlations has such a physical origin that the effect should be independent of the particular geometry of the charged aggregates. Clearly there are quantitative differences when one considers cylindrical, spherical or irregularly shaped or flexible charged colloidal species, but the basic mechanism should operate in the same way. That this is the case can be seen from Fig.8, where the potential of mean force for two charged spherical aggregates has been calculated from an MC simulation. For monovalent counterions there is a monotonic repulsion in accordance with the PB predictions, but with multivalent counterions or a solvent with a low dielectric permittivity, the usual entropic double layer repulsion decreases and eventually the correlation term starts to dominate. The standard way when considering forces between particles of a regular shape is to invoke the Derjaguin approximation [18], which relates the force between curved surfaces to the interaction free energy of planar ones. In this way one can use the results from the planar case also for curved objects, as long as the radius of curvature is larger than both the Debye screening length and the distance of closest approach between the two particles. This clearly demonstrates that the breakdown of the PB equation can occur for any (regular) geometry and there is reason to expect that such a breakdown can occur also in the case of irregularly shaped charged particles, like for example flexible polyelectrolytes (see below).

To illustrate the use of the Derjaguin approximation let us consider the force between two identical spherical charged colloidal particles of radius R. Then the approximation states that

$$F_{sphere}(h) = \pi R \frac{G_{plane}(h)}{area} \tag{24}$$

Now the free energy for the planar case $G_{plane}(h)$ is related to the force

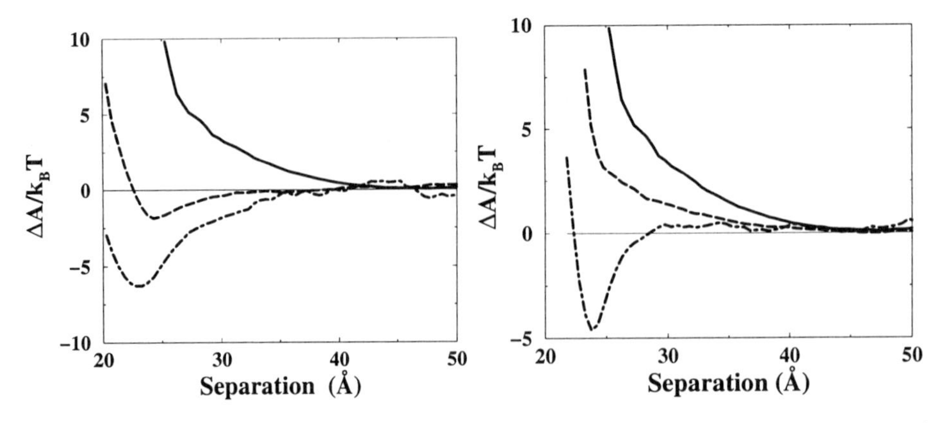

Figure 8. a) The potential of mean force between two spherical aggregates of radius 10 Å and net charge 24. The system contains no salt but only counterions of different valency; solid line=mono-, dashed line=di- and dot dashed line=trivalent ions. The dielectric permittivity is 78 and the temperature 298 K. b)The same as in a), but ϵ_r is varied; solid line=78, dashed line=48 and dot dashed line=18.

$F_{plane}(h)$ by integration

$$G_{plane}(h) = \int_h^\infty dh F_{plane}(h) \tag{25}$$

Asymptotically the force is repulsive due to the fact that the PB equation becomes appropriate and the integral will always contain a repulsive part at large separations. As a consequence one needs to go to somewhat higher charge densities to have an attractive electrostatic force between spherical particles compared to the planar case. Attractive interactions have been demonstrated for parallel cylinders [68, 69, 70] as well as for spheres [71, 72, 73, 74, 6].

If two spherical double layers can correlate and give rise to an attractive interaction and similarly two planar ones, then one should also be able to see the same phenomenon for a single polyelectrolyte chain. That is, the chain should at sufficiently strong coupling contract [75] and eventually take on an end-to-end distance shorter than the corresponding ideal chain. As seen above the ion-ion correlation term can be made important by changing to multivalent counterions. This also works for a single chain and Fig. 13a demonstrates this. Lowering the dielectric permittivity has the same effect as is seen in Fig. 13b.

One would expect that ion-ion correlations become less important at high temperature, since the Coulomb interaction should become relatively weaker. For a dielectric medium with a temperature independent dielectric permittivity this is the case as, for example, seen from the dimensionless parameters of eq.(15). However, when we apply the model to an aqueous system it is necessary to also consider the temperature dependence of ϵ_r. For an ideal dipolar dielectric ϵ_r is inversely proportional to T and the exponent V/k_BT in the partition function becomes for this special case temperature independent in the primitive model. The free energy is then a pure entropy and the force is proportional to the temperature. For water

$$\epsilon_r = 87.84 - 0.3964(T - 273.16) + 7.45 \cdot 10^{-4}(T - 273.16)^2 \qquad (26)$$

and the ratio between the Coulomb interaction and the thermal energy k_BT actually increases with increasing temperature. Thus, the deviations from the PB theory actually *increases* with increasing temperature in the normally studied temperature range according to the primitive model.

INTERNAL DEGREES OF FREEDOM

It is an interesting observation that multivalent atomic ions quite often have disastrous effects on living species. A problem of today for the forest industry is the negative effect of Al^{3+} released from various soils due to the acidic rain. Aluminum ions have also been suggested as a factor in the development of Alzheimer's disease and related conditions. Different water soluble actinide ions form multivalent complexes, which might add to their chemical toxicity. Uranyl complexes, for example, typically carry a rather high net charge. The same is true for many other heavy metal ions, although it is not clear how they act in different situations. Still, biological cells do make use of highly charged systems as for example in the case of DNA. Here a fundamental role is played by small oligoamine ions like spermine and spermidine. Nordenskiöld and co-workers have studied DNA systems extensively in the past and in particular the effect of both multivalent atomic ions and oligoamines [68, 69]. Recently they have also extended their Monte Carlo simulations to include the osmotic pressure in densely packed virus particles [76]. The general result from these, as well as other studies [77, 78] of oligoelectrolytes, is that their effect is very much the same as atomic multivalent ions, although the attraction is slightly reduced.

Polyelectrolytes in electric double layers have been studied by Åkesson *et al.* [79] and by Podgornik [80] and shown to give rise to a net attraction even in a "mean field" description. The attraction is caused by a short range bridging term involving two consecutive monomers along the chain being attached to different surfaces - see Fig. 9. The pressure is conveniently calculated from a modification of eq.(13)

$$p_{osm} = k_BTc(\text{mid-plane}) + p^{corr} + p^{bridge} \qquad (27)$$

(the p_i^{hc} term and the summation have been excluded, since we are restricting the discussion to a salt free system). Since bridging effects are

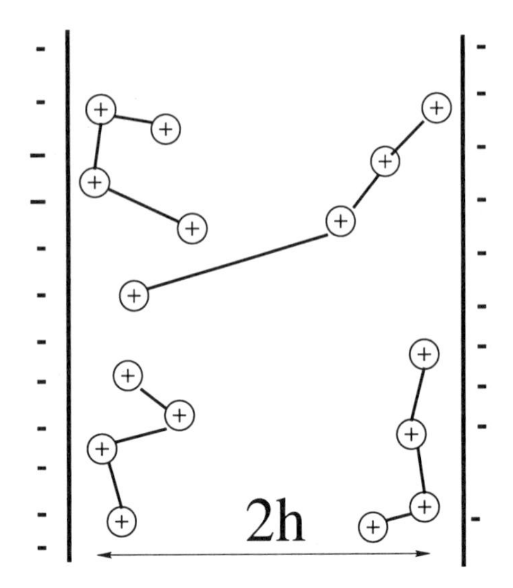

Figure 9. Schematic picture of an electric double layer with neutralizing polyelectrolyte counterions.

appreciable only for separations of the order of a monomer-monomer bond length, the effect should be present also for small oligoelectrolytes like spermin and spermidine. Fig.10 contains a comparison of the osmotic pressure in a planar double layer containing point counterions of valency four and a tetra-electrolyte, respectively. The latter is simply described as point charges connected with harmonic bonds.

The results from Fig.10a can be summarized, that the attraction becomes weaker but more long ranged with oligoelectrolytes as compared to multivalent point counterions. From Fig.10b it is also clear that the bridging term is a major contributor to the net attractive pressure for a flexible tetra-electrolyte.

GENERAL ASPECTS ON CHARGE-CHARGE CORRELATIONS

Charge-charge correlation is a general mechanism for generating attractive interactions in molecular and atomic systems. The textbook example is the dispersion interaction operating between any two atoms or molecules. In a conventional description it is caused by correlations between electrons in the two interacting atoms(molecules) and it invokes a quantum mechanical description of the degrees of freedom.

There exist, however, other well known examples of attractive interactions due to charge-charge correlations involving *classical* degrees of freedom, *e.g.* the Debye interaction between a permanent dipole and a polariz-

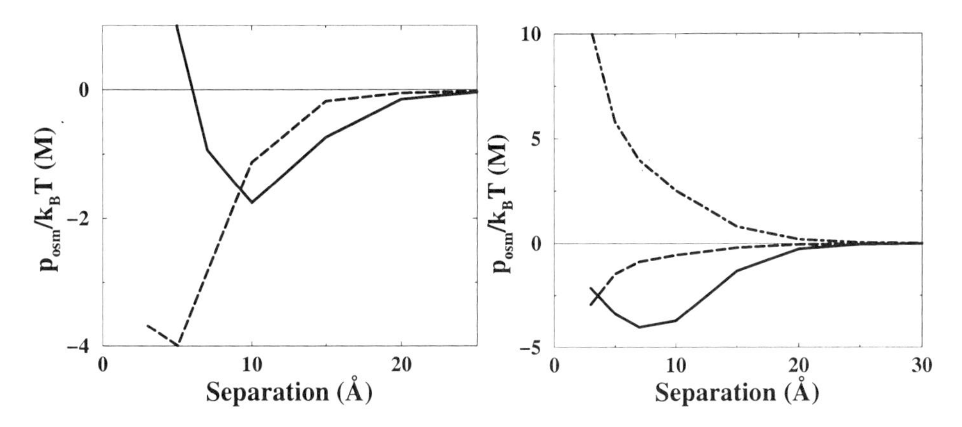

Figure 10. a) The osmotic pressure in a planar electric double layer with tetravalent point counterions (solid line) and with a flexible tetra-electrolyte with average monomer bond lengths of 6 Å (dashed line). b) The total osmotic pressure for the flexible tetra-electrolyte in a) partitioned into the different components given in eq.(27); the bridging term p^{bridge} (solid line), the correlation term p^{corr} (dashed line) and the entropic contribution k_BTc(mid-plane) (dot-dashed line).

ability and the Keesom interaction between two rotating dipoles at a finite temperature [16]. The two latter can be expressed in terms of the electric polarizabilities α of the two species k and l giving

$$V_{ij}(R) = -\frac{3k_BT\alpha_k\alpha_l}{(4\pi\epsilon_0)^2R^6} \tag{28}$$

where R is the separation. In fact, even the basically quantum mechanical dispersion interaction goes asymptotically to this form demonstrating the fundamental connection between these interactions due to charge-charge correlations [81]. Even ion-ion correlation effects can give an interaction of precisely this form if one considers situations where the ions are confined in space [73]. Let us consider two spherical cells, each of radius R_c and each containing one spherical macroion of charge z_M and radius R_M plus neutralizing counterions - see Fig. 11. A distance R separates the macroion centers.

The excess free energy, ΔA, due to the interaction of the two subsystems can be written as,

$$\Delta A = k_BT \ln < \exp[-\Delta V(R)/k_BT] >_0 \approx \frac{< \Delta V(R)^2 >_0}{k_BT} \tag{29}$$

$\Delta V(R)$ is the interaction energy between the cells for a particular configuration of counterions. The angular brackets $<>_0$ represent an average over the counterions in the individual non-interacting cells. The interaction energy $\Delta V(R)$, can be written as a two-center multipole expansion,

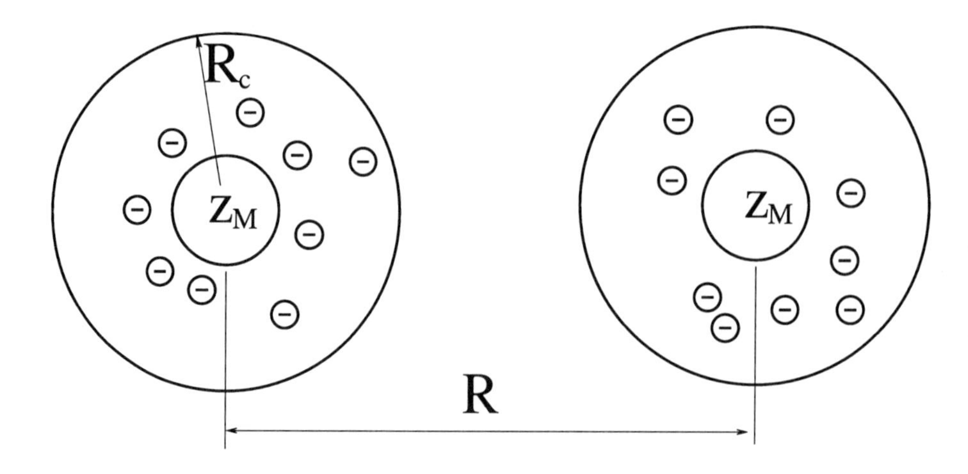

Figure 11. Two interacting, but non-overlapping, charge distributions each with a central macromolecule of charge $z_M e$ surrounded by neutralizing counterions.

where the different terms represent charge-charge, charge-dipole, dipole-dipole interactions etc. For electro-neutral sub-systems the lowest order term contributing to ΔA is due to the dipole-dipole component,

$$\Delta V(R) \approx \sum_i \sum_j \frac{z_i z_j e^2}{4\pi\epsilon_0\epsilon_r R^3}(\mathbf{r}_i\mathbf{r}_j - 3r_{ix}r_{jx}) \tag{30}$$

where the summation over i and$j$ are for the two sub-systems, respectively. Using the spherical symmetry of the sub-systems,

$$< \Delta V(R) >_0 \approx 6(\frac{1}{4\pi\epsilon_0\epsilon_r})^2 < (\sum_i z_i e r_{ix})^2 >_0 < (\sum_j z_j e r_{jx})^2 >_0 \tag{31}$$

where the polarizability, α, can be identified as

$$\alpha = \frac{< (\sum_i z e r_{ix})^2 >_0}{k_B T} \tag{32}$$

If combining eqs.(29-32) then the free energy of interaction can be written as,

$$\Delta A = -\frac{3k_B T \alpha_k \alpha_l}{(4\pi\epsilon_0\epsilon_r)^2 R^6} \tag{33}$$

illustrating the generality of eq.(28) and the analogy with the dispersion force. The only difference between eqs.(28) and (33) is the dielectric permittivity, ϵ_r, of the medium.

Another example of ion-ion correlations in spherical geometry is the case with two droplets formed from an aqueous electrolyte solution, which will interact according to eq.(33) and the polarizabilities will be that of conducting spheres. A similar case is obtained with reversed micelles (or water in oil micro-emulsion droplets) formed by ionic surfactants [82].

Even for net neutral surfaces there is always a lateral charge distribution and correlations with another similar surface will lead to attraction [83]. The in-plane and the out-of-plane dipolar components behave qualitatively differently, since the former can correlate by change of orientation, while the latter requires a translational motion to achieve the attractive effect. The magnitude of these interactions depends on the interactions within the surface, while the distance dependence is generic [84, 85, 86, 87]. A particularly interesting case occurs when one has a structure on a mesoscopic scale on the surface due to surface aggregates or domains. Due to the large size of the correlating entities one can easily reach a direct interaction that clearly exceeds k_BT, meaning that the system is strongly correlated. The interaction is then more long ranged, except asymptotically, and can lead to deviations from the Lifshitz theory at large separations [23, 88, 89]. Studies of such models also reveals a general effect that the classical charge-charge correlations are screened by electrolyte [83, 90, 88]. Due to the large range of the interactions, screening effects become important even at moderate and low electrolyte concentrations. The effective screening length for correlations is half the Debye length [83, 87, 88].

EXPERIMENTAL MANIFESTATIONS OF ION-ION CORRELATIONS

The theoretical discussion has shown that there is an attractive contribution to the force between two similarly charged surfaces or particles. This attraction is short ranged, it decays like h^{-3} for a salt-free planar double layer [91, 92], and it is *e.g.* more significant the higher the valency of the counterion and the lower the dielectric permittivity. It is typically stronger than the van der Waals contribution. Experimentally, the effect can manifest itself as an unexpected attractive force or when the effect is less strong, as a substantial weakening of the repulsive double layer force.

There exist several powerful methods for measuring repulsive forces, while it is much more difficult to measure an attraction, because the system is then inherently unstable. In measurements one has either to balance the attractive force by a known stronger repulsive force or, as in the Surface Force Apparatus (SFA) and the Atomic Force Microscopy, balance the attraction with a sufficiently stiff spring. Neither of these methods is easy to implement for strongly attractive forces and to our knowledge there is no clear-cut experimental measurement of attractive ion-ion correlation forces where one has been able to resolve the distance dependence of the force.

On the other hand, there are numerous experiments indirectly demonstrating the existence of an attractive force that is most likely due to ion-ion correlations.

It is well established for many colloidal systems that di- and tri-valent ions can cause precipitation and/or coagulation [93]. Such observations have usually been interpreted in terms of the DLVO theory even in cases where the quantitative description indicated a discrepancy between theory and experiment. One then concluded that the multivalent ions could form "salt bridges" or coordinate specifically between the surfaces. One common example is the precipitation of soaps by calcium or magnesium ions. It is not easy to separate one effect from the other and different conceptual descriptions in fact overlap.

An attractive force that can be shown to operate between similarly charged surfaces or particles, which has a range that clearly exceeds the scale for molecular contact between surfaces and is stronger than the conventional van der Waals force is a good candidate for an ion-ion correlation force. To further substantiate the case, the force should be independent of chemical details, such as the molecular nature of the surfaces and/or the chemical nature of the ions. In practice, it can be difficult to eliminate all other possibilities for explaining an observed attractive force. However, the success of the PB/DLVO theory for the description of, at least, aqueous systems have given strong support to the applicability of the primitive model and we should be able to trust the theoretical results not only with respect to the mechanism of the ion-ion correlation force but also with respect to the quantitative predictions. Consequently, we argue that when an attractive force is observed under conditions where the theory predicts an attraction, this is a strong indication that one has a manifestation of the effect.

One of the first clear experimental demonstrations of the correlation effect was due to Khan *et al.* [94] who studied the swelling of the lamellar liquid crystalline phase formed by the doubled tailed anionic surfactant AOT. With the normal counterion Na^+, the lamellar phase swells to ca. 80 per cent water. This could be quantitatively modeled using the PB approximation [95], while with divalent counterions such as Ca^{2+} and Mg^{2+} a lamellar liquid crystalline phase was still formed, but it only incorporated around 40 per cent water [94, 96]. The prediction of the PB theory was, that these systems should swell even more than for the monovalent counterions. To quantitatively study the transition in swelling behavior Khan *et al.* [94] prepared the ternary system NaAOT - CaAOT- water and observed that at intermediate mixing ratios there was a transition from strong to weak swelling and simultaneously a coexistence of two lamellar phases. In fact, in the simulations by Guldbrand *et al.* [48] the parameters where chosen to

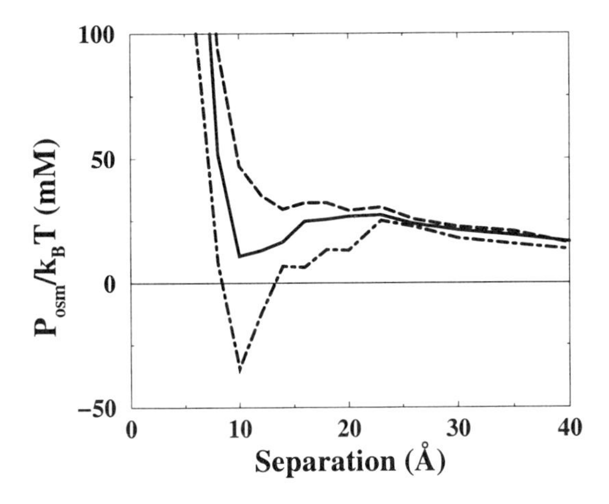

Figure 12. The osmotic pressure between two charged surfaces with variable surface charge density; dashed line = 0.075, solid line = 0.071 and dot-dashed line = 0.080 C/m^2. Divalent point counterions and a Bjerrum length of 7.14 Å are used.

represent the AOT system. The simulations predicted that for monovalent ions there is a small correction to the PB approximation, while for divalent ions attraction dominates. Furthermore, in the transition from attraction to repulsion the force curve shows a non-monotonic behavior on the repulsive side, see Fig.12. This means that two lamellar phases are co-existing. It is interesting to note that the phase separation occurs with a force that is net repulsive at all separations. It is often stated that one needs an attractive region in the force to cause phase separation, but a non-monotonic repulsive force is in fact sufficient.

In subsequent studies of surfactant phase behavior, it has been generically seen that in going from monovalent to divalent counterions the swelling of lamellar phases is drastically decreased [97]. This is true for both anionic and cationic amphiphiles and for a number of different headgroups. Occasionally for less highly charged systems one can observe the coexistence of two lamellar phases. Taken together these studies demonstrates the existence of attractive ion-ion correlation forces. They also show that the results from the primitive model are in reasonable quantitative agreement for aqueous systems. Clearly one should expect the same force to appear in other systems when the physical parameters are such that the theory predicts attractions.

Using the SFA technique one has in several studies found a strong attractive force component at close range [98, 99, 100]. It manifests itself as an instability occurring for longer separations than expected for the conventional van der Waals force. However, the presence of this attractive component does not change the experimental outcome qualitatively, it is still repulsive at long range, and it requires assumptions about the repulsive double layer force to make quantitative statements about the attractive part. Still, the general observation is that when one expects a sizeable extra attractive force it is also observed. Pashley [101] has studied the double layer force between mica surfaces in the presence of trivalent counterions and report a *charge inversion* of the mica surfaces, which is yet another manifestation of ion-ion correlations [59, 102, 103].

For aqueous systems with monovalent ions one expects correlation effects to only give a quantitative correction to the mean field predictions. There are situations, however, where these corrections can be of importance. Soap films formed by ionic surfactants can have a very high charge density, $\sigma \simeq 0.2 - 0.3$ C/m^2. We have earlier suggested [48] that the formation of Newton black films, with their thin aqueous layer, is triggered by ion-ion correlations. A more recent example is found in concentrated emulsions stabilized by an ionic surfactant. Here one has a similar molecular arrangement of a thin aqueous layer separating two charged surfactant films. Sonneville *et al.* [56] observed a discontinuous swelling in such systems and interpreted the attractive component causing this discontinuity as being due to ion-ion correlations.

Above we have concentrated on forces between planar, or nearly planar, surfaces. The first explicit discussion of attractive ion-ion correlation forces due to Oosawa [104], was inspired by experiments in polyelectrolyte systems. In this case the basic models assume a cylindrical geometry. As discussed above there is for these types of systems numerous examples where the addition of divalent and trivalent counterions induce precipitation [105, 106], but it is difficult to separate the role of short range interactions from the typical ion-ion correlation effects operating at a slightly longer range. So far the most studied cases are DNA and virus systems [68, 107, 76] and it has been shown experimentally for a number of cases that multivalent counterions can cause condensation of DNA molecules [108, 109, 110, 111, 112]. Recently, the coil-globule transition of a single DNA molecule has been followed in a fluorescence microscopy study [113, 114, 115, 116]. The transition has been caused by both multivalent counterions and by lowering of the solvent dielectric permittivity see Fig. 13.

Salt effects on ion-ion correlation can be quite spectacular. In a system with multivalent counterions and low salt content attractive forces will dominate. Upon addition of a 1:1 salt, there will be a competition between the original multivalent counterions and the monovalent ones coming from the salt. At sufficiently high salt concentration the attraction can disappear and the system will revert to a "normal" double layer repulsion. Thus, we have the unexpected situation that addition of salt leads to *increased*

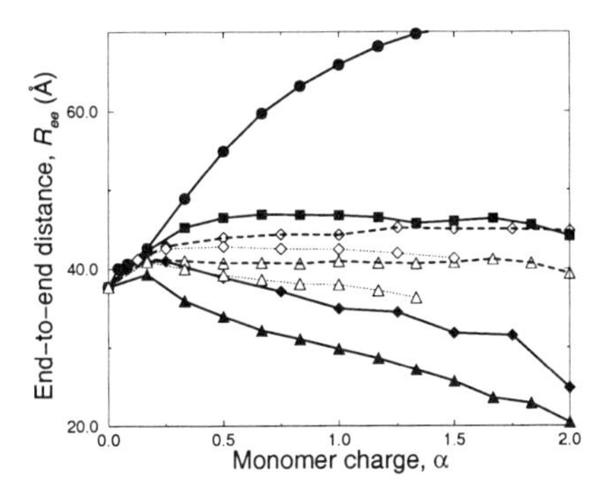

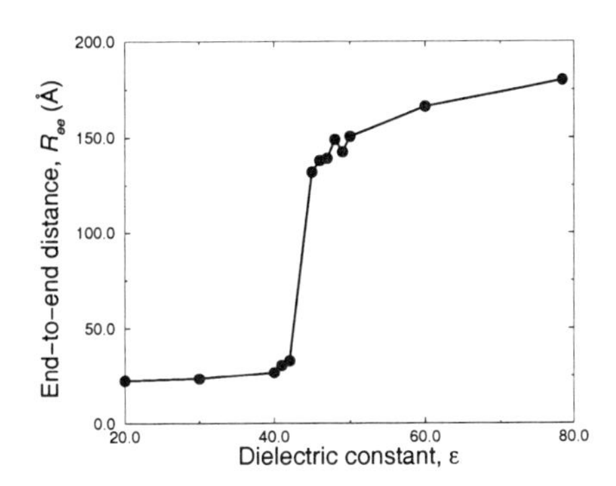

Figure 13. a) The end-to-end distance of a 24 monomer polyelectrolyte as a function of monomer charge. In the MC simulations different counterions have been used. Solid lines are for point counterions, dashed lines are for oligo-electrolytes (connected monovalent ions) with bond length=6 Å and the dotted lines are for oligo-electrolytes with bond length=4 Å. The valencies of the counterions are: circles-monovalent, squares-divalent, diamonds-trivalent and triangles-tetravalent. b) The end-to-end distance of a 60 monomer polyelectrolyte as a function of the dielectric constant. Here the monomer charge is fixed to -1 and simple monovalent counterions are used. All particles have a hard core radius of 4 Å and the bond length is 6 Å and an additional square well between monomers and counterions have been introduced - see also reference [77].

repulsion. The phenomenon can be observed in dilute DNA solutions, where initially compacted DNA (by spermidine or spermine) expands if the salt concentration becomes sufficiently high - see Fig. 14.

For spherical particles it has turned out to be more difficult to find clear demonstrations of the ion-ion correlations. For a given charge density, the electrostatic interactions are weaker between spheres than between planes as, for example, can be seen from the Derjaguin approximation. Consequently, one needs a higher charge density in the spherical system relative to the planar or cylindrical cases in order to see a net attraction. Furthermore, if one uses multivalent counterions, one easily reaches the precipitation limit, where the physical origin of the observed effect is less clear. There are experimental observations of coexistence between concentrated and dilute phases of latex particles [117, 10, 118]. However, for these systems the parameters are such that the observation can not be explained by ion-ion correlation effects. There exists a current debate on the explanations of these observations [119] and in our opinion the suggestion by Sogami and Ise has a very weak foundation [10]. A possible source of the two phase coexistence observed for charged latex particles is charge polydispersity. This may generate a two phase separation with more highly charged particles in a dilute phase and the less charged ones in the concentrated phase.

Parallel to the debate about the "Sogami theory" there is also an ongo-

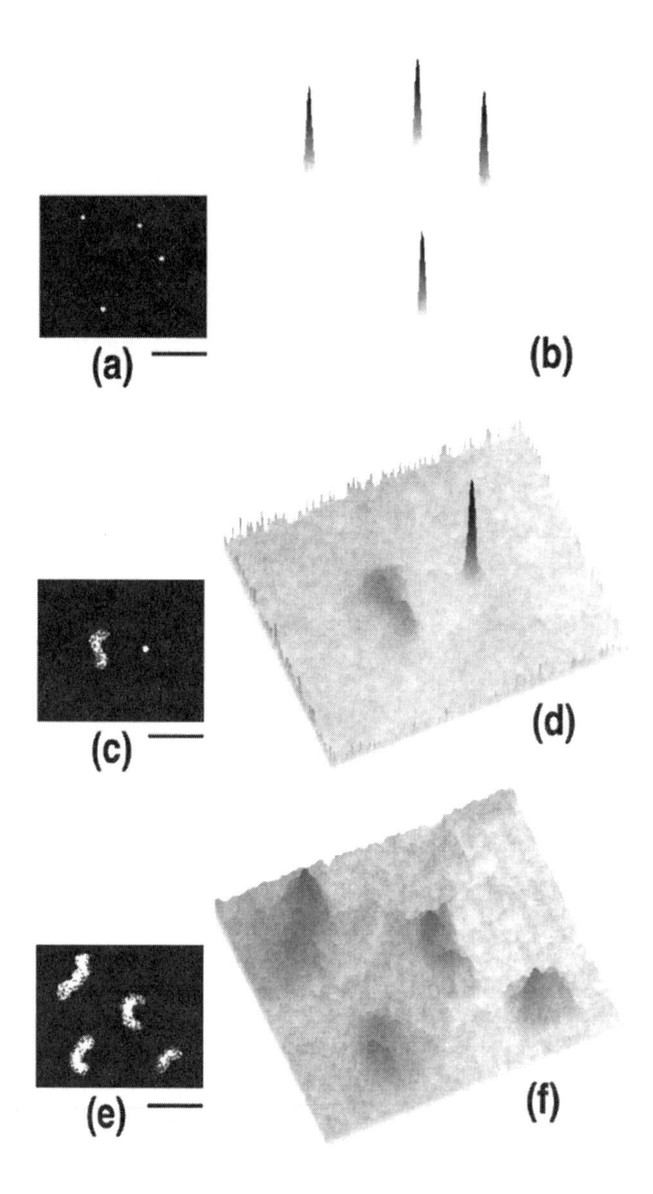

Figure 14. Experimental evidence that simple salt unfolds DNA compacted by spermine. a), c) and e) are video frames from the fluorescence microscopy image of single T4 DNA molecules at [spermine]=2.0×10^{-6} M. The scale bar represents 5 μm. b), d) and f) are the corresponding quasi-three-dimensional representations of the fluorescence intensity. a) and b) shows the salt-free case, [NaCl]=0, where the DNA molecules exhibit a globular conformation. In c) and d), [NaCl]=30 mM, the DNA molecules coexist in both elongated coiled and compacted globular structures. For high salt concentrations, [NaCl]=300 mM, all DNA molecules have a coiled structure.

ing dispute about how to theoretically approach an asymmetric electrolyte using the concept of effective potentials between the macroions. The interested reader is referred to some enlightening papers by Chan [120] and Belloni [121].

Conclusions

In these lecture notes we have tried to describe the phenomenon of ion-ion correlations in as simple terms as possible. We have also pointed out the generality and made connection with quantum mechanical perturbation theory. The generality means that geometry, ion type, details of surface charges etc. are of secondary importance and we have made several attempts to show this by describing different experimental situations, where we believe ion-ion correlations play a major role.

The forces in charged macromolecular systems are always a balance between repulsive forces of entropic origin and attractive forces of energetic origin. For many systems of chemical interest the repulsive forces dominate and the mean field description provided by the Poisson-Boltzmann equation is adequate. There are many ways of changing the balance between entropic and energetic terms, the most efficient way to favor attractive interactions is to reduce the importance of the entropy term by simply decreasing the number of charged particles by using multivalent counterions. The other way to favor attractive forces is by increasing the importance of the energetic term by *e.g.* going to a solvent with a low dielectric permittivity. These observations can be formalized for a pair of planar double layers using the dimensionless parameters K_1 and K_2.

Numerically, the important means for investigating ion-ion correlations have been Monte Carlo simulations and accurate integral equation techniques. One would hope that more simple approaches based on different perturbational schemes should suffice to capture the phenomenon at least semi-quantitatively. This does not seem to be the case and our general conclusion is that whenever ion-ion correlations become important and cause attractive forces to dominate, then we are unfortunately well outside the regime where more simple and analytical tractable approaches are applicable. At the present stage we believe that in order to get further insight into ion-ion correlations, well designed experiments for in particular spherical systems are needed.

References

1. Derjaguin, B. V.; Landau, L. *Acta Phys. Chim. URSS* 1941, **14**, 633.
2. Verwey, E. J. W.; Overbeek, J. T. G. *Theory of the Stability of Lyophobic Colloids;* Elsevier Publishing Company Inc.: Amsterdam, 1948.

3. Gouy, G. *J. de Physique* 1910, **9**, 457.
4. Chapman, D. L. *London and Edinburg Phil. Mag. and J. of Sci.* 1913, **25**, 475.
5. Marčelja, S. *Nature* 1997, **385**, 689.
6. Linse, P. *J. Chem. Phys.* 2000, **113**, 4359–4373.
7. Jönsson, B.; Wennerström, H.; Halle, B. *J. Phys. Chem.* 1980, **84**, 2179.
8. Podgornik, R. *J. Chem. Phys.* 1989, **91**, 5840.
9. McQuarrie, D. A. *Statistical Mechanics;* Harper Collins: New York, 1976.
10. Sogami, I.; Ise, N. *J. Chem. Phys.* 1984, **81**, 6320.
11. Beresford-Smith, B. *Some aspects of strongly interacting colloidal interactions,* Thesis, Australian National University, 1985.
12. Beresford-Smith, B.; Chan, D. Y. C.; Mitchell, D. J. *J. Colloid Interface Sci* 1985, **216**, 216.
13. Overbeek, J. G. *J. Chem. Phys.* 1987, **87**, 4406.
14. Woodward, C. E. *J. Chem. Phys.* 1988, **89**, 5140.
15. Jönsson, B.; Åkesson, T.; Woodward, C. *Ordering and phase transitions in charged colloids, Eds: A. K. Arora and B. V. R. Tata* 1996, 295.
16. Israelachvili, J. *Intermolecular and Surface Forces;* Academic Press: London, 2nd ed.; 1991.
17. Evans, D. F.; Wennerström, H. *The Colloidal Domain - Where Physics, Chemistry, Biology and Technology Meet;* VCH Publishers: New York, 1994.
18. Derjaguin, B. *Kolloid-Z* 1934, **69**, 155.
19. Israelachvili, J. N.; Adams, G. E. *J. Chem. Soc. Faraday Trans. I* 1978, **74**, 975.
20. Pashley, R. M. *J. Coll. Interface Sci.* 1981, **83**, 531.
21. Israelachvili, J. N.; Pashley, R. M. *Nature* 1982, **300**, 341.
22. Israelachvili, J. N.; Pashley, R. M. *J. Coll. Int. Sci.* 1984, **98**, 500.
23. Tsao, Y.-H.; Evans, D. F.; Wennerström, H. *Science* 1993, **262**, 547.
24. Söderman, O.; Engström, S.; Wennerström, H. *J. Coll. Interf. Sci.* 1980, **78**, 110.
25. Sharp, K.; Fine, R.; Honig, B. *Science* 1987, **236**, 1460.
26. Head-Gordon, T.; Brooks, C. L. *J. Phys. Chem.* 1987, **91**, 3342.
27. Gunnarsson, G.; Jönsson, B.; Wennerström, H. *J. Phys. Chem.* 1980, **84**, 3114.
28. Jönsson, B.; Wennerström, H. *J. Coll. Interf. Sci* 1981, **80**, 482.
29. Jönsson, B.; Wennerström, H. *J. Phys. Chem.* 1987, **91**, 338.
30. Ullner, M.; Jönsson, B. *Macromolecules* 1996, **29**, 6645–6655.
31. Gunnarsson, G.; Wennerström, H.; Olofsson, G.; Zacharov, A. *J. Chem. Soc. Faraday Trans. 1* 1980, **76**, 1287.
32. Gilson, M. K.; Rashin, A.; Fine, R.; Honig, B. *J. Mol. Biol.* 1985, **183**, 503–516.
33. Penfold, R.; Warwicker, J.; Jönsson, B. *J. Phys. Chem. B* 1998, **00**, 000.
34. Spassov, V.; Bashford, D. *Prot. Sci.* 1998, **7**, 554–567.
35. Dolar, D.; Skerjanc, J.; Bratko, D.; Crescenzi, V.; Quadrifoglio, F. *J. Phys. Chem.* 1982, **86**, 2469–2471.
36. Carlsson, I.; Fogden, A.; Wennerström, H. *Langmuir* 1999, **15**, 6150.
37. Tanford, C.; Kirkwood, J. G. *J. Am. Chem. Soc.* 1957, **79**, 5333.
38. Warwicker, J.; Watson, H. C. *J. Mol. Biol.* 1982, **157**, 671–679.
39. Fushiki, M.; Svensson, B.; Jönsson, B.; Woodward, C. E. *Biopolymers* 1991, **31**, 1149–1158.
40. Svensson, B.; Jönsson, B.; Woodward, C. E.; Linse, S. *Biochemistry* 1991, **30**, 5209–5217.
41. Wennerström, H.; Jönsson, B.; Linse, P. *J. Chem. Phys.* 1982, **76**, 4665.
42. Kjellander, R.; Marčelja, S. *J. Phys. (France)* 1988, **49**, 1009.
43. Widom, B. *J. Chem. Phys.* 1963, **39**, 2808.

44. Svensson, B.; Åkesson, T.; Woodward, C. E. *J. Chem. Phys.* 1991, **95**, 2717.
45. Greberg, H.; Kjellander, R.; Åkesson, T. *Molec. Phys.* 1996, **87**, 407.
46. Valleau, J. P.; Ivkov, R.; Torrie, G. M. *J. Chem. Phys.* 1991, **95**, 520.
47. Engström, S.; Wennerström, H. *J. Phys. Chem.* 1978, **82**, 2711.
48. Guldbrand, L.; Jönsson, B.; Wennerström, H.; Linse, P. *J. Chem. Phys.* 1984, **80**, 2221.
49. Tiddy, G. J. T. *Phys. Rep.* 1980, **57**, 1.
50. Kjellander, R.; Marčelja, S. *Chem. Phys. Letters* 1984, **112**, 49–43.
51. Kjellander, R.; Marčelja, S. *J. Chem. Phys.* 1985, **82**, 2122.
52. Bratko, D.; Vlachy, V. *Chem. Phys. Lett.* 1982, **90**, 434.
53. Outhwaite, C. W.; Bhuiyan, L. B. *J. Chem. Soc., Faraday Trans. 2* 1983, **79**, 707.
54. Outhwaite, C. W.; Bhuiyan, L. B. *Molec. Phys.* 1991, **74**, 367.
55. Das, T.; Bratko, D.; Bhuiyan, L. B.; Outhwaite, C. W. *J. Phys. Chem.* 1995, **99**, 410–418.
56. Sonneville, O.; Gulik-Krzywicki, V. B. T.; Jönsson, B.; Wennerström, H.; Lindner, P.; Cabane, B. *Langmuir* 2000, **16**, 1566.
57. Kjellander, R.; Åkesson, T.; Jönsson, B.; Marčelja, S. *J. Chem. Phys.* 1992, **97**, 1424.
58. Greberg, H.; Kjellander, R.; Åkesson, T. *Molec. Phys.* 1997, **92**, 35.
59. Tang, Z.; Scriven, L. E.; Davis, H. T. *J. Chem. Phys.* 1992, **97**, 9258.
60. Tang, Z.; Scriven, L. E.; Davis, H. T. *J. Chem. Phys.* 1992, **97**, 494.
61. Penfold, R.; Jönsson, B.; Nordholm, S. *J. Chem. Phys.* 1993, **99**, 497.
62. van Roij, R.; Dijkstra, M.; Hansen, J. P. *Phys. Rev. E* 1999, **59**, 2010.
63. Warren, P. B. *J. Chem phys* 2000, **112**, 4683–4698.
64. Torrie, G. M.; Valleau, J. P. *J. Phys. Chem.* 1982, **86**, 3251.
65. Valleau, J. P.; Torrie, G. M. *J. Chem. Phys.* 1982, **76**, 4623–4630.
66. Bolhuis, P. G.; Åkesson, T.; Jönsson, B. *J. Chem. Phys.* 1993, **98**, 8096.
67. Dubios, M.; Zemb, T.; Belloni, L.; Deville, A.; Levitz, P.; Selton, R. *J. Chem. Phys.* 1992, **96**, 2278.
68. Guldbrand, L.; Nilsson, L.; Nordenskiöld, L. *J. Chem. Phys.* 1986, **85**, 6686.
69. Lyubartsev, A. P.; Nordenskiöld, L. *J. Phys. Chem.* 1997, **101**, 4335–4342.
70. Grønbech-Jensen, N.; Mashl, R. J.; Bruinsma, R. F.; Gelbart, W. *Phys. Rev. Lett.* 1997, **78**, 2477–2480.
71. Patey, G. N. *J. Chem. Phys.* 1980, **72**, 5763.
72. Svensson, B.; Jönsson, B. *Chem. Phys. Lett.* 1984, **108**, 580.
73. Woodward, C. E.; Jönsson, B.; Åkesson, T. *J. Chem. Phys.* 1988, **89**, 5145.
74. Linse, P.; Lobaskin, V. *Phys. Rev. Lett* 1999, **83**, 4208–4211.
75. Stevens, M. J.; Kremer, K. *J. Chem. Phys.* 1995, **103**, 1669–1690.
76. Lyubartsev, A. P.; Tang, J. X.; Janmey, P. A.; Nordenskiöld, L. *Phys. Rev. Letters* 1998, **81**, 5465–5468.
77. Khan, M. O.; Mel'nikov, S. M.; Jönsson, B. *Macromolecules* 1999, **32**, 8836–8840.
78. Rescic, J.; Linse, P. *J. Phys. Chem. B* 2000, **113**, 7852–7857.
79. Åkesson, T.; Woodward, C.; Jönsson, B. *J. Chem. Phys.* 1989, **91**, 2461–2469.
80. Podgornik, R. *J. Phys. Chem.* 1992, **96**, 884–896.
81. Wennerström, H.; Daicic, J.; Ninham, B. W. *Phys. Rev. A* 1999, **60**, 1–4.
82. Bratko, D.; Woodward, C. E.; ; Luzar, A. *J. Chem. Phys.* 1991, **95**, 5318.
83. Ninham, B.; Parsegian, V. A. *J. Theoret. Biol.* 1971, **31**, 405.
84. Attard, P.; Kjellander, R.; Mitchell, D. J. *Chem. Phys. Lett.* 1987, **139**, 219.
85. Attard, P.; Kjellander, R.; Mitchell, D. J.; Jönsson, B. *J. Chem. Phys.* 1988, **89**, 1664.

86. Podgornik, R. *Chem. Phys. Letters* 1988, **156**, 71–75.
87. Podgornik, R. *Chem. Phys. Letters* 1989, **144**, 503–508.
88. Belloni, L.; Spalla, O. *J. Chem. Phys.* 1997, **107**, 465.
89. Forsman, J.; Jönsson, B.; Åkesson, T. *J. Phys. Chem. B* 1998, **102**, 5082–5087.
90. Miklavic, S. J.; Chan, D. Y. C.; White, L. R.; Healy, T. W. *J. Phys. Chem.* 1994, **98**, 9022.
91. Attard, P.; Mitchell, D. J.; Ninham, B. W. *J. Chem. Phys.* 1988, **88**, 4987.
92. Attard, P.; Mitchell, D. J.; Ninham, B. W. *J. Chem. Phys.* 1988, **89**, 4358–4367.
93. Shaw, D. J. *Introduction to Colloid and Surface Chemistry;* Butterworths: London, 4th ed.; 1992.
94. Khan, A.; Fontell, K.; Lindman, B. *J. Colloid Interface Sci.* 1984, **101**, 193.
95. Khan, A.; Jönsson, B.; Wennerström, H. *J. Phys. Chem.* 1985, **89**, 5180.
96. Khan, A.; Fontell, K.; Lindman, B. *Coll. Surf.* 1984, **11**, 401.
97. Wennerström, H.; Khan, A.; Lindman, B. *Adv. Coll. Interface Sci.* 1991, **34**, 433.
98. Marra, J. *Biophys. J.* 1986, **50**, 815.
99. Kjellander, R.; Marčelja, S.; Pashley, R. M.; Quirk, J. P. *J. Chem. Phys.* 1990, **92**, 4399–4407.
100. Petrov, P.; Miklavic, S. J.; Nylander, T. *J. Phys. Chem.* 1994, **98**, 2602–2607.
101. Pashley, R. M. *J. Coll. Interface Sci.* 1984, **102**, 23–35.
102. Sjöström, L.; Åkesson, T.; Jönsson, B. *Ber. Bunsenges Phys. Chem.* 1996, **100**, 889.
103. Shklovskii, B. I. *Phys. Rev. E* 1999, **60**, 5802–5811.
104. Oosawa, F. *Polyelectrolytes;* Marcel Dekker: New York, 1971.
105. de la Cruz, M. O.; Belloni, L.; Delsanti, M.; Dalbiez, J. P.; Spalla, O.; Drifford, M. *J. Chem. Phys.* 1995, **103**, 5781–5791.
106. Wittmer, J.; Johner, A.; Joanny, J. F. *J. Phys II* 1995, **5**, 635–654.
107. Tang, J. X.; Wong, S.; Tran, P. T.; Janmey, P. A. *Ber. Bunsen-Ges. Phys. Chem.* 1996, **100**, 796.
108. Gosule, L. C.; Schellman, J. A. *Biochemistry* 1976, **259**, 333.
109. Wilson, R. W.; Bloomfield, V. A. *Biochemistry* 1979, **18**, 2192–2196.
110. Bloomfield, V. A.; Wilson, R. W.; Rau, D. C. *Biophys. Chem.* 1980, **11**, 339–343.
111. Widom, J.; Baldwin, R. L. *J. Mol. Biol.* 1980, **144**, 431–453.
112. Sen, D.; Cothers, D. M. *Biochemistry* 1986, **25**, 1495–1503.
113. Mel'nikov, S. M.; Sergeyev, V. G.; Yoshikawa, K. *J. Am. Chem. Soc.* 1995, **117**, 2401–2408.
114. Yoshikawa, K.; Takahashi, M.; Vasilvskaya, V. V.; Khokhlov, A. R. *Phys. Rev. Lett.* 1996, **76**, 3029–3031.
115. Kidoaki, S.; Yoshikawa, K. *Biophys. J.* 1996, **76**, 932–939.
116. Mel'nikov, S. M.; Khan, M. O.; Lindman, B.; Jönsson, B. *J. Am. Chem. Soc.* 1999, **121**, 1130–1136.
117. Ise, N.; Okubo, T.; Sugimura, M.; Ito, K.; Nolte, H. J. *J. Chem. Phys.* 1983, **78**, 536.
118. Ise, N.; Ito, K.; Okubo, T.; Dosho, S.; Sogami, I. *J. Am, Chem. Soc.* 1985, **107**, 8074–8077.
119. Tata, B. V. R.; Ise, N. *Phys. Rev. E* 2000, **61**, 983–985.
120. Chan, D. Y. C. *Phys. Rev. E* 2001, **00**, 00 in press.
121. Belloni, L. *J. Phys – Condens Mat.* 2000, **12**, R549–R587.

DIRECT SURFACE FORCE MEASUREMENT TECHNIQUES

PATRICK KÉKICHEFF
Institut Charles Sadron, C.N.R.S., UPR 022
6, rue Boussingault
67083 - Strasbourg Cedex (France)

Abstract. Techniques for the direct measurement of interactions between two bodies as a function of the distance of separation of their surfaces are reviewed. These include methods based on osmotic equilibria, force measurement devices using macroscopic solid surfaces (surface force balances, surface force apparatus, atomic force microscope, etc), and force measurement techniques on individual freely moving particles (total internal reflection microscopy, etc). The advantages and limitations of each of the techniques are discussed and illustrative results are presented.

Introduction

Surface forces play a crucial role in the stabilization of colloidal suspensions, aggregation, and phase behavior in surfactant systems, adhesion, lubrication, coating, and many other commercially important areas (detergency, food industry, cosmetics, drug delivery, enhanced oil recovery, biocompatibility, etc) [1]. Both repulsive and attractive forces can occur between surfaces, and they can extend in range from only a few molecular diameters to micrometers. Surface forces have been inferred from different techniques. The primary concern of this review is with the techniques for the *direct* measurement of the interactions between two bodies as a function of the distance of separation of their surfaces. Here "direct" is intended to signify that the force, or a function of the force, is measured at discrete separations of the surfaces. The space between the bodies may consist of a vacuum, gases or vapors, or pure liquids or solutions. The surface properties of the bodies may be modified by the adsorption of various molecular species, e.g. vapor, liquid condensates (as bridges of capillary-condensed liquid), ions, surface active agents and macromolecules. However, the main emphasis is on the techniques rather than on the results of measurement, regardless of the materials constituting the bodies (crystalline, amorphous, metallic, solid or liquid interfaces, etc) and of the composition of the separating gap.

C. Holm et al. (eds.), Electrostatic Effects in Soft Matter and Biophysics, 205–282.
© 2001 *Kluwer Academic Publishers. Printed in the Netherlands.*

The scope of this review is devoted to normal forces only (the measurement of tangential forces, e.g. friction, is not considered here) with the focus on equilibrium forces (dynamic measurements and hydrodynamic interactions are not addressed). The techniques that are to be discussed are capable of measuring forces somewhere in the range of 10^{-3} to 10^{-15} N and detecting separation (to better than 1 nm) between two surfaces over quite large distances (from several micrometers down to atomic or molecular contact). These ranges in force and separation are sufficient to study most regions of interest.

The interactions between particles dispersed in solution can be measured through four different methods. (1) Measure the spontaneous fluctuations of particle volume fraction, using the scattering of incident radiation (light, X-rays, or neutrons) $I(q)$ as a function of the scattering vector q. The dispersions must be transparent to light, or provide a good contrast for X-rays or neutrons. (2) Use membrane exchanges or apply an external field (gravity, centrifugation, etc) to measure the relation of osmotic pressure, Π, to volume fraction in the dispersions. This method is always applicable and requires little equipment but a fair amount of physical chemistry work ; it measures a behavior which is very close to that observed in applications, i.e. the resistance to forced deswelling. The interpretation of Π in terms of particle interactions requires a statistical treatment over the $\approx 10^{23}$ particles similar to the one carried out to extract interaction profiles from the scattered intensity curve $I(q)$ of the preceding method. (3) Use two macroscopic surfaces of the same nature as the particle surfaces, and measure the force-to-distance relation with a deflection method (force balance or spring device). The main difficulty is to make such surfaces or to stick the microscopic particles, without altering the surface chemistry of their interfaces, to the macroscopic surfaces of these devices. The resulting force law can be identified with the corresponding law for particles immersed in a solution, at constant geometry, material and solvent composition. In all these three methods a potential of mean force is measured ; it depends on the relative positions of all particles inside the solution, in the intervening gap, and on the macroscopic surfaces. (4) Monitor the Brownian movement of an individual particle freely suspended in solution around an equilibrium point where its potential energy is a local minimum. This method is the only one where the measured interaction arises directly from a single particle and not from the collective response averaged on an ensemble of particles. However, for submicronic particles, the direct measurement of pair interactions is generally not possible due to technical limitations in visualization, separation control and force evaluation.

The experimental techniques used to determine surface interaction forces / potentials fall into two complementary categories. Either the interaction profile is measured at constant chemical potential of all species (force balance and spring devices of (3)), or the response to a change of the chemical potential of solvent for example is investigated (all the other methods). As the description of the scattering intensity based methods can be found in many textbooks, the first group of these techniques will not be described, and the review will deal with the other three groups only.

1. Osmotic equilibria based techniques

Methods using osmotic equilibria [2-8] are based on exchanges of solvent molecules between a solution of solutes (neutral or charged particles, macromolecules, etc) and a reservoir. In living bodies, similar processes govern the transport of water and ions between cells and their environment. When a solution is placed in contact with a dialysis membrane it exchanges water, ions, surfactants, etc, with a reservoir containing large amounts of these species. At equilibrium, the chemical potentials of all species able to cross the membrane are identical in the solution and in the reservoir. Therefore it is possible to control the state of a sample through manipulation of these chemical potentials. This is particularly useful for colloidal systems where the adsorption of ions (H^+, OH^-, ...) or molecules (surfactants) from the aqueous solution onto the particles determines the surface charge, and therefore the stability of the dispersions. The same equilibria make it possible to extract water from the dispersion by adding to the reservoir high molar mass species which bind water. At equilibrium, the chemical potentials of water on either side of the membrane are equal, and therefore the osmotic pressure (see definition in next paragraph) of the sample equals that of the polymer in the reservoir. Through this procedure, dispersions may be concentrated at a set value of the osmotic pressure [9].

With such thermodynamic methods, named osmotic stress techniques, one imposes set values to all intensive variables, for instance the osmotic pressure, and one measures the response of the system, which is an extensive variable : the water content or the separation between particles. This remark underlines the encounter of inherent experimental limitations to be overcome :

- concerning the sample : has the isothermal thermodynamic equilibrium been reached together with all the relevant parameters correctly imposed ? Indeed the equilibria are slow since molecule exchanges and solvent transfer must take place ; therefore one must allow enough time for the species to diffuse and for the particle surfaces to reorganize.

- concerning the reservoir : how the non-infinity of its volume can be taken into account ? How is its osmotic pressure known ? This question may be particularly vexing when changes in pH and salinity may affect the osmotic pressure of the stressing polymer solution used.

- concerning the membrane : what type of suitable membrane to be used (material, pore size, resistance to high applied pressure, contamination, etc) ?

- concerning the exchanging solutes : what kinds of species are exchanged through the membranes (for instance, in the case of micellar solutions, surfactant molecules only can permeate) ?

In this section we review the different methods to measure the relation of osmotic pressure to volume fraction in the dispersions. Its determination enables phase diagrams to be built and interactions between particles to be inferred.

1.1. GENERAL DEFINITION OF THE OSMOTIC PRESSURE

Consider a solution comprised of a non-volatile solute and a solvent. When this solution is topped by a layer of pure solvent, the two volumes will progressively mix by diffusion and convection processes until the solute concentration is uniform in the whole system. If the two phases I (pure solvent) and II (solution of a solute dispersed in the solvent) are separated by a semi-permeable membrane, that is a membrane which is permeable to solvent but not to solute preventing the two liquids from mixing, one observes that the volume of phase I diminishes while the volume of phase II increases. This shows a situation out of equilibrium where solvent flows from the pure phase into the solution. By diluting the phase II (the dispersion), the solvent tends to equalize its chemical potential on both sides of the membrane. *Any transport of solvent across a membrane, induced by a concentration gradient is called an* <u>*osmosis*</u> (from the Greek word *ôsmos*, meaning to push or thrust). In the preceding experiment the volume occupied by the phase I can be kept constant if one applies a pressure P' (for instance by means of a movable piston ; see Fig.1). Similarly the volume of phase II will be held constant if one applies a pressure $P' + \Pi$: at equilibrium the excess pressure Π necessary to stop the influx of solvent through the membrane is called the *osmotic pressure* of the dispersion. This is why the solutions used in intra-vein injection, such as physiological serum, must have the osmotic pressure of the blood plasma (about 8 atm at 37 °C ; such a high value is due to the presence of chloride and sodium ions). Indeed, in a hypotonic environment ($P' < \Pi$) red globules swell with water and explode whereas in a hypertonic medium ($P' > \Pi$) they get dehydrated.

Another way of understanding this general definition of osmotic pressure is to consider the arrangement in Fig.2. :

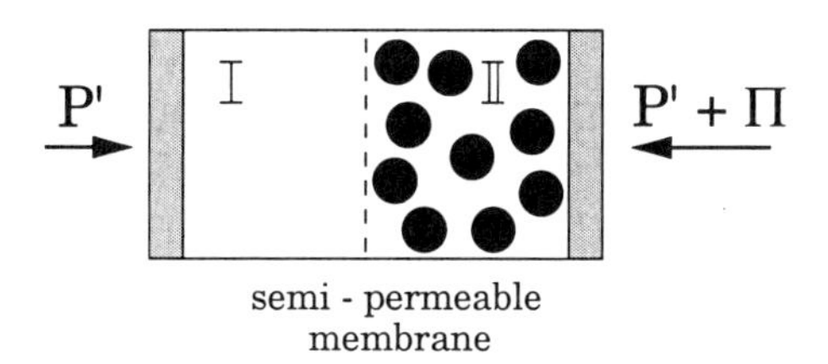

Figure 1. The membrane is permeable to solvent but not to solute. The solvent may be prevented from entering the solution by application of an excess pressure Π, the osmotic pressure of the solution.

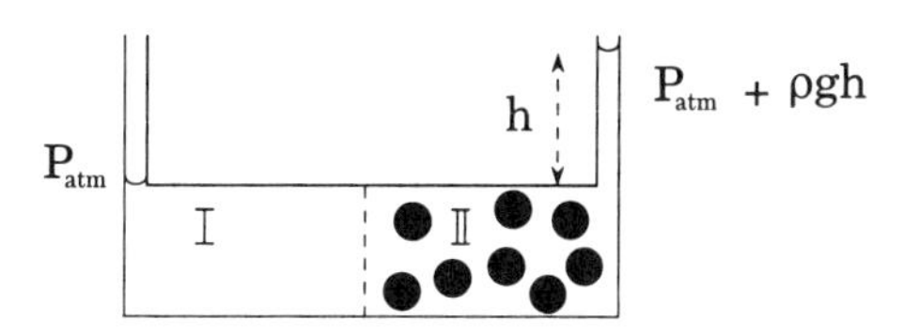

Figure 2. Capillary rise. At equilibrium the hydrostatic pressure developed in the capillary matches the osmotic pressure of the solution and opposes the flow of solvent through the membrane.

Passage of solvent from side *I* to side *II* causes the liquid level of side *II* to rise in the capillary tube, setting up a hydrostatic pressure which *opposes* the flow of liquid through the membrane. When this pressure becomes sufficiently high, that is when the hydrostatic pressure in the column of the solution matches the osmotic pressure, further net flow is prevented and equilibrium is reached. The equilibrium osmotic pressure of the solution is thus given by the difference in pressures between the two sides of the membrane :

$$\Pi = \rho g h$$

where ρ is the mass density of the solution and h the capillary rise[1]. In thermodynamic terms the osmotic pressure counteracts the difference in solvent activity (see § 1-2).

The existence of an osmotic pressure in any solvent-based dispersion of solutes yields to two main effects. First, one can note from the very definition of the osmotic pressure that any dispersion in contact with pure solvent tends to increase its volume : a dispersion possesses a "swelling" pressure. Conversely, upon a forced deswelling process, the osmotic pressure

[1] The complication inherent to this arrangement is that the entry of solvent into the solution results in its dilution, so it is more difficult to treat than the arrangement in Fig.1 in which there is no flow and the concentration remains unchanged. In practice, to prevent too much dilution of the solution as a result of the solvent flow into it, the column in which the pressure head develops is generally of a very narrow diameter.

of dispersions is measured as a resistance to a decrease in the volume available to the particles ; this resistance originates from an increase of energy (interparticle forces) or from a loss of entropy (configurations) or from both.

Classically the osmotic pressure has been mainly used to determine particle mass (macromolecules, proteins, etc) by extrapolation to zero concentration. However, there is a wealth of additional information to be gained by carrying out measurements at higher concentrations. One can obtain insight into the behavior of concentrated dispersions and thereby their stabilization mechanism. In particular, the factors controlling the properties of the solutions can be evidenced and the nature of the interactions between particles can be extracted.

1.2. THERMODYNAMIC TREATMENT

The thermodynamic treatment of osmosis depends on noting that, at equilibrium, the chemical potential of the solvent and of any species crossing must be the same on each side of the membrane. Consider first a two-component system, consisting of solvent and a single non-electrolytic solute. We shall designate with a single prime (e.g. μ') the thermodynamic parameters on side I of the membrane, and by unprimed symbols (μ) for the solution on side II. On the pure solvent side the chemical potential of the solvent, which is at pressure P' (Fig.1), is $\mu_{\text{solvent}}(T, P')$. On the solution side, the chemical potential is lowered by the presence of the solute at a particle concentration ρ_p but is raised on account of the greater pressure P $(= P' + \Pi)$ that the solution experiences. At equilibrium the two are equal :

$$\mu_{\text{solvent}}(T, P') = \mu_{\text{solvent}}(T, P, \rho_p)$$

The presence of solute is taken into account in the normal way via the activity :

$$\mu_{\text{solvent}}(T, P, \rho_p) = \mu_{\text{solvent}}(T, P) + RT \ln(a_{\text{solvent}})$$

The chemical potential of the pure solvent at pressure $P = P' + \Pi$ is related to the chemical potential at pressure P' by the relation

$$\mu_{\text{solvent}}(T, P' + \Pi) = \mu_{\text{solvent}}(T, P') + \int_{P'}^{P'+\Pi} \left(\frac{\partial \mu_{\text{solvent}}}{\partial P}\right)_T dP \qquad (1)$$

The partial derivative $\left(\frac{\partial \mu_{\text{solvent}}}{\partial P}\right)_T$ is equal to the partial molar volume of the solvent $\bar{v}_{\text{solvent}}$ (for pure water this is 1 liter for 55.5 moles). This quantity is virtually independent of the pressure, certainly for the small pressure changes encountered in osmotic pressure work, and may therefore be taken outside the integral sign[2]. The above equation becomes

$$\mu_{\text{solvent}}(T, P' + \Pi) = \mu_{\text{solvent}}(T, P') + \bar{v}_{\text{solvent}}\,\Pi$$

and the condition for equilibrium gives

$$\Pi = -\frac{RT}{\bar{v}_{\text{solvent}}}\ln a_{\text{solvent}} \tag{2}$$

This equation shows that the osmotic pressure is a direct measure of the activity of the solvent in the solution.

(i) ideal solution (dilute solution) :

For an ideal solution (dilute solution), the activity a_{solvent} may be replaced by the mole fraction, i.e. $a_{\text{solvent}} \approx x_{\text{solvent}} \approx 1 - x_p$ and $\ln(1 - x_p) \approx -x_p$ where x_p is the mole fraction of the particles. When the solution is dilute $x_p \approx N_p/N_{\text{solvent}}$. Moreover, because $N_{\text{solvent}}\bar{v}_{\text{solvent}} = V$ the total volume of the solvent, eqn.2 simplifies to the van't Hoff equation :

$$\Pi V = N_p RT \quad \text{or} \quad \Pi = \rho_p k_B T \tag{3}$$

where ρ_p is the number density in particles of the solution. The van't Hoff equation applies to very dilute solutions in which the behavior can be considered as ideal. It shows that the osmotic pressure of a dilute solution, consisting of solvent and a single solute, is equal to the pressure of the perfect gas of solute.

Note that one of the most common applications of osmometry has been to the measurement of molar masses of macromolecules (proteins and synthetic polymers). As these large molecules dissolve to produce solutions that are far from ideal, it is assumed that the van't Hoff equation is only the first term of a virial-like expansion :

$$\Pi = \rho_p k_B T\,(1 + B\rho_p + \ldots)$$

The additional terms take the non-ideality into account. The osmotic pressure is measured at a series of concentrations (it is $c_p = \rho_p M_p$ which is the quantity known experimentally), and a plot of Π/ρ_p against ρ_p is used to find the molar mass of the dissolved molecules.

(ii) real solution (concentrated solution) :

[2] This is equivalent to say that the liquids are almost incompressible in the pressure range considered.

At the thermodynamic equilibrium the Gibbs-Duhem equation expresses the fact that the free energy is minimum :

$$SdT - VdP + N_{\text{solute}}\, d\mu_{\text{solute}} + N_{\text{solvent}}\, d\mu_{\text{solvent}} = 0$$

where S is the entropy of the system. At constant temperature, this equation may be written as :

$$dP = \frac{N_p}{V}\, d\mu_p + \frac{N_{\text{solvent}}}{V}\, d\mu_{\text{solvent}} = \rho_p\, d\mu_p + \rho_{\text{solvent}}\, d\mu_{\text{solvent}}$$

The osmotic pressure is defined as the difference between the absolute pressures in the solution P and in the reservoir P' ($\Pi = P - P'$) and thus $d\Pi = dP$. In addition, the reservoir contains only pure solvent and at the osmotic equilibrium $d\mu_{\text{solvent}} = 0$. One then obtains :

$$d\Pi = \rho_p\, d\mu_p$$

which can also be written as :

$$\left(\frac{\partial \beta\Pi}{\partial \rho_p}\right)_{T,\ \mu_{\text{solvent}}} = \rho_p \left(\frac{\partial \beta\mu_p}{\partial \rho_p}\right)_{T,\ \mu_{\text{solvent}}} \tag{4}$$

by noting $\beta = 1/k_B T$. Introducing the statistical thermodynamics definition of the partial structure factor $S_{\alpha\eta}(0)$ for a mixture of species α, η, ... in the limit of vanishing scattering vector ($Q \to 0$) :

$$S_{\alpha\eta}(0) = \frac{1}{\sqrt{N_\alpha N_\eta}} \left(\frac{\partial N_\alpha}{\partial \beta\mu_\eta}\right)_{T,\ V,\ \mu_\lambda} \tag{5}$$

the partial structure factor for the solute ($\alpha = \eta =$ solute noted p) is

$$S_{pp}(0) = \frac{1}{N_p} \left(\frac{\partial N_p}{\partial \beta\mu_p}\right)_{T,\ V,\ \mu_{\text{solvent}}}$$

which can be written in terms of density as

$$S_{pp}(0) = \frac{1}{\rho_p} \left(\frac{\partial \rho_p}{\partial \beta\mu_p}\right)_{T,\ V,\ \mu_{\text{solvent}}}$$

Because the chemical potential of pure solvent is constant, one gets :

$$\left(\frac{\partial \beta\mu_p}{\partial \rho_p}\right)_{T,\ V,\ \mu_{\text{solvent}}} = \frac{1}{\rho_p\, S_{pp}(0)}$$

Finally, eqn.4 can be written as

$$\left(\frac{\partial \beta \Pi}{\partial \rho_p}\right)_{T,\ \mu_{\text{solvent}}} = \frac{1}{S_{pp}(0)} \tag{6}$$

showing that the value $S_{pp}(0)$ of the structure factor in the limit $Q \to 0$ is closely related to the isothermal osmotic compressibility of the system ($\chi_T = \frac{1}{\rho}\left(\frac{\partial \rho}{\partial \Pi}\right)_T$). Note that this relation is only valid for monodisperse particles and not in a multi-component system [10]. In the absence of interaction, the structure factor is equal to 1, and eqn.6 simplifies to the ideal gas law (eqn.3). One may thus interpret the osmotic pressure of a concentrated solution (or not), consisting of solvent and a single solute, as the pressure of the monodisperse gas of solute particles. This result can be generalized to the case of a solvent and several solutes that cannot cross through the membrane : the osmotic pressure of such a system is the pressure of a polydisperse gas of the solutes[3]. In eqn.6 the solvent does not appear explicitly : its contribution is contained in the structure factors where the correlations between the particles due to the interactions are mediated by the solvent.

Coming back to the van't Hoff's limiting law for dilute solutions (perfect gas equation of state ; eqn.3) the osmotic pressure appears as a colligative property of a solution as it measures the *number* of species[4], whatever is their mass :

$$\Pi = \rho_p k_B T \quad \text{i.e.} \quad \Pi = \frac{c_p}{M_p} RT$$

where c_p is the concentration in particles of molecular weight M_p. A 10 g/l (1 wt.%) solution raises an osmotic pressure of 2500 Pa if the solute has a 10 000 g molar mass, but only 25 Pa (that is 2.5 mm of water) for a 10^6 g/mol solute. Particles of large mass are not measurable and in a polydisperse system

$$\Pi = RT \sum_i \frac{c_i}{M_i}$$

[3] What we are discussing here should, strictly speaking, be called the *colloid osmotic pressure*. The *total osmotic pressure* of a solution is defined as the osmotic pressure in which the membrane separating solution from solvent is impermeable to all solutes, regardless of size (i.e., it might be a vapor barrier if only the solvent is volatile). This type of osmotic pressure, however, is rarely measured. The term *osmotic pressure* here will always be used to signify only colloid osmotic pressure.

[4] So far we have assumed no chemical interaction between solutes and, for the present, we have excluded charged particles, which give rise to the Donnan effect described in the next section. Note that association between solute molecules, or their dissociation, will affect the number of solute particles, and hence the osmotic pressure.

only the lightest particles contribute the most to the osmotic pressure (this is the converse in scattering experiments where the largest species are the most important ones).

This remark is illustrated in Fig.3 for a latex colloidal dispersion for which the measured pressures are orders of magnitude larger than expected [9]. It was shown that the surface of the latices may release charged polymers of low mass in the bulk. Intensive cleaning through dialysis or ion exchange resin cannot eliminate their presence in the solution as the release is an equilibrium between surface and bulk. These additional particles are not small enough to cross a dialysis membrane and remain in the solution between the colloids ; as a consequence they contribute to and even dominate the total number of particles as can be observed in osmotic pressure experiments (Fig.3). On the other hand they are hardly evidenced in any scattering experiments.

(iii) charged systems :

Electrostatic charges on particles and the presence of ions modify the thermodynamic behavior of solutions in twofold. The requirement that solutions of macro-ions must be electrically neutral implies that solutions containing only a single kind of macro-ion must always contain a sufficient number of small ions to neutralize the macro-ion charge. In addition, electrostatic interactions between the charges on the particles and those of other particles and small ions in the vicinity lead to a non-ideality behavior of the solutions. In this section we shall discuss how the osmotic pressure is effected by the presence of ions and charges.

Let us consider charged particles or macro-ions dispersed in a solution at a number density ρ_p. Even in the absence of added salt the solution contains small ions also, since the solution as a whole must be neutral. To keep the treatment simple we shall suppose that the counter-ions present are all univalent ions. Extension to ions of higher charge is straightforward. Notations will be as follows. As before unprimed symbols are used to designate the solution containing particles, whereas primed symbols indicate the side of the membrane free from macro-ions. The charge Z_p of the particles (expressed in electronic charges) is supposed to be negative. In the presence of added salt (number density ρ_s) the number density of counter-ions is $\rho_+ = \rho_s + Z_p\rho_p$, and that of co-ions is $\rho_- = \rho_s$. The semi-permeable membrane which separates this solution from the reservoir allows solvent and all ions but the particles to permeate. In the reservoir the number density of counter-ions and co-ions will be equal by electroneutrality condition for a monovalent electrolyte : $\rho'_+ = \rho'_- = \rho'_s$.

We are interested in answering the following question. Knowing both the ionic strength in the reservoir and the particle concentration in the solution what are the salt content and the osmotic pressure of the solution ?

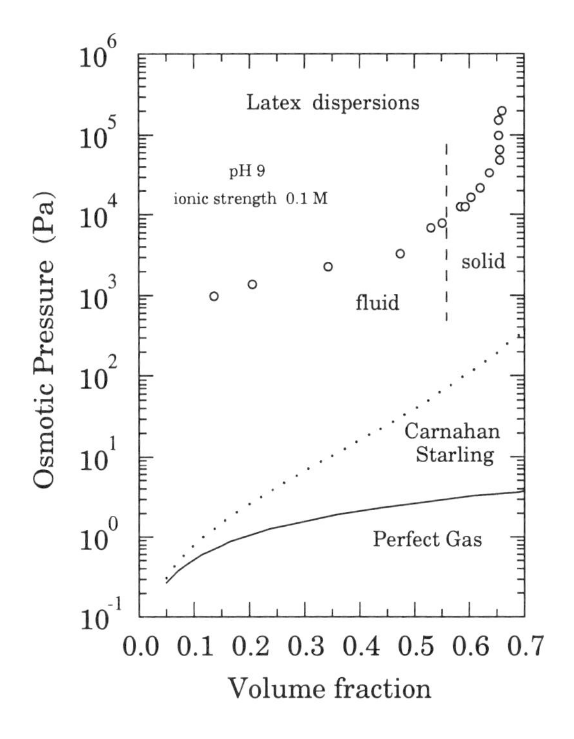

Figure 3. Osmotic pressure of aqueous latex dispersions (particle diameter 115 nm) [9]. In the regime of moderate volume fractions ($0.1 < \phi < 0.5$) the dispersion is expected to behave as a moderately concentrated gas, because the repulsions are short range (no more than 1 nm) and there is plenty of room available (the separation between particles is at least 50 nm) [9]. However, the measured pressures are four to three orders of magnitude higher than the pressure calculated assuming a perfect gas law (solid line) or through the Carnahan-Starling equation that describes a system of particles with excluded volume (dotted line). The experiments indicate that the pressure originates from other species besides the particles. These species must be numerous, since they produce a high pressure proportional to their number density at a low volume fraction (the slope of the compression curve is the same as that of a perfect gas), and of a small molar mass, but still large enough to be excluded by the pores of the dialysis bag. It was shown that the contribution to the pressure was indeed dominated by the presence of small macromolecules with molar mass around 3000 g/mol, that were released by the latices into the solution [9].

The condition for equilibrium is the equality of the chemical potential for the salt in both compartments : $\mu_s = \mu'_s$ (with $\mu_s = \mu_+ + \mu_-$ and $\mu'_s = \mu'_+ + \mu'_-$). Note that this condition applies only to *components* and not to species which cannot be transferred independently from one phase to another : single ion species cannot be transferred from one side to the other side of the membrane without having them accompanied by ions of opposite charge to maintain neutrality. For monovalent electrolytes ions permeate only by neutral pairs. As before the equation is most conveniently transformed into a relation between activities :

$$\mu_s(T, P' + \Pi) = \mu_s^0(T, P') + RT \ln a_s + \int_{P'}^{P'+\Pi} \bar{v}_{\text{electrolyte}}\, dP$$
$$\mu_s'(T, P') = \mu_s'^0(T, P') + RT \ln a_s'$$

The integral in the first equation can be neglected for any ordinary electrolyte[5], and the equality between these two equations gives :

$$a_s' = a_s \tag{7}$$

that is

$$\rho_s'^2 \gamma_+' \gamma_-' = \rho_s \left(\rho_s + Z_p \rho_p\right) \gamma_+ \gamma_- \tag{8}$$

where $\gamma_\pm$ are the ion activity coefficients.

For ideal solutions the ion activity coefficients can be taken equal to 1 and eqn.8, $\rho_s'^2 = \rho_s \left(\rho_s + Z_p \rho_p\right)$, shows that the salt content must be unequally distributed on the two sides of the membrane :

$$\rho_s < \rho_s' < \rho_s + Z_p \rho_p \tag{9}$$

This is the so-called Donnan, salt exclusion effect. It is also called negative adsorption in colloidal systems : at a concentration of a 1 wt.%, colloidal particles of molecular weight 10^6 carrying a charge of 250 ($Z_p \rho_p = 2.5 \times 10^{-3}$ mol/l) equilibrate with three times less salt than the reservoir content of 1 mM added 1:1 electrolyte. This relative concentration difference is levelled off at sufficiently high ionic strength (at 0.01 in the reservoir the salinity in the solution would be only 10% smaller). Eqn.8 shows that this is a general result : if $Z_p \rho_p$ is much smaller than ρ_s, the ion activities on the two sides become equal. This is a very important conclusion, for it is often desirable to suppress the Donnan effect, and one can note that this goal can be achieved in the presence of a sufficiently high concentration of neutral salt.

So far we examined the consequence of the equilibrium condition $\mu_s = \mu_s'$ for the salt. Taking up the equality of the chemical potential of the

[5] This integral amounts to $\approx \bar{v}_{\text{electrolyte}} \Pi \approx -RT(\bar{v}_{\text{electrolyte}}/\bar{v}_{\text{solvent}}) \ln a_{\text{solvent}}$ (by using eqn.2). In a standard solution, there is much more solvent molecules than solute molecules ($N_{\text{solvent}} \gg N_{\text{solute}}$) and $a_{\text{solvent}} \approx 1$ whereas $a_s \ll 1$. The term containing $\ln a_{\text{solvent}}$ is negligible here (even though in dilute regime one had to investigate its asymptotic behavior precisely). More precisely, $\ln a_{\text{solvent}} \approx -x_{\text{solute}} = -N_{\text{ions + particles}}/N_{\text{solvent}}$ is a negligible quantity as long as the total concentration of added ions and particles is much smaller than 55 Mol. This term represents a few percent change in a_s. Thus we may neglect the effect of the osmotic pressure on the chemical potential of the salt.

solvent on both sides of the membrane one is lead to the osmotic pressure relation given by eqn.2. The net osmotic pressure Π of the solution (against the electrolyte reservoir via a membrane permeable to the ions) can be *formally* decomposed into the difference between the pressures of the solution P and of the reservoir P' (P and P' represent the pressures of the gas of solute ; experimentally, there are themselves osmotic pressures against *pure* water via a membrane permeable to solvent only) :

$$\Pi = P_{\text{gas}}(\text{particles } \rho_p + \text{salt } \rho_s) - P_{\text{gas}}(\text{salt } \rho'_s)$$

For ideal solutions this equation becomes :

$$\Pi = k_B T(\rho_p + \rho_+ + \rho_-) - k_B T(\rho'_+ + \rho'_-)$$
$$\text{i.e.} \quad \Pi = k_B T(\rho_p + Z_p\rho_p + 2\rho_s - 2\rho'_s) \tag{10}$$

Note that eqns.10 and 8 (when the ion activity coefficients $\gamma_\pm$ are taken to 1) are approximations of Debye-Hückel like-type. Two limiting situations can be investigated. In excess of salt or at low particle concentration, $Z_p\rho_p$ is much smaller than ρ'_s and $\rho_s \approx \rho'_s - Z_p\rho_p/2$. Therefore eqn.10 reduces, as for non-electrolyte solute, to van't Hoff's limiting law $\Pi \approx \rho_p k_B T + {Z_p}^2{\rho_p}^2/4\rho'_s \approx \rho_p k_B T$ showing that the colloid particles contribute alone to the value of the osmotic pressure. This means that when one particle with its Z_p counter-ions is inserted inside the solution, $Z_p/2$ pairs of salt ions leave the solution through the membrane and the net pressure is increased by only one $k_B T/V$. The other limit is obtained when the solution is concentrated or when there is little salt added. For these salt-free cases ($Z_p\rho_p \gg \rho'_s$), the salinity inside the solution is negligibly small ($\rho_s \approx {\rho'_s}^2/Z_p\rho_p \ll \rho'_s$) and all counter-ions contribute to the perfect gas pressure :

$$\Pi = \rho_p k_B T(1 + Z_p) \quad \text{that is} \quad \Pi = k_B T(\rho_p + \rho_{\text{counter-ions}})$$

This result emphasizes the fact that osmotic pressure counts solute particles. The macro-ion cannot pass through the semi-permeable membrane. In the absence of added salt, its counter-ions will not permeate either since the electroneutrality of the solution must be maintained. Therefore the equilibrium pressure is that associated with $(1+Z_p)$ particles. The presence of increasing salinity leads to a levelling off of the ion concentration on the two sides of the membrane. The effect of the charge on the macro-ion is essentially swamped out with increasing electrolyte.

For non-ideal solutions of electrolyte solutes the real situation is richer than the one encountered with neutral particles in the absence of any exchangeable ions through the membrane. The complexity of behaviors imposes care in the analysis of the experimental osmotic pressure profiles and different theoretical approaches have been proposed [11].

1.3. TECHNIQUES TO MEASURE THE OSMOTIC PRESSURE

(i) Capillary rise :

The method already introduced in § 1-1 is the simplest measurement of osmotic pressure (see Fig.2). All variations in apparatus design involve an inner solution compartment with a relatively large opening at the membrane end and a capillary at the small end (one example is given in [12]). The entire solution chamber is then immersed in a tube, container, or reservoir containing only the pure solvent. Once assembled, the entire assembly is placed in a constant temperature bath for equilibration. A popular set up uses a dialysis bag. The solution to be investigated is enclosed in a bag made of some semi-permeable material. The sealed bag is then placed in a quantity of the solvent. The ensemble — bag topped with a long capillary — resembles a derrick. As the solute particles cannot permeate the membrane the concentration in particles, ρ_p, is known. The thermodynamic equilibrium is reached when the level of the liquid column no longer rises. The height, h, of the capillary rise is measured precisely with a catetometer. The correct pressure required to attain equilibrium may be obtained by observing the rate of rise (or fall) of the liquid level in the capillary at various pressures and determining graphically the pressure required to make this rate zero. At equilibrium the osmotic pressure is given by $\Pi = \Pi(\rho_p) = \rho g h$, where ρ is the mass density of the solution (not to be confused with the number density in particles, ρ_p). Note that capillaries are used to minimize the dilution effects, otherwise corrections for capillary rise must be taken into account. An upper practical limit for h is 50 cm, otherwise enormous volumes must be used as reservoirs. As a consequence the upper limit for the measurement of the pressure is 5000 Pa, while pressures as low as 10 Pa may be measured. Good thermostating is an extremely important practical feature, in addition to the theoretical requirement of isothermal conditions in osmometry experiments. Indeed, the set up consists of a large liquid volume attached to a capillary and therefore has the characteristics of a liquid thermometer : the location of the meniscus is quite sensitive to temperature fluctuations.

Other problems such as poor solvent drainage in the capillaries, adsorption of solute on the membrane, partial permeation of solute through the membrane, gas dissolution changing the pH, etc, can interfere with the attainment of a true osmotic equilibrium. The time taken to reach equilibrium increases with the amount of solvent that must be displaced. The establishment of osmotic equilibrium may take days or weeks.

The membrane is the source of most of the difficulties in osmometry. There is no general way to select a membrane material that will be permeable to one component and impermeable to another for any conceivable combination of chemicals. Very high molecular weight components are generally more easily retained. The membrane must be sufficiently thin to permit equilibration at a reasonable rate. At the same time, it must be strong enough to withstand the considerable pressure differences that may exist across it. Membranes must be free from minute imperfections that would constitute a leak between the two compartments : such a leak would invalidate the experimental results. One peculiarity of membranes is their tendency to display what is known as an asymmetry pressure ; that is, an equilibrium pressure difference may exist across a membrane even when there is solvent on both sides. This must be measured and subtracted as a blank correction.

Membranes may use different mechanisms to retain the solute. Some solutes, especially in the colloidal size range, are retained by a sieve effect : the particles are too large to pass through the pores in the membrane material. Another possibility is a mechanism by which the membrane displays selective solubility, that is the membrane dissolves the solvent but not the solute. In this way the solvent can pass through the membrane, while the solute is retained. An analogous mechanism for charged particles may arise by the membrane repelling (and thus retaining) particles of one particular charge.

A large variety of membrane materials have been used in osmotic pressure experiments, including various forms of cellophane and animal membranes as the most common ones (cellulose ester or regenerated cellulose for organic solvent systems ; cellulose acetate or nitrocellulose (collodium) for aqueous solutions), and various other polymers, such as polyvinyl alcohol, polyurethane, polytrifluorochloroethylene. Because contaminants, such as surfactant molecules, or cellulose fragments, may be released from the membranes, dialysis bags must be conditioned in water at the appropriate ionic strength and pH prior to use.

(ii) Membrane osmometer :

The time required to reach equilibrium is much reduced through the use of commercially available automatic membrane osmometers. If, for example, the capillary height in the solution chamber increases because solvent permeates from the solvent chamber, this is immediately compensated by the application, via a servomechanism, of a pressure on the solution chamber, such that the capillary height remains the same. Since this method involves the transport of only very small amounts of liquid, equilibrium is reached after only a few minutes (< 30 min).

An osmometer is made of two compartments, separated by a semipermeable membrane, which is rigid and cannot be deformed. One compartment has a small volume (≈ 0.05 cm^3) and contains the pure solvent while the other one of volume a few times larger is filled with the solution to be investigated. As soon as the two liquids are in contact some solvent molecules will cross the membrane in order to dilute the dispersion and to equalize the chemical potential of solvent on either side of the membrane. However, the volume of the cell (that of both chambers and the membrane) is constant and the rigid membrane cannot deform. Therefore the crossing of a few solvent molecules induces a depression in the solvent chamber. This negative pressure is measured by the gauge, and its absolute value is equal to the osmotic pressure of the solution. As the solvent volume exchanged is small the equilibrium is reached rapidly, in a few minutes. With conventional osmometers the osmotic pressure of a dispersion can be measured in the range 0.001 to 0.04 atm. The accuracy is not great (> 1 Pa) as the pressure gauge is very sensitive to variations in temperature, to rapid changes in external pressure (air cabinets, etc), and to external vibrations.

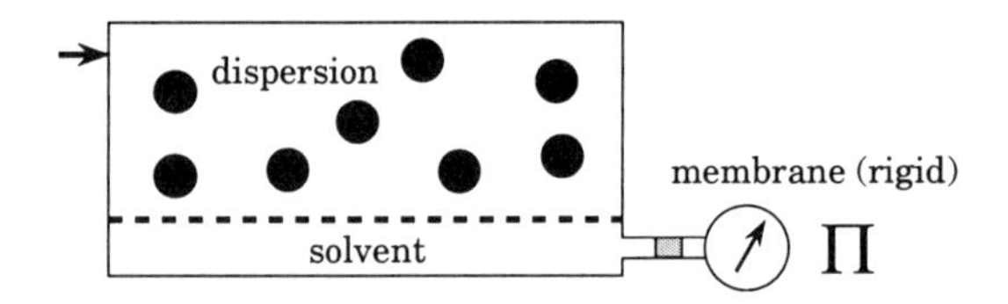

Figure 4. Membrane osmometer.

In the case of a dispersion containing several solutes where only one cannot cross the membrane (for instance a dispersion of particles and salt), it is necessary to modify the nature of the liquid filling the lower compartment, for example by injecting first a salted aqueous solution that was equilibrated osmotically with the dispersion. Equilibrium is obtained when a stable value is observed : the chemical potential of salt in both chambers is the same. Once this has been achieved the measurement of the osmotic pressure of the dispersion relatively to this new "zero" may begin.

(iii) Osmotic stress by dialysis :

Osmotic stress technique enables a controlled removal of water from a macromolecular, a surfactant, or a colloidal system, while keeping the properties of boundary water under thermodynamically well-defined conditions [3-7]. The basis of the method is to let the system of interest come to equilibrium with a polymer solution of known osmotic pressure. Because the chemical potential of solvent in a polymer solution can be tuned by changing the concentration in dissolved macromolecules, the technique provides the possibility also of changing the chemical potential of the solvent in the investigated system ; therefore through this procedure one is able to impose a set value of the osmotic pressure. Coupling the technique with scattering / diffraction of beams (X-rays, neutrons, or light) allows to determine the osmotic pressure profile as a function of the average separation between particles [9, 13-24]. Many systems will form ordered arrays during osmotic equilibration, enabling one to determine the structural consequences of solvent removal. One is thus able to measure directly not only intermolecular separation, but also the chemical potentials or works of condensing the array.

The experimental procedure consists merely of enclosing the solution of interest in a bag made of some semi-permeable material. The sealed bag is then placed in a reservoir to where are added polymers which bind water. High molar mass species are used because the radii of gyration of the macromolecules are the order of tens of nanometers and cannot enter the pores of the dialysis membranes. If the reservoir volume can be considered as infinitely large, the concentration ρ_{polymer} at the thermodynamic equilibrium in the stressing solution is known ($\rho_{\text{polymer}}^{\text{equilibrium}} = \rho_{\text{polymer}}^{\text{injected}}$) and the equation of state is $\rho_p(\Pi)$ (rather than $\Pi(\rho_p)$). When the reservoir is not too large, solvent exchanges may affect ρ_{polymer} and the osmotic pressure Π is not strictly imposed. In practice it is essential that components of the stressing solution be in sufficient excess or replaced often enough to ensure known fixed activity during equilibration. When equilibrium is attained, the macromolecule concentration in the reservoir is determined through size exclusion chromatography (SEC) or organic carbon analysis (TOC) ; from this concentration the osmotic pressure is calculated and thus the osmotic pressure of the sample is known as it equals that of the polymer in the reservoir. An aliquot of the bag solution is also extracted and the volume fraction of the dispersion is determined through drying, titration, etc. From these measurements one point of the equation of state is determined : the osmotic pressure as a function of the volume fraction of the dispersion can thus be obtained by varying the concentration of the stressing polymer solution. Note that one may use very small reservoir volumes, even much smaller than the sample volume to be investigated ! In such a situation, this is the polymer solution which takes up the osmotic pressure imposed by the sample solution (now the equation of state is $\Pi(\rho_p)$). This method enables the osmotic pressure plateaus of phase transitions to be determined precisely (for an illustration see [21] where equilibria between lamellar phases L_α/L'_α have been investigated).

Stressing polymer solutions enable to cover the range from 10-100 Pa up to 10-100 atm in osmotic pressures (higher values are difficult to obtain as concentrated polymer solutions ($>$ 40 wt.%) become too viscous). One delicate aspect of the osmotic stress technique is getting accurate osmotic pressures of the polymer solutions. Variations in polymer sample from different batches are particularly vexing at low polymer concentration :

the presence of small diffusible species (such as antioxidant additives found with commercial polyethylene glycols for preservation purposes) or different molecular weight distributions are of particular concern at polymer osmotic pressures of less than 1 atm. Dialysis of polymer sample is an almost unavoidable step before use. At higher concentrations and thus in the range of large osmotic pressures these concerns are less dramatic, since the macromolecules overlap to the extent that the sizes of individual macromolecules become irrelevant. Furthermore, the chosen stressing polymer must give the same osmotic pressure regardless of ions, pH, and temperature. Dextran polymers fulfill this requirement. These branched D-glucopyranose —$(C_6H_{10}O_5)_n$— polymers are synthesized from sucrose by Lactobacillae bacteria. Their osmotic pressures have been measured by different methods and are reproduced in Fig.5 :

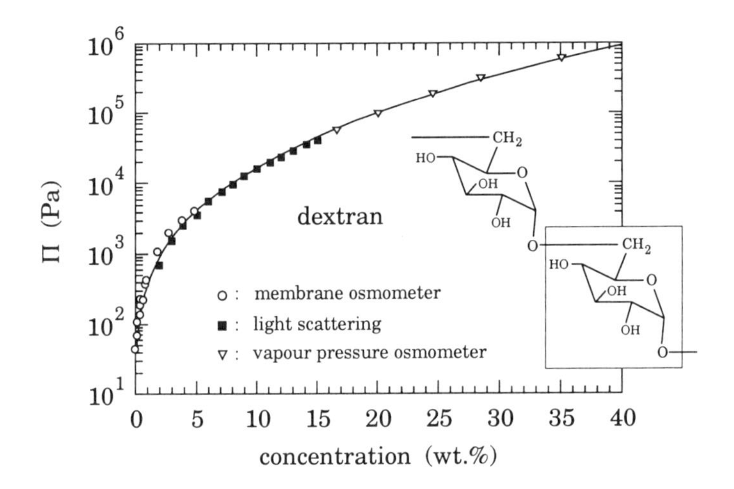

Figure 5. Osmotic pressure of dextran solutions. At low concentrations ($c < 15$ wt.%) the osmotic pressures of dextran solutions (from Fluka, grade T110, with a distribution of molar mass M_n = 70 000 and weight average M_w =150 000) were determined with a Knauer membrane osmometer for pressures up to 4000 Pa ($0.1 < c < 5\%$), at higher pressures from the intensity of scattered light, which gives the osmotic compressibility ($0.2 < c < 15\%$) [23], and in the range 16% to 35% using a vapor pressure osmometer [9]. The results can be fitted (solid line) by the following expressions for pressure Π (in Pa) versus concentration c in weight percent : $\log \Pi = 0.385 + 2.185\, c^{0.2436}$ for $c \leq 10\%$ and $\log \Pi = 0.872 + 1.657\, c^{0.3048}$ for $c \geq 10\%$. The values calculated from this fit also match well the osmotic pressures measured at high pressures using concentrated dextran solutions of molar mass 500 000 from other sources [7, 25] ; the above expressions represent well the osmotic pressures of these dextran solutions in the whole range of concentrations (0.1 to 40 wt.%).

Patience is required to determine a whole curve of osmotic pressure as a function of the volume fraction in solute particles. At each higher polymer concentration, the dispersed system must come to equilibrium with water of lower chemical activity. Attainment of equilibrium is slow (days or weeks) since solvent transfer must take place. During these long lasting experiments caution must be taken as bacteria may develop, damage the bag membrane and eventually induce a leak. To avoid their development, often activated by sun light and heat (a dextran solution may then become opaque) bactericides such as quaternary ammonium based molecules may be added to the stressing polymer solution. However complications in the studied system will inevitably arise. The osmotic pressure may be modified and adsorption onto the membrane or onto the particle surfaces may occur. Other possibilities would involve osmotic experiments to be performed under inert environment (argon, etc). In all cases it is recommended to carry out experiments no longer than during one month.

Mechanically, the polymer solution exerts an osmotic pressure on the dispersed system. At the new equilibrium, and for each polymer concentration, the repulsion between particles creates an osmotic pressure which must equal that of the polymer solution. This is strictly true if the mechanical resistance of the dialysis bags is not a concern upon the swelling or the deswelling processes of the dispersed system. When the initial solution of solute is equilibrated at a lower pressure it takes up large amounts of solvent ; the dispersion expands so much that the dialysis bag may be mechanically stretched (and eventually may rupture owing to the internal pressure buildup developed by the solvent imbibed). Then the pressure inside the bag is no longer equal to the pressure of the stressing solution and the equilibrium is only apparent : the concentration of the dispersion remains larger than the one it should be if imposed by the osmotic pressure of the outer solution. This situation can be easily recognized in the dilute regime. Indeed, as was stated in § 1-2, the osmotic pressure responds to the number of solute particles present and then must vary linearly with concentration following an ideal gas law. To avoid such experimental artefacts, ample air space must be present in the bag at the beginning. Another technique is to leave the dialysis bag open or to mount it at the tip of a glass capillary where the liquid dispersion could rise (Fig.6) ; then the pressure of the dispersion can be read directly from the level difference in an equilibrium against an aqueous solution with or without polymer.

Conversely when the dispersion is equilibrated at a pressure higher than its original pressure, it looses solvent, and if the bag contains too little material, the dispersion may not concentrate as much as it would have done under the action of a large pressure. This is due to the mechanical resistance of the bag upon the deflation process. During the initial deswelling of the dispersions it is necessary to refill the bag with more solution (in order to obtain a sufficient amount of concentrated dispersion) and to exchange

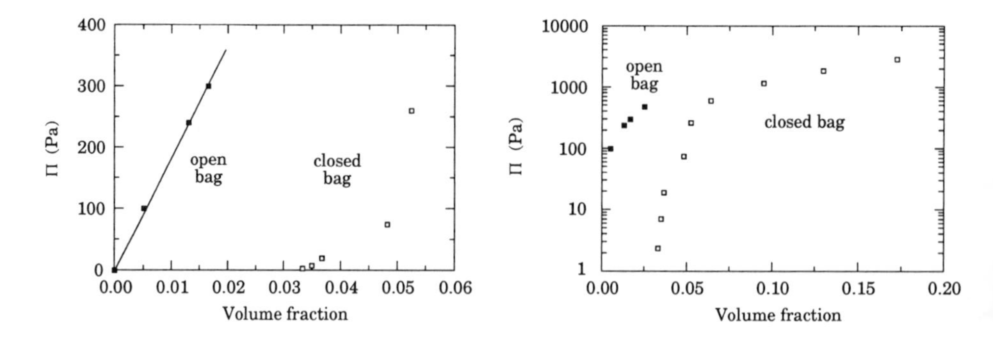

Figure 6. Difference in the measured pressure upon the swelling of a latex dispersion placed in open or closed dialysis bags [9]. The measured pressure is the osmotic pressure of the dextran stressing solution outside the dialysis bag ; it equals the osmotic pressure inside the bag minus a contribution from the stretching of the bag. The latex particles (115 nm of diameter) have a core made of styrene-butyl acrylate copolymers and a surface membrane made of a copolymer of these materials and acrylic acid (pH 9 ; ionic strength : 7.5×10^{-3} M).

the stressing solutions (in order to avoid dilution with water from the dispersion).

In all experiments a rule of thumb is the following : the bag must remain floppy and keep its equilibrium shape resembling that of a red globule (in cross-section) : thin in the center with rolls along the sides.

(iv) Vapor-phase osmometry :

The technique is based on the thermodynamic fundamentals that the vapor pressure of a solvent above a solution, P, is lower than when the solvent is pure, P_0. High osmotic pressures ($10^5 - 10^9$ Pa, that is an upper range much higher than the 10-100 atm conveniently accessible by polymer-induced stress) can be effected by exposing samples to known vapor pressures, of saturated salt solutions [26] for example.

Let us write that the chemical potential of the solvent is the same in all equilibrium phases : for the liquid and vapor phases of the compartment containing only pure solvent,

$$\mu_{\text{solvent}}^{\text{liq}}(T, P_0) = \mu_{\text{solvent}}^{\text{vap}}(T, P_0)$$

and for the equilibrium of the liquid mixture and its vapor in the compartment containing the solution with solute particles (considered as non-volatile) :

$$\mu_{\text{solvent}}^{\text{liq}}(T, P, \rho_p) = \mu_{\text{solvent}}^{\text{vap}}(T, P)$$

Using the definition of the solvent activity in the liquid phase

$$\mu_{\text{solvent}}^{\text{liq}}(T, P, \rho_p) = \mu_{\text{solvent}}^{\text{liq}}(T, P) + RT \ln a_{\text{solvent}}$$

and the definition of relative humidity P/P_0 (vapors are considered as perfect gases)

$$\mu_{\text{solvent}}^{\text{vap}}(T,P) = \mu_{\text{solvent}}^{\text{vap}}(T,P_0) + RT\ln\frac{P}{P_0}$$

one is lead to an equation similar to eqn.2 for the osmotic pressure

$$\Pi = -\frac{RT}{\bar{v}_{\text{solvent}}^{\text{liq}}}\ln\frac{P}{P_0} \tag{11}$$

where $\bar{v}_{\text{solvent}}^{\text{liq}}$ is the partial molar volume of the solvent in the liquid phase. This equation shows that the osmotic pressure is the pressure required to raise the vapor pressure of the solution to that of the pure solvent.

Except at relatively low activities of water, vapor pressure is likely to be prohibitively difficult to use accurately. Indeed, for dilute solutions, the osmotic pressure is $\Pi \approx \bar{v}^{\text{vap}}/\bar{v}^{\text{liq}}(P_0 - P)$ that is $\Pi \approx 1200(P_0 - P)$ for aqueous solutions, and a small variation of the vapor pressure effects the osmotic pressure measurement to a large extent (at 99% relative humidity, Π is some 14 atm, already in the upper range of osmotic stress induced by polymers). Furthermore, the rapid variation with temperature of the vapor pressure of pure water calls for care in the measurements carried out at near saturation. A temperature fluctuation of only 0.2 °C at room temperature leads to an error of more than 14 atm. To get the equivalent 2% accuracy readily achieved in the application of osmotic stress in the range 0-100 atm by polymers would require temperature control to 0.003 °C. In practice it is safest not to carry out measurements much above 85-90% relative humidity.

Note also that the vaporous medium allows equilibration only of volatile species. Not only the activity of the solute of interest, but also the activities of the salt and other nonvolatile small molecules will contribute to the equilibrium state. With vapor-phase osmometry one measures the *total* rather than the net osmotic pressure of the solution (see footnote 3). Unlike the polymer-induced stress method, sample dehydration here will increase the concentration of the salts and molecules that should be kept at constant activity. Any properties that depend on salt concentration will be severely perturbed by this accumulation.

1.4. APPLICATION OF AN EXTERNAL FIELD

The consolidation or concentration of suspended particulate solids under the influence of a body force applied to the particles (e.g. a gravitational force, or an applied pressure gradient developed in a filter) is a problem of widespread practical interest. For instance, it determines the extent to which fine solids can be separated from liquids by mechanical means such as pressure-filtration, gravity-thicknening, or centrifugation.

(i) Constant applied pressure :

The filtering of a solution containing a dissolved solute is possible in an ultrafiltration cell only if the applied external pressure becomes larger than the osmotic pressure of the solution : the solution starts then to flow. Based on this observation, different methods for osmotic measurements have been proposed.

The compression cell is of particular interest for pressures higher than the 10-100 atm conveniently accessible by polymer-induced stress : a hydraulic pressure is exerted on the sample through a rubber membrane (or a piston) and, as pressure is applied, the solvent is squeezed out through a millipore filter (or a strong, supported, semi-permeable membrane) into a calibrated capillary tube [2, 8]. Because of mechanical resistance of the membrane or friction in the piston, this method is unreliable at lower pressures. At equilibrium, when there is no further flow of liquid through the millipore, the applied pressure P_A (measured by means of a pressure gauge) and the pressure due to the rubber membrane P_M (which can be positive, negative, or zero depending on the volume of liquid in the cell) balance the osmotic pressure Π of the solution and the hydrostatic pressure of the liquid P_H (corrected for capillary effects) in the calibrated capillary (Fig.7) :

$$P_A + P_M = \Pi + P_H$$

Because P_A , P_M, and P_H may be readily obtained (P_M is determined as a function of the volume in the cell by doing a blank run with pure water), Π can be determined. Also, because the volume fraction in solute particles can be determined at the completion of the compression run (by knowing the initial volume fraction of the dispersion and by measuring the height of liquid in the calibrated capillary at each equilibrium point), it is then possible to obtain Π as a function of the volume fraction of the solute.

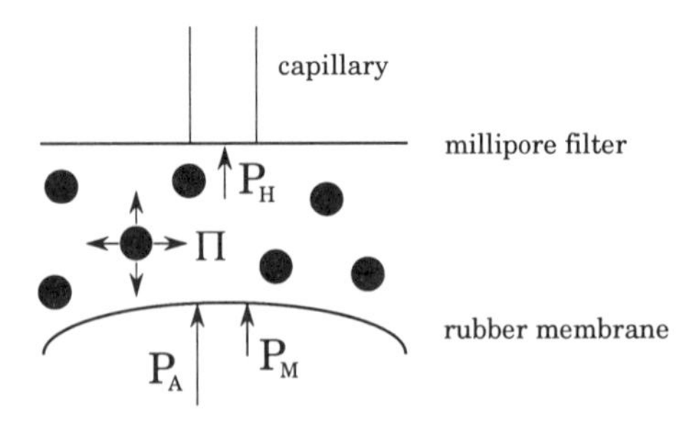

Figure 7. Pressure balance in a compression cell.

Industry takes advantage of the existence of osmotic equilibrium. An application that has attracted a great deal of interest in recent years is the production of potable water from saline water by reverse osmosis. The principle consists to apply a pressure to the salted sea water much larger than its equilibrium osmotic pressure (about 25 atm at 25 °C) : when placed in a compartment sealed by a semi-permeable membrane (that is a membrane which is permeable to the solvent but not to the ions), the sea water concentrates in solute (i.e. in salt) as to reach the thermodynamic equilibrium, and on the other side of the membrane, a debit of salt-free water flows. In this ultrafiltration process the semi-permeable membrane functions like a filter that separates solvent from solute molecules. Since no phase transitions are involved as, for example, in distillation, the method offers some prospect of economic feasibility. However, the rate of reverse osmotic flow depends on the excess pressure across the membrane, and is tied into the technology of developing membranes capable of withstanding high pressures. The method of reverse osmosis may also be used as a method for concentrating solutions. Fruit juices and radioactive wastes, for example, have been concentrated by this method.

(ii) Equilibrium under the influence of gravity :

Consider a suspension of particles settling through a fluid under the influence of gravity. A gradient of concentration builds up for the falling particles of buoyant mass $V_p \Delta\rho$, where V_p is the volume of a particle and $\Delta\rho$ the density difference between particle and solvent phases to take into account the upthrust due to the displaced fluid (particles are considered denser than fluid, i.e. $\Delta\rho > 0$). Interactions influence the spatial distribution of particles and thus have an effect on the local volume fraction $\Phi(z) = \rho_p V_p$ where z is the vertical coordinate measured from the bottom of the sedimenting container and ρ_p the number density in particles of the solution. They also allow energy to be stored elastically within the particle configuration so that in a container of finite depth the accumulation of particles at the bottom of the container sets up a propagating potential gradient which opposes gravity. This latter effect can be accounted for by noting that there is then a transmitted stress or "particle pressure" P that enters the force balance. In the case of colloidally stable suspensions, P is just the osmotic pressure of the particles $\Pi(\Phi)$; for flocculated or coagulated suspensions above their gel-point (i.e. above the concentration where a network forms) P is the elastic stress developed in the particulate network. When equilibrium is attained the particles are acted upon the gravitational force and the opposing osmotic pressure gradient. The force balance performed on a volume element of suspension gives :

$$-\frac{\partial \Pi}{\partial z} = \Delta\rho\,\Phi(z)\,g$$

By integrating this equation, one gets :

$$\Pi(z) = \Pi(z_0) + g\,\Delta\rho \int_{z_0}^{z} \Phi(z')dz' \tag{12}$$

Using the definition of the reduced isothermal osmotic compressibility $\bar{\chi}_T = k_B T(\partial\rho/\partial\Pi)_T$ ($= \chi_T/\chi_T^{\rm ideal}$) the equation for the local particle number density $\rho_p(z)$ can be written as :

$$\frac{d\rho_p}{dz} = -\frac{\bar{\chi}_T}{l_g}\,\rho_p(z) \tag{13}$$

where $l_g = k_B T/V_p\,\Delta\rho\,g$ is the gravitational length (of the order of 0.1 mm for colloidal particles of diameter 0.1 μm). Note that in the dilute gas limit, we obtain from eqn.13 the usual barometric law : $\rho_p(z) = \rho_{p0}\exp(-z/l_g)$ or equivalently $\Pi(z) = \Pi_0\exp(-z/l_g)$.

The osmotic pressure can be determined experimentally as a function of $\Phi(z)$ (or $\rho_p(z)$), according to eqn.12, by a numerical integration of the measured concentration profile. The latter can be inferred from scattering experiments such as light scattering in the beautiful sedimentation experiment performed with model hard-spheres by Piazza *et al.* (Fig.8) [27].

Analysis of a sedimentation profile for heavy enough particles makes it possible to determine not only the equation of state for a wide range of concentrations but also to provide measurements in the range of small osmotic pressure values ($10^{-4} - 1$ Pa).

(iii) Equilibrium under centrifugation :

Since the radial acceleration is proportional to the distance from the axis of rotation an ultracentrifuge provides a way to vary continuously the pressure applied on a dispersed system. The conditions of equilibrium enumerated in the previous section can readily be modified so as to include the influence of a continuously varying applied force. Once centrifugation equilibrium is established, the centrifugal force is balanced by repulsive forces that oppose any further compaction of the particles ; therefore, at all locations in the cell, the centrifugal force is equal to the gradient of the osmotic pressure (also called the disjoining pressure).

Consider a slice of dispersion in the centrifugation tube at a distance z from the axis of rotation. The volume fraction of particles in this slice is $\Phi(z)$. The equilibrium between the centrifugal force and the gradient

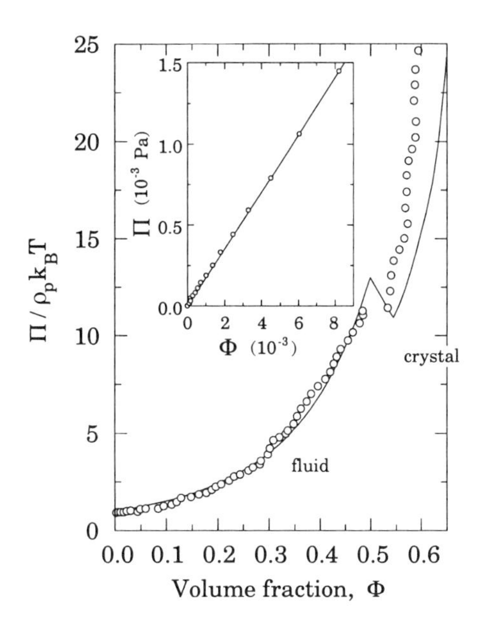

Figure 8. Compressibility factor as a function of the particle volume fraction Φ at equilibrium for a suspension of highly screened 180 nm diameter monodisperse latex spheres [27]. The steady-state sedimentation profile $\Phi(z)$ was obtained after 100 days by measuring the intensity of the depolarized scattered light ; this equilibrium concentration profile was then numerically integrated to give the osmotic pressure according to eqn.12. In the barometric region (ideal gas law at low Φ ; inset) the slope of the profile gives the gravitational length l_g = 0.188 mm. The discontinuity in volume fraction from $\Phi \approx 0.50$ to $\Phi \approx 0.55$ for a value of $\Pi/\rho_p k_B T \approx 12$ indicates the phase transition of the hard-sphere fluid to a crystalline phase. The full curve in the fluid branch follows the Carnahan-Starling equation for hard spheres while the full curve in the solid branch is the Hall empirical fit to molecular dynamics. The deviation from the ideal crystal behavior suggests that the solid might be partially disordered.

of osmotic pressure exerted on this slice by the neighboring slices can be written as follows :

$$-\frac{\partial \Pi}{\partial z} = \Delta\rho\, \Phi(z)\, \omega^2 z$$

where ω is the rotation speed. By integrating this equation, one gets :

$$\Pi(z) = \Pi(z_0) + \omega^2 \Delta\rho \int_{z_0}^{z} z'\, \Phi(z')\, dz' \qquad (14)$$

On centrifugation, the particles may be considered as being forced into a layer at the end of the tube, or conversely the aqueous medium is drained from the interspaces to form a layer of serum at the other extremity. Consequently the centrifugal force is acting to bring the particles into close proximity and to initiate their aggregation. The ultracentrifuge technique can thus be used to determine the minimum pressure required for aggregation or coalescence (in the case of an emulsion) and thereby the stabilization mechanism of colloidal dispersions (latex [28], silica nanoparticles [29]) and of emulsions [30, 31] (Fig.9).

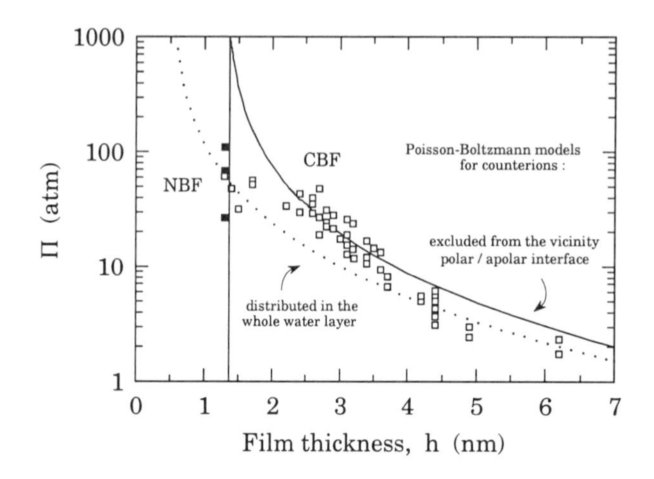

Figure 9. Osmotic pressures measured in an oil (hexadecane) - in - water emulsion stabilized with sodium dodecyl sulfate [31]. In the ultracentrifuge the originally spherical oil drops are flocculated and distorted into polyhedra which approach each other along flat plane-parallel films as in a foam. Once centrifugation equilibrium is reached, i.e. once the centrifugal force is equilibrated by the osmotic pressure gradient, the centrifugation is stopped, and the sample is sliced to provide a set of biliquid foams with different osmotic pressures and water contents. Chemical analysis gives the water and surfactant content. The osmotic pressure is then calculated through eqn.14 and plotted as a function of the thickness of the biliquid foam determined through small angle neutron scattering. At low pressures the films are in the common black film (CBF) state, where the measured disjoining pressure matches the entropic pressure of the counter-ions (solid line). At high pressures, ionic correlations in the counter-ion layer reduce the disjoining pressure and the films jump discontinuously to the Newton black film (NBF). The thickness of the NBF film is stabilized by hydration forces, which resist the dehydration of counter-ions and surfactant headgroups.

2. Force measurement devices between solid surfaces

Of the various methods that have been devised for measuring molecular interactions, the most direct employ macroscopic solid bodies or extended surfaces. In any method of direct measurement of molecular interaction between bodies, the experiment reduces to the measurement of two quantities, namely, the force of interaction between the two bodies and the width of the gap that separates them. Both attractive and repulsive forces can be measured from the deflection of a balance arm or sensitive spring (cantilever) by means of electronics, optics, or interferometric techniques. In static experiments the surface force, F, is simply equated to the restoring force on the balance arm or spring, F_K,

$$F = -F_K \tag{15}$$

Thus the measurement of the force-distance profile is reduced to the determination of the separation and the restoring force (in the case of a cantilever it is given by Hooke's law within the elastic limits of the material, $F_K = -Kz$, where K is the spring constant and z the deflection of the spring)[6]. However, the additional experimental determination of the local radius of curvature of the surfaces, R, is desirable as in most cases the force, F, scales with R. In other words it is not the force, F, which is the most important, but rather the interaction free energy, E, and the latter can be inferred from the former provided the transformation relating the normalized force F/R with the interaction free energy is valid (see § 2-2).

This section will not review particle detachment and peeling experiments which can provide information on interparticle adhesion forces and surface adhesion energy of two solid surfaces in contact. The focus will be rather on the direct measurement of the force between two macroscopic solid surfaces as a function of the surface separation which can provide the full force law. This task presents extreme experimental difficulties. In the first place one wishes to measure a small force — as little as a hundred thousandth of a gram. This in itself is not too difficult ; microchemists routinely deal with far smaller forces. But the interaction to be measured shows up when only the two bodies are extremely close together — within a few tens of nanometers. Furthermore, the force may vary enormously as the distance of separation is changed. It is precisely this variation that one wants to determine. Thus one needs a way to set the width of the gap accurately at any desired distance, and to maintain the separation against the attraction or repulsion force being measured. Not only that ; the equilibrium has to be extremely stable so that chance variations in the gap is corrected before they can build up. The requirements of sensitivity, stability, and rapid response are not easy to reconcile. A sensitive balance tends to swing widely in response to a small change in force.

These experimental difficulties have been overcome successfully by adequate choice of suitable shape and substance for the objects, and by the development of various techniques for measuring the distance dependence of the forces. Direct measurements of the force between surfaces as a function of the separation were first performed in the 1950s. Overbeek and Sparnaay

[6] In dynamic force measurement techniques eqn.15 does not hold as inertial effects arise [32, 33].

[34] and Derjaguin and co-workers [35] developed instruments for measuring these forces in air and vacuum (see § 2-1). Since these pioneering investigations different approaches have been proposed and they will be reviewed in the following sections.

2.1. SURFACE FORCE BALANCE DEVICES

The role ascribed to van der Waals forces in the stability of colloidal dispersions, in the formation of aggregates between particles, and especially the long-range character of these interactions made an independent proof of their existence very desirable. The microbalance developed in the 1950s by Abrikosova and Derjaguin was especially suited for such an investigation, in particular for measuring the long-range part of the attractive dispersion forces acting between solids in air or in vacuum [35]. The experimental results [35-37] conflicted with the classical London-Hamaker theory. The accuracy of the measurements was such that it provided the stimulus for the development, by Lifshitz, of the modern macroscopic theory of dispersion forces. The main difficulty encountered in measuring the molecular attraction between macroscopic bodies is that the forces, F, decrease drastically as the separation, D, between the solids increases. At large distances, they are extremely small, and so the accuracy of measurements is decreased. At small separations, the gradient of the interaction $\partial F/\partial D$ becomes large and positive and this makes it necessary to use rigid cantilevers of stiffness larger than that gradient, otherwise the mechanical stability of the spring device cannot be maintained. Obviously, this problem requires the use of a balance of considerable restoring moment on the one hand and of high sensitivity on the other, these requirements being incompatible in the case of an ordinary balance. These difficulties have been overcome by using negative feedback for the simultaneous solution of two problems : the stabilization of the separation, and the measurement of the attractive interaction.

The bodies whose mutual attraction was measured were a highly polished flat plate and an equally smooth convex lens, both made of quartz (metallic and crystalline materials were also studied). The flat plate rested on one arm of a balance beam three centimeters long, weighing about a tenth of a gram and balanced on a fulcrum made of a wedge-shaped agate bearing (Fig.10). Above the flat plate was mounted the rounded plate in an adjustable bracket. The separation between the flat and the curved lens is calculated from the diameters, d_p, of the Newton's rings measured with a microscope. In the 1950s the gap was illuminated by a cinema lamp through a constant-deflection monochromator. Continuous variation of the illuminating wavelength λ produced by rotation of the monochromator for a constant gap enables the order p of a ring to be determined ($p = 2D_p/\lambda$ where D_p is the air-gap width at the given ring). Having determined p, the separation is then deduced from

$$D = \frac{\lambda}{2}\left(p - \frac{{d_p}^2}{4R\lambda}\right)$$

where R is the radius of curvature of the spherical surface (3-10 cm). Due to roughness of the surfaces, only the region of large distances (> 500 nm) could be investigated.

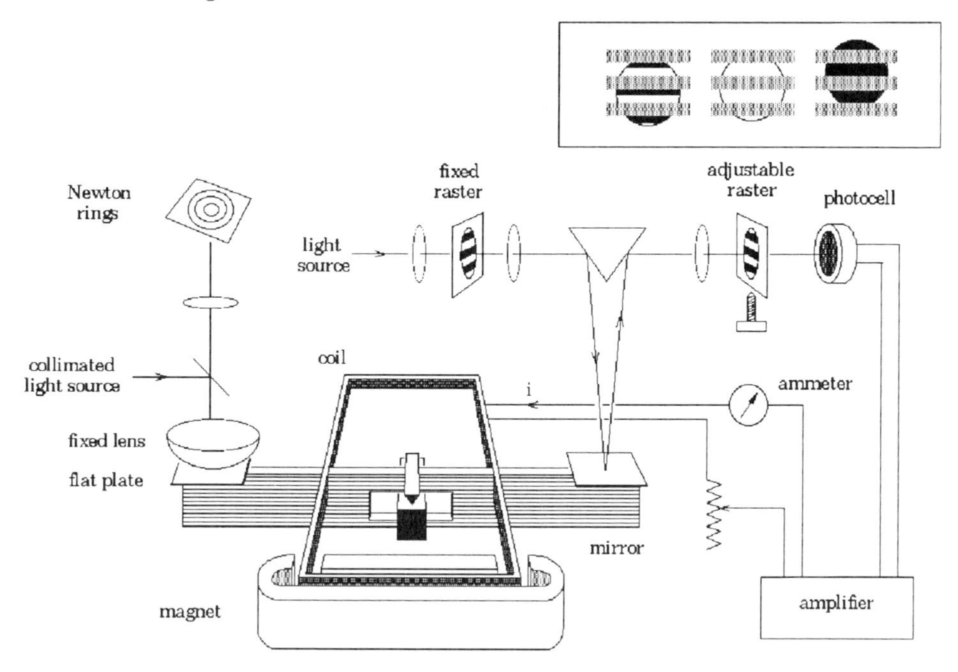

Figure 10. Controlled balance for measuring attractive forces between macroscopic bodies. At the fixed zero position of the balance beam, the current in the coil is equal to zero. Deflection of the balance beam through a small angle alters the passage of the light by the second raster and the illumination of the photo-cell causing a current i. The flow of this current through the coil of the balance situated in the field of the magnet creates an electromagnetic reaction torque which acts on the beam and returns it to the position of equilibrium. *Inset* : Overlapping grids, or rasters, determine the amount of light reaching the photocell. Image of opaque bands (grey bands) on one grid falls on second grid (circle). Light transmitted varies from a maximum when the two sets of the bands coincide (center) to zero when image bands wholly fill transparent spaces (right).

The force of attraction between the two solids is compensated by the balancing torque created by the interaction of a magnetic field with the electric current flowing through a frame rigidly attached to the beam of the microbalance. The current in the coil is preset by a follow-up device comprising a raster-type photorelay. The circuit works as follows. A beam of light shines through a grid, or raster, of alternate opaque and transparent stripes. A mirror on the balance arm reflects the beam through a lens forming an image of the grid. The image falls on a second identical grid, beyond which is a photocell. The quantity of light passing through the second grid and reaching the cell depends on the overlapping of the images of the two rasters. Thus the illumination of the photoelement varies as the mirror turns. The farther the balance arm rotates, the more light passes, and the stronger the current. This electric current is amplified and fed into the balance coil. The passage of the current i in a requisite direction through the frame effects the negative feedback coupling as a result of which the balance beam is subject to a torque $M = ki$ (where k is a constant depending upon the number and shape of the turns in the coil and the magnetic intensity) arranged to be in opposite direction to the turning force on the arm due to molecular attraction. As the measurement of this attractive force is reduced to measure the intensity of the balancing electric current, the device enables the force to be determined with great sensitivity (to within 10^{-9} N).

The force-distance profile is measured as the gap width is varied by bringing the convex surface of the lens and the flat surface of the plate together by means of a fine lifting mechanism. Additional fine control of the separation is obtained by shifting the first raster in a direction perpendicular to its markings with a micrometer. This gives rise to a moment of electromagnetic force (not compensated by gravity), which causes the balance beam to move to a new equilibrium position. In many respects, the feedback circuit resembles a spring since it provides a restoring force (torque) which is proportional to a (detected) displacement. However, unlike a simple mechanical system, it is possible to vary the effective spring constant of the system by altering the gain in the feedback circuit. In other words, the capability exists to alter electronically the stiffness of the force measuring system. This idea will be taken up in the development of recent devices (see § 2-3 and § 2-4).

So far the distances of separation were measured optically. This required that the solids should be transparent. In turn, this meant that the solids should be large (relative to colloidal dimensions). In this regard, Derjaguin and co-workers developed an improved feedback apparatus in which the separation between the solids could be preset and thus no direct measurement of the separation was required. As a consequence, forces between opaque solids were measured and, furthermore, the dimensions of the solids could be reduced considerably to get a point like contact. Use of smaller sample sizes was of considerable advantage as the amount of steric hindrance due to the roughness of the surfaces was concomitantly reduced. Retarded and non-retarded regions of dispersion forces in air were discernable for the interactions measured between crossed-fibers of platinum or quartz with radii of the order 150-500 μm [38, 39].

A variant of this device was also developed for studies in liquid environment between metal filaments with small radii crossing at right angles used as surfaces. The rotation of an elastic torsion suspension relative to a fixed surface was employed to control the distance, and by measuring the angle where the surfaces were brought into contact, the height of the force barrier could be determined, whereas attractive forces could not be measured [40]. Later, this technique was further improved to measure both repulsive and attractive interactions between glass threads in electrolyte solutions by employing a photoelectric mirror galvanometer in a negative feedback arrangement [39].

Particularly productive, the crossed filament technique has been further developed and taken up by other groups [41]. In its latest form the apparatus consists of two L shaped filaments (quartz or other materials) placed with their shorter arms at right angles (Fig.11). The longer arm of one filament is attached to a piezoelectric transducer for fine control of surface separation in series with a coarser positioning control ; that of the other filament has a small mirror attached. The shorter arms can be driven together through the piezo ; where an interaction between the surfaces arises the second filament is deflected and the mirror rotates. This rotation can be followed from the deflection of laser light across a photo-diode array. This method of measurement of separation is similar to that commonly used in atomic force microscopy (see § 2-3) and has the advantage that it can be followed continuously but the disadvantage of not being made at the actual point of contact and not revealing any surface deformation. It is also difficult to obtain a true contact position for the surfaces and the absolute zero of separation is not determined. Valuable results have been obtained for bare surfaces, in electrolyte solutions and with adsorbed polymers [42, 43].

Other workers, while adhering to the basic principles of a balance for measuring the force-distance profile between two macroscopic surfaces solved the problems of operation in different ways, however. In another instrument a balance is used with a quartz surface mounted at one end of a scale-beam, and an iron core dipping into a coil at the other end. Fine adjustment of surface separation is controlled by a piezoelectric tube, and the coil is used both for rough positioning and for compensation of the surface position under the influence of a force as measured with an inductive displacement transducer [44-46]. Capacitors have also been used as force sensors (§ 2-4).

It is remarkable that the essential features of the original designs introduced at the very early stages were retained in the most advanced

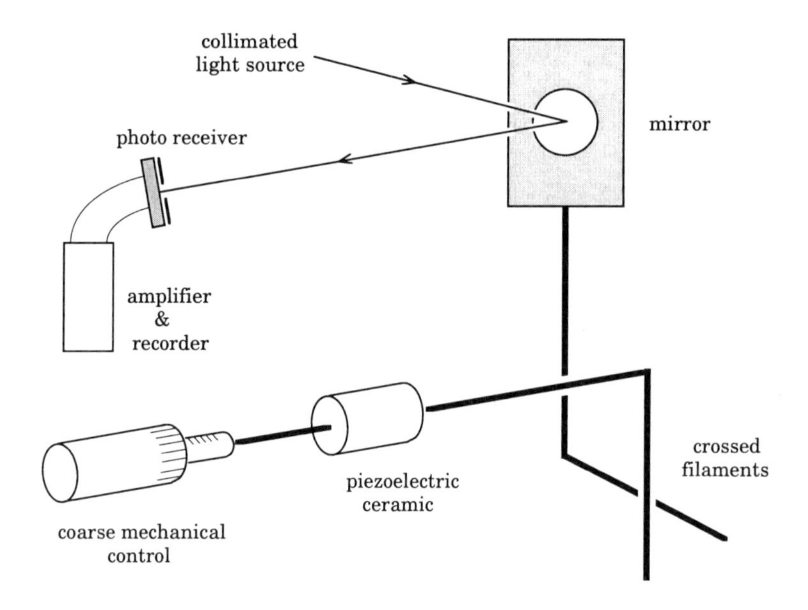

Figure 11. Controlled balance for measuring both attractive and repulsive interactions between crossed filaments.

apparatuses. To measure interactions between macroscopic bodies at molecular separations the geometry of a sphere against a flat is very convenient as it is easier to set the distance between a sphere and a plane than between two planes, which must be held parallel (as in the Dutch device [34]). The crossed-cylinder geometry is even more attractive and this will be the geometry chosen by Tabor and Winterton [47] in the design of the Surface Force Apparatus (§ 2-2). Polished surfaces of large bodies used in the early experiments proved to be insufficiently smooth because the height of their uneveness (50 nm) was commensurate with the width of the gap. To investigate the region of considerably smaller distances requires smooth surfaces as mica sheets or wafers. Methods for determination of the separation, either by interferometry (as in the Soviet device), or by capacitors (as in the Dutch device, not described here) have both been very successful and have been used in the different types of Surface Force Apparatus (the Tabor-Israelachvili type described in § 2-2 and the Ecole Centrale de Lyon device in § 2-4). At last negative feedback automatic control is a powerful arrangement, which accomplishes a quadrupole purpose: (i) it allows one to set the gap at any desired distance, (ii) it maintains the separation by producing a force equal and opposite to the interaction, (iii) through negative feedback it corrects any drift of the surfaces from the preset position and (iv) it provides a means of measuring the opposing force.

2.2. SURFACE FORCE APPARATUS (INTERFEROMETRIC SFA)

The device allows the force to be measured as a function of separation between two macroscopic surfaces in air or immersed in liquids. The separation between the surfaces is determined by optical interferometry and the force is read from the deflection of a cantilever spring on which one of the surfaces is mounted and that can be moved to bring the surfaces to a given separation. The interferometric surface force apparatus (SFA) is available in several different versions such as Mark II [48], Mark III [49], Mark IV [50] and others [51-57]. The principle behind the different instruments is similar, and depicted schematically in Fig.12. It is conceptually similar to earlier models developed by Tabor *et al.* used for measuring van der Waals forces between mica surfaces in air and in vacuum [47, 58].

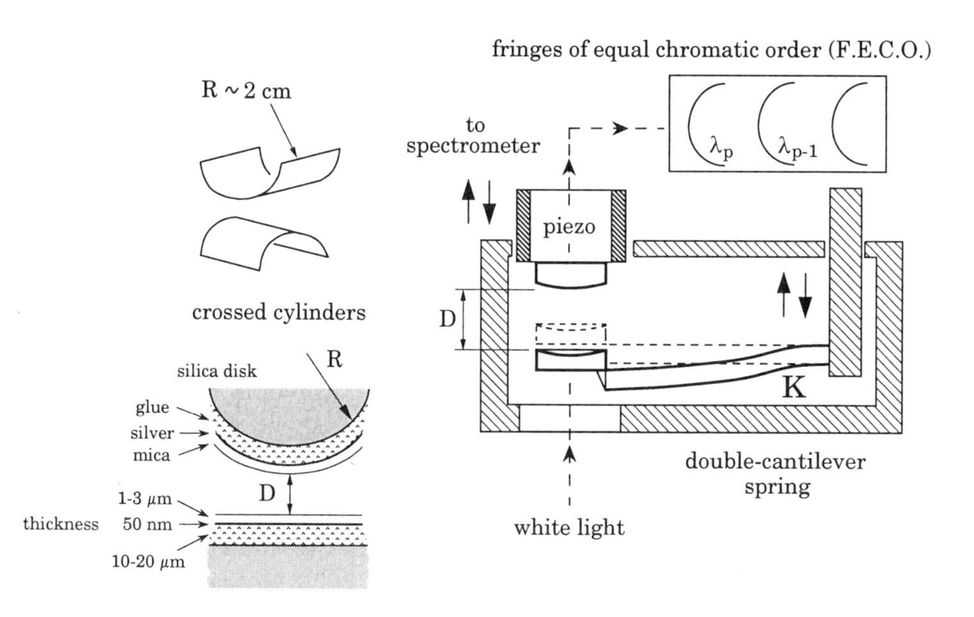

Figure 12. A schematic representation of the interferometric surface force apparatus (SFA). The surfaces are attached to a piezoelectric crystal and on a double-cantilever spring, respectively. The crossed-cylinder geometry is equivalent to that of a sphere against a flat as the radius of curvature, R (about 2 cm), is much larger than the surface separation D (in the drawing the curvature of the surfaces has been exaggerated). The interference FECO pattern set in incident white light gives the real separation between the surfaces (relative to silver backing) and their local shape.

preparation of the surfaces :

The preparation of a pair of surfaces which produce interference fringes yielding resolution of less than a few angstrom is not trivial, requiring a convergence of delicately thin (1-3 μm thick), molecularly smooth and optically transparent substrates, silver layers with a small window of useful reflectivity, and a contaminant free surface. The preferred substrate in the SFA is muscovite mica, although other substrates such as silica [59, 60] and sapphire [61, 62] have also been used. Mica is a naturally occurring mineral consisting of aluminosilicate layers held together by ionic bonds to potassium. When mica is cleaved along crystallographic planes, the ionic interlayer, bonds are broken and large atomically smooth areas (flat to within 0.1 nm) can be easily obtained. Several techniques have been developed to modify the surface of mica and increase the number of surfaces available for study. Surfactant adsorption [63] and Langmuir-Blodgett deposition [64] have been employed to produce both surfactant and bilayer films. Vacuum deposition by thermal evaporation has also been employed to create metallic surfaces suitable for surface force measurement [65]. The mica surface can also be covalently modified by a cold plasma process [66] and a wide variety of different functionalized surfaces can be prepared.

The manual cleaving of mica with pairs of ethanol-rinsed and blow-dried tweezers is carried out in an atmosphere of filtered laboratory air (in a laminar flow cabinet in a clean room). Cleavage is initiated by inserting the tip of a sharp needle into the edge of a thick mica plate and carefully peeling away a thin sheet. Areas of constant thickness are identified by the clarity and uniformity of interference Newton colors observed in reflected light from, for example, a fluorescent tube. Because these colors change abruptly at cleavage steps, with some practice it is possible to recognize regions that are 1-3 μm thick and free of cleavage steps. When such large (typically 10 cm^2) step-free regions are found smaller (1 cm x 1 cm) pieces are cut out of it with a white-hot platinum wire (0.1 mm diameter). The mica melts around the platinum wire and each small sheet ends up being considerably thicker over about half-a-millimeter close to the edge. As each rectangular piece is cut free from the main sheet it is held at one edge with tweezers and immediately placed face down on a freshly cleaved, thicker and larger piece — the backing sheet — onto which it immediately adheres. The molecular contact must take place at once (otherwise contamination may be suspected) except at the edges which have thickened and have been damaged during the melting ; but this is to advantage as it allows the piece to be picked up easily again at a later date with tweezers before gluing it onto a silica support disk. In this way the surface of the pieces in contact with the backing sheet are protected from contamination as long as they remain there. Series of such small rectangular or square pieces may be prepared at the same time and placed on the backing sheet. As working surface contamination is prevented, the backing sheet decorated with these small thin pieces can be transported and silvered by vacuum deposition to a thickness of about 50 nm (reflectivity $> 98\%$ in the green region of the visible spectrum). It is stored in a dessicator under vacuum until used (not longer than 3-4 weeks).

Just prior to an experiment the supporting silica disks (highly polished to lambda to give a cylinder with a radius of 2 cm) are cleaned with acetone and/or ethanol followed by blow-drying in cleaned nitrogen gas, before placing them on a well-cleaned heating plate. Pieces of glue (a thermosetting resin ; Epikote 1004 from Shell Chemical Co.), about 1 mg, are then placed on the disks and the plate heated to 150 °C. As the crystals melt the liquid is distributed and spread evenly over the convex disk surface using a pair of cleaned tweezers or a needle. After removing air-bubbles in the glue layer with the tweezers, and allowing the glue to settle for approximately 20 min. a thin piece of mica is stripped off the backing sheet with tweezers and placed on top of the glue layer, silvered side down. The glue immediately spreads out evenly underneath the mica pulling it into close contact with the glass. The tricky part of the gluing process is to avoid the mica sheet tearing on being stripped off, air bubble trapping, and waving of the thin mica sheet on gluing. The disk is then quickly removed from the heat and allowed to cool off. The mica sheet has in most, if not all, instances a larger area than the area of the curved, glue-covered, surface of the 10 mm diameter disk. Indentation measurements have revealed that the thickness of the glue layer is about 10-20 μm. Epikote resin is very suitable as a glue for three reasons : it is transparent, it does not change its volume as it sets and thus prevents any stress in the glued mica sheet upon cooling, and it is not a source for contaminants. However, care must always be taken as dissolution of the glue may occur in specific solvents, for which other suitable inert glues have been sought [67, 68].

crossed-cylinder geometry :

The two thin sheets of mica (or of silica, etc) freshly attached to the curved supporting silica disks, are mounted in the apparatus facing each other with the cylinder axes at right angles (to each other). With this geometry of two crossed cylinders the precise alignment of two interacting plates and unwanted effects at the edges of the plates are avoided. The geometry of crossed cylinders has an additional advantage in that if the region of contact becomes contaminated or locally damaged (for example due to adhesive properties of the surfaces, etc...) during the course of an experiment, one surface can be moved, first along the axis of one cylinder and then along the other to reveal a new and pristine contact area. Changing contact position several times under the same solution conditions within the same experiment increases the statistics of the measured data and the reliability of the results.

Fortuitously, this arrangement is also convenient for the comparison of experiment with theory. According to Derjaguin the force between crossed cylinders of equal radius R is the same as the force F between a sphere of radius R and a plane flat surface (or between two spheres each of radius $2R$). Further, this force is equivalent to the free energy E of interaction per unit area between two plane parallel surfaces of the same material [69]. Specifically $F = 2\pi RE$. For this reason, forces $F(D)$ measured with the SFA between two crossed-cylinders of mean radius of curvature R are routinely plotted as F/R and are therefore implicitly related to the *interaction free energy* $E(D)$ per unit area of parallel flats (rather than a *force*) :

$$\frac{F}{R} = 2\pi R \tag{16}$$

where R is the geometric mean radius or inverse Gaussian curvature of the surfaces. The validity of this transformation lies on the following requirements : the Derjaguin approximation [69] is valid provided $R \gg D$ and provided R is independent of D (the surfaces must remain undeformed), and so long as $F(D)$ is mathematically well-behaved, i.e. is single-valued and integrable. Furthermore, the range of the force must be small compared to the radius of curvature of the surfaces.

The first condition is always fulfilled in measurements using the SFA : the experimental setup consisting of crossed cylinders may be mapped to a sphere against a plane or to two spheres of twice the radius, with radii of the order of centimeters compared to separations of the order of nanometers or micrometers at the most. However, due to surface deformation effects, the second condition may not always be fulfilled, and when the surfaces are being deformed due to the action of strong surface forces the comparison between experiments and theory no longer becomes straightforward[7]. The last criteria are satisfied in most of all cases. In spite of this, application of the relationship eqn.16 can seem questionable in several cases, in particular when layering is observed upon confinement between macroscopic surfaces. A statistical mechanical derivation of eqn.16 has been given and it has been explicitly shown that it accurately relates the oscillatory structural forces arising between curved surfaces in fluids with short-range repulsions to the interaction free energy between planar walls [70] : the Derjaguin approximation still holds. Similarly, for homeotropically aligned lamellar mesophases, the analysis has included the contribution of the dislocations

[7] We note that for both interferometric and non-interferometric techniques a surface deformation gives rise to an error in the value of F/R, using the radius for the undeformed surfaces. When non-interferometric techniques are used such as in the AFM or in the MASIF (see § 2-3), a surface deformation also gives rise to an error in the deduced surface separation, since the determination of the distance scale relies on a supposed hard wall contact.

and the equilibrium with the surrounding undisturbed multilayer system [71].

distance measurement :

The outer silvered faces of the two mica sheets form an optical cavity (a Fabry-Perot like interferometer) providing the means to measure distances. Collimated white light is directed and impinged normal to the surfaces. Multiple reflections occur between the two reflective films and the light transmitted consists of a spectrum of intensity maxima known as fringes of equal chromatic order (FECO) [72]. A microscope focuses the light emerging from the interferometer on to the entry slit of a grating spectrometer, and the fringes are split up, separated out in space according to wavelength. Analysis of this array makes possible to measure the optical thickness of the film and thus allows the simultaneous determination of both the thickness and the refractive index of each layer in the interferometer.

Note that multiple-beam interferometry requires simply the presence of two highly reflective thin films separated by one or more dielectric materials of total thickness greater than the wavelength of visible light. The spectrum can be accurately predicted using classical electromagnetic theory, and the FECO wavelengths depend on the thicknesses and the optical properties of the media (refractive indices and their dispersions). For a symmetric three-layer interferometer (the usual case of two mica sheets[8] of equal thickness separated by an isotropic homogeneous medium) the separation D between the two surfaces can be calculated from any fringe of wavelength λ_p (p is the fringe order) [73] :

$$\tan\left(\frac{2\pi n D}{\lambda_p}\right) = \frac{2\,\frac{n_m}{n}\,\sin\theta}{\left[1+\left(\frac{n_m}{n}\right)^2\right]\cos\theta \pm \left[\left(\frac{n_m}{n}\right)^2 - 1\right]}$$

$$\text{with} \quad \theta = \frac{\pi\Delta\lambda_p}{\lambda_p(1-\lambda_p^0/\lambda_{p-1}^0)} \tag{17}$$

where n_m and n are the refractive indices of the substrate (mica) and of the medium at the wavelength λ_p ; + and − refers to the fringe order p being odd and even, respectively, and $\Delta\lambda_p = \lambda_p - \lambda_p^0$ represents the shift to longer wavelengths for fringe order p from its contact position at λ_p^0 to λ_p when the surfaces are D apart (λ_p^0 ; λ_{p-1}^0 ; ... are the measured reference positions of the pth ; (p-1)th ; ... fringes when $D = 0$, i.e. the contact wavelengths when the substrates are in molecular contact). Phase changes on reflection yield to corrections in this equation [73]. Other analytic expressions [74] and

[8] Mica is birefringent and each fringe normally appears as a doublet, but if the white light is made to pass through a polarizer one of the vibration directions may be extinguished.

standard multilayer matrix multiplication methods [75] have been proposed to the general case of arbitrary thickness and refractive index for each layer, or when roughness is involved [76], and when the optical media are anisotropic (birefringent or optically active) [77] and/or absorbing [78].

The spatial positions of the observed FECO change continuously as the separation between the surfaces is varied (Fig.13). Note that the separation determined by multiple-beam interferometry is an *absolute* value relative to the predetermined zero separation (measured distance of contact in air, or in an aqueous solution). Hence, with this method the thickness of adsorbed layers can always be measured, which is an important advantage over most of the non-interferometric devices. The uncertainty associated with calculating the central film thickness with multiple-beam interferometry has been discussed in details [73-75, 79]. Under optimum experimental conditions a 0.1 nm resolution can be achieved for surface separations larger than 5 nm, but the method becomes inaccurate for measuring very thin films (< 1 - 3 nm) due to uncertainty in the empirical expressions used for phase changes at the dielectric (mica) - silver interface [79]. Because optical thicknesses through the stratified media are involved when constructive interferences are set, refractive index of the thin sandwiched films can be measured (for instance by use of eqn.17 when the shifts in wavelengths are measured independently for two fringes of different parity). The accuracy is better than 1% at large separations [73] but decreases dramatically when the separation falls below 10 nm [80].

Furthermore, since the gap between the curved surfaces is not uniform, the shape of the observed FECO is a direct representation of the relative geometry of the surfaces (Fig.13). In particular the curvatures of the surfaces can be measured along two perpendicular directions in the plane of the surfaces. This is of great advantage as it may not be assumed that the local radii of curvature of the directions of the principal axes of curvature of the mica surfaces are the same as those of the silica lenses that support them because of the possible significant variation in thickness of the glue that holds the two together. Thus the local geometric mean radius or inverse Gaussian curvature of the surfaces can be determined within 10% accuracy [81] or better [57], offering the possibility of using eqn.16 for the then normalized force. The lateral magnification is set by the microscope that focuses the light emerging from the interferometer to the spectrometer slit (the theoretical limit to the lateral resolution is of the order of a wavelength of visible light) and features of lateral dimensions larger than about 1 μm can be observed[9] : the evolution of liquid bridges and vapor cavities that

[9] Note that the magnification in the radial direction is about 30×, obtained by a low-power microscope, whereas the "magnification" along the normal direction obtained by the optical interference technique is very much greater, typically 30 000×. So while the

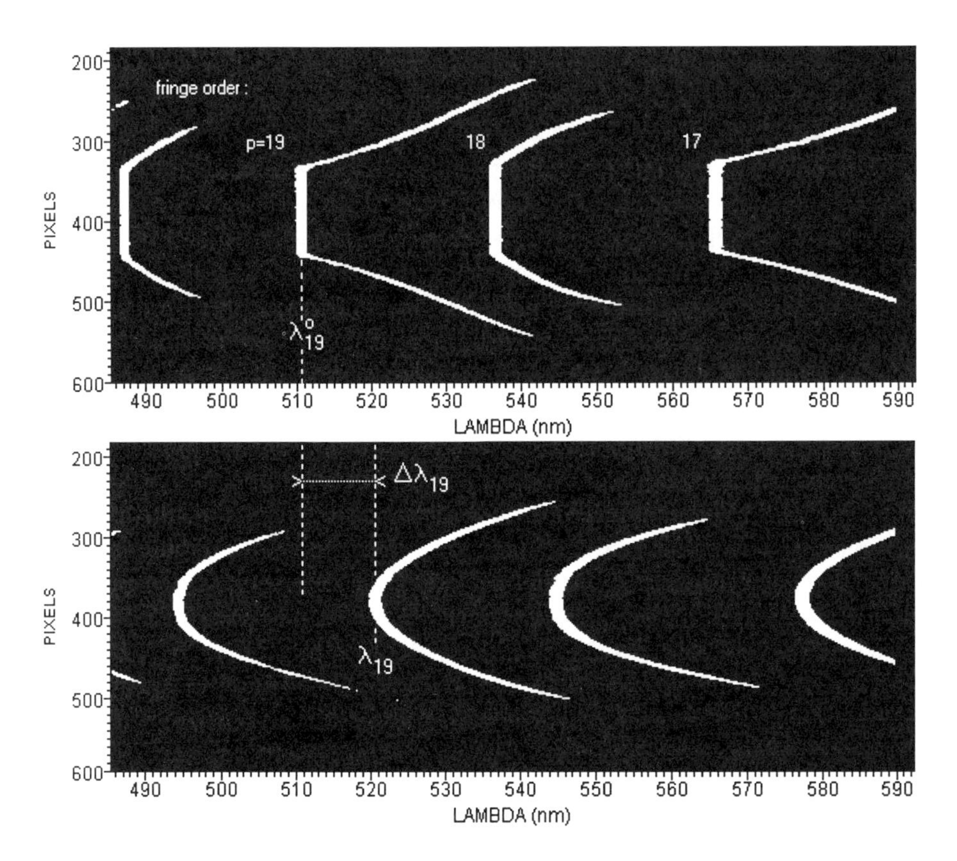

Figure 13. FECO fringe patterns produced by two crossed cylindrical sheets of mica (thickness $\approx 1.58\ \mu$m) in contact (top) and separated by about 70 nm in water (bottom) [57]. Mica is birefringent and each fringe normally appears as a doublet ; here the white incident light is polarized to extinguish one of the vibration directions. Along the radial direction (vertical axis) the magnification is about 32 (one pixel corresponds to about half a micrometer). When surface forces have pulled the two surfaces together a flat region of molecular contact has formed due to the deformation of the underlying glue layer underneath the mica sheets (however, there is probably a layer of water molecules trapped between the surfaces). Note the sharp breakaway at the edges of the contact fringes with alternation of opposite curvature for successive fringes : odd-order fringes have a large gradient $d\lambda/dD$ whereas even-order fringes have a small one (this is due to the alternation between + and − signs in the denominator of eqn.17).

develop between the surfaces [82-85], the changing local shape of the crossed cylinders that may undergo surface deformation due to the action of surface

---

distance resolution in the direction normal to the surfaces is extremely high (≈ 0.2 nm), it is only moderate ($\approx 1\ \mu$m) in the radial direction.

forces [86-88] enabling the deformation of the surfaces about the region of contact to be measured as a function of the applied load upon investigations of adhesion [87-89]. Note that the glue adhering the mica to the discs is rather compliant, and the surfaces deform when they are placed under load. For instance when the surfaces are brought into a strong adhesive contact a "flattened" region forms and is easily observed in the FECO fringe pattern.

force measurement :

Accurate force measurements require precise control of the surface separation. This is usually realized either mechanically, by a system of micrometers and differential springs, which reduce the motion by the ratio of the spring constants, or by piezoelectric devices. The advantage of the latter is that, unlike the mechanical devices, they displace the surfaces smoothly, without causing vibrations. Vibrations may introduce shearing forces and the relative motion of the two surfaces is not then along a straight line. Unfortunately, nonlinearity, hysteresis, and creep are undesirable characteristics common to all piezoelectric positioners limiting their useful range for distance control. These drawbacks may be overcome by suitable electronics [57] such as capacitor insertion to compensate for hysteresis and creep in piezoelectric actuators and a highly accurate, linear, and versatile control of surface displacement without any hysteresis loop can be achieved in the subnanometer range. Alternative ways to change the surface separation have also been proposed (for instance by means of a magnetic field [90]).

In the SFA design schematically pictured in Fig.12 one of the surfaces is attached to a piezoelectric tube while the other one is mounted at the end of a double-cantilever force-measuring spring that can be moved to bring the surfaces to a given separation. A double cantilever is preferable to a single leaf spring because it prevents the curved surfaces from rolling and shearing as the load is varied [91]. Force measurement is begun with the surfaces separated well beyond the range of any surface force, and the piezo crystal (or the dc motor which moves the spring system) is used to step one surface toward the other. After each displacement, the surfaces are allowed to come to rest, and their true distance is measured by the optical method. This process is repeated and the measured separation profile as a function of the displacement of the actuator is recorded. A straight line of slope equal to 1 is obtained at large separations where no force of interaction occurs. This calibration of the motion control is used to infer the spring deflection at separations where a surface force occurs. In this static method, forces between the surfaces of the crossed cylinders are read from the deflection of the cantilever spring system and calculated by Hooke's law knowing the spring constant (of the order of 100 N/m ; the spring constant is calibrated to within 1% after each experiment by placing small weights at the place where the surfaces were contacting and measuring the deflection by a travelling microscope). Provided the drift of the surfaces is negligible during the course of the experiment the force can be determined with a resolution of about 50 nN (when normalized by the radius of curvature of the surfaces ($R \approx 2$cm) this is equivalent to 0.002 mN/m that is a free energy of about 0.3 μJ/m^2)[10].

Repulsive forces are seen as a continuous deflection away from contact and are limited only by the onset of deformation of the surfaces (as the glue between the silver layer and silica support disk is compressed ; this typically occurs for applied loads in the range 8-15 mN/m [86, 87]). This deflection technique is not suitable for the measurement of strongly attractive surface interactions because of the mechanical instability which occurs when the slope of the attractive interaction exceeds the spring constant of the force measuring spring. As a consequence the technique allows forces to be measured only over the regions where the gradient of the force $\partial F/\partial D$ is smaller than the spring constant K. In these measurements only parts of the force curves are directly accessible and the force vs. distance profile appears to be discontinuous, with jumps from unstable (i.e. $\partial(F/R)/\partial D > K/R$) to stable mechanical regimes. Use of stiffer cantilever springs will increase the range of stability, but only at the expense of sensitivity in measuring the force. Another way to overcome this restriction is by means of a spring with a tension that can be varied during the course of an experiment [49, 50, 93]. The device is used to measure the separation at which the spring instability occurs (the jump position) and this is recorded along with the spring constant. In this way a plot of the derivative of the force, which in the Derjaguin approximation is just the pressure between planes, as a function of separation can be constructed.

In the traditional use of the SFA, the spring deflection is inferred from a calibration of the piezo/motor motion control. In other variants it can be measured directly with a force sensor, for instance a piezoelectric bimorph, which permits the spring bending to be measured electronically from the charge developed when the device is strained [94, 95]. With this technique the spring deflection and surface separation can be measured independently. This has allowed to use the force sensor to regulate a magnetic force applied

---

[10] The surface force can also be measured using a dynamic method, one surface being driven toward the other at constant rate. The force-distance profile measured is the sum of both the hydrodynamic force and the static surface force [32]. By recording this profile at different driving speeds, and assuming that the static force is the same at a given separation for each speed, then the static and hydrodynamic components can be extracted. This method has for instance been used when determining the strongly attraction between macroscopic hydrophobic surfaces [92]. It is valid provided the hydrodynamic interaction can be calculated accurately, which means that the viscosity of the medium and the location of the slipping plane should be known.

to the end of the cantilever in a feedback loop [95-97]. With high feedback gains the sensor can be maintained at a constant nominal null deflection regardless of the presence or not of surface forces. This method partly overcomes the spring instability problem and allows continual recording of both repulsive and attractive surface forces [95, 98, 99]. Furthermore, because of the null deflection of the bimorph, the nominal or undeformed separation can be inferred from the motion control without the need to use optical interferometry. This technique has been utilized in a new device known as the MASIF (see § 2-3).

2.3. THE LIGHT LEVER METHOD : ATOMIC FORCE MICROSCOPY AND LIKE-TECHNIQUE

The most common application of the Atomic Force Microscope (AFM) is imaging of surfaces : a three-dimensional image of a surface is obtained by scanning a fine tip over a sample. The design has been derived from the invention of the Scanning Tunneling Microscopy (STM) [100]. In the STM an extremely sharp conducting stylus is brought to within nanometers of a conducting surface. The stylus can be scanned over the surface with the variation in tunneling current monitored as a function of surface coordinate. To overcome the limitation of the STM that the investigated surface must be conducting the Atomic Force Microscope (AFM) was designed [101]. Although similar in concept to existing profilometers, this new instrument provided almost three orders of magnitude in lateral resolution, and one order improvement in vertical resolution.

The force sensing probe in the AFM is simply a cantilevered beam with a sharp tip at the free end. The tip behaves in a way analogous to the pickup on a gramophone. The taper along the length of the stylus acts firstly to resolve tiny lateral distances at the apex, whereupon the broad base of the stylus acts to amplify these displacements to a macroscopic surface. The high vertical resolution comes about from the mechanical gain that the cantilever provides. The weaker the cantilever the more sensitive it is to force. This is counterbalanced by the decrease in resonant frequency gradually making the cantilever sensitive to external vibrations. In practice the surface is scanned passed the stylus and the cantilever is constrained to deflect only in a plane perpendicular to the sample surface. A feedback circuit is designed to maintain the cantilever at a constant deflection by continuously altering the height of the sample surface. Alternatively the cantilever deflection can be recorded as the surface scans at a fixed height underneath the tip. Cantilever deflection can be detected in a number of ways : tunneling current between the cantilever and an STM tip, optical interferometry, capacitance gauges (all reviewed in [102]) ; but the 'light lever' method [103, 104] is by far the most widely used (Fig.14). In this latter operation, the best spatial resolution is usually obtained when the feedback maintains a net deflection of the cantilever against a repulsive force. A common procedure is to use the repulsion arising from the overlap of electronic orbitals (Born repulsion, or 'contact mode'), but any repulsive tip-sample interaction suffices for imaging [105].

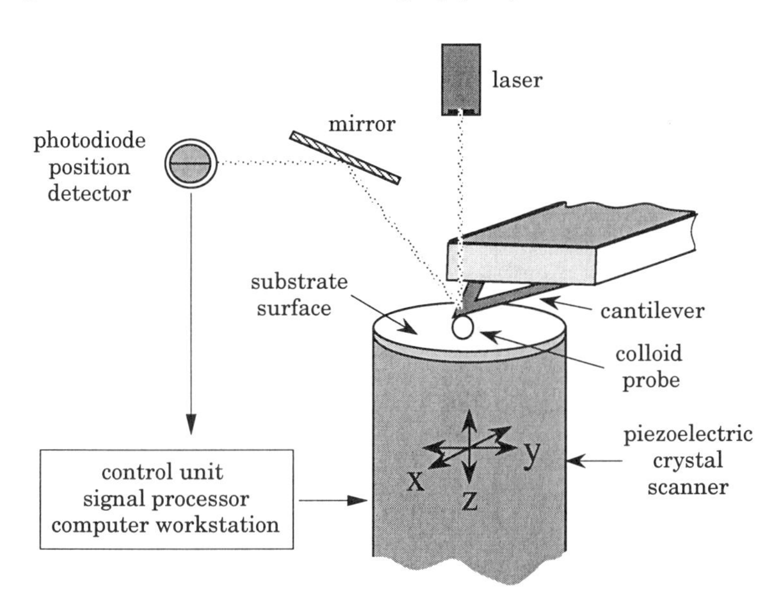

Figure 14. Schematic diagram of the light lever Atomic Force Microscope. A laser is directed onto the tip of a cantilever which reflects onto a split photo-diode. As the cantilever deflects the reflected beam moves across the split diode and the relative strength of the signals from each diode changes, enabling the cantilever deflection, and hence the force, to be obtained. The distance is controlled by the expansion of a piezoelectric ceramic onto which the lower surface is mounted. In this drawing, a spherical "colloid" probe has been attached to the end of the cantilever.

In the light lever method, laser light is focused onto the free end of the cantilever (Fig.14). Reflected laser light from the back of the cantilever initially falls equally across the two photo-diodes (A and B). When the cantilever is deflected, a change in the angle of the reflected laser light creates a greater bias on one diode over the other. The difference of the two photo-diode signals is proportional to the degree of cantilever deflection. Typical lengths of cantilevers can be either 100 or 200 μm long. When this length is compared with the distance from the free end to photo-diodes, approximately 40 mm, a mechanical gain of 200 to 400 times is realized. Thermal noise ultimately limits spatial resolution to around 0.1 nm. Commercially available microfabricated cantilevers are in the form of a flattened 'V'. This geometry reduces the torsional component, as a twist in the cantilever will cause the laser light to reflect out of the plane of the photo-diodes. In direct force measurement the diode difference signal is normalized by the total laser light intensity $(A-B)/(A+B)$, and monitored as a function of vertical piezo displacement, d (Fig.15).

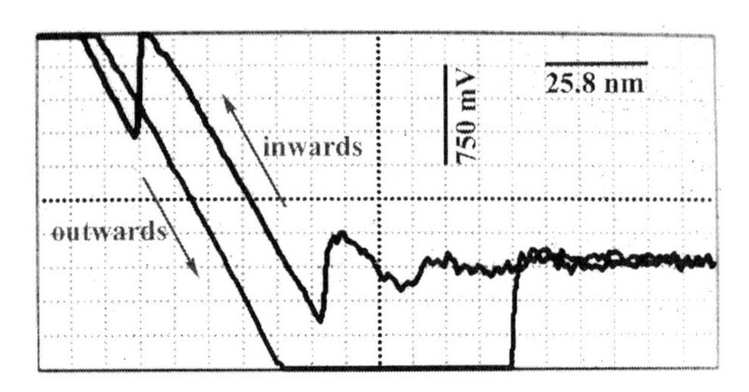

Figure 15. Deflection curves obtained when a glass sphere attached to an AFM cantilever is ramped against a flamed silica flat in an aqueous solution of 3 wt.% CTAB (hexadecyltrimethylammonium bromide) [102]. Note that the spring instabilities appear as near vertical lines. The compliance region is seen as two parallel lines on the far left hand side. The depth of the adhesion has been 'clipped', but may be inferred from the outwards jump distance.

At large surface separations the cantilever is undeflected and the (normalized) photo-diode difference, Z, remains null (region referred to as the *baseline*). When the two surfaces are in contact their movements are coupled so that a piezo displacement, d, causes an equal displacement in the cantilever deflection, z. In this region the slope is constant ($\Delta Z/\Delta d = \Omega$) and has been termed the region of *constant compliance*. That is, the cantilever complies constantly with piezo displacement. The constant compliance region is used to calibrate the optics and hence the gradient of the normalized photo-diode response, Z, into a real deflection in nanometres, $z = \Omega Z$. At mechanical equilibrium the surface force is equal to the opposite spring restoring force given by the Hooke's law $F = -Kz$. Without precise knowledge of spring constants, K (from 2×10^{-2} to 200 N/m for commercially available microcantilevers of various shapes and lengths made in silicon nitride or silicon), no quantitative force measurement is possible ; therefore techniques have been developed and now abound for its determination : calculation by finite element analysis [106] ; resonant frequency measurement [107] ; gravitational field method [108] ; heterodyne interferometry [109] ; methods based on the intrinsic thermal noise of the cantilever [110] or its hydrodynamic drag [111]. As in the SFA, the cantilever becomes mechanically unstable when the gradient of the force exceeds the spring constant. Under these conditions the cantilever jumps to the next stable location where $\partial F/\partial D \leq K$. The spring instabilities appear as near vertical lines in the deflection curves (Fig.15) and the corresponding force-distance profiles become discontinuous with inaccessible parts.

Force-distance profiles can be obtained by ramping continuously the cantilever towards the surface of the sample and apart. Unlike the SFA technique, there is no direct determination of the surface separation, D, since in AFM force measurement, it is inferred from the amount that the sample surface is displaced, d, and the degree the cantilever is deflected, z, in the presence of a surface force, i.e. $D = d+z$. Likewise, the force, although measured directly as an unscaled deflection requires the necessary scaling quantity Ω to renormalize each force curve. The whole technique essentially hinges on the interpretation of the region when the two surfaces come together and their motions become mechanically coupled. Several difficulties arise. In this region it is assumed that the surfaces are incompressible, and consequently only the cantilever is deformed under the loads applied. The observed hysteresis in the position of this 'contact line' due to friction has been the subject of recent discussion [112]. For some systems, especially those where the surfaces have adsorbed species, for example ions or organic molecules, higher applied loads may result in a change in the surface separation or in the slope. This is a result of either the forced desorption, compression, or depletion of species and subsequently another region of constant compliance may be attained (as illustrated in Fig.15: ascribing the region of constance compliance to the very far left hand side was only possible thanks to the comparison with SFA force profiles measured on the same system, Fig.16). Whether or not an adsorbate can be displaced will depend upon the probe radius, as it is the applied pressure which is important. As a consequence not only the magnitude of the force may be in error (since it is transformed via the factor Ω) but also the separation. Since distances are calculated relative to a hard wall contact, the method does not allow the determination of the thickness of firmly adsorbed layers.

A major difficulty associated with the AFM technique is to control the geometry of interaction. This has become increasingly clear as most imaging is now conducted in aqueous environments where the issues of surface chemistry and of geometry cannot be ignored. The act of imaging is derivative to force measurement [105]. Force-distance profiles between a scanning tip and a surface are hard to interpret due to the difficulty of determining the geometry relevant for the interaction zone. The tip interacts with an effective mean curvature depending upon the separation range. Owing to tip roughness, the effective radius, R_{eff}, as seen at large separations (R_{eff} = 50 to 400 nm at D = 10-100 nm [114, 115]) does not necessarily reflect the small scale curvature. When the range of the force

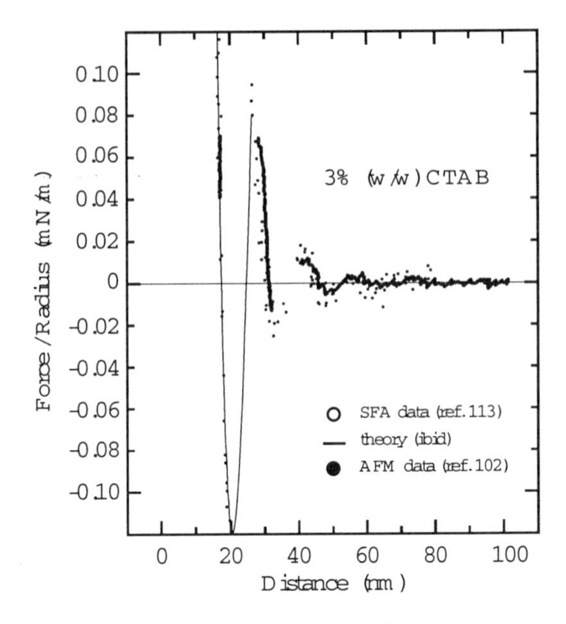

Figure 16. Interaction of two CTAB bilayers supported by a silica sphere and a flamed silica surface across an aqueous solution of 3 wt. % CTAB (the deflection curve for this set of data is shown in Fig.15). The same interaction with mica as the supporting surface [113] is also shown. The comparison of the two sets of data highlights the absence of an absolute scale of separation in AFM experiments.

becomes similar to the scale of the roughness, then the effective radius of the surfaces will decrease.

The inherent deficiency due to the unknown geometry of interaction can be overcome if the fine scanning tip is replaced by a probe of well defined geometry, for example a spherical colloidal particle. Attachment of the probe particle to the cantilever with glue (such as the epoxy resins used in the SFA) proceeds with the help of a XYZ-micromanipulator under a microscope. This 'colloid probe technique' [116] has proved a powerful approach : the surface force can be normalized by the radius of curvature of the attached particle and hence transformed to the surface free energy of the system using the Derjaguin approximation (see § 2-2 ; eqn.16). Particles of different materials, of sizes ranging from 1 μm for colloidal particles to 100 μm or more for glass spheres [102], and of various shapes [117] have been utilized. Whereas colloid probes are useful for large repulsions, particles with large radius of curvature, R, provide access to the range of very small surface forces (down to 5×10^{-4} mN/m) [118] provided the gain in F/R resolution is not offset by difficulties due to the appearance of a large hydrodynamic force. Force measurements must be made at slow speeds so that this additional force is negligible. The reduction in speed eventually becomes victim of thermal drift, that is the baseline drifts appreciably within the duration of the measurement. There are other concerns about the addition of a particle to the end of a microcantilever. As glue is used to attach the probe particle contamination may arise. Another concern is that particle attachment may reduce the length to the point of loading, and hence stiffen the cantilever. Finally, when evaluating the data obtained from the AFM colloidal probe technique it is assumed that the radius of curvature of the particle measured from electron microscopy (in vacuum, after conclusion of the experiment) is the same as the radius of the particle under the conditions of the force measurement (often in a liquid). This may not be correct since the particles may swell in the liquid, may be rough on a small scale that nevertheless is important for the magnitude and range of the interactions (the plane of charge is not as well defined as for an atomically smooth surface), may be porous or built up by aggregated primary smaller colloidal particles. Uncertainty in the effective ratio $(K/R)_{\text{eff}}$ values appears as a major limitation of the AFM technique[11].

Some difficulties encountered in the AFM technique are overcome in the design of a recent device, known as the MASIF (Measurement and Analysis of Surface Interaction Forces) [122]. A bimorph, two piezoelectric slabs glued together with opposite polarization directions and contained in a teflon sheath, constitutes the force sensor ; a charge proportional to its deflection is produced under the action of a surface force and converted to a voltage by an electrometer grade operational amplifier [95]. A well-defined sphere-sphere geometry is used with known surface radii. Usually flame polished glass surfaces are used (radius ≈ 2 mm), prepared by melting the end of a glass rod in a butane-oxygen burner. The surface energy of the glass causes the molten glass to form a highly smooth (RMS roughness of 0.2 nm) spherical drop on the molten end. One of the surfaces is clamped at the end of the bimorph sensor, the other is made to approach using a piezoelectric crystal. An experiment is conducted and analyzed in much the same way as with an AFM when used for force measurements. The surfaces are brought together into contact in a continuous manner, and then separated apart. When the surfaces meet a 'hard wall' contact, the motion of the ramping moving surface is directly transmitted to the bimorph allowing the sensitivity of the force sensor to be calibrated (this assumes of course that the surfaces are incompressible). Thus the deflection of the

---

[11] This issue of force normalization can be successfully addressed by adjusting the radius as a fitting parameter when the measured surface force may be compared to a known interaction. In this method, at the completion of experiments, two surfactant bilayers of known standard long-range interaction [119, 120] are adsorbed to the surfaces and used to scale the magnitude of the absolute force, F, with a calibrated ratio $(K/R)_{\text{eff}}$ [115, 118, 121].

bimorph is known at each instant within 0.05 nm, and the force exerted is obtained using Hooke's law and the spring constant (accurately measured from the resonant frequency of the device ; in the range 100-500 N/m). Assuming that the local radius R at the interaction point does not differ significantly from the macroscopic radius, the resolution in normalized force F/R is about 0.01 mN/m. As the surface separation is calculated from the known expansion of the piezo tube and the measured bimorph deflection, the reported distance in the force profiles is relative to the position of the hard wall reached in each force run.

Despite recent advances in the AFM and MASIF force measurements, the technique of driving continuously one surface toward the other until they come into contact suffers from a main deficiency, which ultimately restricts their use, namely the impossibility of defining an absolute scale of surface separation. The force vs. distance behavior is inferred from a regime assumed to be at 'constant compliance', i.e. when the deflection of the cantilever becomes linear with respect to surface displacement. Furthermore one can note that the use of AFM microcantilevers and MASIF bimorphs, which are both single leaf springs, will lead to an underestimate of the measured adhesion. Indeed, upon separation the motion of the surfaces does not remain parallel and the surface that is mounted on the spring rolls and shears against the other, with a consequent shift in the point (or area) of contact and a concomitant reduction in the pull-off force. By contrast, this drawback is reduced with a double cantilever spring (as used in the SFA ; see § 2-2) because the motion of the spring-mounted surface is constrained to move parallel with the motion of the opposite surface. Finally all these devices suffer from the same spring mechanical instabilities present in the SFA. Feedback schemes are called for. These aspects are addressed in the capacitance based device described in the next paragraph.

2.4. CAPACITANCE BASED DEVICES

Efforts have been made to measure forces between solid surfaces without the reliance on interference fringes as the primary source of information. This means that two main restrictions can be lifted : opaque surfaces can be utilized and solids have no longer to be necessarily large (relative to colloidal dimensions), as shown in the crossed-filaments experiments of Derjaguin and collaborators (see § 2-1). To detect and measure the forces as a function of separation between the solids, the light lever method of the AFM or the use of piezoelectric bimorphs as strain gauge leaf springs to monitor changes in the relative vertical positions, have been very successful (see § 2-3). Another technique is to use capacitors. With the balancing of a type of bridge they can measure spring deflections, which are small, with great accuracy. One of the first uses of the capacitance method was by

Overbeek and Sparnaay who measured, with an accuracy of 1.2 nm for the spring deflection, dispersion forces between glass plates at about 1 μm of separation [34]. This resolution was also exploited in the approach to mapping out the force curve by a dynamic jump method [123], to monitor the dynamic behavior of confined liquid crystals [124], and in the combination of a capacitance sensor and a magnet and coil driving unit called an ultradynanometer by their designers [125, 126]. In addition to measuring spring deflection and distance of separation, the capacitance technique is particularly attractive. By measuring the capacitance between the surfaces, one is probing directly the area of contact between the surfaces, since the measured signal is heavily weighted by the area of least separation. One can thus investigate the dielectric properties of confined materials, and monitor the area of contact of two solids, a technique ideally suited to study the mechanics of contact between two surfaces and adhesion hysteresis with great temporal resolution [56, 127].

Taking advantage of both great sensitivity and temporal resolution in capacitance measurements significant improvement in surface force determination has been achieved in the surface force apparatus designed at the Ecole Centrale de Lyon, France. The main features of the Ecole Centrale de Lyon surface force apparatus (ECL-SFA) [128, 129] are a well-defined sphere - flat geometry for the interacting bodies (although crossed-cylinders may also be used), an accurate control of the surfaces movement together with a high resolution force sensor based on capacitance measurements, and a high rigidity of the device without loss in sensitivity. Static and dynamic measurements can be performed with this device, giving access not only to surface forces, but also to various rheological properties of thin films confined between a rigid sphere and a plate, and to the adhesive properties of many systems. It can also be used as a nanotribometer for nano-rheology studies by replacing the sphere by a diamond tip.

The general principle of the ECL-SFA (Fig.17) is that a macroscopic spherical body can be moved toward and away from, in the three directions $OXYZ$, a planar one using the expansion and the vibration of a piezoelectric crystal (note that a crossed-cylinder geometry can also be used in this apparatus). A sphere of millimetric radius R is firmly fixed to the three axial piezoelectric translator. The plane specimen is supported by double sensors, measuring normal and tangential forces. Each of these is equipped with a capacitive sensor and a double-cantilever spring. The high resolution of the sensor allows a very low compliance to be used for the force measurement (2×10^{-6} m/N). Three capacitive sensors are designed to measure relative displacements in the three directions between the supports of the two solids, with a displacement resolution of 0.01 nm in each direction. Each sensor capacitance is determined by incorporating it in an LC oscillator acting in the range 5-12 MHz. A drop of fluid is introduced between the plane and the sphere to form a meniscus. Two cases must be distinguished. When the size of the meniscus is large (millimetric), the behavior observed at the interface is identical to that of a fully immersed system. Conversely, it is the behavior of microscopic liquid bridges (a few nanometers high) which governs the other phenomena. In the following detailed description of the ECL-SFA principle we will focus on the distance dependence measurement of *normal* forces only.

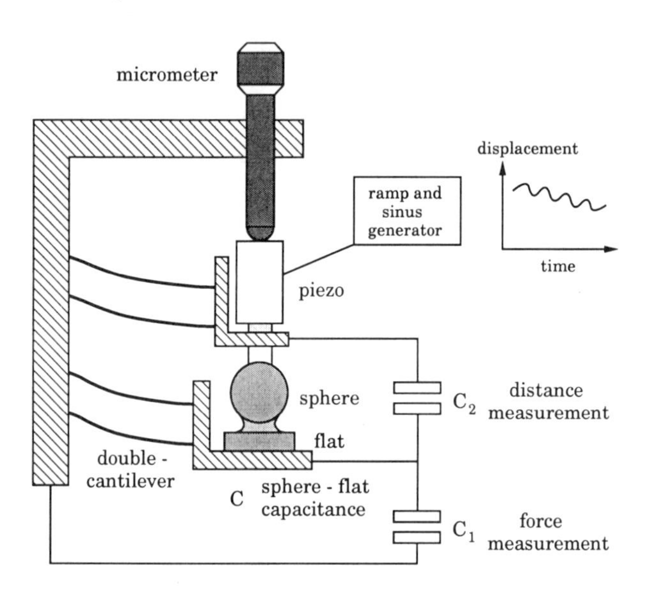

Figure 17. Schematic drawing of the Ecole Centrale de Lyon Surface Force Apparatus. A sphere can be moved towards and away from a plane using a piezoelectric crystal which can expand and vibrate. The flat is supported by a double-cantilever spring, whose stiffness can be adjusted continuously. A first capacitive sensor C_1 measures the elastic deformation of the cantilever and thus the force transmitted to the plane ; a second one C_2 measures the relative displacement between the two solids. Taking advantage of the metallic nature of the bodies, the measurement of the electrical capacitance of the sphere-flat interface allows one to determine an absolute scale of separation.

A first capacitive sensor C_1 measures the elastic deformation of the cantilever and thus the force transmitted through the liquid to the plane. A second sensor C_2 is designed to measure the relative displacement h between the supports of the two solids ($h = D + h_0$ where D is the surface separation and h_0 a constant offset). A feedback loop acting between this measurement and the piezoelectric crystal allows to control h. The normal approach of the sphere to the plane is a superposition of two components : a steady ramp (which gives a constant sphere-plane approach speed between 5 and 0.02 nm/s) and a small amplitude oscillatory motion of about 0.1 nm

RMS with a frequency ranging between 10^{-3} and 500 Hz. This combination allows three different kinds of measurements to be simultaneously recorded :

(i) the "quasi-static" components of the normal force and of the relative displacement D or h, obtained by measuring the mean value of the two signals given by the transducers. The resolution is 0.015 nm for the displacement and 10 nN for the force. The high resolution of the sensor allows high stiffness to be used for force measurement. Not only large applied loads (up to 50 mN) can be measured thanks to the high rigidity available, but the extent of the inaccessible regions of force profiles due to the mechanical spring instabilities (when $\partial F/\partial D > K$) is reduced. As the stiffness of the double-cantilever spring supporting the plate can be adjusted continuously over a wide range (K between 4×10^3 and 6×10^6 N/m), a continual recording of both repulsive and attractive forces as a function of surface separation becomes available [130]..

(ii) the mechanical transfer function. The harmonic components of the two capacitor signals are measured with two-phase lock-in analyzers. So long as static forces remain constant they do not disturb dynamic measurements. This is achieved by vibrating the piezoelectric crystal at low amplitude around an average separation D. The displacement resolution, which is better than 10^{-4} nm, makes possible dynamic experiments with very small amplitudes, as little as 5×10^{-3} nm. This is a very important advantage when performing experiments at high frequencies (> 1 kHz) and small separations with a highly viscous liquid, as the strain must be low enough for the linear viscoelastic condition to be fulfilled (i.e. the hydrodynamic force being proportional to the shear rate). Elastic modulus, shear modulus, and other viscoelastic parameters have been extracted as a function of the confinement [131, 132]. In particular when the confined liquid is a pure viscous fluid (viscosity η), the thickness D_0 of the immobile layer of liquid adjacent to the surface can be extracted from the damping function $A = -F_{\text{hydrodynamic}}/(dD/dt)$ measured during the steady approach of the surfaces : $A = 6\pi\eta R^2/(D - 2D_0)$ (Fig.18) [129].

(iii) the electrical capacitance of the sphere-flat interface (the bodies are conductors), C, obtained by measuring the harmonic component that results from the sinusoidal motion of the displacement. As absolute determination is difficult, the first derivative, $\partial C/\partial D$, is used. This measurement allows to address successfully the delicate question of defining an origin and an absolute scale of surface separation. Theoretical calculations show that $\partial D/\partial C = D/2\pi R\epsilon_0\epsilon_r$ where ϵ_0 is the absolute dielectric constant of vacuum and ϵ_r the relative dielectric constant, as long as $D \ll R$ and the separation between the electrodes is larger than four to five times the peak-to-valley amplitude of roughness [133]. As in such conditions D and h only differ by an unknown but constant offset, an extrapolated point such as $\partial D/\partial C = 0$

corresponds to the zero relative to the separation D (Fig.18). The origin so defined has the advantage of still having a meaning whatever happens throughout the close approach of the solids (e.g. deformation, adhesion).

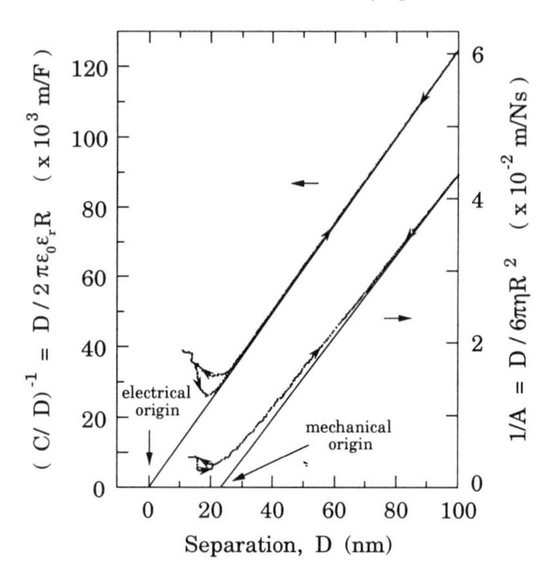

Figure 18. Drainage of a colloidal solution thin film between a sphere (R = 15.4 mm) and a flat (the peak-to-valley roughness of both steel surfaces is about 3 nm) [134]. Both the inverses of the electrical capacitance $(\partial C/\partial D)^{-1}$ and of the damping function $1/A(D)$ are linear with the approach displacement at large separations. Ideally, if the dielectric constant and the viscosity equal the bulk values at all separations, the curves should fall on straight lines passing through the origins. The experimental data agree with this analysis for $D > 12$ nm indicating that the sphere-plane interface is not deformed. The difference between the electrical and mechanical origins (23 nm) indicates that an immobile region of adsorbed colloids has built up on each surface with a thickness comparable to the width of a colloid (10 nm diameter).

The surface force is extracted from the measurement of the total force present between the sphere and the plane, once the hydrodynamic force and the wetting or capillary force, have been subtracted. The hydrodynamic contribution is determined from the dynamic components of the measurement (i.e. the damping function of the interface; see above) ; its contribution can be made negligible at low speeds and thus one is left with quasi-static normal forces only. The wetting force is due to the presence of an annulus of liquid confined between the sphere and the flat. It includes a contribution resulting from the negative Laplace pressure, and a resolved component of the liquid-vapor surface tension γ around the perimeter of the annulus. For *macroscopic* fluid confined drops, the first contribution is equal to $4\pi R\gamma/(1+2DR/r^2)$ where r is the principal radius of the meniscus (see inset of Fig.19). Because r is much smaller than the millimetric radius of the sphere, R, the surface tension contribution is negligible in comparison to the capillary pressure in the annulus ; furthermore, the experiments are conducted in such manner that $D \ll r$, so that the volume of the droplet stays constant. Therefore the wetting force does not vary significantly with the separation and its contribution to the total force can be considered as a constant offset. Conversely, *microscopic* liquid droplets, as those formed by capillary condensation between solid substrates, give a complex behavior to the interaction [135] :

$$F_{\text{wet}} = -4\pi\gamma R \left(1 - \frac{D - 3e}{2r_c}\right) \tag{18}$$

where r_c is the mean radius of curvature of the droplet (here r_c stands for the absolute value of the radius of curvature, which is indeed negative) and $e \ll r_c$, the thickness of the wetting films coating the surfaces far from the liquid bridge (Fig.19). The factor $3e$ in eqn.18 reflects the thickening of the wetting films in the region where they merge into the liquid bridge. The fluid interface has a nanometric equilibrium curvature r_c^{-1}, governed by the principal radius of the meniscus, r ($\approx$ 4 - 100 nm), as the axisymmetric size ρ of the droplets is a few micrometers ($r_c^{-1} = r^{-1} + \rho^{-1} \approx r^{-1}$). When the rate of the change in separation is slow enough, the liquid phase remains in equilibrium with its vapor, and the curvature of the meniscus remains constant : the capillary force is negative and almost linear (Fig.19). In the other limit of negligible liquid-vapor exchanges, the volume of liquid in the bridge remains constant and the curvature varies as D is changed [136].

The location of capacitive displacement transducers in the vicinity of both the sphere and the flat, obliges experiments to be conducted in dry air conditions. To overcome this limitation a variant of the ECL-SFA has been recently designed [137], and aqueous systems can now be investigated. The relative displacement between the supports of the sphere and the flat is still measured by a capacitor, but the deflection of the double-cantilever, and hence the force, is now determined by means of a Nomarski interferometer.

3. Force measurement between an individual freely moving particle and a solid wall or another particle

In none of the techniques described in the previous section, the devices cannot mimic the dynamic behavior of colloidal particles : either the radius of curvature of the macroscopic surfaces used in the balance devices or SFA - like apparatuses (§ 2) is several orders of magnitude larger than that of a typical colloidal particle, or the attached specimen in the colloid probe AFM technique (§ 2-3) cannot move freely as it would do in a bulk environment. These arrangements are well suited for investigating the e-quilibrium interaction of two bodies, but real particles undergo Brownian

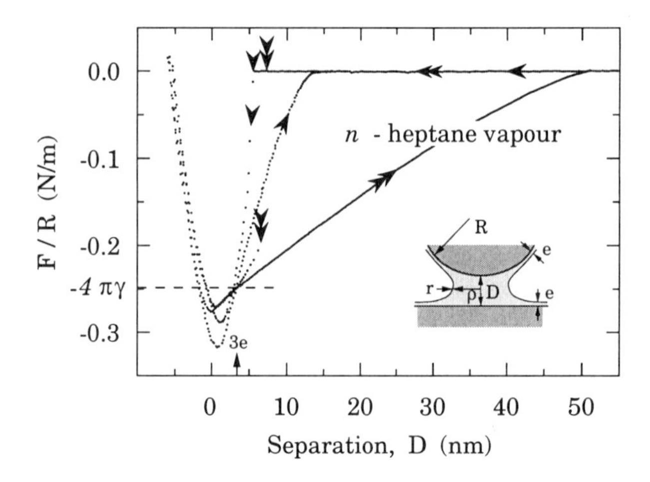

Figure 19. Forces normalized by the radius of the interacting sphere with the plane as a function of their separation in an atmosphere of undersaturated n-heptane vapor (both substrates are made of fire-polished pyrex coated with a 50 nm thick platinum layer) [135]. The arrows on the curves show the direction of the surfaces relative motion, at a constant velocity of 0.1 nm/s. The number of arrows increases with the vapor pressure. When the surfaces advance toward each other, the interaction undergoes a discontinuity due to the condensation of a liquid bridge. The central liquid droplet persists as the surfaces are separated until a separation threshold is exceeded. The trend of larger condensation and evaporation distances, and the fact that the difference in these locations increases as the vapor pressure increases both indicate that there is a hysteresis in the phase transformation signifying a first-order phase transition induced by confinement.

movement, and the distance between them continually fluctuates. Then the equilibrium interaction might not be the relevant one. This is particularly true of colloidal dispersions stabilized by steric repulsion of adsorbed polymer layers, whose conformation can be disturbed by the squeezing flow induced by the relative movement of two interacting bodies. Conversely, osmotic stress techniques maintain the particles in bulk, but the extracted informations result from statistical averages over a large ensemble and the interactions are not directly measured between individual entities. Therefore, for measuring interactions directly between individual entities other techniques must be sought.

One of the first experimental techniques to investigate the interactions between particles and surfaces was the rotating disc method [138-140]. From the measurement of the deposition rate the primary maximum in the potential energy between the particles and the surface could be inferred. In the same way that particle adsorption kinetics provided the first probe for particle - surface interactions, the first studies of particle - particle interactions were through coagulation kinetics experiments, from which the height of the energy barrier preventing coagulation could be extracted.

However, neither the form, nor the range of the potential energy - distance profiles were determined. The last fifteen years have seen the emergence of new techniques which could achieve just this.

3.1. TOTAL INTERNAL REFLECTION MICROSCOPY

Consider a microscopic particle located near a flat surface. Because of Brownian movement the particle - plate separation distance will fluctuate around an equilibrium point where the net force is zero and the potential energy is a local minimum. Assuming that the distribution of separation distances is described by a Boltzmann distribution, profiles of the mean potential energy can be derived from the distance distributions (in contrast surface force balance or spring devices provide a direct measurement of the mean force). Based on this principle, Prieve and Alexander developed a hydrodynamic technique to measure the instantaneous separation distance between a single colloidal sphere and a flat plate [141]. A shear flow tangential to the plate was imposed and the movement of the particle was observed with a microscope. The closer to the wall the more the particle lags behind the fluid due to the hydrodynamic interaction with the wall ; for a constant shear rate the particle velocity decreases monotonically with distance. From the measurement of the distribution in velocities, the distribution in elevations can be deduced, and in turn the potential energy of the particle as a function of the distance from the wall can be extracted. To improve the 10 nm spatial resolution of this shear flow technique, the total internal reflection microscopy (TIRM) was introduced [142]. Also built on the idea of monitoring the Brownian fluctuations of an individual particle located near a surface, the TIRM technique provides a measurement of the instantaneous separation distance with a spatial resolution on the order of 1 nm.

The separation distance is inferred from the intensity of light scattered by the particle as it interacts with an evanescent wave formed upon total internal reflection of light at an interface separating two media of different refractive indices. When an object which scatters light, e.g. a colloidal sphere, approaches the surface close enough to enter the evanescent field frustrated total reflection will occur. Micron-sized dielectric particles located in an evanescent wave scatter radiation at an intensity that decays exponentially by about 1% per nanometer increase in separation between the particle and the reflecting interface [143]. Measuring this scattered light intensity as a function of time thus provides a sensitive and non-intrusive method to determine the separation distance. Because one is dependent on Brownian motion to move the particle against the potential energy gradient, the whole range of the interaction between the particle and the wall is not always accessible. In order to extend the range of the experiment one may apply a radiation pressure to the particle by means of a second laser [144] (Fig.20) : the particle can thus be micromanipulated, pushed against the surface or conversely driven away from the interface, hold at a fixed lateral position or moved to different locations above the plate. By monitoring the separation between the particle and the flat surface over a sufficiently long time (typically a histogram of about 50 000 separation measurements taken at 10 ms intervals is built up), the probability of finding the particle at a given separation is measured, and since this probability is given by the Boltzmann distribution in the potential field, the interaction potential as a function of separation can be evaluated [145]. Using the most probable elevation above the plate as the reference state, potential energy differences up to 10 k_BT and as small as 0.1 k_BT can be resolved. This potential energy usually contains additive contributions from gravitational forces, radiation pressure, and surface forces (Fig.20).

The ability to utilize a freely moving Brownian particle is a unique feature of the TIRM technique which also allows the determination of the mobility and diffusion coefficients of individual particles near surfaces [145]. However, the need for Brownian motion means that measuring forces at very small separations is difficult if not impossible, because the particle may simply deposit onto the plate. Determining absolute separation distances can also be problematic. The most commonly used method is to force the particle to deposit onto the plate (by increasing, e.g., the ionic strength of the solution) and measuring the scattering intensity for a stuck particle. By assuming that the scattering intensity varies exponentially with separation, the distance corresponding to any intensity measurement can be calculated. Although easy to accomplish, roughness on the particle or plate, presence of layers of polymer chains when the surfaces are coated with adsorbed polymer, changes in refractive index upon solution conditions, can all lead to errors in the measured separation. The last drawback to the method that should be mentioned is the validity of the exponential relationship between scattering intensity and separation distance. Valid for micron-sized particles [143], this relationship could be questionable for very large particles and particles with a very high refractive index. Extension of the technique to other systems could require reevaluating this relationship so critical in interpreting the scattering data.

Even with a technique which does not have the angstrom resolution of surface force balance or spring devices, TIRM offers several advantages over these techniques as it does allow one surface to be reduced to colloidal dimensions, with the ability to investigate the interactions of a single, freely moving, Brownian particle, near interfaces (although applications have so far been restricted to particles interacting with solid/fluid interfaces, the technique could easily be adapted to handle liquid/liquid interfaces as well).

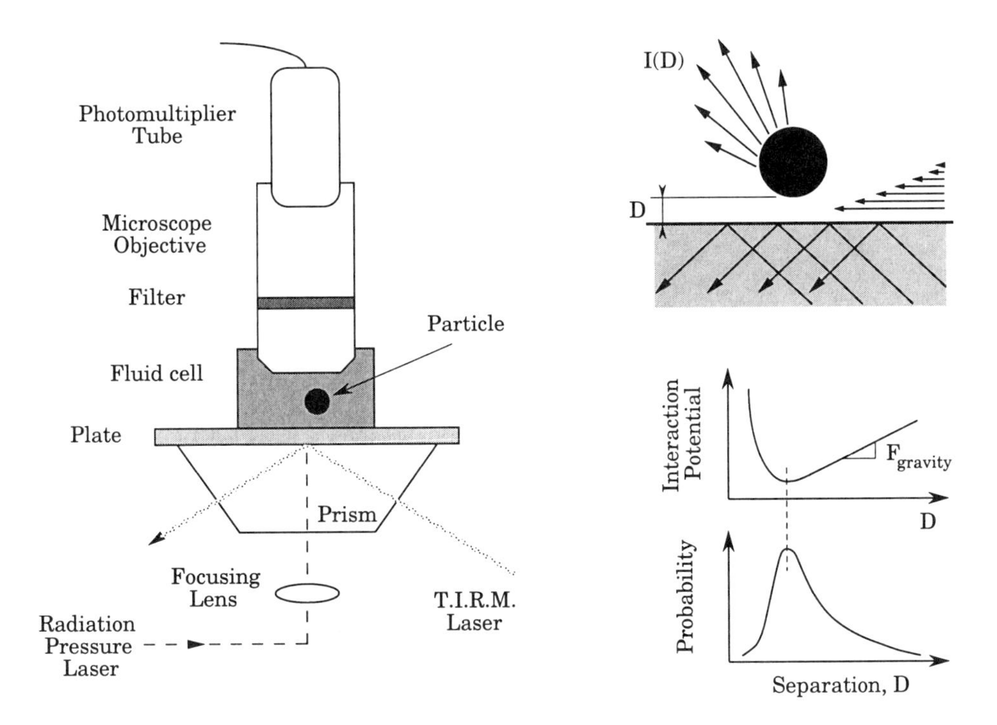

Figure 20. Schematic representation of the combined TIRM/radiation pressure apparatus. The lasers used to form the evanescent wave and radiation pressure are at different wavelengths. A band-pass filter blocks the radiation pressure beam, allowing only the scattered light from the evanescent wave to enter the photomultiplier tube where the intensity is measured. The top right drawing is an expanded view of the beams striking the particle. When illuminated by an evanescent wave (horizontal arrows), the sphere scatters light which is exponentially sensitive to its elevation, D. When only gravity and electrical double layer repulsion act, the potential energy profile of a particle denser than solvent resembles the schematic drawing in the bottom right of the figure. When the particle is far from the plate, gravity is the dominant interaction : the profile increases linearly with separation. When the particle ventures close to the plate, it experiences electrostatic repulsion due to the overlap of the double layers and the potential energy increases as the separation decreases. The most probable location of the particle corresponds to the bottom of the energy profile well, where gravity and double layer repulsion are equal.

Apart from some caution about its use for investigating small separations, the extreme sensitivity of the technique (10^{-14} N in force, i.e. 2-3 orders of magnitude smaller than all other techniques when one normalizes the force by the 2-20 μm particle radius) makes it somewhat ideal for studying long-range interactions. Electrical double layers [146-148], dispersion forces [149, 150], gravity [151], depletion interactions [152-154], and steric repulsion due to adsorbed polymer layers [155] have been studied. Additional size reduction of the second surface where the plate is replaced by a second microscopic particle allows measurement of the interactions between two individual particles. These achievements are presented in the next sections.

3.2. DIGITAL VIDEO MICROSCOPY, AND DYNAMIC OPTICAL TRAPPING

The combination of optical trapping to manipulate individual colloidal particles, high-resolution digital video microscopy, and total internal reflection microscopy, makes possible precise measurements of interactions between particles and dynamics. These techniques have been reviewed in the series of lectures presented by Prof. D. Grier. In Chapter 4 it is described how interactions between isolated or confined colloidal pairs can be determined and what are the effects influencing colloidal motions, including hydrodynamic and electrostatic coupling.

3.3. MAGNETIC CHAINING TECHNIQUE

The technique exploits the fact that the anisotropy of the forces between induced dipoles causes particles to form linear chains [156]. The magnitude of the induced dipole can be controlled by the strength of the applied field, allowing the force to be determined. Moreover, if the particles are uniform in size, light is Bragg scattered from the chains, allowing the separation to be measured. The force-distance profiles can thus be directly determined in a wide variety of materials, including emulsions, foams, sols, polymers or polyelectrolytes [156-161]. Although elegant, the main limitation of this technique is that it can only measure the interaction between magnetic particles.

The technique was employed successively to study electrical double layer and depletion interactions between monodisperse oil-in-water ferrofluid emulsion droplets [156-158, 161]. In this system because the resultant droplets are paramagnetic the droplets align along the applied magnetic field direction (Fig.21). At low droplets volume fraction ($< 0.1\%$) the chains remain well separated and are only one droplet thick. The chaining causes a strong modification of the optical properties : illuminated by a white light source, the emulsion appears colored in the backscattering direction. As the external magnetic field strength is increased, the color changes from brown to red, to yellow, and to green. These colors originate from Bragg diffraction and provide a straightforward measurement of the spacing between droplets within the chains. Indeed, along one chain the center-center distance, d, between droplets can be directly deduced from the spectral distribution of the scattered light at a constant angle. For perfectly aligned particles illuminated by incident white light parallel to the chains, the first order

Bragg condition reduces to

$$2d = \frac{\lambda_0}{n} \tag{19}$$

where n is the refractive index of the suspending medium and λ_0 is the wavelength of the light back scattered at an angle of 180°. This relation gives an accurate measurement of the center-center separation between droplets provided the emulsion polydispersity is reduced to its minimal level. This is one of the main difficulties of the technique.

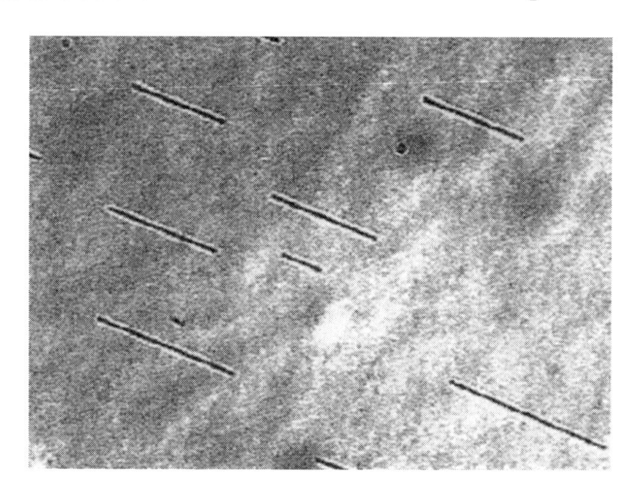

Figure 21. Chains of oil-in-water ferrofluid emulsion droplets (200 nm diameter) aligned with a magnetic field [156].

From eqn.19 the interdroplet surface spacing, D, is then deduced from :

$$D = d - 2R \tag{20}$$

where R is the radius of the droplets (the radius can be inferred from the form factor measured by static light scattering in very dilute systems). Accurate determination of R is of prime importance to get the reference state where the droplets surfaces are in contact ($D = 0$). Inferring the inter-droplet surface spacing from eqn.20 requires also that surface deformation is negligible over the whole range of separation.

The repulsive force between the droplets balances the magnetically induced attractive force between dipoles. For dipoles aligned parallel to the field, the force is proportional to d^{-4} and to the square of the induced magnetic moment of each drop, m :

$$F(d) = \frac{1.202}{2\pi\mu_0} \times \frac{3m^2}{d^4} \tag{21}$$

where μ_0 is the magnetic permeability of free space. Note that only the dipole term has been kept in the multipole expansion of the force. Higher order multipole moments generated by the nonuniform field from the other droplets are usually negligible, and the point-dipole approximation is sufficient for perfect spheres. Within this framework the total applied field H_T is given by the sum of the external applied field H_{ext} and the field from the induced magnetic moments in all neighboring droplets in the chain (considered as infinite) :

$$H_T = H_{\text{ext}} + 1.202 \times \frac{2m}{4\pi\mu_0 d^3} \tag{22}$$

where the induced magnetic moment of the droplet, m, is proportional to its volume and to the total applied field :

$$m = \mu_0 \frac{4}{3}\pi R^3 \chi_S H_T \tag{23}$$

Here χ_S is the susceptibility of a spherical droplet, related to the intrinsic susceptibility, χ, of the ferrofluid by

$$\chi_S = \frac{\chi}{1 + \chi/3} \tag{24}$$

to take into account the demagnetization effect due to polarization. The intrinsic susceptibility, χ, of the ferrofluid follows the classical Langevin form

$$\chi = \frac{\alpha}{H_{\text{ext}}} \left(\coth \beta H_{\text{ext}} - \frac{1}{\beta H_{\text{ext}}} \right) \tag{25}$$

where α and β are two coefficients determined by magnetic susceptibility measurements on the bulk ferrofluid. Expressions (21) to (25) constitute a self consistent set of equations.

Note that this technique can only measure repulsive force profiles and that regions of the force - distance profile are inaccessible whenever the variation of the repulsive force as a function of separation becomes smoother than the variation of the magnetic attraction. This is a situation analog to the one encountered with the spring devices described in § 2 when the gradient of the force becomes smaller than the spring constant. This technique allows one to measure interparticle forces as small as 2×10^{-13} N, corresponding to the minimum force required for forming chains. When normalized by the droplet radius, the ratio F/R corresponding to the onset of chaining up is about 10^{-3} mN/m. The upper limit is set by the saturation of the magnetization. For a ferrofluid, the range of measurable forces is narrow as the largest force measurable is 0.1 mN/m. The typical surface separation ranges from 1 to about 150 nm with an accuracy of $\pm$ 1 nm.

4. Comparison between techniques

Results from three complementary methods of force or energy measurement : the osmotic pressure techniques, the force balance or spring devices, and the single particle techniques, have given insights into the nature of the interactions operating in electrolyte solutions, colloidal dispersions, membrane suspensions, polymer and polyelectrolyte solutions, etc. In each of the methods, one imposes all intensive variables, for instance the osmotic pressure (or the interaction energy between the surfaces in a surface force balance), and one measures the response of the system, which is an extensive variable : the solvent content or the distance of separation between particles or surfaces. The three groups of techniques provide a measurement of the interactions operating between particles or surfaces as a function of their separation, expressed as an osmotic pressure, a surface force, or an interaction energy profile. Hopefully these quantities can be deduced from each other and measurements from the different techniques can be compared : the force normalized by the local radius of curvature of the surface can be transformed into an interaction free energy per unit area provided the Derjaguin approximation is valid (§ 2-2), while the free energy can also be obtained by integration of the pressure vs. separation curve. Nevertheless, there is a fundamental difference between the methods. In force balance or spring devices the measurements are carried out keeping the chemical potential of both solvent and solute constant whereas upon osmotic stress, measurements are performed by changing the chemical potential of solvent for example[12]. Since each of these techniques provides direct measures of some quantities and is limited by inference of extrapolation for estimating others, the best strategy is to use all three to the extent possible. One can note that very different amounts of surfaces are involved in the three methods. The advantages and disadvantages this represents are now discussed with emphasis on the consequences to attain the thermodynamic equilibrium.

[12] An illustrative example is given in § 4-1 for surfactant systems. For macroscopic SFA surfaces coated by adsorbed surfactants the area per surfactant molecule increases as the surfaces are brought together and a repulsive force is experienced (due for example by the overlap of electric double layers). Conversely upon a higher applied osmotic stress, carried out by changing the chemical potential of water, the reverse is true, i.e. this mean area decreases as the interbilayer separation decreases. This means that one would expect the swelling pressure to be slightly more repulsive than the force measured using the surface force technique, even in the absence of undulation force (see Figs.22 and 23).

4.1. SMALL OR LARGE AMOUNT OF SURFACES ? THERMODYNAMIC EQUILIBRIUM

Force balance or spring devices measure the interaction of macroscopic bodies of the order of 1 cm in size, whereas particles or surfaces of interest are several orders of magnitude smaller in size. Of course, one can use Derjaguin's approximation to scale the measurements made there (§ 2) to other geometries like two microscopic spheres or a sphere and a plate (§ 3). However that may be, the area of the surfaces involved in all these force measurement techniques is small (1 cm^2 to a few μm^2) compared to the typical 10 m^2 amount of surface required in techniques based on osmotic pressure measurements (§ 1). There are both advantages and disadvantages in the use of small areas for the surfaces. The surface equilibria for the species that may adsorb or desorb are rapid and the reservoir is allowed to be not necessarily infinitely large as the system investigated is extremely small. Conversely, the presence of any quantity, even in tiny amounts, of species not desorbable may poison the surface and create a non-representative state of the system. In addition it is not always easy to obtain surfaces that resemble the particles to be studied. In particular, due to their large specific surface area colloidal systems are very sensitive to the variation of certain intensive parameters (surface charge area density, electrostatic screening, etc). Therefore it is often desirable to maintain the surface in a given state : with force balance or spring techniques the states of the system are not all assessed.

On the other hand, large sample size allows for accurate definition of composition and favors the detection of small effects. In the techniques based on osmotic equilibria where large specific areas of hundreds of cm^2 per gram are involved, all the preceding limitations are overcome. The sample under study is in a state representative of the system and if molecular rearrangements are not kinetically limited the osmotic pressure measurements will be carried out at thermodynamic equilibrium. In practice the price to pay will be the duration of experiments as the thermodynamic equilibrium will be achieved slowly in a sample containing a large specific area. Indeed one must allow enough time for the species to diffuse, for the molecules to exchange, for the solvent to be transferred, and for the particle surfaces to reorganize, before any measurement is initiated.

An instructive example is the measurement of the interactions between bilayers of surfactants or lipids. In the SFA, one approach has consisted in coating lipids, either by adsorption from suspension [83] or by passage through monolayers (Langmuir-Blodgett deposition) [64, 162] onto mica sheets. Forces between the surfaces of the coated crossed cylinders are measured as a function of their separation as described previously (§ 2-2). Two important differences between the force measurement by means of a SFA and applying osmotic stress to spontaneously forming multilayers are (i) the immobilization of the bilayers that comes from attachment to the mica surface and (ii) the cylindrical versus parallel geometry. To correct for the geometry, the Derjaguin approximation is used and normalized forces F/R measured with the SFA are implicitly related to the *energy* between parallel surfaces [69]. So to compare interactions measured on multilayers with those between crossed cylinders it is necessary either to differentiate the cylinder-cylinder forces or to integrate multilayer osmotic pressures from a hypothetical infinity. For frozen hydrocarbon chains, such as double-chained ammonium acetate surfactants, the bilayers are stiff, and there is no amplifying action of undulatory thermal fluctuations in the bulk. Therefore a good agreement is observed between the SFA results and the interactions measured between bilayers within a multilayer array through the application of osmotic stress to the bulk mesophase : here the multilayers are equilibrated against polyethylene glycol or dextran solutions of known osmotic pressure or against vapors of saturated salt solutions, and X-ray diffraction of the equilibrated phase gives the repeat distance of the surfactant plus water layers (Fig.22).

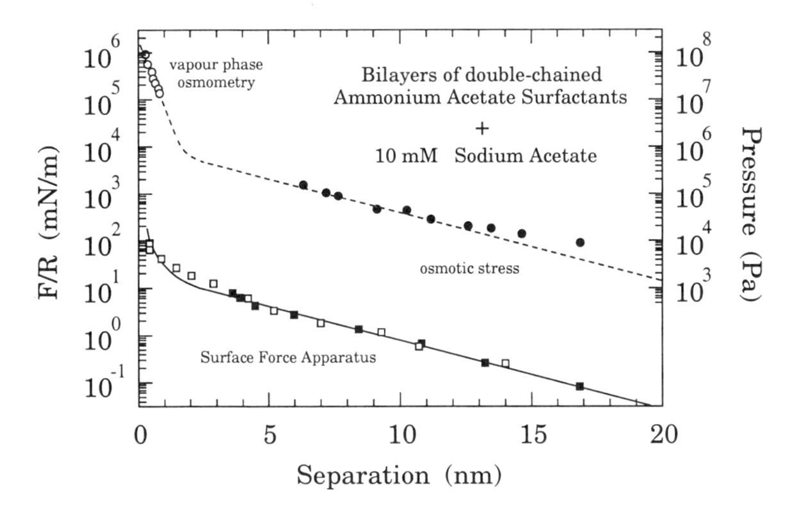

Figure 22. Comparison of forces between DHDAA (dihexadecyldimethylammonium acetate) bilayers in a multilayer using osmotic stress [163] and between bilayers immobilized onto the crossed mica cylinders of the SFA [164]. Osmotic stress data for pressures were fitted (dashed line) as the sum of two exponential repulsions : a long-range electrostatic double layer interaction and a short-range hydration contribution, and then integrated to predict a F/R curve (solid line) that should be obtained from the SFA data. Two SFA series (solid and empty squares) fall remarkably well on that (solid) line.

However, for chains in a liquid-like state, measured interactions between bilayers undulating within a multilayer array deviate from those between adsorbed bilayers immobilized onto rigid mica cylinders of the SFA, where undulations are presumably impossible. Fig.23 shows the force as a function of the separation between bilayers on crossed mica cylinders, differentiated to give the equivalent force per molecule (shaded band), together with measurements of repulsion between bilayers in a multilayer array, also as a force per molecule (points). Both data sets are for phosphatidylcholines with melted hydrocarbon chains. In the regime of strong repulsion the two show similar forces with only a small horizontal shift due probably to differences in the defined zero of separation. But at low pressures there is a distinct divergence between the two data sets ; the limiting spacing of the multilayers is considerably greater than that between adsorbed bilayers.

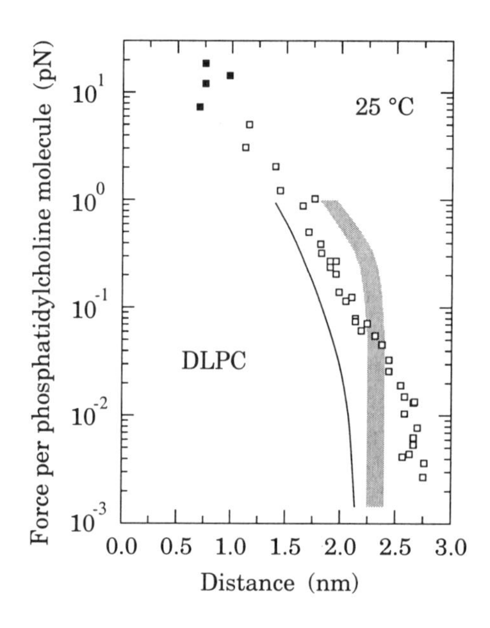

Figure 23. Comparison of forces between bilayers in a multilayer (empty squares) using osmotic stress and between bilayers immobilized onto the crossed mica cylinders of the SFA (shaded band). The data points are for DLPC (dilauroyl phosphatidylcholine) at 25 °C where hydrocarbon chains are melted except at high pressure (solid squares). The solid line is the underlying interbilayer force after subtraction of undulatory fluctuation forces in the multilayer system [165]. This line, remarkably parallel to the fixed-bilayer shaded band of the SFA measurement shows the good agreement between the two techniques once one takes account of the difference in apparent zero of separation (for details see [166]).

These results show that immobilization of bilayers onto macroscopic rigid surfaces modify their interactions, at least by suppressing the undulatory entropic contributions and/or the surfactant head group mobility.

The connection between these measurements to any real interactions between free bilayers or membranes becomes remote and the SFA advantage of directly measuring the interactions is lost. To overcome this drawback another approach has been proposed [71, 167, 168]. In this method, the macroscopic surfaces of the SFA are immersed in a multilamellar mesophase and used as two confining walls. The layers are aligned on the whole parallel to the molecularly smooth walls (homeotropic alignment) contouring the slightly curved surfaces and a network of edge dislocations is formed to accommodate the confinement of the mesophase in the wedge. Note that the alignment of the multilamellar stack would be more difficult with highly curved surfaces such as those involved in the "colloid probe" AFM technique (§ 2-3). The total force across the stack oscillates regularly with surface separations (Fig.24), with a periodicity equals to the bulk reticular spacing. From the parabolic shape of each oscillation (inset Fig.24) elastic properties of the layered structure can be extracted, and thereby the stabilization mechanism of the mesophase can be inferred.

This illustrative example shows that the interactions operating between unsupported amphiphilic aggregates in solution or within the assemblies of the molecular architecture, are not so intractable. The small area of the macroscopic walls involved in force balance or spring devices is not necessarily a penalty and actually there is a wealth of additional information to be gained by application of these techniques and not accessible by others. Beyond the influence of a macroscopic surface on the behavior of complex fluids close to a wall and in thin films confined between two walls (coating, wetting, etc), structural information, such as reticular spacing or order parameter, the nature and dynamics of packing defects [124, 167, 168], the induced phase transformation upon confinement or stress [60, 68, 83, 84, 135, 136, 169], can be obtained.

4.2. RESOLUTION

Now that we have reviewed and seen how the techniques work, let us compare their capabilities. Table 1 gives a quantitative comparison of the techniques. The numbers given represent resolutions in distance and force which have been reported in the literature by the authors who have used each technique.

The first quantity to be measured is the width of the gap that separates the two bodies (macroscopic surfaces, particles, or interfaces of self-assembled entities). All techniques provide a good spatial resolution on the order of 1 nm or better, but they do not necessarily give the *real* separation between the surfaces. Indeed, surface deformations may occur prior to the surfaces contacting each other : the effect of a repulsive surface interaction is to flatten the surfaces and to increase their separation compared to the

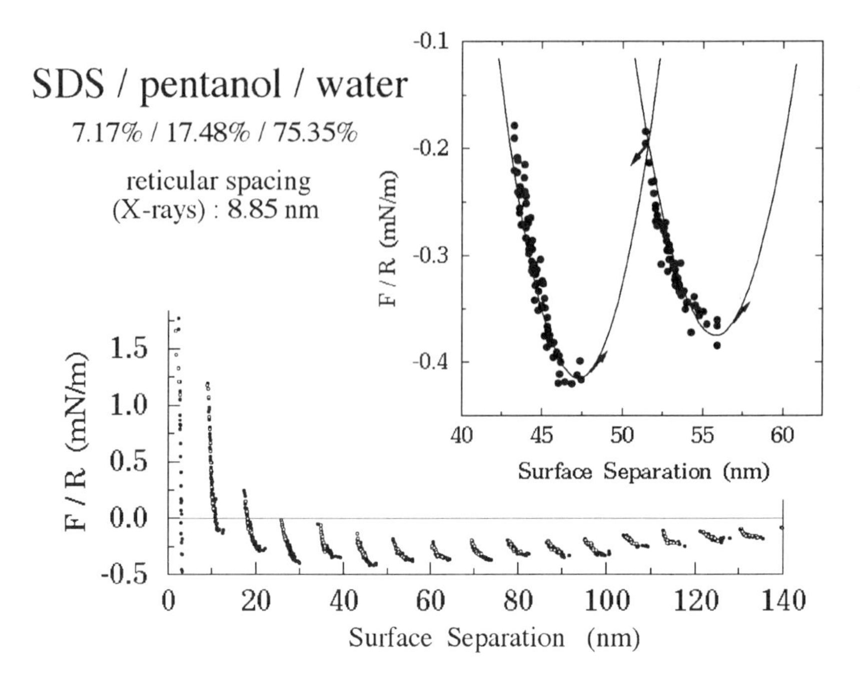

Figure 24. Force F (normalized by the mean radius of curvature of the surfaces) as a function of separation between two mica sheets immersed in a water-swollen lamellar mesophase [168]. The first 16 oscillations from contact are shown. *Top* : enlargement of the forces around the sixth and seventh oscillations. The solid lines are parabolic fits using a layer compressional elastic modulus $B = 1.8 \times 10^5$ J/m^3 and converting the measured forces to free energies between parallel flats according to the Derjaguin approximation. When the separation between the walls changes, the stack experiences an elastic stress induced by the weak normal strain. Every time two adjacent parabola intersect, the elastic energy of the system is lowered by the exclusion or introduction of a layer, via the creation or annihilation of an elementary edge dislocation [71].

undeformed surface shape (Fig.25), whereas the surfaces bulge out toward one another under the influence of attractive forces. These deformations may become significant at short separations (a few nanometers), amounting to the same order of magnitude than the gap width [86, 87]. Only the interferometric SFA (§ 2-2) and the capacitance-based SFA (§ 2-4) give a direct access to the *real* separation between the surfaces, however they are deformed or not. In all the other techniques the separation extracted is the distance between surfaces, as *if* they were not deformed. Indeed, X-ray, neutron, and light scattering, give the repeat or average distance between the centers of neighboring particles or self-assembled entities (§ 1 and § 3-3), and TIRM monitors the instantaneous elevation of a particle wandering above a macroscopic plate (§ 3-1).

TABLE I. Comparison of the resolution provided by the techniques for the direct measurement of interactions between two bodies as a function of the distance of separation of their surfaces.

	Technique	Separation			geometry of interaction	radius of curvature R	Pressure (Pa)	Force		Energy per unit area (J/m^2)
		method of measurement	zero	resolution (nm)				resolution (nN)	F/R (mN/m) range	
PRESSURE	capillary rise[1]	light, neutron, X-ray scattering	yes/no	0.1	any	any	10 - 5000			obtained by integration of the pressure
	osmometer[1]						100 - 4×10^3			
	osmotic stress[1]						10 - 10^7			
	vapour phase osmometry[1]						10^5 - 10^9			
	gravity						10^{-4} - 1			
	centrifugation						10^6 - 10^9			
FORCE	force balance	interferometry[2] electronics	yes	0.1	sphere-plane crossed-cylinders	10-50 mm 100 μm	obtained by derivation of the force or of the energy	10	5×10^{-3} - 100	10^{-8} - 2×10^{-2}
	SFA[3]	interferometry[2]	yes	0.1	crossed-cylinders	20 mm		50	2×10^{-3} - 20	3×10^{-7} - 2×10^{-3}
	ECL-SFA	capacitance	yes	0.015	sphere-plane	2-10 mm		10	2×10^{-3} - 5000	3×10^{-7} - 1
	AFM[3,4,5] (light lever)	optics[2] electronics	no	0.1	sphere-plane	1-100 μm		0.01	5×10^{-4} - 50	10^{-7} - 10^{-2}
	MASIF[4]	electronics (bimorph)	no	0.1	sphere-sphere	2 mm		20	10^{-2} - 100	2×10^{-8} - 2×10^{-2}
	chain magnetic technique[6]	light diffraction	yes/no	1	sphere-sphere	0.1 μm		2×10^{-4}	10^{-3} - 0.1	2×10^{-7} - 2×10^{-1}
ENERGY	TIRM	evanescent laser wave	yes/no	1	sphere-plane	2-30 μm		10^{-5}	10^{-6} - 10^{-5}	10^{-9} - 10^{-8}
	video-microscopy optical trapping	optics	yes/no	1	sphere-sphere	2-30 μm		10^{-5}	10^{-6} - 10^{-5}	10^{-9} - 10^{-8}

[1] solute must be not permeable to membrane.
[2] system transparent in visible light.
[3] glue deformation upon large applied loads is a limitation of the technique.
[4] single cantilever is used.
[5] K/R is not well defined.
[6] magnetic system only.

From these distance measurements a gap width is extracted, provided the size of the particle, the thickness of the membrane, etc, are known[13]. Similarly, in AFM and like-techniques (§ 2-3), there are two potential problems with the procedure of pushing the surfaces into contact to determine the sensitivity of the light lever. Firstly the calibration for the light lever obtained may be incorrect due to deformation of the surface and the probe, and, secondly, the surface separation obtained from the deflection is not equal to the point of closest approach. As a result, except with the SFAs where an *absolute* scale in separation is defined, all other techniques suffer from drawbacks and limitations in defining a zero for the distances. Note also that the resolution in separation distance will be ultimately limited by the roughness on the surfaces, either that on the macroscopic surfaces used in force balance or spring devices, or that on the microscopic particles ; by contrast the use of atomically smooth mica sheet for both surfaces gives to the SFA the best spatial resolution of any technique.

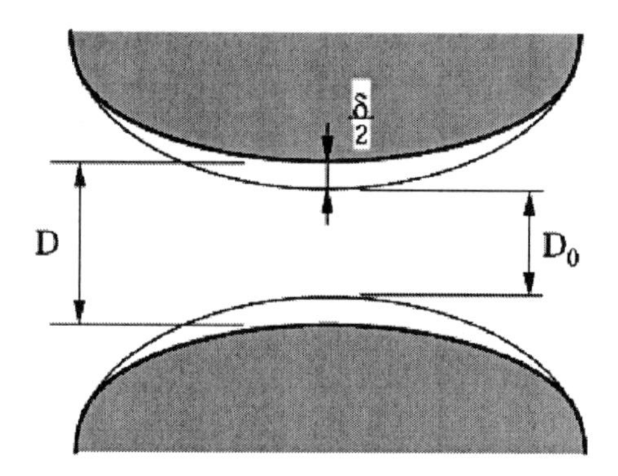

Figure 25. Two surfaces deformed as a result of a repulsive interaction. The thin line is the undeformed shape, while the thick line is the actual shape. In the SFAs, the real separation, D, is directly obtained by interferometry or by measurement of the capacitance between the two surfaces across the gap. All other techniques give the width of the gap, D_0, as if the surfaces were not deformed. The central displacement, $\delta = D - D_0$, is the deformation along the axis of symmetry.

[13] If bilayers were incompressible, then changes in repeat spacing would equal changes in bilayer separation. A force vs. distance relation could easily be constructed from measured pressure vs. repeat spacing reduced by a constant bilayer thickness. But bilayers are laterally compressible, and dissection of the measured repeat distance into a bilayer thickness and a solvent separation becomes a difficult problem : the same isotropic osmotic stress that pushes bilayers together also acts to deform them laterally. There is a predicted decrease in cross-sectional area per molecule and bilayer thickness [4]. Consequently, estimates of these structural changes are required in order to estimate bilayer separation.

On the other hand, both AFM and TIRM have a better resolution in the force than SFA. Of course, that greater sensitivity is really needed, since the forces on a microscopic sphere are orders of magnitude weaker than between two macroscopic bodies. A fairer basis of comparison might be the force divided by the characteristic radius of the interacting bodies : using Derjaguin's approximation, the force divided by the radius equals the energy per unit area between two infinite parallel plates [69]. On this basis, TIRM is two orders of magnitude more sensitive than other techniques. The reason TIRM is so much more sensitive is that it uses a molecular gauge for energy ($k_B T$) rather than a mechanical gauge for force (spring constant). While TIRM is capable of measuring much weaker interactions, it is incapable of measuring interactions as strong as some reported for AFM or SFA. In force balance and spring devices the upper limit depends on the most elastic component of the device and is usually governed by the means of attaching the surfaces (for instance the underlying glue fixing the mica sheet to the silica hemicylinder in the SFA or the microscopic sphere to the cantilever in the AFM). Note that the inherent limitations of all these techniques are related to the fact that under high applied loads surface deformation becomes substantial and measured values of F/R no longer corresponds to the free energy of interaction as related by Derjaguin's approximation. These remarks may also be applied to the techniques where the osmotic pressure is measured. Given the many ways of applying osmotic stress, it is possible to bring structures to virtually complete dehydration at pressures corresponding to over 1000 atmospheres. But the assignment of a measured work of dehydration to a force or energy of interaction between particles requires estimates about the work of rearranging their packing and structural configuration (loss of entropy), but also about the work spent to deform their interfaces. For example, for self-assembled systems in solution the measured pressure is recognized as both an inter-aggregate and a lateral stress (cf. footnote 13).

5. Future outlook

The developments in the area of surface force measurements have been many and important since the early days. Nevertheless progress is needed to overcome the difficulties in this research field. It is helpful to divide the problems associated with the direct measurement of interactions into two groups. The first group is characterized by those problems associated with the design of apparatus to detect and measure the forces as a function of the distance of separation between the particles or the surfaces. The second group is characterized by those problems associated with the choice, nature, and preparation of the surfaces of the bodies, between which the forces are to be measured. The discussion of all this review has been mainly directed towards the problems associated with the first group. The limitations and drawbacks of each of the techniques have been identified. It appears that definition of an absolute scale of separation and surface deformation are the two main issues to be addressed. Improvements of particular importance would be to control the force rather than the distance : this will allow to map out the relaxation of e.g. polymer layers, rearrangement in the Stern layer, etc. Such developments have already been proposed [57]. Another attractive research direction is the study of non-equilibrium forces. This is particularly important for polymer-coated solid surfaces, non-equilibrium effects in electrical double layers, etc. The dynamic method of the capacitance-based SFA on one hand, and on the other, the TIRM and related techniques seem the most promising tools to investigate such issues. Let us now focus to the second type of difficulties, those concerning surfaces and systems.

One of the aims of surface studies is to learn in detail the relation between surface chemical composition, surface heterogeneity, and surface interaction. This is not a simple task since the proper evaluation of the surface force measurements requires that the systems investigated are free of contaminating particles. One way to overcome the problem of contamination is to use samples of large specific areas, as in osmotic pressure based techniques. Another way is to decrease the radius of the solid surfaces, with the additional advantage that surface roughness is concomitantly reduced as well. This is what has been done in the AFM colloid probe technique, but to develop this method further would require the production of microscopic particles with a controlled chemistry and a very well defined geometry. This approach is however inherently limited since surface-altering treatments of the particle arise upon attachment to the cantilever (the interfaces are subjected to air exposure, melting, or vacuum exposure). Attempts to relate the measured force with colloidal stability and behavior of real systems can be of only limited success as the intricate surface chemistry is not equivalent. It has been demonstrated indeed, that not only quantitative but even qualitative (from attraction to repulsion) differences in the interactions may be observed [170]. One must resort to methods where particles are irreversibly adsorbed on macroscopic surfaces directly from bulk dispersions in order to prevent their colloidal interfaces from experiencing any environment outside their native aqueous state. The difficult task is to assemble particles in a compact, homogeneous layered and relatively flat film. As a result, coated macroscopic substrates, between which the interactions are to be measured, will have an interface fully representative of the real system. The substrates may be modified first by standard surface modification techniques but optimum bulk solution conditions may also be found to facilitate anchoring of molecules or adsorption of particles on a surface, where some interactions in their subtle interplay are favored with the oppositely charged [171] or like-charged particles [170, 172] and the mediated ions [173].

Controlling the surface chemistry before any investigation has been one of the recent issues of importance in surface force measurement. The corollary is to investigate the relation between surface heterogeneity and interactions and this will be a topic of importance for future studies. Modelling heterogeneity, either of chemical nature or structural nature could be obtained from controlled inhomogeneous macroscopic surfaces with adsorbed nanoparticles [174]. Again attempts to relate the measured force with real systems can be of only limited success if roughness and heterogeneity of the surface cannot be controlled *in situ*. One can note that the interpretation of surface force measurements has until now been hampered by the lack of a method of characterizing the material in the intersurface gap or along the surface *in situ*. By controlling the nature of the confining surfaces, it would be possible to separate the effects induced by a surface-specific interaction from effects purely due to confinement. New experimental approaches must be developed and are called for. A very promising avenue is to combine different techniques. To this respect, the simultaneous use of spectroscopic tools, such as X-ray diffraction [175-177], fluorescence microscopy [178], or birefringence measured with a laser [179] provide structural and orientational information along the surfaces and within the thin films upon confinement or under stress as the gap between the interfaces is changed.

References

1. Israelachvili, J.N. *Intermolecular and surface forces. With Applications to colloidal and biological systems.* Academic Press, London, 1985.
2. Barclay, L.M. & Ottewill, R.H. *Spec. Discuss. Faraday Soc.* **1**, 138-147 (1970)
3. Hachisu, S. & Kobayashi, Y. *J. Colloid Interface Sci.* **46**, 470-476 (1974).
4. Parsegian, V.A., Rand, R.P. & Fuller, N.L. *Proc. Natl. Acad. Sci. U.S.A.* **76**, 2750-2754 (1979).
5. Horkay, F., Nagy, M. & Zrinyi, M. *Acta Chim. Acad. Sci. Hung.* **103**, 387-395 (1980).
6. Rohrsetzer, S., Kovács, P. & Nagy, M. *Colloid Polym. Sci.* **264**, 812-816 (1986).
7. Parsegian, V.A., Rand, R.P., Fuller, N.L. & Rau, D.C. *Methods in Enzymology*, Biomembranes, Part O, Academic Press (L. Packer, Ed.), Vol. 127, 400-416 (1986).
8. Goodwin, J.W., Ottewill, R.H. & Parentich, A. *Colloid Polym. Sci.* **268**, 1131-1140 (1990).
9. Bonnet-Gonnet, C., Belloni, L. & Cabane, B. *Langmuir* **10**, 4012-4021 (1994).
10. Klein, R. in *Light Scattering, Principles and Development*, Oxford University Press (W. Brown, Ed.), chap. 2, 1996.
11. Belloni, L. *J. Phys. : Condens. Matter* **12**, R549-R587 (2000).
12. Reus, V., Belloni, L., Zemb, T., Lutterbach, N. & Versmold, H. *J. Phys. II France* **7**, 603-626 (1997).

13. LeNeveu, D.M., Rand, R.P. & Parsegian, V.A. *Nature (London)* **259**, 601-603 (1976).
14. LeNeveu, D.M., Rand, R.P., Gingell, D. & Parsegian, V.A. *Biophys. J.* **18**, 209-230 (1977).
15. Cowley, A.C., Fuller, N.L., Rand, R.P. & Parsegian, V.A. *Biochemistry* **17**, 3163-3168 (1979).
16. Lis, L.S., Parsegian, V.A. & Rand, R.P. *Biochemistry* **20**, 1771-1777 (1981).
17. Rand, R.P. *Annu. Rev. Biophys. Bioeng.* **10**, 277-314 (1981).
18. Rau, D.C., Lee, B. & Parsegian, V.A. *Proc. Natl. Acad. Sci. U.S.A.* **81**, 2621-2625 (1984)
19. Rau, D.C. & Parsegian, V.A. *Science* **249**, 1278-1281 (1990).
20. Parsegian, V.A., Rand, R.P. & Fuller, N.L. *J. Phys. Chem.* **95**, 4777-4782 (1991).
21. Dubois, M., Zemb, T., Belloni, L., Delville, A., Levitz, P. & Setton, R. *J. Chem. Phys.* **96**, 2278-2286 (1992).
22. Chang, J., Lesieur, P. Delsanti, M., Belloni, L., Bonnet-Gonnet, C. & Cabane, B. *J. Phys. Chem.* **99**, 15993-16001 (1995).
23. Spalla, O., Nabavi, M., Minter, J. & Cabane, B. *Colloid Polym. Sci.* **274**, 555-567 (1996).
24. Peyre, V., Spalla, O., Belloni, L. & Nabavi, M. *J. Colloid Interface Sci.* **187**, 184-200 (1997).
25. Vérétout, F., Delaye, M. & Tardieu, A. *J. Mol. Biol.* **205**, 713-728 (1989).
26. Wexlwe, A. "Humidity and Moisture : Measurement and Control in Science and Industry. " Reinhold, New York (1965).
27. Piazza, R., Bellini, T. & Degiorgio, V. *Phys. Rev. Lett.* **71**, 4267-4270 (1993).
28. El-Aasser, M.S. & Robertson, A.A. *J. Colloid Interface Sci.* **36**, 87-93 (1971).
29. Ilekti, P., Ph.D Thesis, Université Paris VI (2000).
30. Vold, R.D. & Groot, R.G. *J. Colloid Interface Sci.* **19**, 384-398 (1964).
31. Sonneville-Aubrun, O., Bergeron, V., Gulik-Krzywicki, T., Jönsson, B., Wennerström, H., Lindner, P. & Cabane, B. *Langmuir* **16**, 1566-1579 (2000).
32. Chan, D.Y.C. & Horn, R.G. *J. Chem. Phys.* **83**, 5311-5324 (1985).
33. Attard, P., Schulz, J.C. & Rutland, M.W. *Rev. Sci. Instrum.* **69**, 3852-3866 (1998).
34. Overbeek, J.Th. & Sparnaay, M.J. *Disc. Faraday Soc.* **18**, 12-24 (1954).
35. Derjaguin, B.V., Titijevskaia, A.S., Abricossova, I.I. & Malkina, A.D. *Disc. Faraday Soc.* **18**, 24-41 (1954).
36. Derjaguin, B.V., Abrikosova, I.I. & Leib, E. *Vestnik AN SSSR* **6**, 125 (1951).
37. Derjaguin, B.V. & Abrikosova, I.I. *Zhurn. eksper. teoret. Fisiki* **21**, 945 (1951); **30**, 999 (1956); *Doklady AN SSSR* **90**, 1055 (1953).
38. Derjaguin, B.V., Rabinovich, Y.I. & Churaev, N.V. *Nature* **265**, 520-521 (1977); **272**, 313-318 (1978).
39. Rabinovich, Y.I., Derjaguin, B.V. & Churaev, N.V. Adv. *Colloid Interface Sci.* **16**, 63-78 (1982).
40. Voropaeva, T.N., Derjaguin, B.V. & Kabanov, E.N. in *Research in Surface Forces*, Consultants Bureau, New York (1963), pp. 116-119.
41. Knapschinsky, L., Katz, W., Ehmke, B. & Sonntag, H. *Colloid Polym. Sci.* **260**, 1153-1156 (1982).
42. Götze, Th., Sonntag, H. & Rabinovich, Y. *Colloid Polym. Sci.* **265**, 134-138 (1987).
43. Götze, Th. & Sonntag, H. *Colloid Surf.* **31**, 181-201 (1988).
44. Peschel, G.P. & Belouschek, P. *Prog. Colloid Polym. Sci.* **60**, 108-119 (1976).
45. Peschel, G.P., Belouschek, P., Müller, M.M, Müller, M.R. & König, R. *Colloid Polym. Sci.* **260**, 444-451 (1982).

46. Belouschek, P. & Maier, S. *Prog. Colloid Polym. Sci.* **72**, 43-50 (1986).
47. Tabor, D. & Winterton, R.H.S. *Nature* **219**, 1120-1121 (1968); *Proc. Roy. Soc. A* **312**, 435-450 (1969).
48. Israelachvili, J.N. & Adams, G.E. *J. Chem. Soc. Faraday Trans. 1* **74**, 975-1001 (1978).
49. Israelachvili, J.N. & McGuiggan, P.M. *J. Mater. Res.* **5**, 2223-2243 (1990).
50. Parker, J.L., Christenson, H.K. & Ninham, B.W. *Rev. Sci. Instrum.* **60**, 3135-3138 (1989).
51. Klein, J. *J. Chem. Soc. Faraday Trans. I* **79**, 99-118 (1983).
52. Van Alsten, J. & Granick, S. *Phys. Rev. Lett.* **61**, 2570-2573 (1988).
53. Quon, R.A., Levins, J.M. & Vanderlick, T.K. *J. Colloid Interface Sci.* **171**, 474-482 (1995).
54. Gauthier-Manuel, B. & Gallinet, J.-P. *J. Colloid Interface Sci.* **175**, 476-483 (1995).
55. Grünewald, T. & Helm, C.A. *Langmuir* **12**, 3885-3890 (1996).
56. Frantz, P., Agrait, N. & Salmeron, M. *Langmuir* **12**, 3289-3294 (1996).
57. Kékicheff, P., Iss, J., Courtier, F., Finck, F., Friedmann, P. & Lambour, C. *in preparation.*
58. Israelachvili, J.N. & Tabor, D. *Proc. Roy. Soc. A* **331**, 19-38 (1972).
59. Horn, R.G., Smith, D.T. & Haller, W. *Chem. Phys. Lett.* **162**, 404-408 (1989).
60. Antelmi, D.A., Kékicheff, P. & Richetti, P. *J. Phys. II France* **5**, 103-112 (1995); *Langmuir* **15**, 7774-7788 (1999).
61. Horn, R.G., Clarke, D.R. & Clarkson, M.T. *J. Mater. Res.* **3**, 413-416 (1988).
62. Ducker, W.A., Xu, Z., Clarke, D.R. & Israelachvili, J.N. *J. Am. Ceram. Soc.* **77**, 437-443 (1994).
63. Pashley, R.M. & Israelachvili, J.N. *Colloids Surf.* **2**, 169-187 (1981).
64. Marra, J. & Israelachvili, J.N. *Biochemistry* **24**, 4608-4618 (1985).
65. Smith, C.P., Maeda, M., Atanasoska, L., White, H.S. & McClure, D.J. *J. Phys. Chem.* **92**, 199-205 (1988).
66. Parker, J.L., Cho, D.L. & Claesson, P.M. *J. Phys. Chem.* **93**, 6121-6125 (1989).
67. Kékicheff, P., Richetti, P. & Christenson, H.K. *Langmuir* **7**, 1874-1879 (1991).
68. Freyssingeas, E., Antelmi, D., Kékicheff, P., Richetti, P. & Bellocq, A.-M. *Eur. Phys. J. B* **9**, 123-136 (1999).
69. Derjaguin, B.V. *Kolloid-Z.* **69**, 155-164 (1934).
70. Attard, P. & Parker, J.L. *J. Phys. Chem.* **96**, 5086-5093 (1992).
71. Richetti, P., Kékicheff, P. & Barois, P. *J. Phys. II France* **5**, 1129-1154 (1995).
72. Tolansky, S. *Multiple-Beam Interferometry of Surfaces and Films*, Oxford, Clarendon Press, 1948.
73. Israelachvili, J.N. *J. Colloid Interface Sci.* **44**, 259-272 (1973).
74. Horn, R.G. & Smith, D.T. *Appl. Opt.* **30**, 59-65 (1991).
75. Clarkson, M.T. *J. Phys. D: Appl. Phys.* **22**, 475-482 (1989).
76. Levins, J.M. & Vanderlick, T.K. *J. Colloid Interface Sci.* **158**, 223-227 (1993).
77. Rabinowitz, P. *J. Opt. Soc. Am. A* **12**, 1593 (1995).
78. Mächtle, P., Müller, C. & Helm, C.A. *J. Phys. II France* **4**, 481-500 (1994).
79. Farrell, B., Bailey, A.I. & Chapman, D. *Appl. Opt.* **34**, 2914-2920 (1995).
80. Kékicheff, P. & Spalla, O. *Langmuir* **10**, 1584-1591 (1994).
81. Stewart, A.M. *J. Colloid Interface Sci.* **170**, 287-289 (1995).
82. Christenson, H.K. & Claesson, P.M. *Science* **239**, 390-392 (1988).
83. Horn, R.G. *Biochim. Biophys. Acta* **778**, 224-228 (1984).
84. Christenson, H.K. *J. Colloid Interface Sci.* **104**, 234-249 (1985).

85. Kuhl, T., Ruths, M., Chen, Y.L. & Israelachvili, J.N. *J. Heart Valve Dis.* **3**, S117-S127 (1994).
86. Parker, J.L. & Attard, P. *J. Phys. Chem.* **96**, 10398-10405 (1992).
87. Attard, P. & Parker, J.L. *Phys. Rev. A* **46**, 7959-7971 (1992).
88. Maugis, D. & Gauthier-Manuel, B. *J. Adhesion Sci. Technol.* **8**, 1311-1322 (1994).
89. Horn, R.G., Israelachvili, J.N. & Pribac, F. *J. Colloid Interface Sci.* **115**, 480-492 (1987).
90. Stewart, A.M. & Christenson, H.K. *Meas. Sci. Technol.* **1**, 1301-1303 (1990).
91. Christenson, H.K. *J. Colloid Interface Sci.* **121**, 170-178 (1988).
92. Claesson, P.M. & Christenson, H.K. *J. Phys. Chem.* **92**, 1650-1655 (1988).
93. Israelachvili, J.N. & Pashley, R.M. *J. Colloid Interface Sci.* **98**, 500-514 (1984).
94. Israelachvili, J.N., Kott, S.J. & Fetters, L.J. *J. Polym. Sci. B* **27**, 489-502 (1989).
95. Parker, J.L. *Langmuir* **8**, 551-556 (1992).
96. Parker, J.L. & Stewart, A.M. *Progr. Colloid Polym. Sci.* **88**, 162-168 (1992).
97. Stewart, A.M. & Parker, J.L. *Rev. Sci. Instrum.* **63**, 5626-5633 (1992).
98. Parker, J.L., Richetti, P., Kékicheff, P. & Sarman, S. *Phys. Rev. Lett.* **68**, 1955-1958 (1992).
99. Attard, P. & Parker, J.L. *J. Phys. Chem.* **96**, 5086-5093 (1992).
100. Binnig, G. & Rohrer, H. *Helv. Phys. Acta* **55**, 726-735 (1982).
101. Binnig, G., Quate, C. & Gerber, G. *Phys. Rev. Lett.* **56**, 930-933 (1986).
102. Senden, T.J., Ph.D Thesis, Australian National University, Canberra, Australia (1993).
103. Meyer, G., Amer, N.M. *Appl. Phys. Lett.* **53**, 1045-1047 (1988).
104. Alexander, S., Hellemans, L., Marti, O., Schneir, J., Elings, V., Hansma, P.K., Longmire, M. & Gurley, J. *J. Appl. Phys.* **65**, 164-167 (1989).
105. Senden, T.J., Drummond, C.J. & Kékicheff, P. *Langmuir* **10**, 358-362 (1994).
106. Sader, J.E. & White, L. *J. Appl. Phys.* **74**, 1-9 (1993).
107. Walters, D.A., Cleveland, J.P., Thomson, N.H., Hansma, P.K., Wendman, M.A., Gurley, G. & Elings, V. *Rev. Sci. Instrum.* **67**, 3583-3590 (1996).
108. Senden, T.J. & Ducker, W.A. *Langmuir* **10**, 1003-1004 (1994).
109. Torii, A., Sasaki, M., Hane, K. & Okuma, S. *Meas. Sci. Technol.* **7**, 179-184 (1996).
110. Hutter, J.L & Bechhoeffer, J. *Rev. Sci. Instrum.* **64**, 1868-1873 (1993).
111. Maeda, N. & Senden, T.J. *Langmuir* **16**, 9282-9286 (2000).
112. Hoh, J.H. & Engel, A. *Langmuir* **9**, 3310-3312 (1993).
113. Richetti, P. & Kékicheff, P. *Phys. Rev. Lett.* **68**, 1951-1954 (1992).
114. Burnham, N.A. & Colton, R.J. *J. Vac. Sci. Technol. A* **7**, 2906-2913 (1989).
115. Drummond, C.J. & Senden, T.J. *Colloids Surf. A* **87**, 217-234 (1994).
116. Ducker, W.A., Senden, T.J. & Pashley, R.M. *Nature* **353**, 239-241 (1991).
117. Finot, E., Lesniewska, E., Mutin, J.-C. & Gourdonnet, J.-P. *Langmuir* **16**, 4237-4244 (2000).
118. Kékicheff, P. & Spalla, O. *Phys. Rev. Lett.* **75**, 1851-1854 (1995).
119. Kékicheff, P., Christenson, H.K. & Ninham, B.W. *Colloids Surf.* **40**, 31-41 (1989).
120. Rutland, M.W. & Parker, J.L. *Langmuir* **10**, 1110-1121 (1994).
121. Kékicheff, P., Marcelja, S., Senden, T.J. & Shubin, V.E. *J. Chem. Phys.* **99**, 6098-6113 (1993).
122. Parker, J.L. *Prog. Surf. Sci.* **47**, 205-271 (1994).
123. Lodge, K.B. & Mason, R., *Proc. Roy. Soc. London, Ser. A* **383**, 279 (1982); ibid, **383**, 295 (1982).
124. Herke, R.A., Clark, N.A. & Handschy, M.A. *Science* **267**, 651-654 (1995); *Phys. Rev. E* **56**, 3028-3043 (1997).

125. Steblin, V.N., Shchukin, E.D., Yaminsky, V.V. & Yaminsky, I.V. *Kolloidn. Zh* **53**, 684-687 (1991).
126. Yaminsky, V.V., Steblin, V.N. & Shchukin, E.D. *Pure Appl. Chem.* **64**, 1725-1730 (1992).
127. Frantz, P., Artsyukhovich, A., Carpick, R.W. & Salmeron, M. *Langmuir* **13**, 5957-5961 (1997).
128. Tonck, A., Georges, J.-M. & Loubet, J.-L. *J. Colloid Interface Sci.* **126**, 150-163 (1988).
129. Georges, J.-M., Millot, S., Loubet, J.-L. & Tonck, A. *J. Chem. Phys.* **98**, 7345-7360 (1993).
130. Loubet, J.-L. *et al. in preparation*
131. Montfort, J.-P., Tonck, A., Loubet, J.-L. & Georges, J.-M. *J. Polym. Sci. B* **29**, 677-682 (1991).
132. Pelletier, E., Montfort, J.-P., Loubet, J.-L., Tonck, A. & Georges, J.-M. *Macromolecules* **28**, 1990-1998 (1995).
133. Boyer, L., Houzé, F., Tonck, A., Loubet, J.-L. & Georges, J.-M. *J. Phys. D: Appl. Phys.* **27**, 1504-1508 (1994).
134. Georges, J.-M., Loubet, J.-L. & Tonck, A. *in New Materials : Approaches to Tribology*, ed. Pope, L.E., Fehrenbacher, L.L., and Winer, W.O. (Pittsburg, PA: Materials Research Society, vol. 140), p. 67 (1989).
135. Crassous, J., Charlaix, E. & Loubet, J.-L. *Europhys. Lett.* **28**, 37-42 (1994).
136. Crassous, J., Charlaix, E. & Loubet, J.-L. *Phys. Rev. Lett.* **78**, 2425-2428 (1997).
137. Restagno, F., Ph.D Thesis, Ecole Normale Supérieure de Lyon, France (2000).
138. Marshall, J.K. & Kitchener, J.A. *J. Colloid Interface Sci.* **22**, 342-351 (1966).
139. Hull, M. & Kitchener, J.A. *Trans. Faraday Soc.* **65**, 3093-3104 (1969).
140. Clint, G.E., Clint, J.H., Corkill, J.M. & Walker, T. *J. Colloid Interface Sci.* **44**, 121-132 (1973).
141. Prieve, D.C. & Alexander, B.M. *Science* **231**, 1269-1270 (1986); *Langmuir* **3**, 788-795 (1987).
142. Prieve, D.C., Luo, F. & Lanni, F. *Faraday Discuss., Chem. Soc.* **83**, 297-307 (1987).
143. Prieve, D.C. & Walz, J.Y. *Appl. Opt.* **32**, 1629-1641 (1993).
144. Brown, M.A., Smith, A.L. & Staples, E.J. *Langmuir* **5**, 1319-1324 (1989).
145. Prieve, D.C. *Adv. Colloid Interface Sci.* **82**, 93-125 (1999).
146. Prieve, D.C. & Frej, N.A. *Langmuir* **6**, 396-403 (1990).
147. Flicker, S.G., Tipa, J.L. & Bike, S.G. *J. Colloid Interface Sci.* **158**, 317-325 (1993).
148. Flicker, S.G. & Bike, S.G. *Langmuir* **9**, 257-262 (1993).
149. Suresh, L. & Walz, J.Y. *J. Colloid Interface Sci.* **196**, 177-190 (1997).
150. Bevan, M.A. & Prieve, D.C. *Langmuir* **15**, 7925-7936 (1999).
151. Walz, J.Y. & Suresh, L. *J. Chem. Phys.* **103**, 10714-10725 (1995).
152. Sober, D.L. & Walz, J.Y. *Langmuir* **11**, 2352-2356 (1995).
153. Sharma, A. & Walz, J.Y. *J. Chem. Soc. Faraday Trans.* **92**, 4997-5004 (1996).
154. Rudhardt, D., Bechinger, C. & Leiderer, P. *Phys. Rev. Lett.* **81**, 1330-1333 (1998).
155. Bevan, M.A. & Prieve, D.C. *Langmuir* **16**, 9274-9281 (2000).
156. Leal-Calderon, F., Stora, T., Mondain-Monval, O., Poulin, P. & Bibette, J. *Phys. Rev. Lett.* **72**, 2959-2962 (1994).
157. Mondain-Monval, O., Leal-Calderon, F., Phillip, J. & Bibette, J. *Phys. Rev. Lett.* **75**, 3364-3367 (1995).
158. Mondain-Monval, O., Leal-Calderon, F. & Bibette, J. *J. Phys. II France* **6**, 1313-1329 (1996).

159. Mondain-Monval, O., Espert, A., Omarjee, P., Bibette, J., Leal-Calderon, F., Phillip, J. & Joanny, J.-F. *Phys. Rev. Lett.* **80**, 1778-1781 (1998).
160. Espert, A., Omarjee, P., Bibette, J., Leal-Calderon, F. & Mondain-Monval, O. *Macromolecules* **31**, 7023-7029 (1998).
161. Dimitrova, T.D. & Leal-Calderon, F. *Langmuir* **15**, 8813-8821 (1999).
162. Marra, J. *J. Colloid Interface Sci.* **107**, 446-455 (1985); *Biophys. J.* **50**, 815-825 (1986).
163. Parsegian, V.A., Rand, R.P. & Fuller, N.L. *J. Phys. Chem.* **95**, 4777-4782 (1981).
164. Pashley, R.M., McGuiggan, P.M., Ninham, B.W., Brady, J. & Evans, D.F. *J. Phys. Chem.* **90**, 1637-1642 (1986).
165. Evans, E. & Parsegian, V.A. *Proc. Natl. Acad. Sci.* **83**, 7132-7136 (1989).
166. Horn, R.G., Israelachvili, J.N., Marra, J., Parsegian, V.A. & Rand, R.P. *Biophys. J.* **54**, 1185-1187 (1988).
167. Kékicheff, P. & Christenson, H.K. *Phys. Rev. Lett.* **63**, 2823-2826 (1989).
168. Richetti, P., Kékicheff, P., Parker, J.L. & Ninham, B.W. *Nature (London)* **346**, 252-254 (1990).
169. Richetti, P. Moreau, L., Barois, P. & Kékicheff, P. *Phys. Rev. E* **54**, 1749-1762 (1996).
170. Atkins, D.T., Kékicheff, P. & Spalla, O. *J. Colloid Interface Sci.* **188**, 234-237 (1997).
171. Spalla, O. & Kékicheff, P. *J. Colloid Interface Sci.* **192**, 43-65 (1997).
172. Antelmi, D.A. & Spalla, O. *Langmuir* **15**, 7478-7489 (1999).
173. Machou, D. & Kékicheff, P. *to be published.*
174. Spalla, O. & Desset, S. *Langmuir* **16**, 2133-2140 (2000).
175. Idziak, S.H.J., Safinya, C.R., Hill, R.S., Kraiser, K.E., Ruths, M., Warriner, H. E., Steinberg, S., Liang, K.S. & Israelachvili, J.N. *Science* **264**, 1915-1918 (1994).
176. Koltover, I., Idziak, S.H.J., Davidson, P., Li, Y., Safinya, C.R., Ruths, M., Steinberg, S. & Israelachvili, J.N. *J. Phys. II France* **6**, 893-907 (1996).
177. Kékicheff, P. & Riekel, C. *to be published.*
178. Neuman, R.D., Park, S. & Shah, P. *J. Colloid Interface Sci.* **172**, 257-260 (1995).
179. Kékicheff, P. Maret, G. & Tinland, B. *in preparation.*

COUNTERIONS IN POLYELECTROLYTES

ALEXEI R.KHOKHLOV, KONSTANTIN B.ZELDOVICH
and ELENA YU.KRAMARENKO
Physics Department, Moscow State University
Moscow 117234 Russia

1. Introduction

Polyelectrolytes are macromolecules containing electrically charged monomer units. Such units appear due to the dissociation of ionizable (initially neutral) monomer units. In this process, the monomer unit on the chain becomes charged and a counterion is released to the solution. Therefore, counterions are always present in polyelectrolyte systems. Evidently, the number of charged monomer units is equal to the number of counterions.

The properties of ion-containing polymer systems depend essentially on the spatial distribution of counterions.

The most familiar effect connected with the inhomogeneous spatial distribution of counterions is called counterion condensation and has been first described by Onsager in 1940s and later by Manning [1, 2]. They have considered the distribution of counterions around a rigid-rod macroion and have found that if the linear charge density of the macroion is larger than

$$\tau^* = \frac{\varepsilon k_B T}{e},$$

then the counterions are condensed on the macroion ($k_B T$ is thermal energy, e is the elementary charge, and ε is the dielectric constant of the medium).

In this case, any further increase of the linear charge density of the macroion does not lead to the increase of the charge of the ion together with its counterions, since the counterions become condensed.

Although this result does not hold for non-cylindrical geometries, electrostatic attraction of counterions to the charges on polymer chains often leads to important effects, for example, to ion pair formation. In this respect, one can distinguish between the extreme cases: polyelectrolyte behavior, when counterions are dissociated and move freely in solution,

C. Holm et al. (eds.), *Electrostatic Effects in Soft Matter and Biophysics*, 283–316.
© 2001 *Kluwer Academic Publishers. Printed in the Netherlands.*

and ionomer behavior, when counterions are bound to polymer chains and lose their translational mobility. Intermediate cases, where only part of counterions is bound to polymer chains, result in the interesting phenomena of mixed polyelectrolyte/ionomer behavior. Counterion binding has a major effect on osmotic properties of the system, and we will demonstrate its importance for the cases of polyelectrolyte gel swelling and adsorption of polyelectrolytes on oppositely charged surfaces.

Account for the difference between bound and mobile counterions is not sufficient to describe some of the effects occurring in real inhomogeneous polyelectrolyte systems. For example, osmotic pressure in inhomogeneous charged gels can be much lower than expected, even if Manning condensation on network chains is taken into account. Similarly, the swelling ratio of microgels exhibits non-monotonous behavior with the degree of ionization and molecular mass. All these effects are caused by larger-scale inhomogeneity, which is manifested at length scales larger than the size of the polymer coil. The counterions which are not condensed on the chains are not necessarily free, and may become trapped by large-scale inhomogeneities. These effects will be illustrated using dilute polyelectrolyte solutions and inhomogeneous gels as examples of strongly inhomogeneous systems.

1.1. OUTLINE OF THE PRESENTATION

In the next section we recall the key results of the classical theory of polyelectrolyte gel swelling and collapse behavior. Then, we will demonstrate how the trapping of the counterions by the spatial inhomogeneities influences the osmotic pressure of polyelectrolyte gels and compare the results with the classical theory. In Sec. 4 we present the simplest theory for polyelectrolyte solutions and microgel particles of various molecular masses taking into account the difference in states of counterions inside and outside the space occupied by polyelectrolyte macromolecules. In Sec. 5 this theory is generalized to study the influence of ion pair formation on the swelling behavior of polyelectrolyte chains and gels. We will demonstrate that ion pairing becomes progressively important when macromolecules or gels shrink. In this case the formation of a supercollapsed state is possible, when all counterions are trapped and form ion pairs. In Sec. 6 the theory of adsorption of polyelectrolyte chains on oppositely charged surfaces is presented. We show that the effect of surface overcharging in the course of polyelectrolyte adsorption can be explained by the formation of ion pairs in the adsorbed polymer layer.

2. Polyelectrolyte Gels: Classic results

We start with the description of homogeneous charged polymer gels [3, 4]. It has been shown that the main features of polyelectrolyte gels such as high degree of swelling in good solvent and the existence of abrupt collapse transitions are due to the presence of counterions. The classical theories assume that the number of counterions is independent on the gel state and the counterions are evenly distributed throughout the whole gel sample.

In good solvent the swelling ratio of a polyelectrolyte gel is defined by the competition between subchain elasticity tending to compress the gel sample, and the exerting osmotic pressure of (presumably ideal) gas of low-molecular counterions. If N is the average number of monomer units of the gel subchains and σ is the average number of monomer units between two nearest charged groups, then the number of counterions per subchain is equal to N/σ. The requirement of macroscopic electroneutrality ensures that all counterions remain within the gel sample, and if we assume that their spatial distribution is uniform, the free energy of the counterion gas per one subchain reads

$$\frac{F_{gas}}{k_B T} = \frac{N}{\sigma} \ln\left(\frac{n_0}{\sigma \alpha^3}\right), \tag{1}$$

where α is the swelling coefficient of the gel with respect to the reference state in which gel subchains are most close to Gaussian conformation [5], $\alpha = (n_0/n)^{1/3}$, n and n_0 are the concentrations of monomer units in the gel at equilibrium and in the reference state, respectively. If the chains are flexible with the size of monomer unit a, the value of n_0 can be estimated as $n_0 \sim N^{-1/2}/a^3$ [5].

The dominant term in elastic free energy is due to chain expansion,

$$\frac{F_{el}}{k_B T} = \frac{3}{2}\alpha^2, \tag{2}$$

and the free energy of the homogeneous gel in a good solvent as a function of the swelling ratio α can be written as

$$\frac{F}{k_B T} = \frac{3}{2}\alpha^2 + \frac{N}{\sigma} \ln\left(\frac{n_0}{\sigma \alpha^3}\right). \tag{3}$$

Minimizing with respect to α, we obtain the equilibrium swelling ratio of a spatially homogeneous charged gel in a good solvent:

$$\alpha_{hom,eq} = (N/\sigma)^{1/2}. \tag{4}$$

This value of swelling ratio $\alpha \sim N^{1/2}$ is much larger than the swelling ratio of the uncharged gel $\alpha \sim N^{1/10}$, which is defined by the balance between elastic energy and the energy of repulsion of uncharged monomer units in a good solvent. It is the osmotic pressure of counterions within the gel which leads to the additional expansion of the polyelectrolyte chains.

In a poor solvent the interaction of uncharged monomer units becomes dominant, the osmotic pressure of counterions is negligible, and the system behaves like a neutral one: a globular state is formed. In this case the equilibrium volume of the gel is determined by the balance between attractive pairwise interactions and repulsive interactions of higher orders. The energy of interactions can be written in the form of virial expansion:

$$\frac{F}{k_B T} = BNn + CNn^2, \tag{5}$$

where B and C are the second and the third virial coefficients.

Then the equilibrium value of the swelling ratio of the gel in a poor solvent is $\alpha \sim (C/|B|a^3)^{1/3} N^{1/6}$.

The amplitude of the volume change during the gel collapse is defined by the difference in the swelling ratios in the swollen and collapsed states. While in the collapsed state, polyelectrolyte and uncharged gels behave almost identically, the swelling ratio of a polyelectrolyte gel in a good solvent is much higher than that of a neutral gel. Consequently, the magnitude of the collapse for polyelectrolyte gels is much higher.

The classical theories take into account the existence of counterions and their leading role in the swelling behavior of polyelectrolyte gels. However, a more careful consideration shows that some of the effects can not be described by this approach. In particular, the number of counterions contributing to the osmotic pressure can depend on the structure of the gel or it can change in the course of the gel collapse. In the following sections we will show that these effects are very important and should be taken into account when considering more sophisticated models of polyelectrolyte systems.

3. Swelling behavior of inhomogeneous polyelectrolyte gels

As it has been already mentioned, the real polyelectrolyte gel is never homogeneous. The spatial inhomogeneity of polymer density distribution occurs as early as at the synthesis stage, and is generally preserved in the gel phase [6].

In our theory we take into account the inhomogeneity of the polymer gel using a simple two-phase model [7], see Figure 1.

Our model gel consists of the regions of higher density of polymer chains, caused by locally higher crosslink density, separated by regions with lower polymer content. In this model, the actual continuous distribution of subchain lengths is roughly replaced by an assumption that the gel

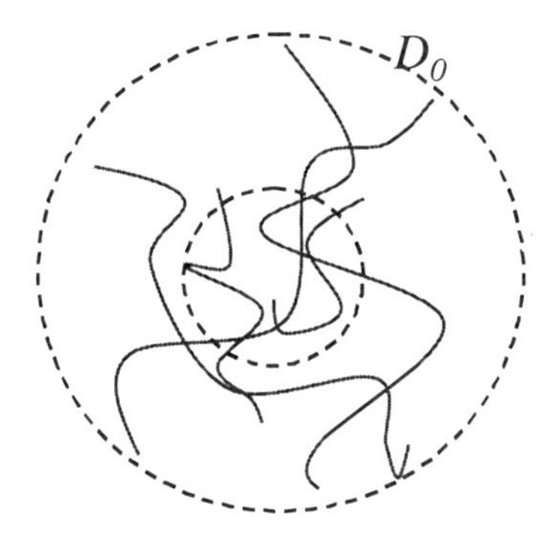

Figure 1. Schematic representation of the core-shell model of inhomogeneity in a polyelectrolyte gel. The core of radius R_0 consists of chains with the degree of polymerization N_1, the degree of polymerization in the shell is N_2.

consists of subchains of two different degrees of polymerization N_1 and N_2, and the spatial structure is depicted as clusters (core) of chains of length N_1 embedded in a matrix of chains of length N_2 (shell). By varying the parameters of the model, namely N_1, N_2, the core size R_0 and the average separation between the cores D_0, we can hope to describe qualitatively any inhomogeneity pattern occurring in real gels.

Let us consider the case when $N_2 > N_1$ (the core is denser than the shell), and $R_0 < D_0$, so that core size is less than the distance between neighboring cores. With this choice of parameters, the model system resembles the presumed picture of the gel microstructure shown in Figure 1: the small dense aggregates in a loosely crosslinked matrix. It is assumed that polymer chains in both core and shell have the same (constant) fraction of quenched charged units $1/\sigma$.

The two-phase model gel is characterized by swelling ratios α_1 and α_2 of the core and the shell, respectively, and the bulk swelling ratio α of the macroscopic sample. Here we define the swelling ratio α as the ratio of end-to-end distance of the polymer chain to its Gaussian size, or, for the gel, the ratio of the size of the gel sample to its size under reference conditions, where the subchains are assumed to be closest to Gaussian.

To derive the free energy of the system and to establish the relation between α_1, α_2 and α we need to agree on the method of calculation of swelling characteristics which does not depend on the details of the mutual positions of denser clusters in the more dilute matrix. Let us assume that the space per one core/shell is of spherical shape and consists of the central highly crosslinked region with subchain length N_1 and radius R_0 (core) surrounded by a spherical layer with radius D_0 consisting of chains of length N_2 (shell) (Fig. 1). The subscript 0 denotes the reference state (normally corresponding to synthesis conditions or swelling of uncharged gels in θ-solvents [3]). Then, the macroscopic swelling ratio α is given by

$$\alpha = \frac{\alpha_1 R_0 + \alpha_2 (D_0 - R_0)}{D_0}, \tag{6}$$

We will consider the case when $D_0 > R_0$ or even $D_0 \gg R_0$, i.e. when the core size is small as compared to the distance between nearest aggregates.

The free energy of the macroscopic gel is proportional to the free energy of the core together with its shell surrounding. The simplification represented by Fig. 1 is in the spirit of our simplest model and does not lead to any loss of generality.

The non-ionic part of the free energy of the core can be expressed as follows [3]:

$$\frac{F_1}{k_B T} = \frac{4\pi}{3} R_0^3 \nu_{0,1} \left\{ \frac{3}{2} \left(\alpha_1^2 + \alpha_1^{-2} \right) + \frac{B n_{0,1}}{\alpha_1^3} + \frac{C n_{0,1}^2}{\alpha_1^6} \right\}, \tag{7}$$

where $\nu_{0,1} \sim 1/(a^3 N_1^{3/2})$ is the average crosslink density in the reference state, $n_{0,1} \sim 1/a^3 N_1^{1/2}$ is the monomer unit density in the reference state, and B and C are the second and third virial coefficients of the non-Coulombic interaction of monomer units, respectively.

Similarly, for the free energy of the shell we can write

$$\frac{F_2}{k_B T} = \frac{4\pi}{3} (D_0^3 - R_0^3) \nu_{0,2} \left\{ \frac{3}{2} \left(\alpha_2^2 + \alpha_2^{-2} \right) + \frac{B n_{0,2}}{\alpha_2^3} + \frac{C n_{0,2}^2}{\alpha_2^6} \right\}, \tag{8}$$

with $\nu_{0,2} \sim 1/(a^3 N_2^{3/2})$ and $n_{0,2} \sim 1/a^3 N_2^{1/2}$.

We are interested chiefly in the properties of the swollen state of the inhomogeneous gel with hydrophilic monomer units, and therefore we suppose that the gel is immersed in a good or θ-solvent. Since a polyelectrolyte gel is highly swollen under these circumstances, we can also neglect all compression-related ($\sim \alpha^{-2}$) elastic terms in the free energy. Therefore we retain only the leading terms in powers of α, neglecting the second and the fourth terms in eqs 7 and 8.

Thus, our final expression for the non-ioinic part of the free energy takes into account the elastic expansion of a polymer chain, and the excluded volume repulsion of monomer units.

The ionic part of the free energy consists of the translational entropy of the counterions and of their electrostatic interactions. Within our two-phase framework, this energy per aggregate can be written as

$$\frac{F_{ion}}{k_BT} = \mathcal{N}_1 \ln \frac{\mathcal{N}_1 a^3}{V_1} + \mathcal{N}_2 \ln \frac{\mathcal{N}_2 a^3}{V_2} + l_B(\mathcal{N}_1 - Q_1)^2 \left(\frac{1}{R_0\alpha_1 + (D_0 - R_0)\alpha_2} - \frac{1}{R_0\alpha_1} \right). \tag{9}$$

The first two terms represent the translational entropy of small ions, while the third term is the electrostatic energy of the system, considered as a spherical capacitor (see also [8]). In these expressions, $l_B = e^2/\epsilon k_B T$ is the Bjerrum length, $\mathcal{N}_1$ and $\mathcal{N}_2$ are the numbers of counterions residing in the core and the shell respectively, and Q_1 is the immobilized charge on polymer chains in the core. The volumes V_1 and V_2 of core and shell are given by

$$V_1 = \frac{4\pi}{3} R_0^3 \alpha_1^3,$$

$$V_2 = \frac{4\pi}{3}((R_0\alpha_1 + (D_0 - R_0)\alpha_2)^3 - R_0^3\alpha_1^3).$$

For subsequent calculations we will also introduce the fraction of counterions β within the core, defined as $\beta = \mathcal{N}_1/Q$,

Thus, the free energy of the system F is given by

$$F = F_1 + F_2 + F_{ion}. \tag{10}$$

The electroneutrality condition is satisfied by setting

$$\mathcal{N}_1 + \mathcal{N}_2 = \frac{1}{\sigma}\nu_{0,1} V_{0,1} N_1 + \frac{1}{\sigma}\nu_{0,2} V_{0,2} N_2, \tag{11}$$

where $V_{0,1}, V_{0,2}$ are the volumes of the corresponding phases in the reference state.

The expression (10) for the free energy depends essentially on the parameters α_1, α_2 and β. It is worth noting that these parameters are not independent: they are connected via a boundary condition of elastic equilibrium on the boundary of the core and the shell. Physically the expansion of chains in the shell will induce a mechanical stress on chains in the core and vice versa. The difference of counterion osmotic pressure on the different sides of the boundary surface also contributes to this effect.

In its general form, this boundary equilibrium condition can be written as

$$\pi_{1,elastic} + \pi_{1,osmotic} = \pi_{2,elastic} + \pi_{2,osmotic}, \tag{12}$$

where π is pressure, and indices 1 and 2 correspond to the core and the shell, respectively.

To get the explicit form of this equation in our case, note that the force of elastic expansion of a single polymer chain is $\sim k_B T\alpha/a\sqrt{N}$, and the unperturbed size of the chain is $\sim aN^{1/2}$, so its cross-section is a^2N.

We assume that the chains of the core and the shell are grafted to core-shell boundary surface with superficial density $\sim 1/(a^2 N_i)$ chains per unit area ($i = 1$ for the core and $i = 2$ for the shell) on the corresponding sides. Thus, the force per one side per unit area (pressure from each side) is

$$P = \frac{k_B T}{a}\frac{\alpha}{\sqrt{N}} \times \frac{1}{a^2 N} = \frac{k_B T}{a^3}\frac{\alpha}{N^{3/2}}.$$

Therefore, for two uncharged gels consisting of chains of length N_1 and N_2 we get

$$\alpha_1 N_1^{-3/2} = \alpha_2 N_2^{-3/2}. \tag{13}$$

The physical meaning of this relation is very clear: in a dense gel the chains will expand only slightly to yield a given total force, whereas in a lightly crosslinked gel consisting of fewer long chains, the swelling ratio will be much higher.

In the case of a polyelectrolyte gel, one should subtract the osmotic pressure of small ions from eq. 13, which gives the final relationship between equilibrium values of α_1 and α_2:

$$\frac{\alpha_1}{N_1^{3/2}} - \frac{\mathcal{N}_1 a^3}{V_1} = \frac{\alpha_2}{N_2^{3/2}} - \frac{\mathcal{N}_2 a^3}{V_2}. \tag{14}$$

Now, to find out the equilibrium state of our model gel, we should minimize the free energy with respect to one of the swelling ratios (say, α_1) and the fraction of counterions within the core β. The swelling ratio α_2 then can be found from the boundary condition (14).

An important issue that we should treat here is the proper choice of the parameters of our model, so that the resulting physical picture would be the closest to a real inhomogeneous gel. We have chosen $l_B = 0.7$ nm, which corresponds to an aqueous system at room temperatures, $R_0 = 20$ nm, and $D_0 = 50$ nm. For the monomer size a we have set $a = l_B$, suggesting that polyelectrolyte chains are flexible, which is the usual assumption in this kind of calculation.

However, it is more difficult to figure out the reasonable ratio of subchain lengths N_1 and N_2 that would represent the experimental situation. We have chosen $N_1 = 100$, and we varied $N_2/N_1 > 1$ as a free parameter. While it is not easy to say which value of N_2/N_1 corresponds to real gels, we can still think that the ratio of maximum and minimum average polymer densities in a real statistically inhomogeneous gel is not very high, namely,

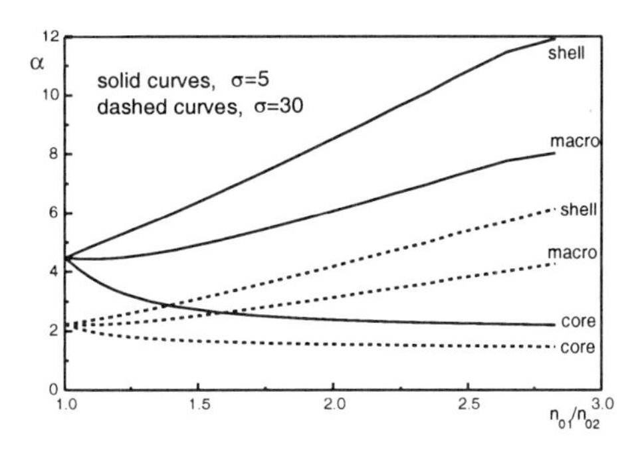

Figure 2. Swelling ratios of the core and the shell and the macroscopic swelling ratio of the inhomogeneous polyelectrolyte gel with different degree of charging σ for $R_0 = 20$ nm, $D_0 = 50$nm. Solid curves: $\sigma = 5$, dashed curves: $\sigma = 30$.

it is of order of unity (2 can be chosen as an estimate). As it is known, the polymer density in the reference state as a function of the degree of polymerization N can be estimated [3] as $n_0\sigma^{-3}N^{-1/2}$. Accordingly, in our model, the ratio of the maximum polymer density in the core and the minimum one in the shell, is equal to

$$n_{0,1}/n_{0,2} = \sqrt{N_2/N_1}. \tag{15}$$

Therefore we expect that this kind of inhomogeneities in real gels correspond to the values of N_2/N_1 not far from unity. As we will show later, in this regime, the dependence of gel macroscopic parameters on N_2/N_1 is not very strong, so the particular value of N_2/N_1 seems to be not very important.

We have performed the calculations for the values of σ equal to $5, 15, 30$, which corresponds to charged units fraction usually encountered in weakly charged polyelectrolyte gels.

The swelling ratios of the core α_1 and of the shell α_2 and the macroscopic swelling ratio α are shown in Fig. 2. In the limit $n_{0,1}/n_{0,2} \to 1$ (homogeneous gel) all these values coincide with each other and with the swelling ratio of the homogeneous polyelectrolyte gel (see below). This proves the self-consistency of our model. We can see that α and α_2 increase with the increase of the ratio $n_{0,1}/n_{0,2}$. There are primarily two reasons for this phenomenon: first, counterions escaped from the core increase the osmotic pressure inside the shell, and, second, at a given counterion concentration, the shell swelling ratio increases with the length of shell subchains just as it should be in the homogeneous gel (see below). It is worth noting, however, that the core swelling ratio α_1 is decreasing with $n_{0,1}/n_{0,2}$. This is caused, first, by decreased counterion concentration in the core resulting in less osmotic pressure, and, second, by the elastic tensions that are developed in

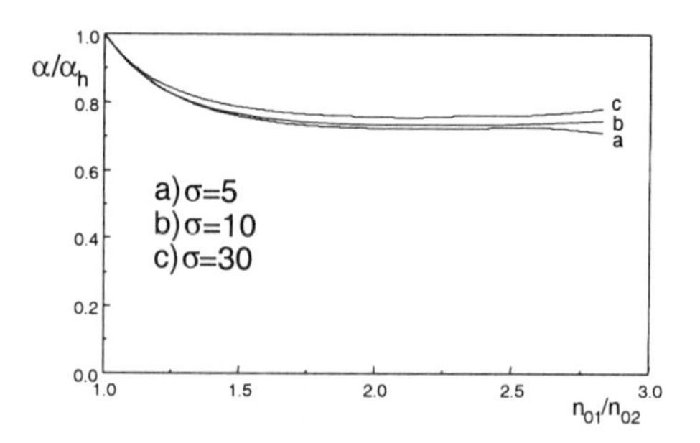

Figure 3. Comparison of the swelling ratio of the inhomogeneous gel to the swelling ratio of the equivalent uniform gel as a function of gel inhomogeneity $n_{0,1}/n_{0,2}$ for different values of σ.

the inhomogeneous gel. All three swelling ratios increase with the increase of the gel charge (decreasing σ) (Fig. 2), in the same way as in the case of an ordinary polyelectrolyte gel.

It is interesting to compare the macroscopic swelling ratio of an inhomogeneous gel with that of a homogeneous polyelectrolyte gel. For this purpose, we introduce an equivalent homogeneous gel with the same charge and the same number of monomer units in the volume as in the inhomogeneous gel under consideration. The total number of monomer units S_{inh} in the part of the inhomogeneous gel shown in Fig. 1 is

$$S_{inh} = \frac{4\pi}{3}\nu_{0,1}R_0^3 N_1 + \frac{4\pi}{3}\nu_{0,2}(D_0^3 - R_0^3)N_2$$

For a homogeneous gel,

$$S_{homo} = \frac{4\pi}{3}\nu_{0,homo}D_0^3.$$

Substituting $\nu_0 \sim 1/a^3N^{3/2}$ and requiring $S_{inh} = S_{homo}$, we obtain the following equation for the length of subchains of an equivalent gel N_{homo}:

$$\frac{D_0^3}{\sqrt{N_{homo}}} = \frac{D_0^3 - R_0^3}{\sqrt{N_2}} + \frac{R_0^3}{\sqrt{N_1}}. \tag{16}$$

The free energy of an equivalent gel in the same approximation as we made for an inhomogeneous gels is expressed by eq. 3 and the equilibrium swelling ratio α_{homo} is given by eq. 4 where $N = N_{homo}$ is determined from eq. 16.

In Fig. 3 we plot the ratio of the macroscopic swelling ratio α of an inhomogeneous gel to the swelling ratio α_{homo} of the equivalent homogeneous gel. The x-axis is $n_{0,1}/n_{0,2}$. First, note that $\alpha/\alpha_{homo} < 1$, i.e. the inhomogeneous gel is less swollen than an equivalent homogeneous gel.

The higher is the gel charge, the less is the swelling degree of the gel in comparison to its homogeneous analogue.

The reason for this effect is simple: some counterions are entrapped by the regions with high polymer charge density, therefore they are osmotically passive, and do not contribute to the full extent to the exerting osmotic pressure.

¿From Figure 3 it can be seen that, as expected, the more is the inhomogeneity in the distribution of charges in the gel (roughly defined as $n_{0,1}/n_{0,2}$), the more is the depression of the gel swelling ratio in comparison with the homogeneous gel (i.e. the more is the fraction of osmotically passive counter ions). This effect is due to the greater depth of entrapping potential wells, so that a smaller part of counterions can escape from the core into the shell. This leads to the decrease of the osmotic pressure and the gel swelling ratio with respect to that of a homogeneous gel.

4. Solutions of polyelectrolyte chains or microgel particles

In the previous section we demonstrated the importance of the effect of the counterion trapping by the inhomogeneously distributed charges within the gel on the swelling behavior of gels. Another example of the systems where large-scale inhomogeneity in counterion distribution plays an essential role are dilute polyelectrolyte solutions.

In this section we consider two cases: solutions can be solutions of linear polyelectrolyte chains and solutions of microgel particles. The conformational changes of microgels have attracted much attention recently [10]. Such particles are intermediate between linear chains and macroscopic gels. They normally can be represented as a collection of elastic chains, some of these chains forming a gel framework, while the other chains are pendant. With an increase in the total number of chains the fraction of pending chains is expected to decrease.

We present an unified model to describe and to compare the swelling properties of single polyelectrolyte chains in a dilute solution, microgel particles of various molecular masses and as a limiting case - macroscopic gels [8]. We paid main attention to one key effect, namely the redistribution of counterions between the interior of a polymer coil and the outer solution. We wanted to concentrate on this main effect in the framework of a simplest model not going into the details of the polymer chain structure.

4.1. THE MODEL

Let us consider a dilute solution of weakly charged microgel particles (linear macromolecules and macroscopic gels being two different limiting cases of a microgel). We suppose that the polymer concentration in the solution is very low, so that microgels do not overlap and do not interact with each other, and we shall consider the swelling behavior of a single microgel particle. The amount of the solvent per one particle is defined by polymer concentration, c.

Each microgel consists of ν subchains of N monomeric units with the characteristic size a. The limiting case of $\nu = 1$ corresponds to a single polymer chain and it will be considered separately in Sec.4.2.

The microgel particles are weakly charged. We assume that the subchains of the microgels are formed by copolymers containing neutral monomer units (for instance, acrylamide units) and a small fraction of ionized units (such as sodium methacrylate units) carrying a charge e. Thus, the position of charges along the subchains is fixed. The total charge of a microgel does not depend on pH and is defined by the number of charged units in the subchains and the number of subchains. As in Sec. 2 N/σ is the total number of charges in each subchain. We consider the case of a salt-free solution, thus, the only mobile ions in the system are the counter ions which appear due to dissociation of monomeric units of the microgel subchains. The number of counterions per microgel particle is equal to the number of charged groups, i.e. to $N\nu/\sigma$.

The free energy of a single polymer microgel can be written as a sum of four terms:

$$F = F_{el} + F_{int} + F_0 + F_{el-st}. \tag{17}$$

The first term in this expression, F_{el}, is the elastic free energy. We write it in the Flory form [11], modified by Birshtein and Pryamitzyn [12] to take into account the entropy loss due to expansion of the microgel subchains in good solvents as well as due to compression of the subchains in poor solvents (cf. eq. 2):

$$F_{el} = \frac{3}{2}\nu k_B T(\alpha^2 + 1/\alpha^2) \tag{18}$$

The next term in eq. 17, F_{int}, is the free energy of interaction of the monomer units of the microgel. We write it in the form of virial expansion in powers of the concentration of monomer units within the particle, n, see eq. 5.

We assume that the polymer chains are flexible and a is the only characteristic length and the following estimates for the virial coefficients are valid not far from the θ-point [3]:

$$B \sim a^3\tau,\ C \sim a^6$$

where τ is the relative temperature deviation from the θ-point, $\tau = \frac{T-\theta}{\theta}$.

It should be mentioned that the expression 5 is valid at low polymer concentrations. This is indeed the case in good solvents and near the transition point to the collapsed state in slightly poor solvents. Far from the θ-point in the extremely poor solvent the density can be rather high and one needs to use more exact expressions, for instance, this contribution can be written in the framework of Flory-Huggins approximation [3]. However, the specific form of the expression for the contribution F_{int} does not influence the qualitative conclusions of the present paper. Thus, below we will use the approximate expression 5 for all values of τ.

Finally, the last two terms in the expression 17, F_0 and F_{el-st}, account for the main contributions connected with the charges on the microgel subchains. The total number of counterions per one microgel particle is equal to the number of charges, i.e., $N\nu/\sigma$. Let us denote the fraction of counterions which are inside the particle as β, their number is equal to $\beta N\nu/\sigma$. Then the number of counterions which are in the region outside the microgel is $(1-\beta)N\nu/\sigma$. In the free energy we have to take into account the contribution from the translational entropy of counterions both inside and outside the polymer. Let us denote this contribution as F_0. Besides, movement of counterions from the interior of the microgels to the solution leads to the appearance of non-compensated charge on the particles, and hence to an additional energy of Coulombic interactions, F_{el-st}. The simplest way to express this last contribution is in terms of the energy of a spherical capacitor with charge $(1-\beta)N\nu e/\sigma$. The radius of the inner electrode of this capacitor is of the order of the mean size of the microgel, $R \sim (N\nu/n)^{1/3}$, and the radius of the outer electrode can be estimated as $R_{out} \sim V_{out}^{1/3}$, V_{out} is the volume of the solvent per one particle, $V_{out} = N\nu/c$.

Taking into account all these contributions we obtain for F_0 and F_{el-st} the expressions:

$$\frac{F_0}{k_BT} = \frac{N\nu}{\sigma}\beta \ln\left(\frac{n_0\beta}{\sigma\alpha^3}\right) + \frac{N\nu}{\sigma}(1-\beta)\ln\left(\frac{n_0(1-\beta)}{\sigma(1/\gamma^3-\alpha^3)}\right) \tag{19}$$

$$\frac{F_{el-st}}{k_BT} = \left(\frac{N}{\sigma}(1-\beta)\right)^2 \nu^{5/3}uN^{-1/2}\left[\frac{1}{\alpha}-\gamma\right] \tag{20}$$

where $u = \frac{e^2}{\epsilon a k_B T}$ and the parameter γ defines the degree of dilution of polymer solution, $\gamma = (V_0/V_{out})^{1/3}$, V_0 is the volume of the microgel in the reference state, $V_0 = N\nu/n_0$. For very dilute solutions $\gamma \ll 1$.

The equilibrium value of the swelling ratio of the microgel and the fraction of counter ions inside the microgel particle can be obtained by minimizing the total free energy expressed by eqs. 17-20 with respect to α and β.

4.2. SWELLING AND COLLAPSE OF SINGLE CHAINS IN A DILUTE SOLUTION

Let us first consider the behavior of linear polyelectrolyte chains in dilute solution and let us compare it to the swelling behavior of polyelectrolyte gels with the same structure of the gel subchains (see Sec. 2). In theories describing collapse transitions in polyelectrolyte gels the condition of total electroneutrality of gels is usually used, i.e. it is assumed that in salt-free solutions all the mobile counterions of a network remain inside the network sample and their number is therefore equal to the number of charges on the gel subchains. In the case of a singe polyelectrolyte macromolecule in a dilute solution a large fraction of counterions of the chain leave for the outer solution this fact leading to a significant difference in the swelling and collapse behavior of linear chains and microscopic gels.

The free energy of the polyelectrolyte macromolecule in a dilute solution is described by eqs. 17-20 with $\nu = 1$. The analysis of the free energy shows that at low polymer concentrations when $\gamma\alpha < 1$ the parameter β is close to zero. It means that the condition of electroneutrality of the single chain is totally violated: the counterions are not held within the polymer coil; the entropic contribution to the free energy of the mobile counterions dominates over the electrostatic interactions and practically all the counterions exit the coil.

In Fig. 4 we show the dependence of the swelling ratio α on the parameter τ for $\gamma = 0.001$ and two different degrees of ionization of the chain, $\sigma = 100$ and $\sigma = 50$. For comparison we also show in Fig. 4 the corresponding behavior of α of a polyelectrolyte gel with the same subchain structure.

One can see that removal of the counterions of the chain leads to a significant decrease in its dimension in good solvents. The swelling ratio of the chain is significantly smaller than that for corresponding subchains of a gel for which $\beta = 1$.

This effect can be explained as follows. Since all the counterions leave the chain, the conformation is defined by electrostatic repulsion between charges along the chain. In this case the chain can be represented as a sequence of "blobs" [3, 13]. Indeed, the analysis of the free energy (for $\beta = 0$ and $\alpha \gg 1$) leads to the following estimate for the swelling ratio in good solvents:

$$\alpha \sim N^{1/2} u^{1/3} \sigma^{-2/3}. \tag{21}$$

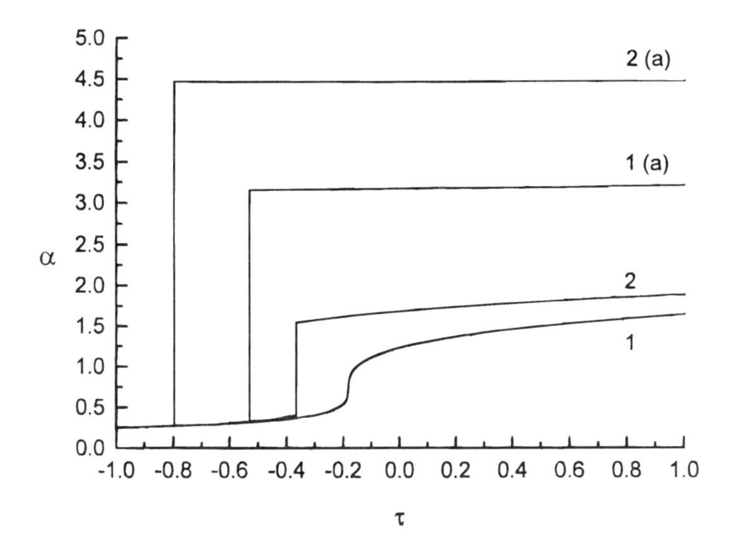

Figure 4. Dependence of the swelling ratio α in the linear polymer on the relative temperature deviation from the Θ-point τ for the following values of the parameters: $N = 1000, u = 1, C/a^3 = 1, \gamma = 0.001$, and $\sigma = 100$ (curve 1) and $\sigma = 50$ (curve 2). Curves 1a and 2a correspond to the case $\beta = 1$ and $\sigma = 100$ and $\sigma = 50$, respectively.

This result is consistent with the "blob" picture [3, 13] and gives an increase of the macromolecular dimensions due to electrostatic repulsion.

On the other hand, for the case of the swelling of macroscopic gels the main reason for the increase of α is the osmotic pressure of counterions rather than the direct Coulomb repulsion. For this case we obtain the following estimate for the swelling ratio and for the equilibrium radius of the chain:

$$\alpha \sim N^{1/2}\sigma^{-1/2} \tag{22}$$

One can see that for $\sigma \gg 1$ the swelling ratios of linear macromolecules and macroscopic gels differ significantly, the osmotic pressure leads to much higher degree of swelling than electrostatic interactions.

On the other hand, in poor solvents the value of the swelling ratio of the chain is defined by non-Coulomb interactions of uncharged monomer links described by the virial coefficient B and C, and the swelling degree α does not depend strongly on the value of β. Therefore the amplitude of the collapse transition becomes smaller.

In addition it should be mentioned that the transition point between the swollen and collapsed states of the chain and the character of this transition depends essentially on whether counterions are inside the polymer coil or if they have left for the outer solution region. When $\beta \sim 0$ the

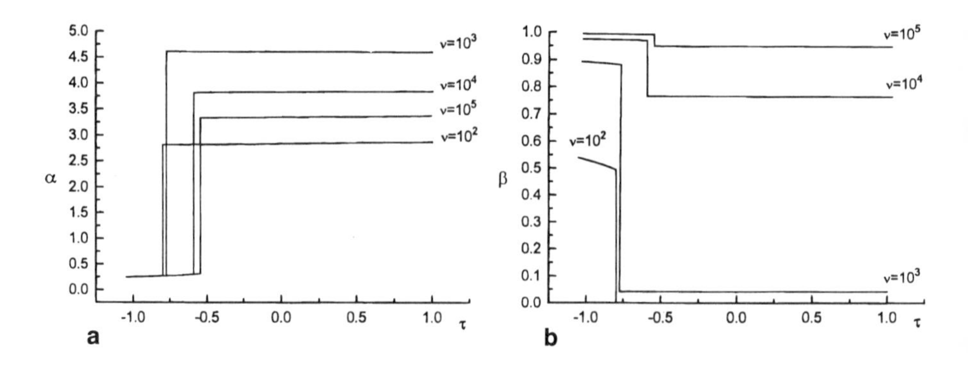

Figure 5. Dependence of the swelling ratio α (a) and the fraction of counterions inside the microgel β (b) on τ for $N = 1000, u = 1, C/a^3 = 1, \sigma = 100, \gamma = 0.001$ and different numbers of microgel subchains ν.

transition occurs at higher temperatures than for the case $\beta = 1$, and also the amplitude of the collapse is smaller and in some cases the character of the transition becomes continuous in contrast to a jumplike first order phase transition for $\beta = 1$ (cf. the curves 1 and 1a in Fig. 4).

4.3. SWELLING OF MICROGEL PARTICLES

Let us consider now the behavior of microgels in dilute solutions. It appears that the distribution of counterions and, hence, the swelling properties of microgel particles depend essentially on molecular mass of the particles.

The swelling ratio, α, and the fraction of counterions inside the microgel, β, for microgels with different number of subchains, ν, as functions of τ are shown in Fig. 5. One can see that if we increase the value of ν the polymer retains inside itself more and more counterions, and the value of β increases and tends to unity, so for rather large particles the condition of electroneutrality holds very well.

The value of β is higher in the globular state and the collapse of the gel is thus accompanied by some kind of "counterion condensation" on the microgel: the fraction of counterions inside the gel increases in a jumplike fashion when the gel shrinks.

It should be mentioned that the swelling ratio of a gel in a good solvent changes non-monotonously with an increase of the number of the subchains (Fig. 5). This fact is clearly seen also in Fig. 6 where we present the functions $\alpha(\nu)$ and $\beta(\nu)$ for $\tau = 1$, with a constant polymer concentration $\gamma = 0.001$ and three different degrees of ionization. While the value of β increases monotonously with an increase in the number of the gel subchains, the dependence of the swelling ratio, α, on ν has a maximum at some value

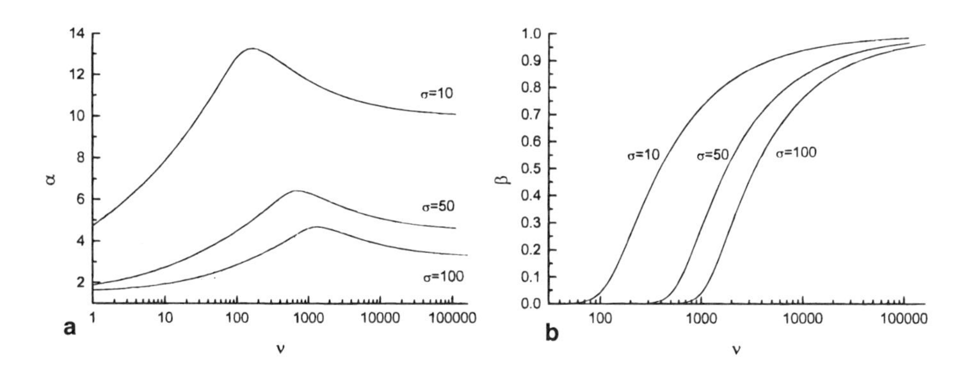

Figure 6. Dependence of the swelling ratio α (a) and the fraction of counterions inside the microgel β (b) on the number of microgel subchains ν for different degrees of charging of the gel $\sigma = 10, 50, 100$.

of ν. This maximum shifts to smaller values of ν with an increase of the ionization degree of the gel chains.

This non-monotonous behavior of the swelling ratio with change of ν is due to the different distribution of counterions between the interior of the microgel and the outer solution for microgel particles of different molecular masses. In the case of small particles, i.e. particles with small number of subchains, the value of β is close to zero (Fig. 6,b). This means that most of counterions leave the gel for the outer solution. Thus, the initial growth of the swelling ratio at small values of ν is connected with an increase in the uncompensated electrostatic repulsion for larger microgel particles. In this case the swelling ratio of the microgel is defined by the interplay of elastic free energy and the energy of Coulombic repulsion between the charges on the subchains and we obtain the following estimate for α:

$$\alpha \sim N^{1/2} u^{1/3} \sigma^{-2/3} \nu^{2/9} \tag{23}$$

On the other hand, with increasing ν, more and more counterions are trapped within the gel. Thus the electrostatic repulsion between the gel subchains decreases and the gel volume begins to decrease as well at large values of ν (Fig. 6,a). It is worth noting that at large ν the value of β is close to unity and we reach the regime of macroscopic gel swelling. In this case the behavior of the gel is described by the swelling of one subchain and α is independent of ν and is defined by eq. 22.

5. Polyelectrolyte/ionomer behavior of polyelectrolytes in poor solvents

In the previous sections we supposed that all potentially charged groups on polyelectrolyte chains are dissociated and the total number of counterions in the system is fixed. Now we consider the influence of ion pair formation on the behavior of polyelectrolyte systems. For this purpose the theory of the polyelectrolyte solutions described in Sec. 4 will be generalized for the case of three possible states of counterions [9].

As before σ is the number of monomeric units between two nearest potentially charged groups. In reality not all of these potentially charged groups are in fact dissociated, since some of the counterions stay bound forming ion pairs with the charges of microgel subchains. Let us denote as θ the fraction of bound counterions (i.e. the fraction of ion pairs). As we will show below the value of θ depends essentially on the dielectric constant, ϵ, of the medium inside the microgel [14, 15].

On the other hand, nonbound counterions can either remain within the microgel or float in the outer solution. Let us denote as β the fraction of counterions kept within polymer (including those forming ion pairs), then $1-\beta$ is the fraction of counterions moving freely outside the microgel.

Thus, in the general case considered here counterions are distributed between three possible states: free counterions in the solution, free counterions within the microgel, and bound counterions forming ion pairs, and $m\nu(1-\beta)/\sigma$, $m\nu(\beta-\theta)/\sigma$ and $m\nu\theta/\sigma$ are the total numbers of counterions in these states, respectively.

The free energy of a single microgel particle can be written as a sum of six terms:

$$F = F_{el} + F_{int} + F_{ion} + F_0 + F_{el-st} + F_{comb}. \tag{24}$$

The first term in this expression is the elastic free energy, it is described by eq. 18. The next term in eq. 24, F_{int}, is the free energy of interaction of the monomer units of the microgel. We write it in the framework of Flory-Huggins approximation [11]. The volume fraction of monomer units inside the microgel is equal to $na^3 = n_0a^3/\alpha^3$ and

$$\frac{F_{int}}{k_BT} = \left(\alpha^3\frac{N\nu}{n_0a^3} - N\nu\right)\ln\left(1 - \frac{n_0a^3}{\alpha^3}\right) - \chi N\nu\frac{n_0a^3}{\alpha^3}, \tag{25}$$

where χ is the polymer-solvent interaction parameter.

The next term, F_{ion}, accounts for the energy gain from ion pairing. We assume that the characteristic distance between ions in ion pairs is of order

a, then each ion pair gives the contribution to energy gain of order of $\frac{e^2}{\epsilon a}$ and therefore

$$\frac{F_{ion}}{k_B T} = -A \frac{N\nu}{\sigma} \theta \frac{e^2}{\epsilon a k_B T}, \tag{26}$$

where A is a numerical constant of order of unity. We will assume $A = 1$ in the subsequent numerical calculations.

As it has been shown in [14] the value of the dielectric constant of polymer medium ϵ influences significantly the number of ion pairs formed by the charges of the gel chains, on the other hand the value of ϵ can change itself during collapse transition of polymer particles. Indeed, the volume fraction of polymer inside a swollen particle is very low, $na^3 \ll 1$, and the value of ϵ is close to that of the pure solvent (water), ϵ_0, while in the collapsed state it is close to that of the dry polymer, ϵ_1. As in ref. [14] we suppose here that the value of ϵ depends linearly on the fraction of polymer within the microgel, i.e.,

$$\epsilon = \epsilon_0 - (\epsilon_0 - \epsilon_1) n_0 a^3 / \alpha^3 \tag{27}$$

and hence

$$\frac{F_{ion}}{k_B T} = -A \frac{N\nu}{\sigma} \theta \frac{u_0}{1 - \frac{(\epsilon_0 - \epsilon_1)}{\epsilon_0} \frac{n_0 a^3}{\alpha^3}}, \tag{28}$$

where $u_0 = \frac{e^2}{\epsilon_0 a T}$.

The contribution F_0 is connected with the translational entropy of mobile counterions inside microgel and in the outer solution. It has the following usual form [14]:

$$\frac{F_0}{k_B T} = \frac{N\nu}{\sigma}(\beta - \theta) \ln \left[\frac{n_0}{\sigma} \frac{\beta - \theta}{\alpha^3} \right] + \frac{N\nu}{\sigma}(1 - \beta) \ln \left[\frac{n_0}{\sigma} \frac{1 - \beta}{1/\gamma^3 - \alpha^3} \right]. \tag{29}$$

The term F_{el-st} is the free energy of Coulombic interactions of non-compensated charge of the coil, it has the form of eq. 20 .

Finally, the last term in eq. 24, F_{comb}, is the free energy connected with the combinatorial entropy of distribution of counterions between three possible states:

$$\frac{F_{comb}}{k_B T} = 2\frac{N\nu}{\sigma}\theta \ln\left[\frac{N\nu}{\sigma}\theta\right] + \frac{N\nu}{\sigma}(\beta-\theta)\ln\left[\frac{N\nu}{\sigma}(\beta-\theta)\right] + \\ +\frac{N\nu}{\sigma}(1-\beta)\ln\left[\frac{N\nu}{\sigma}(1-\beta)\right] + \frac{N\nu}{\sigma}(1-\theta)\ln\left[\frac{N\nu}{\sigma}(1-\theta)\right] \tag{30}$$

Thus, eqs. 24, 18, 25, 28, 29, 20, 30 define completely the free energy of a polyelectrolyte microgel in a dilute salt-free solution.

5.1. SUPERCOLLAPSED IONOMERIC STATE OF POLYELECTROLYTES

Let us consider first the swelling behavior of polyelectrolyte gels taking into account the possibility of ion pair formation and compare it with the swelling of gels without ion pairing. The free energy of gels is described by eqs. 24, 18, 25, 28, 29, 20, 30 with the value of $\beta = 1$.

Analysis shows that the behavior of the gel depends essentially on the ratio ϵ_1/ϵ_0 of the dielectric constants of the dry gel and the solvent. If the ratio ϵ_1/ϵ_0 is not very small accounting for the ion pair formation does not lead to a qualitative change of the behavior of the system in comparison with that described in Section 2. The formation of a small fraction of ion pairs leads to a decrease of the number of free counterions and their osmotic pressure. This fact results in a small decrease of α in comparison with what was obtained in the classical theories.

The behavior of the system changes qualitatively if the ratio ϵ_1/ϵ_0 is smaller than some critical value much less than unity. In this case, the accounting for the dependence of ϵ on the volume fraction of polymer in the network leads to the appearance of third minimum of the free energy. This minimum appears when

$$\epsilon_1/\epsilon_0 < (\epsilon_1/\epsilon_0)_{cr} \sim (Au_0/\sigma)^{1/2} \tag{31}$$

and corresponds to the completely collapsed gel very close to the densely-packed state. In this state practically all the ions are associated into ion pairs, i.e. $\theta \sim 1$.

The occurrence of such supercollapsed state is a manifestation of the following avalanche-type process: the decrease of network volume causes the decrease of ϵ and additional ion pair formation, thus, exerting osmotic pressure of counter ions decreases which in turn leads to the further shrinking of the gel and so on. This loop of positive feedback brings the gel quite close to the most dense conformation.

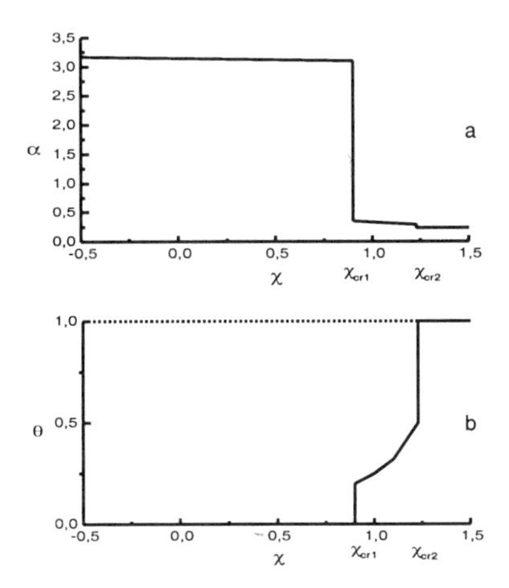

Figure 7. Dependences of the swelling ratio α (a) and the fraction of ion pairs θ on the Flory-Huggins parameter χ for the following values of the parameters: $N = 500, n_0 = 0.01, \sigma = 50, A = 1, u_0 = 1, \epsilon_1/\epsilon_0 = 0.07$.

Since for small values of ϵ_1/ϵ_0 the free energy generally has three minima, by changing external conditions it is possible to induce the conformational transitions between all three states of the network which correspond to these minima (extended gel, collapsed gel, supercollapsed densely packed gel). In particular, the collapse of the network can be realized as the sequence of two phase transitions. The corresponding dependences of α and θ are shown in Fig.7. When the solvent quality becomes poorer the network first collapses into intermediate state that corresponds to an ordinary collapsed gel and then the avalanche process sets in and the gel further collapses in the supercollapsed state. Both transitions are jump-like first-order phase transitions. The gel volumes in the collapsed and supercollapsed state are in fact close to each other. However, two collapsed states are clearly distinguished by the values of θ: in the usual collapsed gel $\theta < 0.5$ (see Fig.7b), i.e. more than a half of all the counter ions are mobile, while in the supercollapsed gel $\theta \sim 1$, i.e. practically all the counter ions form ion pairs. The latter distinction between collapsed and supercollapsed gel is characteristic for most of the cases for which our calculations have shown simultaneous existence of these two states.

The network can also bypass the intermediate state and collapse directly from very extended to the supercollapsed state. This situation takes place for

$$\epsilon_1/\epsilon_0 < (\epsilon_1/\epsilon_0)'_{cr} \sim Au_0/\sigma \tag{32}$$

i.e. for the range of values of ϵ_1/ϵ_0 even smaller than the range given by

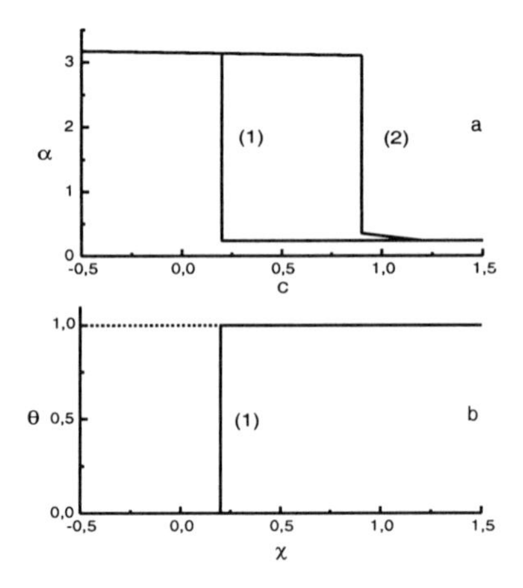

Figure 8. Dependences of the swelling ratio α (a) and the fraction of ion pairs θ on the Flory-Huggins parameter χ for the following values of the parameters: $N = 500, n_0 = 0.01, \sigma = 50, A = 1, u_0 = 1, \epsilon_1/\epsilon_0 = 0.02$. Curve 2 corresponds to disregard of ion pair formation.

the inequality 31. Characteristic dependences $\alpha(\chi)$ and $\theta(\chi)$ for this case are shown in Fig.8. It can be seen that due to the formation of ion pairs the collapse transition point shifts significantly to lower values of χ, i.e. the region where the collapsed conformation is favorable increases.

5.2. COLLAPSE INDUCED BY EXTRA IONIZATION

Several experimental facts indicate that the collapse transition in weakly charged polyelectrolyte gels or single macromolecules can take place with an increase in the degree of ionization (e.g. during titration of a weak polyacid).

One system of this kind (a solution of partially ionized poly(acrylic acid) (PAA) in methanol at room temperature) was investigated in detail in ref. [16]. Ionization was accomplished by titration with CH_3ONa. It was observed that the reduced viscosity at first grows and then decreases rapidly in a small interval of degree of dissociation of PAA. A thorough investigation of the titration behavior of PAA in methanol using different techniques (elastic light scattering, osmotic, potentiometric, conductance and UV spectrophotometric measurements) has clearly demonstrated that after the initial swelling PAA molecules undergo a conformational transition which results in small compact particles where the macromolecular chain is collapsed. The same non-monotonic conformational behavior has been found for poly(methacrylic acid) (PMAA) linear chain solutions and PMMA gels titrated in methanol [17, 18]. At first sight, this behavior is

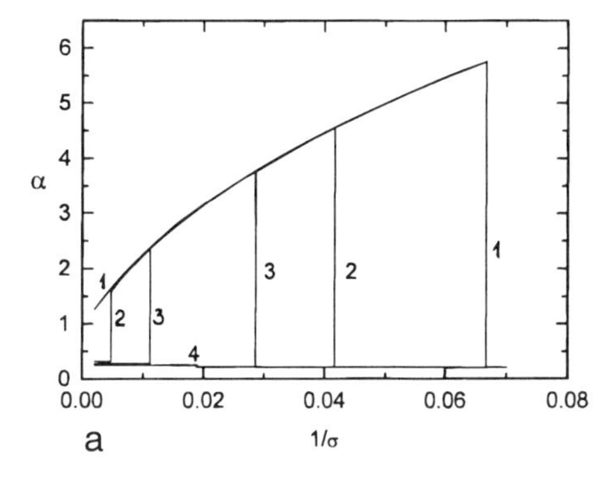

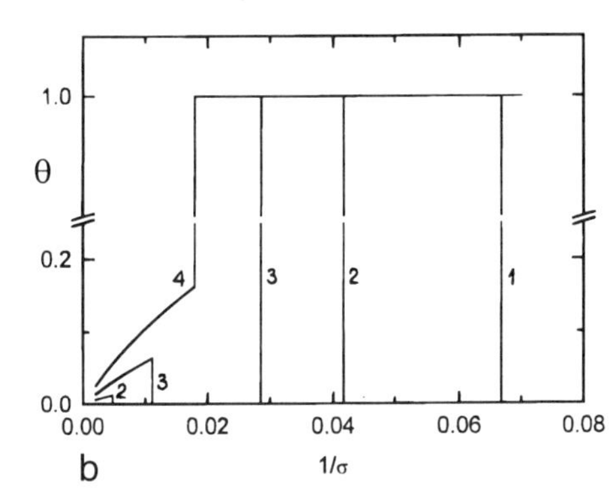

Figure 9. Dependence of the swelling ratio of the gel α and of the fraction of ion pairs θ on the parameter $1/\sigma$ for different values of $\chi = 0.5(1), 0.7(2), 0.8(3), 0.9(4)$.

rather surprising: with an increase of the degree of ionization the chain first swells (which is natural) but then rapidly shrinks.

It appears that this unusual behavior can be explained by the energy gain from the formation of ion pairs in the collapsed state of low polarity, which competes with the swollen state where most of the counter ions are dissociated. An increasing degree of ionization increases the thermodynamic advantages of the collapsed state with an ionomeric multiplet structure over the swollen polyelectrolyte state. In some regimes, this effect can lead to reentrant phase transition behavior, i.e. to an initial decollapse of the gel upon charging with subsequent jump-like collapse at a higher fraction of charged monomer units.

The dependence of the swelling coefficient on the degree of charging of the gel is presented in Fig. 9. It can be seen that incorporation of a small amount of charges in the network chains during titration at first leads to the swelling of the network. In a Θ-solvent this swelling is continuous, while in poor solvents, where the neutral gel is in globular state, it is accompanied by a discontinuous first-order phase transition. With the increase of $1/\sigma$ the transition point shifts to higher value of the ionization degree, in accordance with general expectations. In the swollen state practically all counter ions are mobile (θ is close to zero, see Fig.9,b); it is the osmotic pressure of mobile counter ions that defines the volume of the network in this state. Thus, the value of θ depends little on χ for highly swollen gels but grows with the increase of $1/\sigma$.

The swelling of the gel continues until $1/\sigma$ reaches the critical value $(1/\sigma)_{cr} \sim Au_0\epsilon_1/\epsilon_0$. At this degree of ionization the gel suddenly shrinks and the value of θ becomes close to unity (i.e. all counter ions form ion pairs). The reason for this is clear: the minimum of the free energy corresponding to the "supercollapsed" state becomes more favorable than the

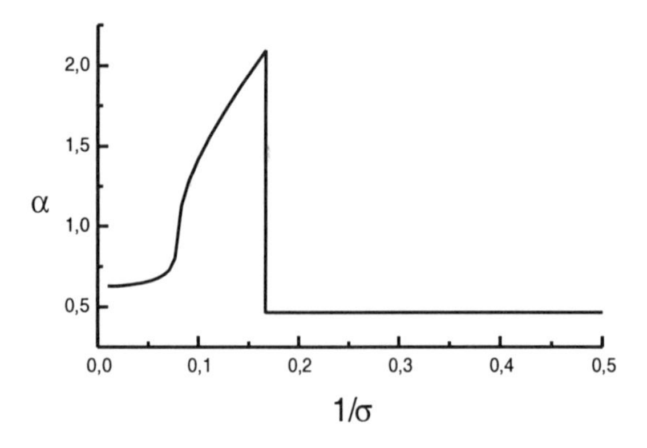

Figure 10. Dependence of the swelling ratio, α, on the degree of ionization of the chain $1/\sigma$ for the following values of the parameters: $N = 100$, $n_0 = 0.1$, $u_0 = 1$, $\epsilon_1/\epsilon_0 = 0.06$, $\gamma = 0.001$, $\chi = 0.75$.

minimum corresponding to the swollen state, and the system switches from the swollen polyelectrolyte to the "supercollapsed" ionomeric behavior. The transition is accompanied by the avalanche-type process described at the end of Section 5.1. It is interesting to note that in the "supercollapsed" state the network is more compact than the initial neutral gel. The reason for this is the tendency to form a denser gel to minimize the dielectric constant and maximize the energy gain from ion pairing. Thus, in agreement with the observations of ref. [18] the change of the volume of the gel in the process of the titration is non-monotonous: initial swelling is followed by sudden collapse and switch from polyelectrolyte to ionomer behavior.

With the increase of the value of χ (i.e.when the solvent becomes poorer) the initial decollapse occurs at higher ionization degrees of the network chains, while the reverse transition to "supercollapsed" state takes place at lower values of $1/\sigma$. Thus, the poorer the solvent, the smaller is the region of stability of the swollen state.

The same behavior is typical also for polyelectrolyte chains in dilute solutions. In Fig. 10 we present the dependence of the swelling ratio of the chain on the parameter $1/\sigma$. The initial swelling of the chain is due to increasing role of electrostatic interactions of the uncompensated charges on the chain, at this stage practically all counterions escape from the space occupied by the coil, and the values of β and θ is close to zero. On the other hand an increasing degree of ionization of the chain enhances the thermodynamic advantages of the supercollapsed state of the chain when practically all counterions are bound and form ion pairs. As in the case of polyelectrolyte gels at some critical value of σ the transition to this state takes place. This non-monotonous behavior of the swelling ratio of

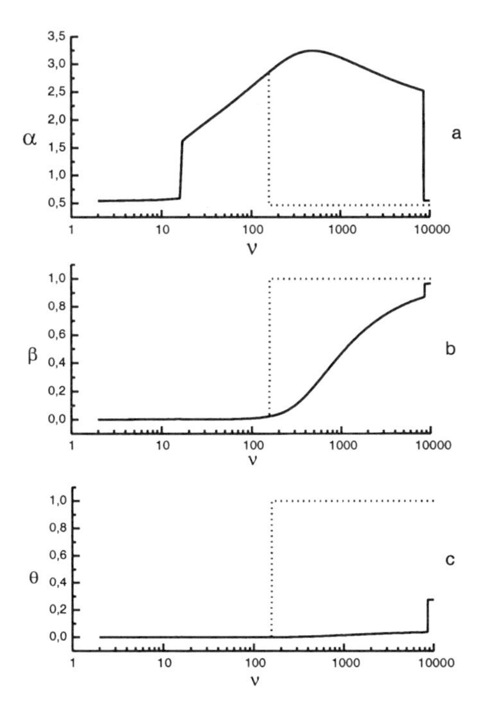

Figure 11. Dependence of the swelling ratio, α (a), the fraction of counterions kept inside the microgel β (b), and the fraction of ion pairs θ (c) in the microgel particle, on the number of microgel subchains ν for the following values of the parameters: $N = 100, n_0 = 0.1, u_0 = 1, \gamma = 0.001, \sigma = 20, \chi = 1$ and for different values of the ratio ϵ_1/ϵ_0.

charged macromolecules upon increase of their degree of ionization is in good agreement with experimental observations [16, 17].

5.3. SWELLING OF MICROGEL PARTICLES IN POOR SOLVENTS

It is interesting to analyze the behavior of microgel particles in poor solvents as function of their molecular masses.

In Fig. 11 the dependences of α, β and θ on the number of microgel subchains are shown. At the same external conditions some microgel particles are in the swollen state while others appear to be collapsed (depending on the molecular weight of these particles). One can see that small particles are in the ordinary collapsed state, in this state practically all counterions are floating in the outer solution, this fact leading to the existence of an uncompensated charge on the particles. With an increase of the molecular mass of the particles the value of this uncompensated charge increases while the change in the particle volume is minor. At some value of $\nu = \nu^*$ the electrostatic energy loss becomes dominant over non-electrostatic volume interactions between uncharged monomer units and it becomes more favorable for the particles to adopt swollen conformations at $\nu > \nu^*$ (the first jump on the dependence $\alpha(\nu)$). On the other hand, with further increase of ν more and more counterions concentrate within polymer particles, the value of β grows significantly. The electrostatic interactions become screened and the collapsed phase is again the most favorable. In our three-state model for counterions the energy gain from ion pairing makes the large particles adopt the collapsed state with the formation of ion pairs. One should mention that taking into account the dependence of the dielectric constant on the volume fraction of polymer we obtain that at smaller values of the ratio ϵ_1/ϵ_0 the process of ion pairing sets in for the particles of smaller molecular masses bringing them to the supercollapsed state.

6. Adsorption of Polyelectrolytes on Oppositely Charged Surfaces

Adsorption of polyelectrolytes on oppositely charged surfaces is an important tool for surface modification [19, 20] and presents substantial fundamental interest. The phenomenon of surface charge reversal and formation of stable multilayers has attracted much attention from both theoretical and experimental sides.

We will demonstrate that ion pairs can play an important role in determining the polymer density profile and the magnitude of charge of the adsorbed layer.

We will employ the formalism similar to that presented above, however, instead of two-phase approximation for counterion distribution we will use a continuum model, where polymer and counterion concentrations and the fraction of ion pair θ are continuous functions of the distance from the plane on which polyions are adsorbed.

6.1. MODEL

We consider a flexible polyelectrolyte molecule interacting electrostatically with an oppositely charged plane in the presence of low-molecular salt. We assume that the charged units fraction of the polymer p is low enough, so that the effects of electrostatic persistence length can be neglected. We are considering the case of a good solvent, and we do not distinguish between polymer counterions and salt ions of the same sign.

The free energy of the system has the contributions corresponding to the (neutral) polymer, translational entropy of low-molecular salt ions and polymer counterions, electrostatic interactions, the coupling with a reservoir, and can be written as follows:

$$F = \int [f_{polymer} + f_{ions} + f_{el-stat} + f_{reserv}]\, d^3\vec{r} \tag{33}$$

We will assume that the system is homogeneous in $y-z$ plane parallel to the surface, and therefore we will characterize the system by monomer unit concentration $c(x)$, electrostatic potential $\psi(x)$, and positive and negative small ion concentrations $c^+(x)$ and $c^-(x)$ as functions of the distance x from the surface. For convenience, we introduce the polymer order parameter ϕ by the relation $\phi^2(x) = c(x)$.

Since we are dealing with a system having an inhomogeneous spatial distribution of polymer concentration, the polymer part of the free energy should contain a gradient term [21], which can be expressed as

$$\frac{f_{inhom}}{k_B T} = \frac{a^2}{6}|\nabla\phi|^2, \tag{34}$$

where a is the Kuhn segment length of the polymer chain, k_B is the Boltzmann constant, and T is the temperature.

The non-ionic interactions of the polymer are accounted using a Flory–Huggins form of the free energy within the framework of a lattice model [11] with cell size a_{FH} and polymer-solvent interaction parameter χ. The use of the full Flory–Huggins form of the free energy instead of a virial expansion [21] is due to the expected high polymer volume fraction within the adsorption layer, when the virial expansion is not valid. Therefore, the polymeric part of the free energy can be expressed as follows:

$$\frac{f_{polymer}}{k_B T} = \frac{a^2}{6}|\nabla\phi|^2 + \frac{1}{a_{FH^3}} \Big[a_{FH}^3\phi^2\chi(1 - a_{FH}^3\phi^2) +$$

$$(1 - a_{FH}^3\phi^2)\ln(1 - a_{FH}^3\phi^2)\Big]. \tag{35}$$

The entropic contribution of small ions f_{ions} depends on the concentrations of positive and negative ions and the ion redistribution between free ions and ions forming ion pairs. We account for ion pairs by introducing the ion pair fraction θ, which is, by definition, the fraction of polymer charged groups forming ion pairs, $0 < \theta < 1$. Since all small ions of the same sign are equivalent, the corresponding contribution can be expressed as

$$\frac{f_{redist}}{k_B T} = \theta\ln\theta + (1-\theta)\ln(1-\theta) \tag{36}$$

per ion, and since in our model the ion pairs can be formed only between polymer and small ions, the expression (36) should be multiplied by $p\phi^2$ to get the volume density of this part of the free energy. The total entropic part of the free energy therefore reads

$$\frac{f_{ions}}{k_B T} = c^+ \ln c^+ a_{FH}^3 - c^+ + c^- \ln c^- a_{FH}^3 - c^-$$

$$+p\phi^2[\theta \ln \theta + (1-\theta)\ln(1-\theta)]. \tag{37}$$

The electrostatic energy can be expressed as follows (e is the elementary charge)

$$f_{el-stat} = p\phi^2(1-\theta)e\psi + c^+ e\psi - c^- e\psi - p\phi^2\theta\Delta U - \frac{\epsilon(\phi)}{8\pi}|\nabla\psi|^2 \tag{38}$$

The first term takes into account the energy of charges on polymer chains in the electrostatic potential ψ (only charged units not forming ion pairs feel the electric field). The next two terms describe the energy of positive and negative small ions in the electric filed, respectively. The fourth term is the energy gain due to the formation of ion pairs, where ΔU is the energy of the formation of one ion pair. The last term corresponds to energy of the electric field created by the potential ψ in the medium with local dielectric constant $\epsilon(\phi(x))$. The expression (38) (apart from the ion pair contribution, fourth term) is equivalent to the usual expression for the electrostatic energy,

$$W = \int \frac{\epsilon \vec{E}^2}{8\pi} dV,$$

where $\vec{E} = -\vec{\nabla}\psi$ is the electric field.

Indeed, if we denote the volume charge density by ρ, then

$$W = \frac{1}{2}\int \rho\psi dV = \int \frac{\vec{E}^2}{8\pi} dV = \int \left(\rho\psi - \frac{\vec{E}^2}{8\pi}\right) dV.$$

Here the last expression corresponds to the equation (38). This form of expression for electrostatic free energy was proposed in ref. [22]. It allows to derive easily the Poisson-Boltzmann equation by minimization of the total free energy of the system.

We will use the simplest possible assumption for the dependence $\epsilon(\phi)$, namely the linear decrease of ϵ with polymer volume fraction,

$$\epsilon(\phi) = \epsilon_w - \phi^2 a_{FH}^3(\epsilon_w - \epsilon_p), \tag{39}$$

where ϵ_w is the dielectric constant of the pure solvent (water, $\epsilon_w = 81$), and ϵ_p is the dielectric constant of the pure polymer. We have chosen $\epsilon_p = 2$ for subsequent calculations.

Finally, the term f_{reserv} couples the system with a reservoir, and ensures that all the concentrations in the limit $x \to \infty$ are equal to their bulk values.

$$f_{reserv} = -\mu_P \phi^2 - \mu^+ c^+ - \mu^- c^- - \mu^- p \phi^2 \theta \tag{40}$$

Here μ_P, μ^+, μ^- are the chemical potentials of the polymer, positive and negative small ions respectively.

The electrostatic interaction with the surface is introduced via the boundary condition

$$\left.\frac{d\psi}{dx}\right|_{x=0} = \frac{4\pi\sigma_S}{\epsilon}, \tag{41}$$

where σ_S is the surface charge.

The equilibrium profiles of polymer and small ions concentration can be obtained by minimizing the functional F with respect to $\phi(x)$, $\psi(x)$, $c^+(x)$, $c^-(x)$ and θ.

The boundary conditions at $x = 0$ are

$$\frac{d\psi}{dx} = \frac{4\pi\sigma_S}{\epsilon}, \qquad \phi = 0, \tag{42}$$

while infinitely far from the plane all the derivatives should vanish, and the concentrations should be equal to their bulk values:

$$\frac{d\psi}{dx} = 0, \quad \frac{d\phi}{dx} = 0, \quad c^+ = c_{salt},$$

$$c^- = c_{salt} + p\phi_\infty^2(1 - \theta_\infty), \quad \phi^2 = \phi_\infty^2. \tag{43}$$

The problem reduces to finding the chemical potentials μ^+, μ^-, and μ_P such that the concentration profiles are not uniform and satisfy the boundary conditions (42),(43). ¿From the formal viewpoint, we have an eigenvalue problem, where the eigenvalues are the chemical potentials μ^+, μ^-, μ_P, and the eigenfunctions correspond to the equilibrium concentration profiles. This problem has been solved numerically using a relaxation method, and the results are presented below.

The choice of parameters for the calculations is expected to approximate the experimental situation of adsorption of flexible polyelectrolytes. It is known that for this case at low concentrations, where virial expansion is applicable, the second virial coefficient of the interaction v is of order a^3, $v/a^3 \sim 1$ [3]. For this reason we have chosen the following parameters for the subsequent calculations: $\chi = 0$, $a = 10$Å, and a_{FH}=5.85Å, corresponding to $v = 200\text{Å}^3$. The experimental values of polymer concentration of the polyelectrolyte solution are in the range of 10^{-4} to 10^{-2} M, see [19, 23,

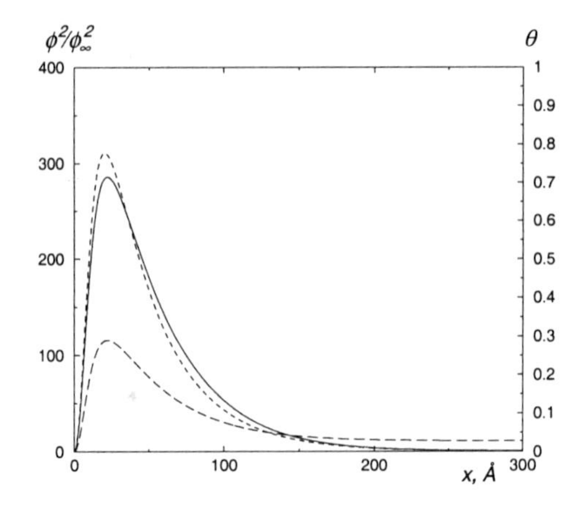

Figure 12. Polymer concentration profiles and fraction of ion pairs θ as function of the distance from the plane x for the case of weak electrostatic interaction. Solid line, actual polymer concentration profile, long dash, fraction of ion pairs θ (right ordinate axis), short dash, polymer concentration profile in the approximation $\theta = 0$ (no ion pairs).

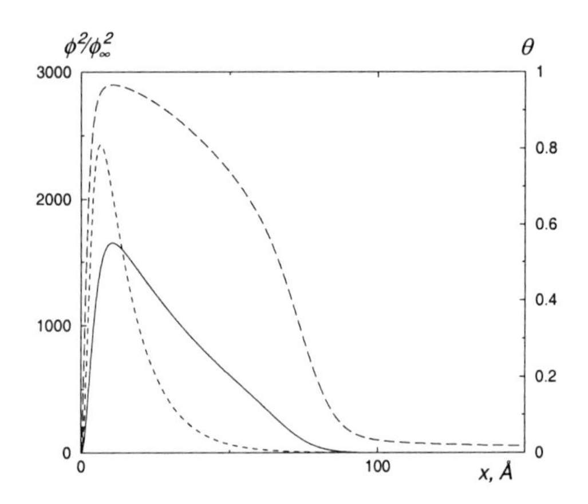

Figure 13. Polymer concentration profiles and fraction of ion pairs θ as function of the distance from the plane x for the case of strong electrostatic interaction. Solid line, actual polymer concentration profile, long dash, fraction of ion pairs θ (right ordinate axis), short dash, polymer concentration profile in the approximation $\theta = 0$ (no ion pairs).

24]. We have chosen $\phi_\infty^2 = 10^{-6}\text{Å}^{-3}$, which corresponds to 1.67 mM. The magnitude of the substrate charge was of order of 0.01 $e/\text{Å}^2$. The energy gain of ion pair formation ΔU has been set to 10 k_BT.

6.2. POLYELECTROLYTE CONCENTRATION PROFILES AND OVERCHARGING

The profiles of polymer concentration ϕ^2 and fraction of ion pairs for weak and strong electrostatic interactions are shown in Fig. 12 and Fig. 13. In the case of weak electrostatic interaction, characterized by low polymer charge, $p = 0.05$, low surface charge density, $\sigma_S = 0.01e/\text{Å}^2$, and high salt concentration, $c_s = 2mM$, the profile of polymer concentration is smooth, and the spatial distribution of ion pairs qualitatively follows the polymer profile. This regime (without accounting for ion pairing) has been extensively investigated, e.g. in ref. [22]. Our results show that even in this regime the fraction of ion pairs θ in the maximum is relatively high, about 0.5, and their influence on the properties of the adsorption layer is significant, although probably not changing the qualitative picture. The polymer profile calculated for the same system, but with the assumption $\theta = 0$ (no ion pairs) is shown in Fig. 12 by short-dash curve. The differences are small; the shapes of the profiles are quite similar, but accounting for ion pairs reduces the charge on the polymer, and thus leads to weaker attraction and consequently to lower peak polymer concentration. It can be shown that the polymer volume fraction in this case does not exceed 0.1; thus a virial expansion should hold in the polymer part of the free energy.

For stronger electrostatic interactions, $p = 0.2, \sigma_S = 0.1e/\text{Å}^2$, and lower screening by salt, $c_S = 0.01\text{mM}$ (Fig. 13), the concentration profiles are much different. First, the peak polymer concentration reaches about 2.7M ($\sim$ 1600 times higher than the bulk concentration) due to increased electrostatic attraction of charged polymer chains to the surface. Second, the fraction of ion pairs within the adsorbed layer is close to unity and their profile (Fig. 12, long-dashed line) has a rather sharp outer boundary which may be used as a definition of the adsorbed layer boundary. These findings support our initial conjectures about the structure of the adsorbed layer. The sharp spatial boundary between ionomer and polyelectrolyte regimes should have a surface charge, and probably can serve as a substrate during multilayer formation.

In this case ion pairs play an important role in the adsorbed layer, and the polymer profile (Fig. 13, solid line) is very different from the one calculated without accounting for ion pairs (Fig. 13, short-dashed line): the adsorption layer is thicker and decays much slower. The magnitude of polymer volume fraction in the adsorbed layer in this regime depends on substrate charge, polymer charge, and salt concentration; in this particular case it is equal to 0.4, i.e. we have a rather dense packing of the polymer in the inner regions of the layer. This fact proves the necessity of usage of the Flory-Huggins type of free energy instead of a virial expansion.

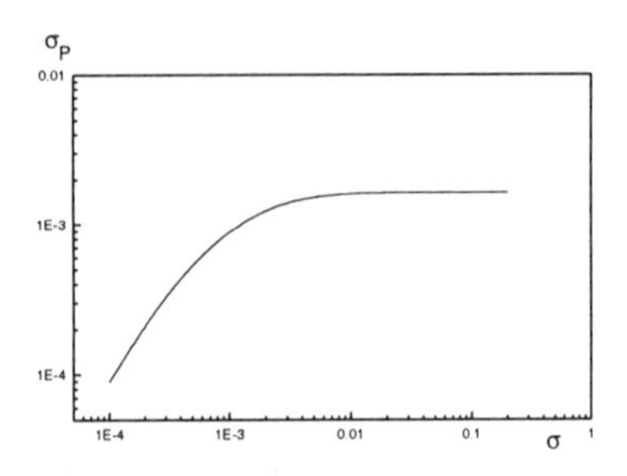

Figure 14. Dependence of the polymer layer charge on the substrate charge.

Thus, we can expect that the surface charge is indeed neutralized, the substrate charge density is of little importance, and the layer charge is defined mostly by outer regions of the layer. Below we will present numerical evidences of this fact.

The charge of polymer layer can be defined as

$$\sigma_P = \int_0^\infty \left\{ p\phi(x)^2(1-\theta(x)) - p\phi_\infty^2(1-\theta_\infty) \right\} dx, \tag{44}$$

i.e. the excess charge of the polymer layer over the bulk polymer charge density. The typical dependence of σ_P on the surface charge σ_S is shown in Fig. 14. We can see that when the substrate surface charge exceeds some threshold value, the charge of adsorbed polymer layer is practically independent of the substrate charge ("saturation regime").

Therefore, multilayer assembly by alternating adsorption of polycations and polyanions becomes possible in the following way: polyelectrolyte A adsorbs onto a surface with charge density σ_1. This charge density had formed due to a layer of a previously adsorbed oppositely charged polyelectrolyte B; if σ_1 is high enough, then overcharging and saturation occur, and the topmost A-layer develops charge density $-\sigma_2$, which is independent of σ_1 and is a characteristic property of polyelectrolyte A under given conditions (solvent quality, salt concentration, etc.). If $|\sigma_2|$ is high enough, polyelectrolyte B can adsorb, resulting in the surface with charge density σ_1, and so on, which allows formation of a multilayer of arbitrary thickness.

7. Conclusions

In this article we have considered various effects connected with the role of counterions in polyelectrolyte systems. We have demonstrated that counterions can be present in such systems in different states, from totally bound (condensed) to polymer chain to completely free. The proper account for the various states of counterions is important for the correct theoretical description of physical effects in polyelectrolyte systems.

References

1. Oosawa, F., *Polyelectrolytes*, Marcel Dekker, New York 1971.
2. Manning, G. S. (1972) *Annu. Rev. Phys. Chem.*, **Vol. 23**, p. 117.
3. Grosberg, A.Yu. and Khokhlov, A.R. *Statistical Physics of Macromolecules*, AIP Press, NY, 1994.
4. Khokhlov, A.R., Starodubtzev, S.G., and Vasilevskaya, V.V. (1993) *Adv. Polym. Sci.*, **Vol. 109**, p. 123.
5. Khokhlov, A.R. (1980) *Polymer*, **Vol. 21**, p. 376.
6. Dusek, K. and Prins, W. (1969) *Adv. Polym. Sci.*, **Vol. 6**, p. 1.
7. Zeldovich, K.B., and Khokhlov, A.R. (1999) *Macromolecules*, **Vol. 32**, p. 3488.
8. Kramarenko, E.Yu, Khokhlov, A.R., and Yoshikawa, K. (1997) *Macromolecules*, **Vol. 30**, p. 3383.
9. Kramarenko, E.Yu, Khokhlov, A.R., and Yoshikawa, K. (2000) *Macromol. Theory Simul.*, **Vol. 9**, p. 249.
10. Antonietti, M., Basten, R., and Lohmann, S. (1995) *Macromol. Chem. Phys.*, **Vol. 196**, p. 441.
11. P. J. Flory *Principles of Polymer Chemistry*; Cornell University Press: Ithaca, NY, 1953.
12. Birshtein, T.M. and Pryamitsyn, V.A. (1987) *Vysokomol.Soed. A*, **Vol. 29**, p. 1858.
13. de Gennes, P.-G. *Scaling Concepts in Polymer Physics*; Cornell University, Ithaca, 1979.
14. Khokhlov, A.R. and Kramarenko, E.Yu., (1994) *Macromol. Theory Simul.*, **Vol. 3**, p. 45.
15. Khokhlov, A.R. and Kramarenko, E.Yu., (1996) *Macromolecules*, **Vol. 29**, p. 681.
16. Klooster, N.Th.M., van der Touw, F., and Mandel, M. (1984) *Macromolecules*, **Vol. 17**, p. 2070, p. 2078, p. 2087.
17. Morawetz H., and Wang Y. (1987) *Macromolcules*, **Vol. 20**, p. 194.
18. Philippova, O.E., Sitnikova, N.L, and Demidovich, G.B., and Khokhlov, A.R. (1996) *Macromolecules*, **Vol. 29**, p. 4642.
19. Decher, G., (1997) *Science,* **Vol. 277**, p. 1232.
20. Lvov, Y.M., and Decher, G. (1994) *Crystallography Reports*, **Vol. 39**, p. 628.
21. Lifshitz, I.M., Grosberg, A.Yu., and Khokhlov, A.R. (1978) *Rev. Mod. Phys.*, **Vol. 50**, p. 683.
22. Borukhov, I., Andelman, D., and Orland, H. (1998) *Macromolecules*, **Vol. 31**, p. 1665.
23. Schlenoff, J.B., Ly, H., and Li, M. (1998) *J. Am. Chem. Soc.*, **Vol. 120**, p. 7626.
24. Dubas, S.T., and Schlenoff, J.B. (1999) *Macromolecules*, **Vol. 32**, p. 8153.

$\Delta C^{(1)}$

DISTRIBUTION FUNCTION THEORY OF ELECTROLYTES AND ELECTRICAL DOUBLE LAYERS

Charge Renormalisation and Dressed Ion Theory

ROLAND KJELLANDER
Dept. of Chemistry, Physical Chemistry
Göteborg University
SE-412 96 Göteborg
Sweden

Abstract. The exact charge renormalisation scheme of Dressed Ion Theory (DIT) for bulk electrolytes and double layer systems is presented. A brief introduction to some aspects of distribution function theory for homogeneous and inhomogeneous fluids is also given, including linear response theory (with a non-linear extension), integral equation theories, polarisation response functions and the dielectric function of bulk electrolytes.

DIT provides a general and exact formalism for electrolyte systems in which entities like effective (renormalised) charges of particles and surfaces, the (proper) Screened Coulomb potential and the effective permittivity of electrolyte solutions acquire precise and physically transparent definitions. In DIT the renormalised charges are defined such that the theory of electrolyte systems can be written in terms of these charges rather than the bare ionic charges. This results in a substantial conceptual simplification. The renormalised charge constitutes the appropriate source charge when the screened Coulomb potential is used to calculate the electrostatic potential. DIT also provides a convenient framework for analysing the decay behaviour of various functions like the charge density and mean electrostatic potential around each particle and the pair correlations and potential of mean force between particles. Links to some commonly used approximations in the theory of electrolytes are given.

1. Introduction

In the application of liquid state theory to soft matter systems like colloid or surfactant dispersions and solutions of polyelectrolytes or other macromolecules one has several different length scales to consider, for example the small size of the solvent molecules and dissolved salt ions, the appreciably larger size of the macroparticles or aggregates and the average separation between the particles. The range of the electrostatic interactions in the system may span either length scale depending on the salt concentration of the

C. Holm et al. (eds.), *Electrostatic Effects in Soft Matter and Biophysics*, 317–366.
© 2001 *Kluwer Academic Publishers. Printed in the Netherlands.*

solution. To treat, for example, the interaction between the macroparticles while at the same time include the molecular details on the length scale of the small molecules is cumbersome and computationally expensive. Still, what happens molecularly on the microscale often has important consequences on the larger scale. Thus, even when the main interest concerns phenomena on the large scale, the effects of the molecular details somehow have to be taken into account.

For such reasons one has turned to effective interaction potentials between the macroparticles where the molecular degrees of freedom of for example the solvent and dissolved salt are included implicitly. Furthermore what happens on the microscale close to the macroparticles can be included in quantities like effective charges of the latter. A classical example of an effective interaction potential is the Coulomb interaction in a dielectric medium, where the molecular degrees of freedom of the solvent are contained the dielectric constant. This is appropriate to do for interactions across distances that are sufficiently large compared to the molecular size of the solvent. Another example is the so–called Screened Coulomb (Yukawa) potential from the Debye–Hückel (DH) theory that implicitly incorporates the ionic degrees of freedom of the dissolved salt. The presence of the small ions leads to a screening that is manifested as exponentially decaying electrostatic potentials and expressed by the Screened Coulomb potential.

When using effective interaction potentials one has simplified the problem so that one can focus on what is happening on the length scales of the macroparticles without explicitly treating the solvent and small ion degrees of freedom. Such a separation of the degrees of freedom can be done without invoking any approximations as demonstrated in the McMillan–Mayer theory [1, 2], but the price one then has to pay is to use a complicated many–body effective potential (potential of mean force) between the macroparticles. It is also possible to define exact effective pair potentials [3, 4] in which case both macroparticle and small particle degrees of freedom contribute to the effective potential. In practice one may use approximations based on either kind of effective potentials or something similar.

It is very common to use the screened Coulomb potential $\phi^{0,\mathrm{DH}}$ from the DH theory (defined below in eq. (61)) to calculate the effects of electrostatic interactions between charged macroparticles dispersed (or dissolved) in an electrolyte solution. Thereby, one often assumes that the interaction energy between two charges q_{I} and q_{J} immersed in the electrolyte is equal to $q_{\mathrm{I}}q_{\mathrm{J}}\phi^{0,\mathrm{DH}}(r)$, where r is the separation between the charges, i.e. one uses the bare charges of the particles when calculating their interaction energy. There are several problems with this. The charged groups of the macroions are surrounded by ion clouds, the nature of which depends, for example, on the charge, size and shape of the group and on the composition of the electrolyte solution. The presence of the ion clouds affects the interaction energy between the macroions. If similarly charged groups are close together on the macroion they will enhance the development of each other's ion clouds in a non–linear manner. This will also affect the interaction energy. The function $\phi^{0,\,\mathrm{DH}}$ depends on the overall concentration of small ions in the dispersion and do not contain any information about the local ion clouds. Thus one should not use the bare charges combined with $\phi^{0,\,\mathrm{DH}}$ to calculate the electrostatic energy between the macroions.

A solution would be to make use of effective charges of the macroions instead of the bare charges. These effective charges should contain information about the local ion clouds and depend on the nature of the macroions and the electrolyte composition. The question is then how to define the effective charge such that the resulting electrostatic interaction is the correct one, i.e. the one obtained when treating the small ion degrees of freedom explicitly and correctly. This raises a further problem, namely that $\phi^{0,\,\mathrm{DH}}$ is an approximate entity that is useful provided the electrolyte is sufficiently dilute (the DH theory is strictly valid in the limit of infinite dilution). If one wants to obtain the correct potentials, the screened Coulomb potential one uses must be some other function than that predicted by the DH approximation. Is it possible to find such a potential and can one define effective charges that satisfy these demands?

These questions and their positive answers is a major subject in Dressed Ion Theory (DIT) and the main part of the present article is devoted to this theory. The article also includes a more general but brief background which covers some essential liquid state theory needed for DIT. This background material may, however, be of more general interest since it gives an exposition of several important concepts and relationships in distribution function theory and gives, for example, a basis for several integral equation approximations for homogeneous and inhomogeneous fluids. For a more thorough account of distribution function theory the reader may, for example, consult [5–10]; for electrolyte systems see in particular [6, 8–10].

The DIT formalism is an exact reformulation of the classical statistical mechanics for electrolyte systems where the solvent is modelled as a dielectric continuum, e.g. the primitive model. In this formalism each particle (small and large ones) acquires a diffuse effective charge, a dress, which contains a well–defined part of the ion cloud charge density around the particle. The dressed particles replace the bare particle charges in the theory, which thereby becomes simplified and expressed in a physically transparent and appealing manner. Each dressed particle constitutes the effective source charge for the electrostatic potential that arises due to the particle. The electrostatic screening of the potential due to the surrounding electrolyte is thereby governed by the "proper" Screened Coulomb potential ϕ^0 of the system, which goes over to $\phi^{0,\mathrm{DH}}$ in the limit of infinite dilution of the electrolyte.

DIT gives a framework to analyze the decay behaviors of ϕ^0, the electrostatic potential and various correlation functions. We shall see how these behaviors are connected to the physical interpretation of DIT and how this motivates the definitions of, for example, effective point charges of ions, effective surface charges of macroions and effective permittivities of electrolyte solutions. It is found that for 1:1 electrolytes in aqueous solution at room temperature the dominant screening mode in bulk is close to that predicted by DH theory up at least 0.1 M concentration, while this is not the case for electrolytes with divalent or higher valencies. The effective surface charges of macroions differ substantially from the bare charges unless the surface charge density *and* the electrolyte concentration is small.

The DIT parts of the current article is to a large extent based on the original publications [11–24]. DIT is closely related to the so–called Hypervertex formalism of Stell (see e.g. [25]), which provides a general formalism in liquid state theory. The precise link between the theories is shown in [11]. DIT is also related to some approaches in field theory [26].

2. Basic Theory for Inhomogeneous Fluids

2.1. EQUILIBRIUM CONDITION; BOLTZMANN DISTRIBUTION FOR DENSITY

Let us consider a fluid consisting of particles that interact with the pair potential $u(\mathbf{r}_1, \mathbf{r}_2)$ where $\mathbf{r}_1$ and $\mathbf{r}_2$ are the co–ordinates of two of the particles. The fluid particles can also interact with an external potential, e.g. due to a body immersed in the fluid, and we denote this potential $\nu(\mathbf{r})$. In absence of the external potential $[\nu(\mathbf{r}) = 0]$ we have a bulk fluid with constant number density $n = n_{\text{bulk}}$. In presence of this potential the fluid is inhomogeneous and its number density is co–ordinate dependent, $n = n(\mathbf{r})$. We shall for simplicity in notation initially only consider the case of a one–component fluid; the results are easily generalized to the many–component case.

The chemical potential μ for a bulk fluid can be split in the ideal and excess parts [2]

$$\mu = \mu_{\text{bulk}}^{\text{ideal}} + \mu_{\text{bulk}}^{\text{excess}} = \mu^0 + k_B T \ln n_{\text{bulk}} + \mu_{\text{bulk}}^{\text{excess}} \tag{1}$$

where $\mu_{\text{bulk}}^{\text{ideal}}$ is the chemical potential of a non–interacting fluid (ideal gas) of number density n_{bulk}, $\mu_{\text{bulk}}^{\text{excess}}$ is the contribution due to intermolecular interactions, μ^0 is the standard chemical potential, k_B is Boltzmann's constant and T is the absolute temperature. The chemical potential is the total change in free energy (= reversible work done) when adding a new particle to the system; the excess part is the work against intermolecular interactions when placing the particle in the fluid at an arbitrary but fixed point and the ideal part is the contribution when releasing it (due to ideal entropy of mixing). For monatomic particles $\mu^0 = k_B T \ln \Lambda^3$, where Λ is the de Broglie thermal wave length. The excess chemical potential is in general a complicated quantity to evaluate since it includes contributions due to the reorganization of the particles in the neighborhood of the new particle when adding the latter.

In presence of the external potential $\nu(\mathbf{r})$, the work done against interactions when inserting a new particle is different depending on where we put it. In this case we have [27]

$$\mu = \mu^{\text{ideal}}(\mathbf{r}) + \mu^{\text{excess}}(\mathbf{r}) = \mu^0 + k_B T \ln n(\mathbf{r}) + \mu^{\text{excess}}(\mathbf{r}) \tag{2}$$

where $\mu^{\text{excess}}(\mathbf{r})$ is the work against interactions when adding a new particle at $\mathbf{r}$ and keeping it fixed there. The excess contribution is due to both the external and internal interactions $\mu^{\text{excess}}(\mathbf{r}) = \nu(\mathbf{r}) + \mu^{\text{int}}(\mathbf{r})$, where the first term gives the work against the external potential and $\mu^{\text{int}}(\mathbf{r})$ is defined as the work due to interactions with the fluid particles. In the bulk fluid where $\nu(\mathbf{r}) = 0$, e.g. far from the source of the external potential, we have $\mu^{\text{excess}}(\mathbf{r}) = \mu^{\text{excess}}_{\text{bulk}} = \mu^{\text{int}}_{\text{bulk}}$ and $n(\mathbf{r}) = n_{\text{bulk}}$. At equilibrium the total chemical potential is equal everywhere, i.e. μ is independent of $\mathbf{r}$ and we obtain from eqs. (1) and (2)

$$n(\mathbf{r}) = n_{\text{bulk}} e^{-\beta[\nu(\mathbf{r}) + \mu^{\text{int}}(\mathbf{r}) - \mu^{\text{int}}_{\text{bulk}}]}. \tag{3}$$

where $\beta = (k_B T)^{-1}$. In general, eq. (3), the Boltzmann distribution for density, applies for an inhomogeneous fluid in equilibrium with a bulk fluid of density n_{bulk} (infinitely large reservoir of particles).

Let us apply this result to an inhomogeneous system where we have placed a fluid particle fixed at co-ordinate $\mathbf{r}'$. We maintain equilibrium with the same bulk fluid of density n_{bulk}. The interaction $u(\mathbf{r}, \mathbf{r}')$ between this particle and the other particles of the fluid (which are mobile) can be included in the external potential, which we denote $\nu(\mathbf{r} \mid \mathbf{r}') \equiv \nu(\mathbf{r}) + u(\mathbf{r}, \mathbf{r}')$, i.e. the external potential at $\mathbf{r}$ when there is a particle fixed at $\mathbf{r}'$. The density of this system is denoted $n(\mathbf{r} \mid \mathbf{r}')$. In analogy with eq. (3) we obtain

$$n(\mathbf{r} \mid \mathbf{r}') = n_{\text{bulk}} e^{-\beta[\nu(\mathbf{r}|\mathbf{r}') + \mu^{\text{int}}(\mathbf{r}|\mathbf{r}') - \mu^{\text{int}}_{\text{bulk}}]} \tag{4}$$

where $\mu^{\text{int}}(\mathbf{r} \mid \mathbf{r}')$ is the work against interactions with the mobile particles in this system when adding a new particle at $\mathbf{r}$ and keeping it fixed there.

For the original inhomogeneous system (in presence of the external potential $\nu(\mathbf{r})$ alone) $n(\mathbf{r})$ is the obtained by averaging the instantaneous density distribution over *all* possible particle configurations (a grand canonical ensemble average). One can show that $n(\mathbf{r} \mid \mathbf{r}')$ is equal to the density distribution one obtains for the *same* system by averaging over the subset of particle configurations in which there happen to be a particle located at $\mathbf{r}'$ (a subset of the previous set of configurations). By definition of the pair distribution function $g(\mathbf{r}, \mathbf{r}')$, we have $n(\mathbf{r} \mid \mathbf{r}') = n(\mathbf{r})g(\mathbf{r}, \mathbf{r}')$. From this relation and eqs. (3) and (4) follow

$$g(\mathbf{r}, \mathbf{r}') = e^{-\beta[u(\mathbf{r},\mathbf{r}')+\mu^{\text{int}}(\mathbf{r}|\mathbf{r}')-\mu^{\text{int}}(\mathbf{r})]} \tag{5}$$

and we emphasize that $g(\mathbf{r}, \mathbf{r}')$ is the pair distribution of the original inhomogeneous system with external potential $\nu(\mathbf{r})$.The pair distribution function is symmetric, $g(\mathbf{r}, \mathbf{r}') = g(\mathbf{r}', \mathbf{r})$, although this is not immediately apparent from eq. (5).

We now introduce the *singlet (one-point) direct correlation function* $c^{(1)}$ defined as

$$c^{(1)}(\mathbf{r}) \equiv -\beta\mu^{\text{int}}(\mathbf{r}), \tag{6}$$

which implies

$$\mu^{\text{excess}}(\mathbf{r}) = \nu(\mathbf{r}) - k_B T c^{(1)}(\mathbf{r})\,. \tag{7}$$

We can write eqs. (3) and eq. (5) as

$$n(\mathbf{r}) = n_{\text{bulk}} e^{-\beta\nu(\mathbf{r})+\Delta c^{(1)}(\mathbf{r})} \tag{8}$$

and

$$g(\mathbf{r}, \mathbf{r}') = e^{-\beta u(\mathbf{r},\mathbf{r}')+\Delta c^{(1)}(\mathbf{r}|\mathbf{r}')} \tag{9}$$

respectively, where $\Delta c^{(1)}(\mathbf{r}) \equiv c^{(1)}(\mathbf{r}) - c^{(1)}_{\text{bulk}}$ and $\Delta c^{(1)}(\mathbf{r} \mid \mathbf{r}') \equiv c^{(1)}(\mathbf{r} \mid \mathbf{r}') - c^{(1)}(\mathbf{r})$ in obvious notation. Equation (8) gives the density distribution function in terms of the change of $c^{(1)}$ in absence and presence of the external potential and eq. (9) gives the pair distribution function in terms of the change of $c^{(1)}$ in absence and presence of a particle fixed at $\mathbf{r}'$. Note that in both cases we need to know how $c^{(1)}$ changes when we change the external potential.

2.2. LINEAR RESPONSE THEORY

We first investigate what happens with $c^{(1)}(\mathbf{r})$ when we change the external potential slightly. From eq. (7) we see that this can be obtained from the corresponding change in excess chemical potential. Remember that $\mu^{\text{excess}}(\mathbf{r})$ is the work against interactions when inserting a new particle at $\mathbf{r}$ and keeping it fixed there. When we do this insertion the density distribution at other points $\mathbf{r}''$ changes from $n(\mathbf{r}'')$ to $n(\mathbf{r}'' \mid \mathbf{r})$, i.e. by

$$\Delta n(\mathbf{r}'' \mid \mathbf{r}) \equiv n(\mathbf{r}'' \mid \mathbf{r}) - n(\mathbf{r}'') = n(\mathbf{r}'')h(\mathbf{r}'', \mathbf{r}) \tag{10}$$

where $h(\mathbf{r}'', \mathbf{r}) \equiv g(\mathbf{r}'', \mathbf{r}) - 1$ is the pair correlation function (also called the total correlation function).

We now change the external potential $\nu(\mathbf{r}'') \longrightarrow \nu(\mathbf{r}'') + \delta\nu(\mathbf{r}'')$, where $\delta\nu(\mathbf{r}'')$ is small everywhere. The excess chemical potential then changes $\mu^{\text{excess}}(\mathbf{r}) \longrightarrow \mu^{\text{excess}}(\mathbf{r}) + \delta\mu^{\text{excess}}(\mathbf{r})$, where $\delta\mu^{\text{excess}}(\mathbf{r})$ is the extra work required due to the additional potential $\delta\nu(\mathbf{r}'')$ when inserting the particle at $\mathbf{r}$. In the limit of infinitesimally small $\delta\nu(\mathbf{r}'')$ the change in $\mu^{\text{excess}}(\mathbf{r})$ is linearly related to $\delta\nu(\mathbf{r}'')$ and one can show that

$$\delta\mu^{\text{excess}}(\mathbf{r}) = \delta\nu(\mathbf{r}) + \int \delta\nu(\mathbf{r}'')\Delta n(\mathbf{r}'' \mid \mathbf{r})\, d\mathbf{r}''\,. \tag{11}$$

This relation can be rationalized as follows. The first term is obviously the extra interaction with the external potential felt directly by the inserted particle at $\mathbf{r}$. The second term gives the indirect effect due to the interaction between $\delta\nu(\mathbf{r}'')$ and the other (mobile) particles. The number of particles in volume element $d\mathbf{r}''$ is changed by $\Delta n(\mathbf{r}'' \mid \mathbf{r})\, d\mathbf{r}''$ when the new particle is inserted at $\mathbf{r}$, i.e. particles are moved in to or out from this volume element (the probability that they are in $d\mathbf{r}''$ increases or decreases). The integrand equals the work due to the additional potential $\delta\nu(\mathbf{r}'')$ when the number of particles is changed in $d\mathbf{r}''$ and the integral is the work from the change in density in the entire volume.

Eq. (7) implies

$$\delta\mu^{\text{excess}}(\mathbf{r}) = \delta\nu(\mathbf{r}) - k_B T \delta c^{(1)}(\mathbf{r}) \tag{12}$$

and eq. (11) thus give

$$\delta c^{(1)}(\mathbf{r}) = -\beta \int \delta\nu(\mathbf{r}'')n(\mathbf{r}'')h(\mathbf{r}'', \mathbf{r})\, d\mathbf{r}'', \tag{13}$$

where we have inserted eq. (10). In the formalism of functional derivatives this equation is (by definition) equivalent to the statement

$$\frac{\delta c^{(1)}(\mathbf{r})}{\delta\nu(\mathbf{r}'')} = -\beta n(\mathbf{r}'')h(\mathbf{r}'', \mathbf{r}), \tag{13'}$$

but we shall not use this formalism here. Instead we shall write variational relationships like eq. (13) which are exact for infinitesimally small variations. For variations that are not infinitesimally small one must add non–linear terms to the right hand side (rhs); the equations give the linear response only.

We can now also determine how much the density distribution changes when the external potential changes by $\delta\nu(\mathbf{r}'')$. Assume that the density changes $n(\mathbf{r}) \longrightarrow n(\mathbf{r}) + \delta n(\mathbf{r})$. Since we hold μ constant (equilibrium with the same bulk fluid), eq. (2) implies that $k_B T\delta[\ln n(\mathbf{r})] + \delta\mu^{\text{excess}}(\mathbf{r}) = 0$. To linear order we have $\delta[\ln n(\mathbf{r})] = \delta n(\mathbf{r})/n(\mathbf{r})$ and hence

$$\frac{1}{n(\mathbf{r})}\delta n(\mathbf{r}) = -\beta\delta\mu^{\text{excess}}(\mathbf{r}) \tag{14}$$

Thus eqs. (10) and (11) give the change in density

$$\delta n(\mathbf{r}) = -\beta n(\mathbf{r})\left[\delta\nu(\mathbf{r}) + \int \delta\nu(\mathbf{r}'')n(\mathbf{r}'')h(\mathbf{r}'',\mathbf{r})\,d\mathbf{r}''\right] \tag{15}$$

This equation (or the equivalent functional derivative statement) is called the *first Yvon equation* [7].

We now know how much $\mu^{\text{excess}}(\mathbf{r})$, $c^{(1)}(\mathbf{r})$ and $n(\mathbf{r})$ changes when we make an infinitesimally small change $\delta\nu(\mathbf{r}'')$ in external potential. Since these changes are related by linear relationships it should be possible to express, for example, $\delta c^{(1)}$ in terms of δn, i.e. to eliminate $\delta\nu$ from eqs. (13) and (15). To this end we define the *pair (two-point) direct correlation function* $c^{(2)}(\mathbf{r}'',\mathbf{r})$ from

$$\delta c^{(1)}(\mathbf{r}) = \int \delta n(\mathbf{r}'')c^{(2)}(\mathbf{r}'',\mathbf{r})\,d\mathbf{r}'' \tag{16}$$

and it should be possible to express $c^{(2)}$ in terms of n and h from eqs. (13) and (15). If one combines eqs. (13), (15) and (16) and use the fact that $\delta\nu(\mathbf{r}'')$ is an infinitesimally small but otherwise arbitrary variation, one finds after some algebra that pair direct correlation function must satisfy the equation

$$h(\mathbf{r}',\mathbf{r}) = c^{(2)}(\mathbf{r}',\mathbf{r}) + \int h(\mathbf{r}',\mathbf{r}'')n(\mathbf{r}'')c^{(2)}(\mathbf{r}'',\mathbf{r})\,d\mathbf{r}'' \tag{17}$$

which is the famous *Ornstein–Zernike (OZ) equation* [6, 7]. This equation is often used as the definition of $c^{(2)}$ from n and h and then eq. (16) follows as a consequence of this definition. Here we have instead used eq. (16) as the definition and then the OZ equation follows. The two points of view are equivalent. The physical meaning of $c^{(2)}$ is, according to eq. (16) and the definition (6) of $c^{(1)}$, that $-k_B Tc^{(2)}(\mathbf{r}'',\mathbf{r})$ tells how much a small change in density at $\mathbf{r}''$ influences μ^{int} at $\mathbf{r}$ [recall that $\mu^{\text{int}}(\mathbf{r})$ is the work against interactions with the fluid particles when inserting a new particle at $\mathbf{r}$].

Eq. (15) gives the change in density $\delta n(\mathbf{r})$ when we have exposed our system to a given small change in external potential $\delta\nu(\mathbf{r}'')$. We are now in a position where we can answer the following "reverse" question: What change in potential $\delta\nu(\mathbf{r})$ is necessary in order to induce a given small change in density $\delta n(\mathbf{r}'')$? Eqs. (12) and (14) give

$$\delta\nu(\mathbf{r}) = -\frac{1}{\beta}\left[\frac{1}{n(\mathbf{r})}\delta n(\mathbf{r}) - \delta c^{(1)}(\mathbf{r})\right] \tag{18}$$

and by using the definition (16) of $c^{(2)}$ we obtain the *second Yvon equation* [7]

$$\delta\nu(\mathbf{r}) = -\frac{1}{\beta}\left[\frac{1}{n(\mathbf{r})}\delta n(\mathbf{r}) - \int \delta n(\mathbf{r}'')c^{(2)}(\mathbf{r}'',\mathbf{r})\,d\mathbf{r}''\right] \tag{19}$$

which is the answer to our question, i.e. it gives $\delta\nu$ in terms of δn. Note that it is the OZ equation that makes eq. (19) follow from eq. (15) and vice versa.

2.3. NON–LINEAR RESPONSE; INTEGRAL EQUATIONS

If the change in external potential is not infinitesimally small we must complement the linear response results above with non–linear terms. This is the case when we want to obtain $\Delta c^{(1)}$ in eqs. (8) and (9), which are due to changes in external potential from 0 to $\nu(\mathbf{r})$ and from $\nu(\mathbf{r})$ to $\nu(\mathbf{r} \mid \mathbf{r}')$ respectively. As we have seen, the density distribution changes from n_{bulk} to $n(\mathbf{r})$ and from $n(\mathbf{r})$ to $n(\mathbf{r} \mid \mathbf{r}')$ in these two cases.

Let Δn be the change in density and $\Delta c^{(1)}$ the change in direct correlation function when the external potential is varied by $\Delta\nu$, which is not small. If we apply eq. (16) to this case we have

$$\Delta c^{(1)}(\mathbf{r}) = \int \Delta n(\mathbf{r}'')c^{(2)}(\mathbf{r}'',\mathbf{r})\,d\mathbf{r}'' + \text{ terms non–linear in } \Delta n \tag{20}$$

where $c^{(2)}$ is evaluated for the system *before* we change the external potential, i.e. as it is in linear response theory. In the case of eq. (8) we have $\Delta n(\mathbf{r}'') = n(\mathbf{r}'') - n_{\text{bulk}}$ and we can write eq. (20) as

$$\Delta c^{(1)}(\mathbf{r}) = \int \Delta n(\mathbf{r}'')c^{(2)}_{\text{bulk}}(\mathbf{r}'',\mathbf{r})\,d\mathbf{r}'' + e^{(1)}(\mathbf{r}) \tag{21}$$

where $e^{(1)}(\mathbf{r})$, which is called the *singlet bridge function*, is the sum of all non–linear terms in Δn. One can, in fact, express $e^{(1)}(\mathbf{r})$ explicitly in terms of $\Delta n(\mathbf{r}'')$, n_{bulk} and $h_{\text{bulk}}(\mathbf{r}'',\mathbf{r})$ (an infinite sum of complicated multicenter integrals of products of these functions; so–called "bridge diagrams" or "elementary diagrams" [6, 28]), but this is in practice not an accessible route to obtain $e^{(1)}(\mathbf{r})$. Instead one approximates the bridge function in various ways (see below).

If we apply eq. (20) to calculate $\Delta c^{(1)}(\mathbf{r} \mid \mathbf{r}')$ in eq. (9), in which case the change in density is $\Delta n(\mathbf{r}'' \mid \mathbf{r}) = n(\mathbf{r}'' \mid \mathbf{r}) - n(\mathbf{r}'') = n(\mathbf{r}'')h(\mathbf{r}'', \mathbf{r}')$, we obtain

$$\Delta c^{(1)}(\mathbf{r} \mid \mathbf{r}') = \int \Delta n(\mathbf{r}'' \mid \mathbf{r}')c^{(2)}(\mathbf{r}'', \mathbf{r})\, d\mathbf{r}'' + e^{(2)}(\mathbf{r}, \mathbf{r}') \tag{22}$$

where $e^{(2)}(\mathbf{r}, \mathbf{r}')$ is the sum of all non–linear terms in Δn for this case and is called the *pair bridge function.* One can express $e^{(2)}(\mathbf{r}, \mathbf{r}')$ explicitly in terms of $n(\mathbf{r})$ and $h(\mathbf{r}'', \mathbf{r})$ as an infinite sum of complicated integrals [28], but again one has to resort to approximations in practice (see below). By inserting eq. (22) into eq. (9) and using the OZ equation (17) one obtains

$$g(\mathbf{r}, \mathbf{r}') = e^{-\beta u(\mathbf{r},\mathbf{r}')+h(\mathbf{r},\mathbf{r}')-c^{(2)}(\mathbf{r},\mathbf{r}')+e^{(2)}(\mathbf{r},\mathbf{r}')} \tag{23}$$

which is the most commonly used form of the relationship between g, u, h, $c^{(2)}$ and $e^{(2)}$.

So far all equations have been exact, but we will finish this section by introducing some commonly used approximations in liquid state theory. Let us start with bulk fluids where $n(\mathbf{r}) = n_{\text{bulk}}$ and the pair functions are isotropic, e.g. $h(\mathbf{r}'', \mathbf{r}) = h(|\mathbf{r}'' - \mathbf{r}|)$. If one uses eq. (23) with an approximate $e^{(2)}$ (expressed in terms of, for example, h and/or $c^{(2)}$) together with the OZ equation (17), there are only two unknowns, h and $c^{(2)}$. These two equations can therefore be solved (at least numerically). The approximation is called a closure approximation because it results in a closed set of equations. This is the basis of the approximate *integral equation theories* of bulk fluids [6].

The simplest approximation for the bridge function is simply to set $e^{(2)}(\mathbf{r}, \mathbf{r}') = 0$, which is the so–called *Hypernetted Chain (HNC) approximation.* Despite that the non–linear contributions in eq. (22) thereby are neglected, the HNC approximation is a non–linear approximation since eq. (23) gives a non–linear relationship between g and the exponent

$$g(\mathbf{r}, \mathbf{r}') = e^{-\beta u(\mathbf{r},\mathbf{r}')+h(\mathbf{r},\mathbf{r}')-c^{(2)}(\mathbf{r},\mathbf{r}')} \qquad \text{(HNC)} \tag{24}$$

A linear approximation for h and $c^{(2)}$ is obtained by linearizing the factor $\exp[h - c^{(2)}]$ in eq. (24) whereby

$$g(\mathbf{r}, \mathbf{r}') = e^{-\beta u(\mathbf{r},\mathbf{r}')}[1 + h(\mathbf{r}, \mathbf{r}') - c^{(2)}(\mathbf{r}, \mathbf{r}')] \qquad \text{(PY)} \tag{25}$$

This is the *Percus–Yevick (PY) approximation.* Alternatively one can express this as eq. (23) with the approximation $e^{(2)} = \ln[1 + h - c^{(2)}] - h + c^{(2)}$.

Another linear approximation is the *Mean Spherical Approximation (MSA)*, which is used for particles with a hard core (diameter a). In this approximation one sets $g(\mathbf{r}, \mathbf{r}') = 0$ for $|\mathbf{r} - \mathbf{r}'| < a$, as required, and

$c^{(2)}(\mathbf{r},\mathbf{r}') = -\beta u(\mathbf{r},\mathbf{r}')$ for $|\mathbf{r}-\mathbf{r}'| \geq a$. The latter Ansatz arises from the general property of $c^{(2)}$ that $c^{(2)}(\mathbf{r},\mathbf{r}') \longrightarrow -\beta u(\mathbf{r},\mathbf{r}')$ when $|\mathbf{r}-\mathbf{r}'| \longrightarrow \infty$ and one assumes that this limiting property holds as an equality all the way down to contact between the particles. Alternatively one can express MSA as eq. (23) with the approximation $e^{(2)} = \ln[1+h] - h$ for $|\mathbf{r}-\mathbf{r}'| \geq a$. In the *Generalized Mean Spherical Approximation (GMSA)*, one improves this approximation by adding some empirical function to $e^{(2)}$ (and thereby subtract it from the Ansatz above for $c^{(2)}$). This empirical function can, for example, be an exponentially decaying function with some fitting parameters, which can be selected such that some self–consistency criteria holds (e.g. the equality between the pressure calculated via the virial and the compressibility equations).

An approach that often gives particularly accurate results is to approximate $e^{(2)} \approx e^{(2)}_{\mathrm{Ref}}$, where $e^{(2)}_{\mathrm{Ref}}$ is the bridge function from another system, the reference system (usually a hard sphere fluid), for which $e^{(2)}$ is known with reasonably high accuracy. This is the so–called *Reference Hypernetted Chain (RHNC) approximation.* One can select the diameter of the hard spheres of the reference system such that some self–consistency criterion holds (cf. GMSA above). Then the approximation is often called the *Modified Hypernetted Chain (MHNC) approximation.*

For inhomogeneous fluids [10] one needs also an equation for the density distribution $n(\mathbf{r})$. In the vast majority of integral equation theories for inhomogeneous fluids one uses eqs. (8) and (21) with an approximate $e^{(1)}(\mathbf{r})$. These are the *singlet level integral equation* theories, since they use an approximation for the singlet distribution function of the inhomogeneous fluid. A common choice is to set $e^{(1)}(\mathbf{r}) = 0$, the *singlet HNC approximation.* In addition one needs an approximation to calculate the pair correlation functions [like $c^{(2)}_{\mathrm{bulk}}$ in eq. (21)] for the bulk fluid that is in equilibrium with the inhomogeneous one. If, for example, the PY approximation is used for the latter, the resulting approximation is called the *HNC/PY approximation* and likewise for other combinations of approximations for the singlet and bulk functions.

A more refined kind of theory for inhomogeneous fluids are *pair level integral equation* theories where one does an approximation on the pair level, e.g. to use the HNC approximation (24) for the anisotropic pair functions of the inhomogeneous fluid, the *Anisotropic Hypernetted Chain (AHNC) approximation* [29, 30]. To obtain the density distribution $n(\mathbf{r})$ one can then use eq. (8), since in the AHNC approximation one can, in fact, express $\Delta c^{(1)}(\mathbf{r})$ explicitly in terms of $h(\mathbf{r}'',\mathbf{r})$, $c^{(2)}(\mathbf{r}'',\mathbf{r})$ and $n(\mathbf{r})$ without any further approximation. The set of equations thus obtained is then solved numerically for h, $c^{(2)}$ and n in a self–consistent manner. Other approximations at this level are the analogous *Anisotropic RHNC and PY approximations* etc. [31-33] where one uses the respective closure approximation above for the pair distributions of inhomogeneous fluids. In these cases one utilizes some exact equation for the density distribution $n(\mathbf{r})$, e.g.

$$\nabla n(\mathbf{r}) = -\beta n(\mathbf{r}) \left[\nabla \nu(\mathbf{r}) + \int \nabla \nu(\mathbf{r}'') n(\mathbf{r}'') h(\mathbf{r}'', \mathbf{r})\, d\mathbf{r}'' \right], \tag{26}$$

which can be derived in a straight–forward manner [7] from the first Yvon equation (15).

2.4. MANY–COMPONENT SYSTEMS

The results above are easily generalized to many–component systems. We use subscripts to indicate species, for example $u_{ij}(\mathbf{r}, \mathbf{r}')$ is the pair potential between species i and j, $\nu_i(\mathbf{r})$ is the external potential for species i and $n_i(\mathbf{r}) g_{ij}(\mathbf{r}, \mathbf{r}')$ is the density of species i at $\mathbf{r}$ when a particle of species j is located at $\mathbf{r}'$.

For example, the first Yvon equation becomes

$$\delta n_j(\mathbf{r}) = -\beta n_j(\mathbf{r}) \left[\delta \nu_j(\mathbf{r}) + \sum_i \int \delta \nu_i(\mathbf{r}'') n_i(\mathbf{r}'') h_{ij}(\mathbf{r}'', \mathbf{r})\, d\mathbf{r}'' \right] \tag{27}$$

which gives the change in density $\delta n_j(\mathbf{r})$ of species j when the external potentials for all species i are changed slightly by $\delta \nu_i(\mathbf{r}'')$ (infinitesimally small change). The converse relationship, the second Yvon equation, becomes

$$\delta \nu_i(\mathbf{r}) = -\frac{1}{\beta} \left[\frac{1}{n_i(\mathbf{r})} \delta n_i(\mathbf{r}) - \sum_j \int \delta n_j(\mathbf{r}'') c_{ij}^{(2)}(\mathbf{r}'', \mathbf{r})\, d\mathbf{r}'' \right] \tag{28}$$

and the OZ equation

$$h_{ij}(\mathbf{r}', \mathbf{r}) = c_{ij}^{(2)}(\mathbf{r}', \mathbf{r}) + \sum_l \int h_{il}(\mathbf{r}', \mathbf{r}'') n_l(\mathbf{r}'') c_{lj}^{(2)}(\mathbf{r}'', \mathbf{r})\, d\mathbf{r}'' . \tag{29}$$

3. Polarization of Bulk Electrolyte and Dressed Ion Theory

In this section we shall deal with bulk electrolytes, so all pair functions are isotropic, e.g. $h_{ij}(\mathbf{r}', \mathbf{r}) = h_{ij}(|\mathbf{r}' - \mathbf{r}|)$. We shall use the notation n_i for the bulk density of species i, i.e. we drop the subscript "bulk" in what follows. The charge of an ion of species i is denoted q_i. We shall only consider primitive model electrolytes, so the ions are hard spheres with a central charge and the solvent is a dielectric continuum with dielectric constant ε_s. This means that the pair potential used is

$$u_{ij}(r) = \begin{cases} u_{ij}^{\mathrm{Coul}}(r), & r \geq a_{ij} \\ \infty, & r < a_{ij} \end{cases}$$

where $u_{ij}^{\mathrm{Coul}}(r) = q_i q_j/(4\pi\varepsilon_s\varepsilon_0 r)$ is the Coulomb potential, ε_0 is the permittivity of vacuum and a_{ij} is the distance of closest approach of the ion centers for species i and j (the "ionic size" parameter). However, the results below are of more general validity; the ions can be interacting with any short range interaction potential in addition to the Coulomb potential, provided the former has finite range or at least decays substantially faster than the exponentially decaying distribution functions we will consider below.

Let us expose a bulk electrolyte solution to an applied electrostatic potential $\Psi^{\mathrm{appl}}(\mathbf{r}')$ (we assume that it has this value in vacuum). The dielectric polarization of the continuum solvent makes the ions to experience the external electrostatic potential $\Psi^{\mathrm{ext}}(\mathbf{r}') = \Psi^{\mathrm{appl}}(\mathbf{r}')/\varepsilon_s$. This potential induces a polarization charge density of the electrolyte which we denote $\rho^{\mathrm{pol}}(\mathbf{r})$ (the polarization charge of the dielectric continuum solvent is *not* included in ρ^{pol}). The total electrostatic potential $\Psi(\mathbf{r}')$ is the sum of $\Psi^{\mathrm{ext}}(\mathbf{r}')$ and the potential from $\rho^{\mathrm{pol}}(\mathbf{r})$ i.e.

$$\Psi(\mathbf{r}') = \Psi^{\mathrm{ext}}(\mathbf{r}') + \frac{1}{\varepsilon_s}\int \phi(|\mathbf{r}' - \mathbf{r}|)\rho^{\mathrm{pol}}(\mathbf{r})\, d\mathbf{r} \tag{30}$$

where $\phi(r) = 1/(4\pi\varepsilon_0 r)$ is the electrostatic potential in vacuum from a unit charge.

3.1. LINEAR REGIME

3.1.1. *Polarization response*

Let us first expose a homogeneous electrolyte to an infinitesimally small external electrostatic potential $\delta\Psi^{\mathrm{ext}}(\mathbf{r}')$ which gives rise to a polarization charge $\delta\rho^{\mathrm{pol}}(\mathbf{r})$. The latter arises because ionic species j in the electrolyte experiences an external potential $\delta\nu_j(\mathbf{r}') = q_j\delta\Psi^{\mathrm{ext}}(\mathbf{r}')$ which makes the ion density change by $\delta n_j(\mathbf{r})$. (In absence of the external potential $\delta\Psi^{\mathrm{ext}}(\mathbf{r}')$ the ion density is uniform and the charge density is zero.) From the first Yvon equation (27) we have

$$\delta n_j(\mathbf{r}) = -\beta n_j \left[q_j\delta\Psi^{\mathrm{ext}}(\mathbf{r}) + \sum_i \int q_i\delta\Psi^{\mathrm{ext}}(\mathbf{r}')n_i h_{ij}(\mathbf{r},\mathbf{r}')\, d\mathbf{r}' \right]. \tag{31}$$

The polarization charge is $\delta\rho^{\mathrm{pol}}(\mathbf{r}) = \sum_j q_j\delta n_j(\mathbf{r})$ and by inserting eq. (31) we obtain

$$\delta\rho^{\mathrm{pol}}(\mathbf{r}) = -\beta\sum_j q_j^2 n_j\delta\Psi^{\mathrm{ext}}(\mathbf{r}) + \int A(\mathbf{r},\mathbf{r}')\delta\Psi^{\mathrm{ext}}(\mathbf{r}')\, d\mathbf{r}' \tag{32}$$

where $A(\mathbf{r},\mathbf{r}') = -\beta\sum_{ij} q_i q_j n_i n_j h_{ij}(\mathbf{r},\mathbf{r}')$. The first term gives the local and the second term the non–local polarization response to $\delta\Psi^{\rm ext}(\mathbf{r}')$.

Conversely, the second Yvon equation (28) implies that for all i

$$\beta q_i \delta\Psi^{\rm ext}(\mathbf{r}') = -\frac{1}{n_i}\delta n_i(\mathbf{r}') + \sum_j \int \delta n_j(\mathbf{r}) c_{ij}(\mathbf{r},\mathbf{r}')\, d\mathbf{r} \tag{33}$$

provided $\delta n_j(\mathbf{r})$ is the same as before [for a general $\delta n_j(\mathbf{r})$, we may have $\delta\nu_i(\mathbf{r}') \neq q_i\delta\Psi^{\rm ext}(\mathbf{r}')$, i.e. a $\delta\Psi^{\rm ext}(\mathbf{r}')$ common for all species may not exist]. Here we have dropped superscript (2) on $c_{ij}^{(2)}(\mathbf{r},\mathbf{r}')$ since it is obvious that it is a pair function. The total electrostatic potential $\delta\Psi(\mathbf{r}')$ is given by (cf. eq. (30))

$$\delta\Psi(\mathbf{r}') = \delta\Psi^{\rm ext}(\mathbf{r}') + \frac{1}{\varepsilon_{\rm s}}\int \phi(|\mathbf{r}'-\mathbf{r}|)\delta\rho^{\rm pol}(\mathbf{r})\, d\mathbf{r}. \tag{34}$$

Eqs. (33) and (34) yields

$$\beta q_i \delta\Psi(\mathbf{r}') = -\frac{1}{n_i}\delta n_i(\mathbf{r}') + \sum_j \int \delta n_j(\mathbf{r}) c_{ij}^0(\mathbf{r},\mathbf{r}')\, d\mathbf{r} \tag{35}$$

where

$$c_{ij}^0(\mathbf{r},\mathbf{r}') = c_{ij}(\mathbf{r},\mathbf{r}') + \beta u_{ij}^{\rm Coul}(|\mathbf{r}-\mathbf{r}'|) \tag{36}$$

and we have used that $u_{ij}^{\rm Coul}(r) = q_i q_j \phi(r)/\varepsilon_{\rm s}$. The function $c_{ij}^0(\mathbf{r},\mathbf{r}')$ is the short–range part of the direct correlation function, which follows from the asymptotic relationship $c_{ij}(\mathbf{r},\mathbf{r}') \sim -\beta u_{ij}^{\rm Coul}(|\mathbf{r}-\mathbf{r}'|)$ when $|\mathbf{r}-\mathbf{r}'| \longrightarrow \infty$ [in general we have $c_{ij}(r) \sim -\beta u_{ij}(r)$ when $r \longrightarrow \infty$].

Eq. (35) gives the total potential $\delta\Psi(\mathbf{r}')$ in terms of the density change. Note that c_{ij}^0 here plays the same role vis–à–vis $\delta\Psi(\mathbf{r}')$ as c_{ij} plays vis–à–vis $\delta\Psi^{\rm ext}(\mathbf{r}')$ in eq. (33). Just like eq. (31) is the inverse relationship to eq. (33) (which, as we have seen, is a consequence of the OZ equation (29) being satisfied by h_{ij} and c_{ij}), we can form the inverse relationship to eq. (35) provided we introduce the function h_{ij}^0 that together with c_{ij}^0 satisfies the OZ equation

$$h_{ij}^0(\mathbf{r}',\mathbf{r}) = c_{ij}^0(\mathbf{r}',\mathbf{r}) + \sum_l \int h_{il}^0(\mathbf{r}',\mathbf{r}'') n_l c_{lj}^0(\mathbf{r}'',\mathbf{r})\, d\mathbf{r}''. \tag{37}$$

(h_{ij}^0 is thereby defined by eq. (37)). By analogy with eq. (31) we obtain

$$\delta n_j(\mathbf{r}) = -\beta n_j \left[q_j \delta\Psi(\mathbf{r}) + \sum_i \int q_i \delta\Psi(\mathbf{r}') n_i h_{ij}^0(\mathbf{r},\mathbf{r}')\, d\mathbf{r}' \right] \tag{38}$$

which gives the change in ion density in terms of the total electrostatic potential. Consequently, the polarization charge is given by

$$\delta\rho^{\mathrm{pol}}(\mathbf{r}) = -\beta \sum_j q_j^2 n_j \delta\Psi(\mathbf{r}) + \int A^0(\mathbf{r}, \mathbf{r}')\delta\Psi(\mathbf{r}')\, d\mathbf{r}' \tag{39}$$

where $A^0(\mathbf{r}, \mathbf{r}') = -\beta \sum_{ij} q_i q_j n_i n_j h_{ij}^0(\mathbf{r}, \mathbf{r}')$. Note that while eq. (32) expresses the polarization charge in terms of the external electrostatic potential, eq. (39) expresses it in terms of the *total* electrostatic potential. Eqs. (32) and (39) are equivalent for infinitesimally small external potentials.

Eqs. (32) and (39) motivates the introduction of the response functions B and B^0, that gives the polarization response in terms of the applied and the total electrostatic potentials respectively:

$$\delta\rho^{\mathrm{pol}}(\mathbf{r}) = \int B(|\mathbf{r} - \mathbf{r}'|)\delta\Psi^{\mathrm{appl}}(\mathbf{r}')\, d\mathbf{r}' \tag{40}$$

and

$$\delta\rho^{\mathrm{pol}}(\mathbf{r}) = \int B^0(|\mathbf{r} - \mathbf{r}'|)\delta\Psi(\mathbf{r}')\, d\mathbf{r}' \tag{41}$$

From eq. (32) and $\delta\Psi^{\mathrm{ext}} = \delta\Psi^{\mathrm{appl}}/\varepsilon_{\mathrm{s}}$ follow

$$B(r) = -\frac{\beta}{\varepsilon_{\mathrm{s}}} \sum_j q_j n_j [q_j \delta^{(3)}(r) + \rho_j^{\mathrm{cloud}}(r)] \tag{42}$$

where $\delta^{(3)}(r) = \delta(x)\delta(y)\delta(z)$, $\delta(x)$ is Dirac's delta function and

$$\rho_j^{\mathrm{cloud}}(r) = \sum_i q_i n_i h_{ij}(r) = \sum_i q_i n_i g_{ij}(r) \tag{43}$$

is the charge density of the ion cloud around a j ion. We have here used the fact that the pair functions of the bulk electrolyte are isotropic.

Likewise eq. (39) yields

$$B^0(r) = -\beta \sum_j q_j n_j [q_j \delta^{(3)}(r) + \rho_j^{\mathrm{dress}}(r)] \tag{44}$$

where

$$\rho_j^{\mathrm{dress}}(r) = \sum_i q_i n_i h_{ij}^0(r) = \sum_i q_i n_i g_{ij}^0(r) \tag{45}$$

and $g_{ij}^0 = h_{ij}^0 + 1$. The superscript "dress" will be motivated later (see section 3.2.1). Note that the square bracket in eq. (42) is the total charge density of a j ion and its ion cloud

$$\rho_j^{\mathrm{tot}}(r) = q_j \delta^{(3)}(r) + \rho_j^{\mathrm{cloud}}(r). \tag{46}$$

Analogously we introduce the following notation for the square bracket in eq. (44):

$$\rho_j^0(r) = q_j \delta^{(3)}(r) + \rho_j^{\text{dress}}(r), \tag{47}$$

which will be referred to as the charge density of a *dressed ion* (this name is motivated later), i.e. the charge density of a j ion and its dress.

3.1.2. *The Dielectric Function*

The response of a dielectric continuum to a small applied potential $\delta\Psi^{\text{appl}}(\mathbf{r}')$ (the latter as measured in vacuum) is such that the resulting total potential $\delta\Psi$ is proportional to the applied potential: $\delta\Psi(\mathbf{r}') = \delta\Psi^{\text{appl}}(\mathbf{r}')/\varepsilon_s$. Thus, the value of $\delta\Psi$ at each point is obtained from the value of $\delta\Psi^{\text{appl}}$ at this same point. This is not true for other media; $\delta\Psi$ at each point $\mathbf{r}$ is obtained from the value of $\delta\Psi^{\text{appl}}$ in an entire neighborhood of $\mathbf{r}$ since there are non–local contributions due to the charge distribution induced by $\delta\Psi^{\text{appl}}$. This non–locality is easiest to express in Fourier space and is customarily described by the *dielectric function* $\hat{\varepsilon}(k)$ defined below, where $k = |\mathbf{k}|$ and $\mathbf{k}$ is the wave vector. (In what follows we restrict ourselves to bulk fluids consisting of particles that interact with spherically symmetric potentials.) The Fourier transform $\hat{f}(\mathbf{k})$ of a function $f(\mathbf{r})$ is given by $\hat{f}(\mathbf{k}) = \int f(\mathbf{r}) \exp(-\mathrm{i}\mathbf{k}\cdot\mathbf{r})\, d\mathbf{r}$, where i is the imaginary unit (do not confuse this with the species index i).

For a dielectric continuum we have $\delta\hat{\Psi}(\mathbf{k}) = \delta\hat{\Psi}^{\text{appl}}(\mathbf{k})/\varepsilon_s$ and its dielectric function defined by the denominator, $\hat{\varepsilon}_{\text{cont}}(k) \equiv \varepsilon_s$, is constant (subscript 'cont' stands for continuum). This means that $\varepsilon_{\text{cont}}(r)$ is a delta function in ordinary space, which implies that it is entirely local. In general, we define the dielectric function $\hat{\varepsilon}(k)$ (more precisely the static, longitudinal dielectric function) from the relationship [6, 8]

$$\delta\hat{\Psi}(\mathbf{k}) = \frac{\delta\hat{\Psi}^{\text{appl}}(\mathbf{k})}{\hat{\varepsilon}(k)}. \tag{48}$$

Recall that $\delta\Psi$ is the total electrostatic potential that results from the infinitesimally small applied potential $\delta\Psi^{\text{appl}}$ and that it includes the potential from the polarization charge induced by the latter. In general $\hat{\varepsilon}(k)$ is not constant and eq. (48) implies that the attenuation of the applied potential is different for different wave lengths ($\hat{\varepsilon}$ does not depend on the direction of the wave vector $\mathbf{k}$ since the solution is isotropic). Eq. (48) also implies that $\delta\Psi(\mathbf{r})$ is given by the convolution of $\delta\hat{\Psi}^{\text{appl}}(\mathbf{r}')$ and the inverse Fourier transform of $1/\hat{\varepsilon}(k)$, which in general is non–local since $\hat{\varepsilon}(k)$ is not constant.

We can obtain explicit expressions of $\hat{\varepsilon}(k)$ from the results in the previous section. The Fourier transform of eq. (34) with $\delta\Psi^{\text{ext}} = \delta\Psi^{\text{appl}}/\varepsilon_s$

inserted is equal to

$$\delta\hat{\Psi}(\mathbf{k}) = \frac{1}{\varepsilon_s}[\delta\hat{\Psi}^{\text{appl}}(\mathbf{k}) + \hat{\phi}(k)\delta\hat{\rho}^{\text{pol}}(\mathbf{k})] \tag{49}$$

and $\delta\hat{\rho}^{\text{pol}}$ can be can obtained from the transform of eq. (40) or (41), i.e. $\delta\hat{\rho}^{\text{pol}} = \hat{B}\delta\hat{\Psi}^{\text{appl}}$ and $\delta\hat{\rho}^{\text{pol}} = \hat{B}^0\delta\hat{\Psi}$. By inserting $\delta\hat{\rho}^{\text{pol}}$ into eq. (49) we obtain

$$\delta\hat{\Psi}(\mathbf{k}) = \frac{1}{\varepsilon_s}[1 + \hat{\phi}(k)\hat{B}(k)]\delta\hat{\Psi}^{\text{appl}}(\mathbf{k}) \tag{50}$$

and

$$[\varepsilon_s - \hat{\phi}(k)\hat{B}^0(k)]\delta\hat{\Psi}(\mathbf{k}) = \delta\hat{\Psi}^{\text{appl}}(\mathbf{k}) \tag{51}$$

respectively. By comparing these equations with eq. (48) we can now identify $\hat{\varepsilon}(k)$ of the electrolyte solution as (cf. [34])

$$\hat{\varepsilon}(k) = \frac{\varepsilon_s}{1 + \hat{\phi}(k)\hat{B}(k)} \tag{52}$$

and

$$\hat{\varepsilon}(k) = \varepsilon_s - \hat{\phi}(k)\hat{B}^0(k). \tag{53}$$

Eqs. (52) and (53) are equivalent. The former is a standard result [35, 36] while the latter is seldom used. Eq. (53) is, however, a very useful relationship as we shall see. We can write it in the following form by using eqs. (44) and (47)

$$\hat{\varepsilon}(k) = \varepsilon_s + \frac{\beta}{\varepsilon_0 k^2}\sum_j n_j q_j \hat{\rho}_j^0(k) \tag{54}$$

where we have utilized that $\hat{\phi}(k) = 1/(\varepsilon_0 k^2)$. The first term in eq. (54) is the contribution to $\hat{\varepsilon}(k)$ from the solvent and the second term is that from the electrolyte.

It is illustrative to examine what eq. (54) becomes in the Debye–Hückel (DH) approximation. This approximation (which is the same as MSA for point particles) is obtained by setting $c_{ij} = -\beta u_{ij}^{\text{Coul}}$, i.e. $c_{ij}^0 = 0$, for all i and j, which implies that $\rho_j^{\text{dress}} = 0$ for all j and hence $\hat{\rho}_j^0 = q_j$. If we insert this in eq. (54) we obtain

$$\hat{\varepsilon}^{\text{DH}}(k) = \varepsilon_s\left[1 + \frac{\kappa_D^2}{k^2}\right] \tag{55}$$

where superscript DH indicates that it is valid in the DH approximation and κ_{D} is the Debye parameter ($1/\kappa_{\mathrm{D}}$ is the Debye length) defined from

$$\kappa_{\mathrm{D}}^2 = \frac{\beta}{\varepsilon_s \varepsilon_0} \sum_j n_j q_j^2. \tag{56}$$

We see that the contribution to the dielectric function from the electrolyte is equal to $\varepsilon_s \kappa_{\mathrm{D}}^2/k^2$, which diverges when $k \longrightarrow 0$, i.e. when the wave length $\longrightarrow \infty$. This is a consequence of the fact that the electrolyte is a conductor. Since $[\hat{\varepsilon}(k)]^{-1} \longrightarrow 0$ like k^2 when $k \longrightarrow 0$, eq. (48) implies that the total electric field in the electrolyte becomes zero when the system is exposed to an applied uniform field, which is appropriate for a conductor. The latter features are retained in the general case, for which we can write

$$\hat{\varepsilon}(k) = \hat{\varepsilon}^{\mathrm{DH}}(k) + \frac{\beta}{\varepsilon_0 k^2} \sum_j n_j q_j \hat{\rho}_j^{\mathrm{dress}}(k), \tag{57}$$

but otherwise the behavior is at least quantitatively different since $\rho_j^{\mathrm{dress}} \neq 0$. We see that ρ_j^{dress} contains everything that is left out from the DH approximation.

3.1.3. *The Proper "Screened Coulomb Potential"* ϕ^0

Let us consider the special case where the applied potential originates from an infinitesimally small point charge δq immersed in the bulk electrolyte, i.e.

$$\delta\Psi^{\mathrm{appl}}(r) = \frac{\delta q}{4\pi\varepsilon_0 r} = \delta q \phi(r). \tag{58}$$

Eq. (48) implies that the total potential, which we shall denote $\delta\psi$ in this case, equals

$$\delta\hat{\psi}(k) = \frac{\delta q \hat{\phi}(k)}{\hat{\varepsilon}(k)} = \delta q \hat{\phi}^0(k) \tag{59}$$

where we have introduced

$$\hat{\phi}^0(k) = \frac{\hat{\phi}(k)}{\hat{\varepsilon}(k)}. \tag{60}$$

While $\delta q \phi(r)$ in eq. (58) is the ordinary (unscreened) Coulomb potential from the point charge δq, the potential $\delta q \phi^0(r)$ contains all effects of the screening due to the surrounding electrolyte solution. We therefore nominate $\phi^0(r)$ as the *Screened Coulomb potential.*

The term "Screened Coulomb potential" is, in fact, commonly used in electrolyte theory and its applications to denote the Yukawa potential obtained in the Debye–Hückel approximation when calculating the screened potential due to a unit point charge immersed in an electrolyte, i.e.

$$\phi^{0,\,\mathrm{DH}}(r) = \frac{e^{-\kappa_{\mathrm{D}} r}}{4\pi\varepsilon_{\mathrm{s}}\varepsilon_0 r}. \tag{61}$$

In fact, this potential is obtained by inserting eq. (55) into eq. (60)

$$\hat{\phi}^{0,\,\mathrm{DH}}(r) = \frac{1}{\varepsilon_{\mathrm{s}}\varepsilon_0[k^2 + \kappa_{\mathrm{D}}^2]} \tag{62}$$

and taking the inverse Fourier transform. Thus the commonly used "Screened Coulomb potential" is the special case of eq. (60) in the DH approximation. To make a distinction we will say that $\phi^0(r)$ from eq. (60) is the "proper" Screened Coulomb potential, since it *involves no approximations* (provided the appropriate $\hat{\varepsilon}(k)$ is used), while $\phi^{0,\,\mathrm{DH}}(r)$ in eq. (61) will be called the DH Screened Coulomb potential and is always approximate. (We shall show below in section 3.2.3 that ϕ^0 is an electrostatic Green's function of the electrolyte solution. Likewise, $\phi^{0,\,\mathrm{DH}}$ is the corresponding Green's function in the DH approximation.)

Note that in the general, exact case $\phi^0(r)$ is not the potential in the electrolyte due to an immersed unit charge. Such a charge is not infinitesimally small, so linear response theory is not applicable and eq. (48) does not give the total potential (non–linear contributions must also be included). Therefore, we can only apply $\delta\phi(r) = \delta q \phi^0(r)$ to cases where the source charge δq is small (for other cases see section 3.2.1). The DH approximation is, on the other hand, a linear theory which means that the total potential is always proportional to the source charge, but the accuracy of this approximation becomes of course worse when the magnitude of source charge is increased.

The potential from an infinitesimally small point charge δq in an electrolyte solution is *never* given by the DH result $\delta q \phi^{0,\,\mathrm{DH}}(r)$, but this is an approximation that becomes more and more accurate at increasing dilution; in the limit of infinite dilution it becomes asymptotically correct. Thus, the proper Screened Coulomb potential is not a simple Yukawa potential; it is only in the DH approximation it has this simple form. Nevertheless, $\phi^0(r)$ decays in most cases like a Yukawa potential for large r values. Provided the electrolyte concentration is not too high we have, as we shall see,

$$\phi^0(r) \sim \frac{e^{-\kappa r}}{4\pi E r}, \quad r \longrightarrow \infty, \tag{63}$$

where κ and E are two constants. The decay length $1/\kappa$ is, in general, not equal to the Debye length and the prefactor is not equal to that of eq. (61), i.e. $E \neq \varepsilon_{\mathrm{s}}\varepsilon_0$. However, since the DH approximation becomes accurate at infinite dilution $\kappa \longrightarrow \kappa_D$ and $E \longrightarrow \varepsilon_s\varepsilon_0$ when the concentration goes to zero.

To see how eq. (63) can be obtained, let us first make the following observations for the DH case. From eq. (55) follows that $\hat{\varepsilon}^{DH}(k) = \varepsilon_s[k^2 + \kappa_D^2]/k^2$. We note that $\hat{\varepsilon}^{DH}(k)$ has simple zeros at $k = \pm i\kappa_D$, where i is the imaginary unit. From eq. (60) follows that $\hat{\phi}^0(k)$ has a singularity (a pole) when $\hat{\varepsilon}(k)$ is zero; in eq. (62) we can clearly see that $\hat{\phi}^{0,DH}(k)$ has simple poles at $k = \pm i\kappa_D$. One can show that the Yukawa form of the potential in eq. (61) is a consequence of the occurrence of these simple poles (it is, in fact, sufficient to consider one of the poles, e.g. $i\kappa_D$, since the other one must exist for reasons of symmetry of the transform). Similar considerations can be done in the general case, to which we now return.

An elegant mathematical method to determine the decay behavior of a function for large r values is based on the investigation of the singularities of the Fourier transform for complex k values. Let us consider the function $f(r)$ and its Fourier transform $\hat{f}(k)$. In general, one can show mathematically (using contour integration in complex k space) that each singularity of $\hat{f}(k)$ corresponds to a term in the asymptotic decay of $f(r)$ when $r \longrightarrow \infty$ (here $\hat{f}(k)$ is analytically continued from the real k axis into complex k space and we restrict ourselves to the upper half space). The singularity closest to the real axis in k space gives the leading asymptotic term of $f(r)$ for large r, while singularities further away give terms that decay faster. Each simple pole gives rise to a term of the Yukawa form, while a singularity of some other kind give a term of different functional form in r space. The DH Screened Coulomb potential is a pure Yukawa function since $\hat{\phi}^{0,DH}(k)$ in eq. (62) only has the poles $k = \pm i\kappa_D$ and no other contributions.

In the general case $\hat{\phi}^0(k)$ has many singularities for complex k values, so $\phi^0(r)$ is not a simple function. The leading contribution for large r originates, however, from a simple pole, i.e. a simple zero of $\hat{\varepsilon}(k)$ in eq. (60). One can show that no other singularity is closer to the real axis in k space (at least when the electrolyte concentration is not too high). Let us denote this zero as $i\kappa$, where κ in general can be a complex number. At low concentrations κ is, however, a real number and this zero of $\hat{\varepsilon}(k)$ gives rise to the Yukawa term shown in eq. (63). We can thus conclude that the decay parameter κ in eq. (63) satisfies

$$\hat{\varepsilon}(i\kappa) = 0 \tag{64}$$

and one can show by residue analysis of eq. (60) in complex k space that E in the denominator of eq. (63) is given by

$$E = \frac{\varepsilon_0}{2}\left[k\frac{d\hat{\varepsilon}(k)}{dk}\right]_{k=i\kappa}. \tag{65}$$

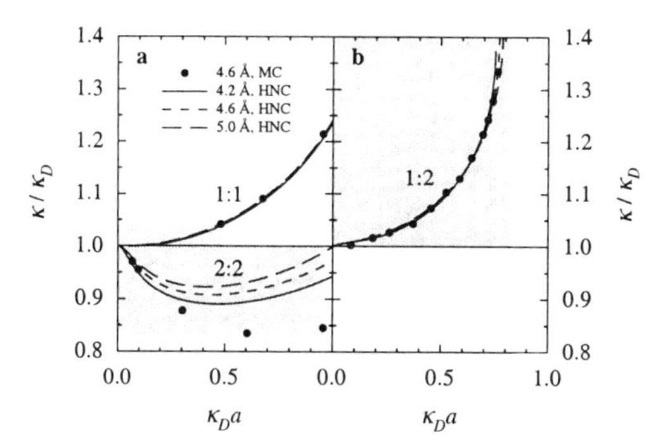

Figure 1. The deviation of the inverse decay length κ from the inverse Debye length κ_D (plotted as κ/κ_D) shown as functions of $\kappa_D a$ for aqueous bulk electrolytes ($\varepsilon_s = 78.4$, $T = 298$ K) with various ion diameters a and valencies (1:1, 2:2 and 1:2). The symbols show Monte Carlo (MC) simulation results for $a = 4.6$Å and the curves HNC results for $a = 4.2$, 4.6 and 5.0 Å. The grey areas indicate the part of the diagram that is relevant for concentrations up to 0.1 M when a = 4.6 Å for the 1:1, 2:2 and 1:2 electrolytes respectively. (The data are taken from [15, 21, 24].)

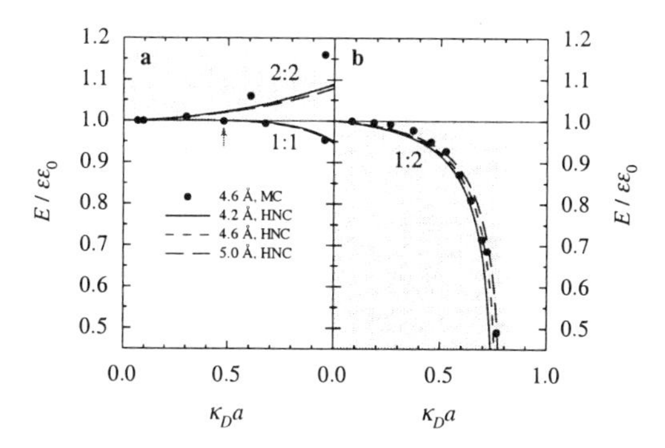

Figure 2. The deviation of the effective permittivity E of the electrolyte solution from the solvent permittivity $\varepsilon_s\varepsilon_0$ (plotted as $E/\varepsilon_s\varepsilon_0$) shown as functions of $\kappa_D a$ for the same cases as in fig. 1. The grey areas have the same meaning as in fig. 1; that for the 1:1 case is, however, to thin to be seen in this figure and the arrow shows where it ends. (The data are taken from [15, 21, 24].)

(Note that eqs. (55) and (65) yields $E = \varepsilon_s\varepsilon_0$ in the DH approximation.)

By using the results in sections 3.1.1 and 3.1.2 one can calculate $\hat{\varepsilon}(k)$ from the pair distribution functions of the electrolyte solution. These functions can be calculated e.g. by computer simulation or integral equation approximations. From $\hat{\varepsilon}(k)$ one can obtain κ and E from eqs. (64) and (65). Fig. 1 shows calculated κ values for various aqueous electrolyte solutions at room temperature. The anions and cations have the same diameter a. The plot shows κ/κ_D as a function of $\kappa_D a$, which is proportional to the square root of concentration, as obtained from Monte Carlo (MC) simulations for $a = 4.6$ Å and from the HNC approximation for this and two other a values. The HNC and MC results agree very well for 1:1 and 1:2 electrolytes. For 2:2 electrolytes HNC results are qualitatively correct but differ quantitatively from the MC results. The grey areas indicate the parts of the diagrams that are relevant for concentrations up to 0.1 M in each case when $a = 4.6$ Å. We see that κ is not very different from κ_D for 1:1 electrolytes in this concentration range, while the difference is appreciable for 2:2 electrolytes and quite large for 1:2 electrolytes.

The deviation of κ from κ_D is described by the following exact expansion valid in the low concentration limit [13, 37, 38]:

$$\frac{\kappa}{\kappa_D} = 1 + \alpha_3^2 \frac{\Lambda \ln 3}{8} + \alpha_4^2 \frac{\Lambda^2 \ln \Lambda}{12} + \mathcal{O}\left(\Lambda^2\right) \tag{66}$$

where $\Lambda = \kappa_D L_B$ is a coupling parameter, $L_B = \beta q_e^2/(4\pi\varepsilon_s\varepsilon_0)$ is the Bjerrum length, q_e the elementary charge and

$$\alpha_\gamma = \frac{\sum_j \nu_j z_j^\gamma}{\sum_j \nu_j z_j^2}$$

($\gamma = 3$ or 4) are constants where z_j is the valency and ν_j is the stoichiometric coefficient of species j. The remainder term, $\mathcal{O}\left(\Lambda^2\right)$, in eq. (66) implies that the higher order terms go to zero at least as fast as Λ^2 (which is proportional to the concentration) when the concentration goes to zero.

The two terms in eq. (66) that contain α_3 or α_4 originate from non–linear, purely electrostatic contributions to the ion–ion correlations in the electrolyte. This implies that linear approximations like MSA and GMSA do not include these contributions, i.e. they do not give the correct leading terms at low concentrations. On the other hand, one can show that the HNC approximation *does* include these leading terms correctly. This is a consequence of the fact that all contributions from the bridge function (neglected in the HNC approximation) are contained in the remainder (in $\mathcal{O}\left(\Lambda^2\right)$) [13, 37]. The ionic size dependent contributions are also contained entirely in the remainder.

Note that eq. (66) implies that for binary electrolytes at sufficiently low concentrations we necessarily have $\kappa < \kappa_D$ for charge–symmetric (like 1:1 or 2:2) electrolytes, for which α_3 is identically zero, and $\kappa > \kappa_D$ for charge–asymmetric (like 1:2) electrolytes. We can clearly see in fig. 1 that this is fulfilled for 2:2 and 1:2 electrolytes. For the 1:1 case, the curve actually goes below 1 at very low concentrations, but this cannot be seen on the scale of the figure (for instance, when $a = 4.2$ Å κ/κ_D first decreases to around 0.999 before it rises above 1). In fact, simple linear approximations like MSA give reasonable values of κ for 1:1 aqueous electrolytes in a wide range of concentrations, while this is not the case for 2:2 and 1:2 electrolytes for which the non–linear effects not included in MSA are very important.

In practice we usually need not to consider deviations of κ from κ_D for aqueous 1:1 electrolytes even for 0.1 M concentrations where the deviation is a few percent, while, as we have seen in fig. 1, the picture is quite different for divalent ions (or higher valencies, cf. [39]). The analogous conclusion holds for the effective permittivity E of the electrolyte. Fig. 2 shows how much E deviates from the permittivity $\varepsilon_s\varepsilon_0$ of the solvent for the same cases as in fig. 1. For aqueous 1:1 electrolytes E is very close to $\varepsilon_s\varepsilon_0$ up to more than 0.1 M concentration, while the deviation is appreciable for 2:2 and quite large for 1:2 electrolytes (the sharp decrease of $E/\varepsilon_s\varepsilon_0$ for the 1:2 case will be discussed in more detail later; E eventually goes to zero at increasing concentrations for all three cases). Again we see that the HNC approximation is accurate for 1:1 and 1:2 electrolytes, while the MC and HNC results agree only qualitatively for the 2:2 case.

The leading term in the decay of $\phi^0(r)$ [eq. (63)] that we have investigated so far only tells a part of the story. In fact, $\mathrm{i}\kappa$ is not the only zero of $\hat{\varepsilon}(k)$. The next zero, which lies further away from the real axis in complex k space, will be denoted $\mathrm{i}\kappa'$, i.e. $\hat{\varepsilon}(\mathrm{i}\kappa') = 0$. It gives rise to one further Yukawa term in $\phi^0(r)$ and we have

$$\phi^0(r) \sim \frac{e^{-\kappa r}}{4\pi E r} + \frac{e^{-\kappa' r}}{4\pi E' r}, \quad r \longrightarrow \infty, \tag{67}$$

where E' is given by eq. (65) but with the square bracket evaluated at $k = \mathrm{i}\kappa'$. Both κ and κ' are real at low concentrations and from the fact that $\mathrm{i}\kappa$ and $\mathrm{i}\kappa'$ are consecutive zeros of $\hat{\varepsilon}(k)$ on the imaginary k axes it follows that E and E' have different signs due to the factor $d\hat{\varepsilon}/dk$ in eq. (65). Since $E \longrightarrow \varepsilon_s\varepsilon_0$ when the concentration goes to zero it follows by continuity that $E > 0$ and hence $E' < 0$. Thus the first term in eq. (67) is positive and the second term is negative. (Other kinds of singularities of $\hat{\phi}^0(k)$ exists, which give rise to terms of other functional form in the decay of $\phi^0(r)$ when r is increased, but we shall not be concerned much about these terms in this presentation.)

Fig. 3 shows both κ and κ' (plotted as κa and $\kappa' a$ as functions of $\kappa_D a$) for the 1:1 and 1:2 electrolytes. For reference, the curves for 2κ and κ_D are also plotted. For the 1:1 electrolyte κ' is much larger than κ at low concentrations, so the second term in eq. (67) decays much faster than the first and is unimportant for large r values. When the concentration is increased κ' and κ approach each other and finally they merge, i.e. the two roots $\mathrm{i}\kappa$ and $\mathrm{i}\kappa'$ of $\hat{\varepsilon}(k) = 0$ coincide. The point of merger is shown as a large

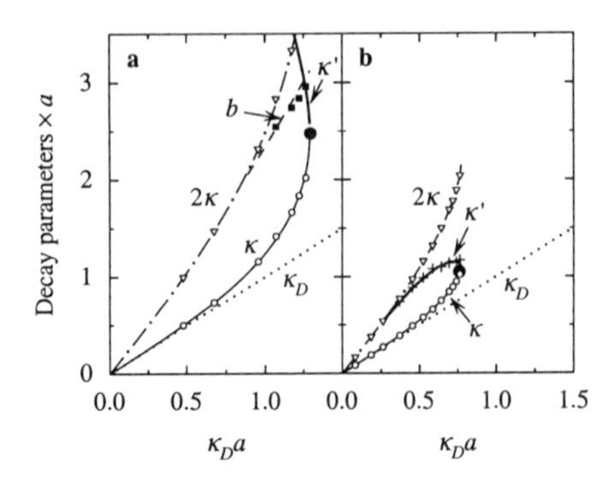

Figure 3. Various decay parameters (inverse decay lengths) multiplied by ionic diameter a and plotted as functions of $\kappa_D a$ for (a) 1:1 and (b) 1:2 electrolytes with $a = 4.6$ Å (same as the corresponding systems in Fig. 1). The symbols show MC results and the curves HNC results: thin full line and open circles, κa; thick full line and plusses, $\kappa' a$; dash–dotted line and open triangles, $2\kappa a$; dashed line and filled squares, ba [only in (a)]; dotted line, $\kappa_D a$. The large filled circle shows the point of merger of κ and κ'. (The data are taken from [21, 24].)

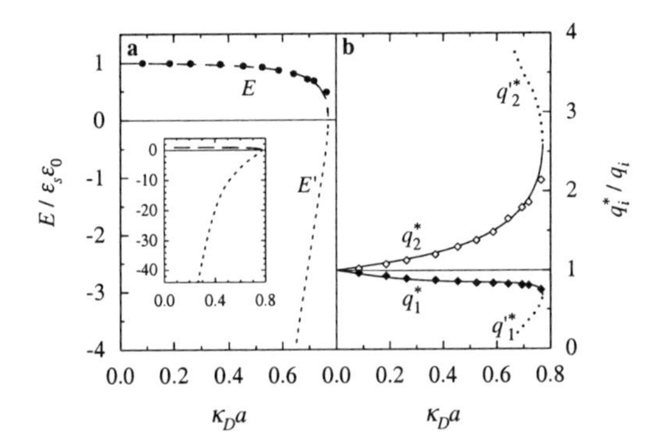

Figure 4. (a) Primary effective permittivity E (long dashes and filled circles) and secondary effective permittivity E' (short dashes) divided by $\varepsilon_s \varepsilon_0$ for a 1:2 electrolyte with $a = 4.6$ Å (same as the corresponding system in Fig. 1). The symbols show MC results and the curves HNC results. The inset shows the same curves on an extended ordinate scale.
(b) The primary effective charge q_i^* (full lines and diamonds) and secondary effective charge q'^*_i (dotted lines) divided by the bare ionic charge q_i for the same system as in (a). The symbols show MC results and the curves HNC results. Monovalent ions $i = 1$ and divalent ions $i = 2$. (The data are taken from [21, 24].)

filled circle in the figure. In a small concentration range below this point the two terms in eq. (67) are about equally important. For aqueous 1:1 electrolytes this is, however, of mainly "academic" interest since the decay length at these concentrations is about the same as the size of the water molecules, so the solvent is not well represented by a dielectric continuum.

The behavior of κ' for 1:2 electrolytes is quite different. Here κ' is bounded above by 2κ so the decay length of the second term in eq. (67) is never smaller than half the decay length of the first term. When the concentration is increased κ' and κ merge in this case too, but since $\kappa' < 2\kappa$ the two decay parameters κ' and κ have similar magnitude in a quite wide concentration range below this point. This range lies below 0.1 M when $a = 4.6$ Å. The quantities E and E' for this case are shown in fig. 4(a). When the point of merger of κ' and κ is approached we have $E' \sim -E \longrightarrow 0$, which follows from the fact that the derivative $d\hat{\varepsilon}(k)/dk$ in eq. (65) becomes zero at the point where the two roots of $\hat{\varepsilon}(k)$ coincide (this will, of course, happen at the point of merger for all other cases too). When the concentration is decreased $\kappa' \longrightarrow 2\kappa$ and $E' \longrightarrow -\infty$ exponentially fast, which means that the second term in eq. (67) becomes very small at low concentrations. Then the next asymptotic term of $\phi^0(r)$ [not shown in eq. (67)] becomes the second leading term in practice (cf. [21]). It decays like $\exp(-2\kappa r)/r^2$ times a slowly varying function, i.e. it has decay length $1/(2\kappa)$ (this term corresponds to a branch point singularity of $\hat{\phi}^0(k)$ at $k = \mathrm{i}2\kappa$).

When concentration is increased beyond the point of merger, κ and κ' become complex (they are each other's complex conjugate). The two terms in eq. (67) then combine into an exponentially damped oscillatory function with decay length $1/K$, where $K = \mathrm{Re}(\kappa) = \mathrm{Re}(\kappa')$ [Re denotes the real part]. Thus, the point of merger of κ and κ' is the transition point to oscillatory decay of $\phi^0(r)$. Since $E' \sim -E \longrightarrow 0$ when this point is approached, each of the terms in eq. (67) diverges but their sum remains finite (as required for physical reasons).

To conclude this section, we see that the electrostatic potential $\delta\psi(r) = \delta q\phi^0(r)$ from an infinitesimally small point charge δq immersed in an electrolyte solution decays like

$$\delta\psi(r) \sim \frac{\delta q e^{-\kappa r}}{4\pi E r} + \frac{\delta q e^{-\kappa' r}}{4\pi E' r} \tag{68}$$

when $r \longrightarrow \infty$. The physical interpretation of this result (when κ and κ' are real) is that E is the (primary) *effective permittivity* of the electrolyte solution, which, as we have seen, goes to the permittivity of the solvent $\varepsilon_s\varepsilon_0$ when the electrolyte concentration goes to zero. The second term has the same functional form and we will denote E' as the *secondary effective permittivity* of the solution. Both E and E' are entirely determined by the dielectric function $\hat{\varepsilon}(k)$ [see eq. (65)] and they describe important aspects of the propagation of electrical disturbances in the electrolyte: the factors $1/E$ and $1/E'$ express the strength (and sign) of the "modes" of propagation that decays with the leading decay lengths $1/\kappa$ and $1/\kappa'$ respectively. These decay lengths are also entirely determined by $\hat{\varepsilon}(k)$ [see eq. (64)].

3.2. NON–LINEAR REGIME

3.2.1. *Charge Renormalization; Diffuse Effective Charge*

In the previous section we investigated the electrostatic potential from an infinitesimally small point charge immersed in a bulk electrolyte. We must now determine what is the electrostatic potential due to a particle with finite charge and/or size. Then linear response theory is not sufficient, we must also include the non–linear contributions. The total potential is given by eq. (30) which we write as

$$\Psi(\mathbf{r}') = \frac{1}{\varepsilon_s}\left[\Psi^{\mathrm{appl}}(\mathbf{r}') + \int \phi(|\mathbf{r}' - \mathbf{r}|)\rho^{\mathrm{pol}}(\mathbf{r})\, d\mathbf{r}\right]. \tag{69}$$

From eq. (41) we have the linear part of the polarization response

$$\rho^{\mathrm{lin}}(\mathbf{r}) \equiv \int B^0(|\mathbf{r} - \mathbf{r}'|)\Psi(\mathbf{r}')\, d\mathbf{r}' \tag{70}$$

and to determine $\Psi(\mathbf{r}')$ from eq. (69) we need to include the non–linear contributions to $\rho^{\mathrm{pol}}(\mathbf{r})$ as well.

Let us consider an ion of species j immersed in the bulk electrolyte. The applied potential due to this ion is $\Psi^{\mathrm{appl}}(\mathbf{r}') = q_j\phi(r')$ and we can identify the polarization charge as the charge density of the surrounding ion cloud, i.e. $\rho^{\mathrm{pol}}(\mathbf{r}) = \rho_j^{\mathrm{cloud}}(r)$. Note that we thereby have included the effects of finite size of the ion as well as finite charge. The resulting potential Ψ will be denoted ψ_j, i.e. the total electrostatic potential around an ion of species j, and we note that ψ_j is identical to what is commonly called the average potential around the ion. If we insert these quantities in eq. (69) and take the Fourier transform we obtain

$$\hat{\psi}_j(k) = \frac{1}{\varepsilon_s}[q_j + \hat{\rho}_j^{\mathrm{cloud}}(k)]\hat{\phi}(k). \tag{71}$$

We now split the polarization charge into the linear and non–linear parts

$$\hat{\rho}_j^{\mathrm{cloud}}(k) = \hat{\rho}_j^{\mathrm{lin}}(k) + \hat{\rho}_j^{\mathrm{nonlin}}(k) \tag{72}$$

where

$$\hat{\rho}_j^{\mathrm{lin}}(k) = \hat{B}^0(k)\hat{\psi}_j(k), \tag{73}$$

which follows from eq. (70). By inserting this in eq. (71) we obtain

$$[\varepsilon_s - \hat{\phi}(k)\hat{B}^0(k)]\hat{\psi}_j(k) = [q_j + \hat{\rho}_j^{\text{nonlin}}(k)]\hat{\phi}(k). \tag{74}$$

From eq. (53) and the definition of the proper Screened Coulomb potential ϕ^0, eq. (60), we see that eq. (74) implies

$$\hat{\psi}_j(k) = \frac{[q_j + \hat{\rho}_j^{\text{nonlin}}(k)]\hat{\phi}(k)}{\hat{\varepsilon}(k)} = [q_j + \hat{\rho}_j^{\text{nonlin}}(k)]\hat{\phi}^0(k). \tag{75}$$

The only difference between eq. (75) and corresponding result for an infinitesimally small point charge δq, eq. (59), is that δq is replaced by $q_j + \hat{\rho}_j^{\text{nonlin}}$. Thus the latter quantity acts as an effective charge (a charge density, a "*diffuse effective charge*" in ordinary space) that replaces the bare charge q_j.

For a finite point charge (e.g. the unit point charge discussed in the previous section) the non–linear contribution cannot be neglected, but in the limit of infinitesimally small charge the effective charge becomes equal to the bare charge as shown by eq. (59). For a particle with finite size and charge $\hat{\rho}_j^{\text{nonlin}}$ is non–zero, which means that the effective charge is not equal to the bare charge. This may also be the case for an uncharged particle with finite size, which accordingly has a non–zero effective charge, i.e. it gives rise to a non–zero electrostatic potential when it is immersed in the electrolyte soluution. This happens, for example, when anions and cations of the electrolyte have different non–electrostatic interactions with the particle or when the electrolyte is asymmetric.

So, what is $\hat{\rho}_j^{\text{nonlin}}$? From the OZ equations (17) and (37) together with the definitions (36), (43), (45), (47) and (71), all transformed to Fourier space, it follows after some algebra [11, 13] that

$$\hat{h}_{ij}(k) = \hat{h}_{ij}^0(k) - \beta\hat{\rho}_i^0(k)\hat{\psi}_j(k) \tag{76}$$

and by using eqs. (43), (44), (45) and (47) in Fourier space we obtain from eq. (76)

$$\hat{\rho}_j^{\text{cloud}}(k) = \hat{\rho}_j^{\text{dress}}(k) + \hat{B}^0(k)\hat{\psi}_j(k). \tag{77}$$

The last term is by eq. (73) equal to $\hat{\rho}_j^{\text{lin}}(k)$, so eqs. (72) and (77) imply

$$\hat{\rho}_j^{\text{nonlin}}(k) = \hat{\rho}_j^{\text{dress}}(k). \tag{78}$$

This mean that the effective charge above is equal to $q_j + \hat{\rho}_j^{\text{dress}}$ or in other words (cf. eq. (47)) that $\rho_j^0(r)$ *constitutes the diffuse effective charge.* This is the reason for concept of a "*Dressed Ion*" with charge density $\rho_j^0(r)$. The dress of the ion is the charge density given by eq. (45) and it constitutes the charge, apart from the bare charge q_j, that is contained in the diffuse effective charge. Usually $\rho_j^{\text{dress}}(r)$ has much shorter range than $\rho_j^{\text{cloud}}(r)$ (exceptions may occur at high concentrations).

These results imply that eq. (75) can be written

$$\hat{\psi}_j(k) = \hat{\rho}_j^0(k)\hat{\phi}^0(k) \tag{79}$$

and in ordinary space

$$\psi_j(r) = \int \phi^0(|\mathbf{r} - \mathbf{r}'|)\rho_j^0(r')\, d\mathbf{r}'. \tag{80}$$

This has the same appearance as Coulomb's law in vacuum, $\psi(r) = \int \phi(|\mathbf{r} - \mathbf{r}'|)\rho(r')\, d\mathbf{r}'$, which gives the potential ψ from a charge density ρ. Eq. (80) says that when one wants to use the proper screened Coulomb potential ϕ^0 to calculate the potential in an electrolyte solution, one should use the diffuse effective charge, the dressed ion charge density ρ_j^0, as the source. For an infinitesimally small point charge δq the effective charge is the bare charge and we recover $\delta\psi(r) = \delta q\phi^0(r)$, cf. section 3.1.3. In most applications of screened Coulomb potential one uses the latter kind of formula even when the charged particle is not a small point charge and then the non–linear contributions are left out. If eq. (80) with the diffuse effective charge is used instead, *the non–linear contributions are included* which can be crucial in many cases.

By combining eqs. (76) and (79) we obtain

$$\hat{h}_{ij}(k) = \hat{h}_{ij}^0(k) - \beta\hat{\rho}_i^0(k)\hat{\phi}^0(k)\hat{\rho}_j^0(k) \tag{81}$$

If we introduce the convolution notation $f_1 * f_2(r) = \int f_1(|\mathbf{r} - \mathbf{r}'|)f_2(r')\, d\mathbf{r}'$ we can write this in ordinary space

$$h_{ij}(r) = h_{ij}^0(r) - \beta\rho_i^0 * \phi^0 * \rho_j^0(r) \tag{82}$$

The factor $\rho_i^0 * \phi^0 * \rho_j^0$ in last term is equal to the electrostatic interaction energy between the diffuse effective charges of an i and a j ion interacting via the proper screened Coulomb potential. From eq. (79) we see that we can write this interaction energy as

$$\rho_i^0 * \phi^0 * \rho_j^0 = \psi_i * \rho_j^0 = \rho_i^0 * \psi_j \tag{83}$$

which implies that there is perfect symmetry in the roles of the effective diffuse charge – it acts both as a source of the potential and as the charge that interacts with this potential. The factor $-\beta$ in the last term of (81) originates from the same factor in exponent of the Boltzmann factor, eq.

(4), and is typical for linear response contributions to distribution functions, cf. the first Yvon equation (27).

Apart from the $\rho_i^0 * \phi^0 * \rho_j^0$ contribution, the pair correlation function $h_{ij}(r)$ in eq. (82) only contains $h_{ij}^0(r)$, which precisely gives the dress of each ion according to eq. (45) and thereby the diffuse effective charge from eq. (47). Thus, when we calculate the charge density of the ion cloud using eq. (43), the last term of eq. (82) gives rise to the part which is not contained in the dress. Since ϕ^0 can also be expressed in terms of ρ_j^0 [eqs. (54) and (60)], all parts of the pair correlation function can be written in terms of $h_{ij}^0(r)$ and the associated charges, i.e. the diffuse effective charges ρ_j^0 (the dressed ions) for all species. We can thus conclude that the decomposition of the ion cloud into the dress and the rest is self–contained.

This concludes the first part of our exposition of DIT for bulk electrolyte solutions. As we have seen it is a reformulation of the exact statistical mechanics, which does not introduce any approximations; the resulting theory remains exact. Accordingly, we do an *exact charge renormalization* when we replace the bare ion charge with the dressed ion charge and consider the interaction between these effective charges using the proper screened Coulomb potential ϕ^0.

3.2.2. *Effective Point Charge*

As we have seen in section 3.1.3 the potential from a sufficiently small point charge δq is proportional to the charge because the non–linear effects are then negligible. This fact is also reflected in eq. (68) which gives the two leading terms in the decay of the potential from δq. Both terms in eq. (68) are proportional to the charge δq. When the charge is increased sufficiently much the proportionality will not hold because non–linearities cannot be neglected any longer. This is the case for the potential around a particle with charge q_j that is not small e.g. a real ion. Then the coefficients of the leading terms are not proportional to the charge and we have instead

$$\psi_j(r) \sim \frac{q_j^* e^{-\kappa r}}{4\pi E r} + \frac{q'^*_j e^{-\kappa' r}}{4\pi E' r}, \quad r \longrightarrow \infty, \tag{84}$$

where q_j^* and q'^*_j in general are different from the bare ionic charge q_j.

To see how this result can be obtained for ions of arbitrary size and charge let us investigate eq. (79) which can be written (using eq. (60))

$$\hat{\psi}_j(k) = \frac{\hat{\rho}_j^0(k)\hat{\phi}(k)}{\hat{\varepsilon}(k)} \tag{85}$$

In section 3.1.3 we saw how the leading contributions to the decay of $\phi^0(r)$ for large r can be determined by investigating leading poles of $\hat{\phi}^0(k)$ in complex k space, which originate from zeros of $\hat{\varepsilon}(k)$ [cf. eq. (60)]. The same strategy can be used to determine the decay of $\psi_j(r)$. We do not need to be concerned about poles of $\hat{\rho}_j^0(k)$ in eq. (85) since they are cancelled by the same kind of pole in $\hat{\varepsilon}(k)$ which contain $\hat{\rho}^0$ terms [see eq. (54)]. (Other kinds of singularities contribute only to higher order terms in the cases of interest here.)

The leading poles of $\hat{\psi}_j(k)$ therefore occur for the same k values as those of $\hat{\phi}^0(k)$, i.e. at the zeros $\mathrm{i}\kappa$ and $\mathrm{i}\kappa'$ of $\hat{\varepsilon}(k)$. As a consequence $\psi_j(r)$ and $\phi^0(r)$ have the same dominating decay lengths for large r. The coefficients of the two terms in eq. (84) can be obtained by residue calculus and it follows that the contribution from $\hat{\varepsilon}(k)$ is the same as for $\phi^0(r)$, i.e. E and E' given by eq. (65) evaluated at $k = \mathrm{i}\kappa$ and $\mathrm{i}\kappa'$ respectively. What is new here is the factor $\hat{\rho}_j^0(k)$ in eq. (85). It gives rise to the factors q_j^* and q'^*_j in eq. (84). Their values are given by

$$q_j^* = \hat{\rho}_j^0(\mathrm{i}\kappa) \tag{86}$$

and the analogous expression for q'^*_j (evaluated at $\mathrm{i}\kappa'$). From the definition of the Fourier transform it follows that this equation can be written explicitly as

$$q_j^* = \int \rho_j^0(r) \exp(-\kappa \mathbf{e} \cdot \mathbf{r})\, d\mathbf{r} = \frac{4\pi}{\kappa} \int_0^\infty \rho_j^0(r) \sinh(\kappa r) r\, dr \tag{87}$$

where $\mathbf{e}$ is a unit vector pointing in the (arbitrary) direction in which the exponential function decays with decay length $1/\kappa$ (this function has a constant value on each plane perpendicular to $\mathbf{e}$). We see that q_j^* and q'^*_j in general are different since κ and κ' are different. The only exception is at the transition point to oscillatory decay when κ and κ' are equal. When κ and κ' are complex q_j^* and q'^*_j are complex too. The two terms in eq. (84) then combine to an exponentially damped oscillatory function.

How to interpret eq. (84) physically? Let us consider cases of low electrolyte concentrations where κ and κ' are real. At sufficiently large distances from the j ion, the electrostatic potential is practically equal to the first term in eq. (84). We now compare with eq. (68), which is also dominated by the first term for large r values. The only difference is that δq in $\delta\psi(r)$ is replaced by q_j^* in $\psi_j(r)$. We see that q_j^* for a real ion has the same role as the bare charge δq has for an infinitesimally small point charge. Thus we can interpret q_j^* as the (primary) *effective point charge* of the j ion, which is generally different from the bare charge q_j because of the effects contained in the diffuse effective charge ρ_j^0 as expressed by eq. (87). From this equation follows that q_j^* is entirely determined by ρ_j^0 and κ. Note that the effective point charge describes how the charge of the ion is perceived at large distances (via the strength of the potential), while the diffuse effective charge describes what happens close to the ion.

Likewise, by comparing the second terms of eqs. (68) and (83) we see that q'^*_j too has the same role as the bare point charge δq. We will refer to q^*_j as the *secondary effective point charge* of the j ion and it is entirely determined by ρ^0_j and κ'. The second order term starts to contribute substantially at smaller r values than the first order term.

An alternative, but related physical interpretation is the following. The potential from the j ion is given by eq. (80). As we have seen, the screened Coulomb potential $\phi^0(r)$ propagates electrical disturbances over large distances in two dominating "modes" with exponential decay lengths $1/\kappa$ and $1/\kappa'$ respectively. The intrinsic strengths of these modes are given by the effective permittivities of the solution $1/E$ and $1/E'$ (cf. eq. (67)). These propagation modes couple with the effective diffuse charge ρ^0_j; the strength of the coupling is given by q^*_j and q'^*_j respectively [as expressed by eq. (87)]. These two parameters, which are effective charges of the ion as "seen" by each mode, summarize how various complicated effects in the charge distribution near each ion (as described by the diffuse effective charge ρ^0_j) influence the potential far away from the ion as propagated via each mode.

Conversely, q^*_j and q'^*_j also describe how strongly an ion interact with each mode of a potential that originates from some other source; as we have seen in eq. (83) ρ^0_j of two ions have symmetrical roles in this respect (and so do the effective point charges). The first equality of eq. (87) can be written

$$q^*_j \exp(-\kappa \mathbf{e} \cdot \mathbf{r}') = \int \rho^0_j(|\mathbf{r} - \mathbf{r}'|) \exp(-\kappa \mathbf{e} \cdot \mathbf{r})\, d\mathbf{r} \tag{88}$$

which can be interpreted in the following way. Consider a potential that decays exponentially with decay length $1/\kappa$. Assume that it interacts with the diffuse effective charge ρ^0_j of a j ion placed at $\mathbf{r}'$. The electrostatic interaction energy is then given by an integral like the rhs of eq. (88). If we want to replace ρ^0_j with a point charge placed at the ion center, the point charge must have the magnitude q^*_j in order to give the same electrostatic interaction energy with the potential as expressed by eq. (88). The corresponding statement holds also for the second propagation mode with decay length $1/\kappa'$.

There exists an intimate relationship between the decay length of each term in eq. (84) and the effective point charges in the corresponding coefficients for all ion species present in the solution. If we insert eq. (54) into eq. (64) and make use of eq. (86) we obtain

$$\kappa^2 = \frac{\beta}{\varepsilon_s \varepsilon_0} \sum_j n_j q_j q^*_j \tag{89}$$

and the analogous expression for $\kappa'$and q'^*_j. This relationship is very similar to the definition of the Debye parameter κ_D in eq. (56). It has also been obtained in approximate theories [40, 41], but it is an exact relationship of general validity. For symmetric binary electrolytes eqs. (56) and (89) imply $q^*/q = [\kappa/\kappa_\mathrm{D}]^2$, where $q = q_+ = -q_-$ and $q^* = q^*_+ = -q^*_-$. Therefore, for the 1:1 and 2:2 electrolytes in fig. 2(a) we obtain q^*/q as a function of $\kappa_\mathrm{D} a$ by taking the square of the curves in the figure.

For asymmetric electrolytes q^*_j and q'^*_j have to be calculated directly from ρ^0_j via eq. (87). The results for a 1:2 electrolyte are shown in fig. 4(b), where index 1 implies monovalent and index 2 divalent ions. At the point of merger of κ and κ' we see that q^*_j and q'^*_j also merge [this is an immediate consequence of eqs. (86) and (87)]. When the concentration is decreased q'^*_1 can change sign. Since q'^*_2 is quite large the corresponding term in eq. (84) is important for divalent ions, provided $|E'|$ is not too large.

Analogously to eq. (66), the deviation of q^*_j from q_j is described by an exact expansion valid in the low concentration limit [13]

$$\frac{q^*_j}{q_j} = 1 + z_i \alpha_3 \frac{\Lambda \ln 3}{4} + z_i^2 \alpha_4 \frac{\Lambda^2 \ln \Lambda}{6} + \mathcal{O}\left(\Lambda^2\right), \tag{90}$$

where all symbols have the same meaning as in eq. (66). This equation implies that for binary electrolytes at low concentrations we necessarily have $|q^*_j| < |q_j|$ for charge–symmetric electrolytes. For charge–asymmetric electrolytes it follows that $|q^*_j| > |q_j|$ for the ion of highest valency and $|q^*_j| < |q_j|$ for the ion of lowest valency. This can be seen in fig. 4(b).

It is of interest to investigate the decays of the pair correlation function $h_{ij}(r)$ and the pair potential of mean force $w_{ij}(r)$ defined from

$$g_{ij}(r) = e^{-\beta w_{ij}(r)}. \tag{91}$$

These decays are related since $h_{ij}(r) = \exp(-\beta w_{ij}(r)) - 1 \approx -\beta w_{ij}(r)$ for small $w_{ij}(r)$, e.g. for large r. The potential of mean force gives the interaction free energy between two ions at distance r from each other in a bulk electrolyte solution, which means that its derivative (with inverted sign) gives the total interaction force between the ions.

The decay behavior can be analyzed using eq. (81), which can be written (by inserting eq. (60))

$$\hat{h}_{ij}(k) = \hat{h}^0_{ij}(k) - \beta \frac{\hat{\rho}^0_i(k)\hat{\phi}(k)\hat{\rho}^0_j(k)}{\hat{\varepsilon}(k)} \tag{92}$$

In most cases it is the last term that gives the leading terms in the decay for large r and, again, it is the zeros of $\hat{\varepsilon}(k)$ that matters. Compared to

$\hat{\psi}_j(k)$ in eq. (85) we have an extra factor $\hat{\rho}_i^0(k)$ in this term. The factor simply gives rise to a factor q_i^* in the leading term of $h_{ij}(r)$ and we have (cf. the first term in eq. (84))

$$h_{ij}(r) \sim -\beta \frac{q_i^* q_j^* e^{-\kappa r}}{4\pi E r}, \quad r \longrightarrow \infty. \tag{93}$$

This implies that $w_{ij}(r) \sim q_i^* \psi_j(r)$ and likewise $w_{ij}(r) \sim q_j^* \psi_i(r)$ as required by symmetry. For the interaction between two equal particles we can conclude that

$$w_{ii}(r) \sim \frac{(q_i^*)^2 e^{-\kappa r}}{4\pi E r} > 0 \tag{94}$$

where the inequality follows by the fact that $E > 0$ and the occurrence of the square of the effective charge. Thus two identical particles repel each other at large separations in an electrolyte solution, provided the effective point charge is non–zero and the concentration is such that κ is real. Since we have made no assumptions about the sizes of the particles, this applies, for example, for the interaction between two spherical colloidal particles in a colloidal dispersion.

A more complete picture is obtained if the two leading terms of $w_{ij}(r)$ are included. A detailed analysis [13, 15, 21] yields the following results. Symmetric and asymmetric electrolytes behave quite differently. Let us start with the latter. For asymmetric electrolytes (like 1:2 electrolytes) we have in analogy with the results for the electrostatic potential

$$w_{ij}(r) \sim \frac{q_i^* q_j^* e^{-\kappa r}}{4\pi E r} + \frac{q'^*_i q'^*_j e^{-\kappa' r}}{4\pi E' r}, \quad r \longrightarrow \infty, \tag{95}$$

(provided the concentration is not too high). For two equal particles of species i we see that the coefficients contain the factors $(q_i^*)^2/E$ and $(q'^*_i)^2/E'$ respectively. Since $E > 0$ and $E' < 0$ the first term is repulsive and the second attractive. It is possible for the effective point charge to be zero (this happens at a point of effective charge reversal) and then the second term dominates in the interaction, which therefore turns attractive at long range in this case.

As we have seen in Figs. 3(b) and 4(a), for 1:2 electrolytes we have $\kappa' < 2\kappa$ and when the concentration goes to zero $\kappa' \longrightarrow 2\kappa$ and $E' \longrightarrow \infty$ exponentially fast, which means that the second term in eq. (95) is unimportant for low concentrations. The next term in eq. (95) (not shown) then takes over in practice as the second leading term (it has the decay length $1/(2\kappa)$ and is a squared Yukawa function times a slowly varying function, cf. discussion of eq. (67) in section 3.1.3). At higher concentrations both terms in eq. (95) are important, just like we found for $\phi^0(r)$. This is the case in a quite wide range at intermediate concentrations where κ and κ' have similar magnitudes. The second term is then particularly important for correlations involving a divalent ion since, as we have seen in fig. 4(b), q'^*_2 can be quite large. At even higher concentrations, after the point of merger of κ and κ', the two terms in eq. (95) combine into an exponentially damped oscillatory function.

The ion–ion correlation functions $h_{ij}(r)$ can be combined linearly into charge–charge, charge–density and density–density corrrelations [8]. For example, for a binary electrolyte the latter is given by $(n_+^2 h_{++} + 2n_+n_-h_{+-} + n_-^2 h_{--})/(n_+ + n_-)^2$. For an asymmetric electrolyte one can show [21, 42] that all three combinations normally decay with the same decay lengths $1/\kappa$ for the leading term and $1/\kappa'$ for the second leading term like in eq. (95). (Exceptions can occur at isolated points e.g the critical point.)

Let us now turn to completely symmetric binary electrolytes (i.e. both size *and* charge symmetric), which behave quite differently. In such systems the charge–density correlations are identically zero due to symmetry and the charge–charge and density–density correlations are decoupled to linear order and decay with different decay lengths $1/\kappa$ and $1/b$ respectively. In this case eq. (95) is replaced by (provided the concentration is not too high)

$$w_{ij}(r) \sim \frac{q_i^* q_j^* e^{-\kappa r}}{4\pi E r} + \frac{\tau_i \tau_j e^{-br}}{4\pi D r}, \qquad r \longrightarrow \infty, \tag{96}$$

where τ_l ($l = i$ and j), D and b are constants. The second term, which gives the leading term in the density–density correlations, comes from h_{ij}^0 in eq. (82). This term is cancelled in the charge–charge correlations and therefore it does not contribute to the charge density and the electrostatic potential. There is also a κ' term like in eq. (95), but κ' is usually much larger than b so this term decays very quickly. It is, however, possible at high concentrations that $\kappa' < b$, in which case the κ' term is the second leading one in $w_{ij}(r)$ for large r. Eq. (96) must then be complemented with such a term.

From symmetry we have $q_+^* = -q_-^* = q^*$ and $\tau_+ = \tau_- = \tau$, which means that we can write eq. (96) as

$$w_{+\pm}(r) \sim \pm\frac{(q^*)^2 e^{-\kappa r}}{4\pi E r} + \frac{\tau^2 e^{-br}}{4\pi D r}, \qquad r \longrightarrow \infty, \tag{97}$$

(valid if the concentration is not too high). The values of κ, b and κ' are plotted in fig. 3(a) for a 1:1 aqueous electrolyte. When the concentration goes to zero $b \longrightarrow 2\kappa$ and $D \longrightarrow \infty$ exponentially fast. This behavior is general for all completely symmetric electrolytes [13] and implies that the second term of eq. (97) is unimportant at low concentrations. The next term of eq. (97) (not shown) then is in practice the second leading term.

It is of the same kind as that mentioned above for eq. (95) (approximately a squared Yukawa function) and contributes only to the density–density correlations. (Note the similarity in behavior of b and D for the current case and κ' and E' for 1:2 electrolytes at low concentrations.)

In fig. 3(a) we see that the b curve crosses the κ' curve close to the transition to oscillatory behavior where κ and κ' are similar. After the cross–over $b > \kappa'$ which implies that a κ' term is the second leading one in $w_{ij}(r)$ rather than the b term as mentioned above. When the electrostatic coupling is increased, e.g. for 2:2 aqueous electrolytes [15], the b values are smaller at high concentrations and then the b curve may cross the κ curve instead. After such a cross–over the b term in eq. (96) is the leading term in $w_{ij}(r)$ for large r instead of the κ term, i.e. the density–density correlations have longer range than the charge–charge correlations. At low concentrations, before this cross–over, $w_{+-}(r) \sim -w_{++}(r)$ for large r, while $w_{+-}(r) \sim w_{++}(r)$ after the cross–over.

3.2.3. *Poisson–Boltzmann-like formulations; ϕ^0 as a Green's function*

Although the formulation of Dressed Ion Theory in Fourier space is compact, elegant and physically appealing, it may hide the links to some approximations commonly used for electrolytes. Here we shall in particular see how DIT can shed some light on the Poisson–Boltzmann approximation.

Poisson's equation for $\psi_j(r)$ is

$$-\varepsilon_s \varepsilon_0 \nabla^2 \psi_j(r) = \rho_j^{\text{tot}}(r) \tag{98}$$

where ρ_j^{tot} is the total charge density of a j ion and its ion cloud (defined in eq. (46)). Let us use the decomposition of the ion cloud in eq. (72) to make the corresponding decomposition of ρ_j^{tot}. By using eqs. (47), (73) and (78) we obtain (as written in ordinary space)

$$\rho_j^{\text{tot}}(r) = \rho_j^0(r) + \int B_0(|\mathbf{r} - \mathbf{r}'|)\psi_j(r')\, d\mathbf{r}'. \tag{99}$$

The first term of the rhs corresponds to ρ^{nonlin} in eq. (72) and is the charge density of the dressed ion. The second term equals ρ^{lin} and gives the part of the charge density of the ion cloud that is not contained in the dress of the ion. B_0 can be expressed in ρ^0 via eqs. (44) and (47), so by inserting eq. (99) into eq. (98) we obtain

$$-\varepsilon_s \varepsilon_0 \nabla^2 \psi_j(r) + \beta \sum_i q_i n_i \int \rho_i^0(|\mathbf{r} - \mathbf{r}'|)\psi_j(r')\, d\mathbf{r}' = \rho_j^0(r). \tag{100}$$

(The Fourier transform of this equation is equivalent to eq. (85).) Eq. (100) constitutes an *exact version of the Linearized Poisson–Boltzmann (LPB) equation.* To see this, let us investigate how the equation looks like in the DH approximation.

As we have seen, in the DH approximation the ions in the electrolyte are undressed, i.e. $\rho_i^{0,\mathrm{DH}}(r) = q_i\delta^{(3)}(r)$. If we insert this in eq. (100) we obtain after simplification

$$-\varepsilon_s\varepsilon_0\nabla^2\psi_j^{\mathrm{DH}}(r) + \beta\sum_i q_i^2 n_i\psi_j^{\mathrm{DH}}(r) = q_j\delta^{(3)}(r) \tag{101}$$

which is the ordinary LPB equation (the DH equation) with a source term (the rhs). For $r > 0$ this equation can be written $-\nabla^2\psi_j^{\mathrm{DH}}(r) + \kappa_{\mathrm{D}}^2\psi_j^{\mathrm{DH}}(r) = 0$, which is the appropriate DH equation for a j ion with no size (a point ion) at the origin. To obtain the corresponding result for a central ion with finite size from eq. (100) in the DH approximation, we have to modify the argument slightly.

For a central j ion with a non–zero size a the DH equation should be applied outside the core region ($r > a$), while in the electrolyte free region inside the core the Laplace equation, $\nabla^2\psi_j^{\mathrm{DH}}(r) = 0$, holds for $0 < r < a$. To obtain this result from eq. (100) one has to acknowledge that the central ion is treated differently from the ions in the surrounding ion cloud in the DH theory. We take $\rho_i^{0,\mathrm{DH}}(r) = q_i\delta^{(3)}(r)$ for all ions in the surrounding electrolyte (as before) and insert this in the integral in eq. (100). For the central ion (rhs in eq. (100)) we take $\rho_j^{0,\mathrm{DH}}(r) = q_i\delta^{(3)}(r) + \varepsilon_s\varepsilon_0\kappa_{\mathrm{D}}^2\psi_j^{\mathrm{DH}}(r)$ inside the core and zero outside. The role of the added term is to cancel the second term of eq. (100) in the core region, so the Laplace equation is obtained there (for $0 < r < a$). Outside the core eq. (101) applies as before. Thereby we have obtained the appropriate equations both inside and outside the core.

Furthermore, if this $\rho_j^{0,\mathrm{DH}}(r)$ for the central ion is inserted in eq. (87) one obtains (after some calculus)

$$q_j^* = \frac{q_j e^{\kappa_{\mathrm{D}}a}}{1+\kappa_{\mathrm{D}}a} \qquad \text{(DH, central ion only)} \tag{102}$$

which is the appropriate prefactor in ψ_j^{DH} for an ion with diameter a and we can write $\psi_j^{\mathrm{DH}}(r) = q_j^*\phi^{0,\mathrm{DH}}(r)$ [with $\phi^{0,\mathrm{DH}}$ as given by eq. (61)]. This is identical to the standard result of the DH theory. Thus a central ion with nonzero size has a dress and an effective point charge $q_j^* \neq q_j$ in the DH approximation. Now, it is the ions in the surrounding electrolyte that determine the screening properties, so if q_i^* would be different from q_i for the ions in the ion cloud too, eq. (89) would give $\kappa \neq \kappa_{\mathrm{D}}$. However, we treat the ions in the ion cloud as point ions and therefore $q_i^* = q_i$ for them in the DH approximation. If we insert $q_i^* = q_i$ in eq. (89) it reduces to eq. (56) and we have $\kappa = \kappa_{\mathrm{D}}$. Thus *it is the inconsistent treatment of the central* *ion and the ions in the ion cloud that make $\kappa = \kappa_D$ in the DH theory*, while $\kappa \neq \kappa_D$ in more accurate theories that treats all ions in the same manner.

MSA is one of the simplest examples of an approximation that is consistent in this respect and for ions with size it consequently predicts that $\kappa \neq \kappa_D$. Since the DH approximation is the same as MSA for point ions, MSA can be described as a natural generalization of the DH theory for ions with size and it arises when all ions are treated in the same manner. However, since MSA is a linear approximation it does not give the correct leading terms for the deviation of κ from κ_D shown in eq. (66). In MSA the leading term is, in fact, proportional to the density (i.e. proportional to Λ^2). As we have seen this has little consequence in practice for 1:1 aqueous electrolytes at room temperature, but cannot be ignored at higher electrostatic coupling and MSA then ceases to be a reasonable approximation for bulk electrolytes.

It may appear odd that there are contributions to the dress inside the core region in this example. This is however a general feature of DIT and is a consequence of the condition that $\rho_j^{\text{cloud}}(r) = 0$ in the core. We can see this easily from eq. (99). The electrostatic potential $\psi_j(r')$ and hence the value of the integral is nonzero inside the core. Therefore, $\rho_j^0(r)$ must be equal to minus the integral in the core region (except at $r = 0$ where there is a delta function), otherwise $\rho_j^{\text{tot}}(r)$ would not be zero there. To have a dress in the core is accordingly the manner in which DIT assures that the core condition is satisfied. Note that this contribution to the dress is linear in the potential, which explains why a linear theory like the DH approximation can imply a nonzero dress.

Let us return to the exact eq. (100), which has $\rho_j^0(r)$ as the source term. As seen explicitly in eq. (80), $\rho_j^0(r)$ acts as a source when the proper screened Coulomb potential ϕ^0 is used to calculate ψ_j. In fact ϕ^0 is the *Green's function* associated with eq. (100), i.e. it satisfies

$$-\varepsilon_s\varepsilon_0\nabla^2\phi^0(r) + \beta\sum_i q_i n_i \int \rho_i^0(|\mathbf{r}-\mathbf{r}'|)\phi^0(r')\,d\mathbf{r}' = \delta^{(3)}(r), \qquad (103)$$

which is easily verified by taking the Fourier transform. Eq. (80) simply shows the standard way to form the solution for the source $\rho_j^0(r)$ using the Green's function. These properties of ϕ^0 emphasizes its role as being the propagator of electrical disturbances in the electrolyte solution. Likewise, the DH screened Coulomb potential, $\phi^{0,\text{DH}}$ (eq. (61)), is the Green's function associated with eq. (101) – a well known fact.

We have now explored the connection between DIT and the DH approximation (the LPB approximation). It is of interest to see if the full (non–linear) PB approximation has an equally interesting connection. Unfortunately it does not, but DIT nevertheless shed some light on some aspects of the PB approximation.

Let us assume that a j ion is the central ion. In the PB approximation the pair distribution function is then approximated as follows

$$g_{ij}(r) = \begin{cases} e^{-\beta q_i \psi_j(r)}, & r \geq a_{ij} \\ 0 & , \quad r < a_{ij} \end{cases} \qquad \text{(PB)} \qquad (104)$$

with $\psi_j(r)$ calculated according to eq. (98) with eqs. (43) and (46) inserted. Thus we have taken (cf. Eq. (91)) $w_{ij}(r) = q_i \psi_j(r)$ for $r \geq a_{ij}$. From eq. (80) follows that this can be written

$$w_{ij}(r) = q_i[\phi^0 * \rho_j^0(r)], \quad r \geq a_{ij} \qquad \text{(PB)} \qquad (105)$$

with all quantities calculated using the approximation (104) [ϕ^0 is the same as in the DH approximation]. Here we clearly see an intrinsic inconsistency. We should have $w_{ij}(r) = w_{ji}(r)$, but this is not generally fulfilled in the PB approximation – a well known fact. The central ion and the ions in the ion cloud are treated differently. Eq. (105) implies that the former ion is treated as a dressed ion with charge density ρ_j^0 while latter are undressed point ions, just as they are in the DH case. As a consequence $\kappa = \kappa_{\text{D}}$ in this case too. Since the central ion has a dress, its effective point charge $q_j^* \neq q_j$ in the PB approximation. (There is, however, no secondary effective charge since there is no decay parameter κ' in this case.)

A symmetrical version of eq. (105) would be [16]

$$w_{ij}(r) = \rho_i^0 * \phi^0 * \rho_j^0(r), \quad r \geq a_{ij} \qquad \text{(Nonlocal PB)} \qquad (106)$$

in which all ions are dressed and are treated in the same way. Since this implies that $w_{ij}(r) = \rho_i^0 * \psi_j^0(r)$, the interaction between the i ions in the ion cloud and the potential from the central ion is non–local, while it is local in the PB approximation where $w_{ij}(r) = q_i \psi_j^0(r)$. Therefore, the approximation (106) may be called the *Nonlocal PB approximation.*

In the literature there exist several other symmetrical generalizations to the PB approximation, but they are entirely different from eq. (106) (see e.g. [43, 44]). As an example of the latter we may take $w_{ij}(r) = [q_i\psi_j(r) + q_j\psi_i(r)]/2$ which is one of the variants of the so–called *Symmetrical PB approximation.* Obviously, it does not deal with the issue of treating all ions in the same manner in a proper fashion. In the *Symmetrical Modified PB approximation* [44] there are additional contributions from the so–called fluctuation potential (due to triplet correlations) and a core contribution.

The exact $w_{ij}(r)$ can be determined from eq. (23). Using eqs. (36) and (82) we obtain

$$w_{ij}(r) = \rho_i^0 * \phi^0 * \rho_j^0(r) + w_{ij}^{\text{corr}}(r), \quad r \geq a_{ij} \qquad (107)$$

where

$$w_{ij}^{\mathrm{corr}}(r) = k_B T[c_{ij}^0(r) - h_{ij}^0(r) - e_{ij}^{(2)}(r)]. \tag{108}$$

We may consider w_{ij}^{corr} as the exact correction term to eq. (106) and it contains all contributions to w_{ij} that are not propagated by ϕ^0. Due to the occurrence of the bridge function one has in practice to resort to some approximation to evaluate w_{ij}^{corr} (cf. section 2.3) or use simulations. Incidentally, we note that the HNC approximation is to take $w_{ij}^{\mathrm{corr}} = k_B T[c_{ij}^0 - h_{ij}^0]$. If the ions are not charged hard spheres but have other short ranged interactions, eq. (107) is simply replaced by

$$w_{ij}(r) = u_{ij}^{\mathrm{short}}(r) + \rho_i^0 * \phi^0 * \rho_j^0(r) + w_{ij}^{\mathrm{corr}}(r) \tag{109}$$

where $u_{ij}^{\mathrm{short}}(r)$ is the total non–electrostatic part of the interaction potential (including the hard core potential, if any). Eq. (109) is valid for all r values. (All functions have, of course, also an implicit dependence on u_{ij}^{short} for all i and j.)

To conclude this section, it is of some interest to note that one can show that the Nonlocal PB approximation gives the correct leading terms in the exact expansions (66) and (90) for $\kappa/\kappa_{\mathrm{D}}$ and q_j^*/q_j (as mentioned above this is also the case for the HNC approximation). The PB approximation on the other hand only gives the correct leading terms for q_j^*/q_j of the central ion, while it predicts $\kappa/\kappa_{\mathrm{D}} = 1$. As mentioned earlier, linear theories do not give these terms which originate from non–linear electrostatic effects.

4. Electrical Double Layers; Dressed Surfaces

The theory presented in section 3 is restricted to bulk electrolytes, but since it is exact it applies also to highly asymmetrical cases like colloidal dispersions of spherical particles. Therefore, it is valid for spherical double layers formed by the ion cloud around a large spherical particle. When the colloidal particles have non–zero concentration they contribute to the screening of the electrostatic potential as expressed by $\phi^0(r)$; for example, their effective point charges enter in eq. (89) for κ. In this section we shall, however, be concerned with a finite number of large particles immersed in an electrolyte solution that contains only small ions. The concentration of large particles is therefore zero and only the small ions contribute to the properties of $\phi^0(r)$. The large particles can have arbitrary geometry and we will consider the double layers around them. We assume throughout that the system is in equilibrium with a bulk electrolyte solution of known composition. The results in section 3 are quite easy to generalize to double layers and we shall only give a brief summary of the most important results.

The interactions between the ions and a large particle can be included in the external potential (cf. section 2). The electrolyte is inhomogeneous near the particle; this inhomogeneity constitutes the diffuse part of the double layer. Far away from the particle we have a homogeneous bulk electrolyte. We may regard the action of the external potential as a strong disturbance of the bulk electrolyte that results in the double layer. Thus, in analogy to what we found above, the propagation of the electrostatic disturbance from the particle is governed by the proper screened Coulomb potential $\phi^0(r)$ of the bulk electrolyte. The immersed large particle (I) acquires a diffuse effective charge (charge distribution of dressed particle) $\rho_{\rm I}^0(\mathbf{r})$ and the electrostatic potential $\psi_{\rm I}(\mathbf{r})$ is given by [cf. eq. (80)]

$$\psi_{\rm I}(\mathbf{r}) = \int \phi^0(|\mathbf{r}-\mathbf{r}'|)\rho_{\rm I}^0(\mathbf{r}')\,d\mathbf{r}' \tag{110}$$

which is equivalent to [cf. eq. 100)]

$$-\varepsilon_{\rm s}\varepsilon_0\nabla^2\psi_{\rm I}(\mathbf{r}) + \beta\sum_i q_i n_i \int \rho_i^0(|\mathbf{r}-\mathbf{r}'|)\psi_{\rm I}(\mathbf{r}')\,d\mathbf{r}' = \rho_{\rm I}^0(\mathbf{r}) \tag{111}$$

where $\rho_i^0(r)$ is the dressed charge distribution of the small ions of species i in the bulk electrolyte.

The interaction free energy (potential of mean force) between two large particles, I and J, immersed in the bulk electrolyte is given by [cf. eq. (109)]

$$w_{\rm IJ}(\mathbf{R}_{\rm IJ}) = u_{\rm IJ}^{\rm short}(\mathbf{R}_{\rm IJ}) + \rho_{\rm I}^0 * \phi^0 * \rho_{\rm J}^0(\mathbf{R}_{\rm IJ}) + w_{\rm IJ}^{\rm corr}(\mathbf{R}_{\rm IJ}) \tag{112}$$

where $\mathbf{R}_{\rm IJ}$ is a separation vector of the two particles [for simplicity in notation we assume that they have fixed orientations in space], $u_{\rm IJ}^{\rm short}$ is the nonelectrostatic interaction between the particles and $w_{\rm IJ}^{\rm corr}$ is defined analogously to eq. (108) [it is a short ranged function]. The functions $\rho_L^0(\mathbf{r})$, L = I or J, only depends on particle L and the species with non–zero concentrations, i.e. the ions of the bulk electrolyte. Thus $\rho_L^0(\mathbf{r})$ is the diffuse effective charge of a single particle L immersed in the electrolyte. The leading term in $w_{\rm IJ}(\mathbf{R}_{\rm IJ})$ for large $|\mathbf{R}_{\rm IJ}|$ originates from the second term in the rhs of eq. (112). It has the same decay length as $\phi^0(r)$, i.e. $1/\kappa$, at least if the electrolyte concentration is not too high. The magnitude of the leading term is proportional to the product of the effective charges of particles I and J, which are determined by $\rho_{\rm I}^0$ and $\rho_{\rm J}^0$ respectively. We see that the formalism is analogous to that of section 3.

To be specific, let us for simplicity restrict ourselves to planar double layers, either the double layer outside an infinitely large wall with uniform surface charge density $\sigma_{\rm I}$ or a double layer system between two such walls with surface charge densities $\sigma_{\rm I}$ and $\sigma_{\rm J}$ in equilibrium with a bulk electrolyte. Let $\psi_{\rm I}(x)$ denote the electrostatic potential at distance x from a single charged wall (I). From eq. (110) follows that the Fourier transform of $\psi_{\rm I}(x)$ satisfies [cf. eqs. (79) and (85)]

$$\overline{\psi}_{\rm I}(k) = \overline{\rho}_{\rm I}^{0}(k)\hat{\phi}^{0}(k) = \frac{\overline{\rho}_{\rm I}^{0}(k)\hat{\phi}(k)}{\hat{\varepsilon}(k)} \tag{113}$$

where $\hat{\varepsilon}(k)$ is the dielectric function of the bulk electrolyte and where we use the notation $\overline{f}_1(k)$ for the one–dimensional and $\hat{f}_2(k)$ for the three–dimensional Fourier transforms of the functions $f_1(x)$ and $f_2(r)$ respectively. The diffuse effective charge of the surface is $\rho_{\rm I}^{0}(x)$, i.e. the charge density of the "dressed surface".

The leading singularities of $\overline{\psi}_{\rm I}(k)$ in eq. (113) originate from the leading zeros of $\hat{\varepsilon}(k)$. This implies (by residue analysis) that

$$\psi_{\rm I}(x) \sim \frac{\sigma_{\rm I}^{*}}{\kappa E}e^{-\kappa x} + \frac{\sigma'^{*}_{\rm I}}{\kappa' E'}e^{-\kappa' x}, \quad x \longrightarrow \infty \tag{114}$$

where κ, κ', E and E' are the same quantities as previously (they are properties of the bulk electrolyte),

$$\sigma_{\rm I}^{*} = \frac{\overline{\rho}_{\rm I}^{0}(\mathrm{i}\kappa)}{2} \tag{115}$$

and $\sigma'^{*}_{\rm I} = \overline{\rho}_{\rm I}^{0}(\mathrm{i}\kappa')/2$ (the factor of 1/2 originates from the fact that the electrolyte is located only on one side of the infinitely large surface [19]). We interpret $\sigma_{\rm I}^{*}$ as the (primary) *effective surface charge density* of the wall surface. This quantity tells how strongly the charge distribution close to the surface influences the potential far away from the wall or, expressed in other words, it gives the strength of the coupling between the diffuse effective charge $\rho_{\rm I}^{0}(x)$ of the wall and the propagation mode of ϕ^0 with decay length $1/\kappa$. Likewise, $\sigma'^{*}_{\rm I}$ is the *secondary effective surface charge density* which is associated with the propagation mode that has decay length $1/\kappa'$. We can compare eq. (114) with the corresponding result in the LPB approximation

$$\psi_{\rm I}^{\rm LPB}(x) = \frac{\sigma_{\rm I}}{\kappa_{\rm D}\varepsilon_{\rm s}\varepsilon_0}e^{-\kappa_{\rm D}x}, \quad x > 0 \tag{116}$$

and in this case the effective surface charge is equal to the bare surface charge. In the general case, eq. (116) gives the correct decay behavior for large x in the limits $\sigma_{\rm I} \longrightarrow 0$ *and* electrolyte concentration $\longrightarrow 0$. In all other cases it is an approximation.

Let us consider the (non–linear) PB approximation. As we have seen, the "central particle" (in this case the wall) has a dress in this case and therefore the effective surface charge density will be different from the bare one. We have

$$\psi_{\rm I}^{\rm PB}(x) \sim \frac{\sigma_{\rm I}^{*\,\rm PB}}{\kappa_{\rm D}\varepsilon_{\rm s}\varepsilon_0}e^{-\kappa_{\rm D}x}, \quad x \longrightarrow \infty. \tag{117}$$

(The analogous definition of an effective charge from the magnitude of the potential in the PB approximation was first suggested by Alexander and co–workers [45] for spherical double layers.) The value of $\sigma^{*\,\mathrm{PB}}$ can be determined from the solution to the PB equation. At least for symmetric electrolytes and for electrolytes with ionic charge ratios 1:2 (i.e. for 1:2, 2:4, etc. electrolytes) one can derive explicit expressions for $\sigma^{*\,\mathrm{PB}}$ as functions of the surface charge σ and the bulk electrolyte concentration. In the both cases we have

$$\frac{\sigma^{*\,\mathrm{PB}}}{\sigma_{\mathrm{Ref}}} = F\left(\frac{\sigma}{\sigma_{\mathrm{Ref}}}\right) \tag{118}$$

where $\sigma_{\mathrm{Ref}} = \beta q_{\mathrm{low}}/(\varepsilon_{\mathrm{s}}\varepsilon_0\kappa_{\mathrm{D}})$ is a reference surface charge that depend on properties of the bulk electrolyte, q_{low} is the ionic charge of the species with the lowest valency and F is a function that *only* depends on the ratio of the ionic charges of the binary electrolyte. For symmetric electrolytes F is given by

$$F(t) = \frac{8}{t}\left(\left[\frac{t^2}{4}+1\right]^{1/2} - 1\right). \tag{119}$$

For electrolytes with charge ratios 1:2 $F(t)$ is a more complicated, but still elementary function [46] that can be obtained from the analytic solution [47] of the PB equation for this case. It follows that if $\sigma^{*\,\mathrm{PB}}/\sigma_{\mathrm{Ref}}$ is plotted as a function of $\sigma/\sigma_{\mathrm{Ref}}$ one obtains a universal curve for all planar double layers with symmetric electrolyte irrespectively of the ion concentration, valency (i.e. 1:1, 2:2 etc. electrolytes), temperature and ε_{s}. The same applies for electrolytes with charge ratios 1:2, but the universal curve is different. These curves are plotted in fig. 5.

An important property of the PB approximation is the well–known fact that the potential far away from the surface is insensitive to the value of the surface charge when the latter is large. This saturation phenomenon, which is due to the build up of a high concentration of counterions near the surface, is reflected in the properties of $F(t)$ for large t values. For symmetric electrolytes eqs. (118) and (119) yield $|\sigma^{*\,\mathrm{PB}}/\sigma_{\mathrm{Ref}}| \longrightarrow 4$ when $\sigma \longrightarrow \pm\infty$. For 1:2 electrolytes the corresponding limiting value depends on the sign of the surface charge, i.e. whether the counterions to the surface charge are monovalent or divalent. In the former case $|\sigma^{*\,\mathrm{PB}}/\sigma_{\mathrm{Ref}}| \longrightarrow 6$ and in the latter $|\sigma^{*\,\mathrm{PB}}/\sigma_{\mathrm{Ref}}| \longrightarrow 6(\sqrt{3}-1)/(\sqrt{3}+1) \approx 1.608$ when $|\sigma| \longrightarrow \infty$.

If we return to the general, exact case $\sigma^{*}/\sigma_{\mathrm{Ref}}$ will *not* be a universal function of $\sigma/\sigma_{\mathrm{Ref}}$ and the saturation, if any, will not be maintained for very large values of $|\sigma|$. The levelling off for $|\sigma^{*\,\mathrm{PB}}/\sigma_{\mathrm{Ref}}|$ when $|\sigma| \longrightarrow \infty$ is possible since the ions in the double layer are assumed to be point–like in the

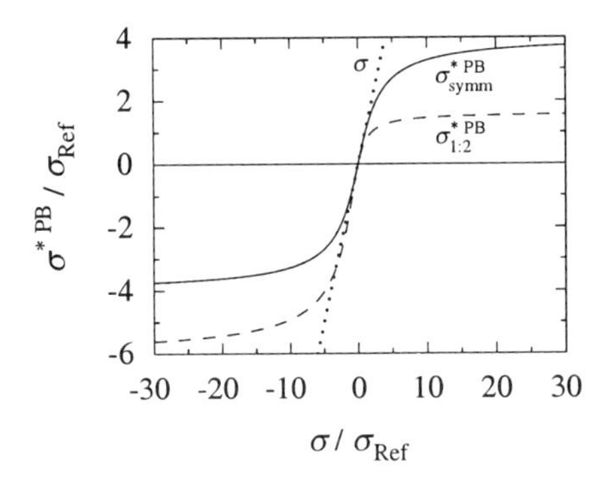

Figure 5. The effective surface charge density $\sigma^{*\,\mathrm{PB}}$ predicted by the PB approximation as functions of the bare surface charge density σ (both in units of σ_{Ref}, see text) for a planar surface in contact with charge symmetric electrolytes (full line) and electrolytes with ionic charge ratio 1:2 (dashed line). The dotted line shows the bare surface charge density σ. For 1:2 electrolytes a negative $\sigma/\sigma_{\mathrm{Ref}}$ implies that the counterions are monovalent and for a positive ratio they are divalent.

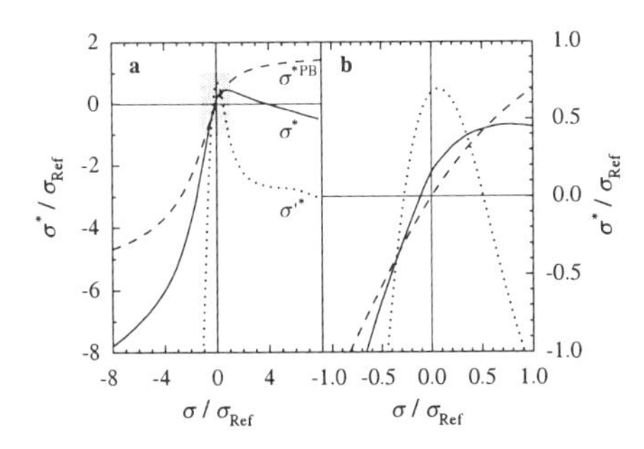

Figure 6. (a) The primary effective surface charge density σ^* (full line) and the secondary effective surface charge density σ^* (dotted line) from AHNC calculations plotted as functions of the bare surface charge density σ (all in units of σ_{Ref}) for a planar surface in contact with a 65 mM 1:2 aqueous electrolyte with a = 4.6 Å (ε_s = 78.7, T = 298 K). For reference $\sigma^{*\,\mathrm{PB}}$ is also shown (dashed line). (b) A detailed view of the part of (a) indicated by the grey rectangle. (The data are taken from [23].)

PB approximation, so there is no upper bound of the density of counterions near the surface. Therefore, an increase in surface charge can always be compensated by a corresponding increase in counterion concentration in the immediate neighborhood of the surface. For ions with finite size the ion density cannot be higher than what corresponds to close packing, so an increase in σ will eventually lead to the formation of more than one layer of counterions near the wall. Thereby the saturation will be less pronounced or, for large counterions, even virtually non–existent.

Fig. 6 shows σ^* and σ'^* calculated in the AHNC approximation for a planar surface with different surface charge densities and in contact with a 65 mM 1:2 electrolyte solution with $a = 4.6$ Å. The PB prediction $\sigma^{*\,\mathrm{PB}}$ is also plotted for reference. Let us start with a discussion of σ^*. For monovalent counterions (negative $\sigma/\sigma_{\mathrm{Ref}}$) we see that the saturation of the effective surface charge mainly follows the qualitative behavior predicted by the PB approximation, but that $|\sigma^*| > |\sigma^{*\,\mathrm{PB}}|$ for large $|\sigma|$, i.e. the shielding of the surface charge is less efficient than in the PB case. This is due to the packing constraints near the surface as discussed above. The same picture holds in systems with 1:1 electrolytes [18] and the shielding becomes much less efficient when the counterion size is increased.

For small σ the qualitative picture for 1:2 electrolytes (see fig. 6(b)) is completely different from that predicted in the PB approximation. While $\sigma^{*\,\mathrm{PB}} = 0$ when $\sigma = 0$, this is not the case in the accurate calculations and we see that $\sigma^*/\sigma_{\mathrm{Ref}} > 0$ when $\sigma = 0$. Thus the uncharged surface has an effective charge and thereby gives rise to an electrostatic potential with the same sign as the monovalent ions. This is a consequence of the ion–ion correlations in the double layer close to the surface, which are entirely neglected in the PB approximation. The divalent ions in contact with the surface feel a stronger attraction to its anisotropic ion cloud (in the direction away from the surface) than a monovalent ion does, which results in higher charge density from monovalent ions than from divalent ones close to the surface. This gives rise to non–zero electrostatic potential and effective surface charge. When $\sigma/\sigma_{\mathrm{Ref}}$ is decreased below zero $\sigma^*/\sigma_{\mathrm{Ref}}$ must pass through zero (since it is negative for large negative $\sigma/\sigma_{\mathrm{Ref}}$), i.e. there exists a point of zero effective surface charge for a small negative value of $\sigma/\sigma_{\mathrm{Ref}}$, see fig. 6(b).

A further qualitative difference compared to the PB case occurs for divalent counterions, i.e. positive $\sigma/\sigma_{\mathrm{Ref}}$ values in fig. 6. Instead of levelling off as $\sigma^{*\,\mathrm{PB}}/\sigma_{\mathrm{Ref}}$ does for large $\sigma/\sigma_{\mathrm{Ref}}$, the accurate $\sigma^*/\sigma_{\mathrm{Ref}}$ first increases, passes through a maximum and then changes sign when $\sigma/\sigma_{\mathrm{Ref}}$ is increased. Thus we have a reversal in sign of the effective surface charge, i.e. the sign of σ^* becomes opposite to that of σ. Then the electrostatic potential far away from the wall has according to eq. (114) opposite sign to the surface charge. This charge reversal is due to an overcompensation of the surface charge by counterion charge in the double layer closest to the wall, i.e. the amount of counterions there is more than what is needed to neutralize the surface charge. The same kind of charge inversion occurs in systems with 2:2 electrolytes [18]; it is the valency of the counterions that is most important. Effective surface charge reversals in presence of divalent counterions in agreement with the theoretical predictions of AHNC theory have been observed experimentally [48].

The potential of mean force between a surface (I) and a j ion at the distance x from it decays like

$$w_{\mathrm{I}j}(x) \sim q_j^* \psi_{\mathrm{I}}(x) \sim \frac{q_j^* \sigma_{\mathrm{I}}^*}{\kappa E} e^{-\kappa x}, \quad x \longrightarrow \infty. \tag{120}$$

Thus when the effective surface charge has reversed sign, the coions of the surface charge are attracted towards the surface for large x (provided, of course, q_j^* has the same sign as q_j, which is normally the case for small ions).

In fig. 5 we see that the secondary effective charge, σ'^*, follows the same general pattern as σ^*, but the changes occur sooner and more rapidly as $\sigma/\sigma_{\mathrm{Ref}}$ is varied. At a point of charge reversal $\sigma_{\mathrm{I}}^* = 0$ and then the second term in eq. (114) determines the sign of $\psi_{\mathrm{I}}(x)$ for large x (the sign is opposite to that of σ'^*_{I} since E' is negative). Otherwise it is the interplay between the two terms in eq. (114) that gives the behavior of $\psi_{\mathrm{I}}(x)$ away from the surface; in fact the sum of these two terms is close to the value of the potential for x values greater than a couple of ionic diameters away from the surface [49]. Thus the asymptotic formulae are not only valid for very large x values, but can give good estimates of the functions even quite close to the surface.

From eq. (112) follows (by residue analysis) that the interaction free energy between two planar walls, I and J, decays for large surface separations R like

$$w_{\mathrm{IJ}}(R) \sim \frac{2\sigma_{\mathrm{I}}^* \sigma_{\mathrm{J}}^*}{\kappa E} e^{-\kappa R}, \quad R \longrightarrow \infty. \tag{121}$$

This can be compared with the corresponding result in the PB approximation

$$w_{\mathrm{IJ}}^{\mathrm{PB}}(R) \sim \frac{2\sigma_{\mathrm{I}}^{*\,\mathrm{PB}} \sigma_{\mathrm{J}}^{*\,\mathrm{PB}}}{\kappa_{\mathrm{D}} \varepsilon_{\mathrm{s}} \varepsilon_0} e^{-\kappa_{\mathrm{D}} R}, \quad R \longrightarrow \infty. \tag{122}$$

In both cases the interaction between two identical surfaces are repulsive for large R (provided the primary effective charge is non–zero). In the PB case this also applies to unequal surfaces with equal sign of charge, while in the exact case it can happen that they attract provided one surface has reversed its effective surface charge and the other has not.

Acknowledgements

I would like to thank my collaborators in various phases of the development and application of DIT, namely D. John Mitchell, Jonathan Ennis–King and Johan Ulander.

References

1. Mayer, J. E. (1968) *Equilibrium Statistical Mechanics*, Pergamon, Oxford.
2. Hill, T. L. (1956) *Statistical Mechanics*, McGraw–Hill, New York.
3. Adelman, S. A. (1976) The Effective Direct Correlation Function, an Approach to the Theory of Liquid Solutions: A New Definition of the Effective Solute Potential, *Chem. Phys. Lett.* **38**, 567–570.
4. Adelman, S. A. (1976) The Effective Direct Correlation Function: An Approach to the Theory of Liquids, *J. Chem. Phys.* **64**, 724–731.
5. McQuarrie, D. A. (1976) *Statistical Mechanics*, Harper & Row, New York.
6. Hansen, J. P. and McDonald, I. R. (1986) *Theory of Simple Liquids*, Academic Press, London.
7. Rowlinson, J. S. and Widom, B. (1989) *Molecular Theory of Capillarity*, Clarendon Press, Oxford.
8. March, N. H. and Tosi, M. P. (1984) *Coulomb Liquids*, Academic Press, London.
9. Martynov, G. A. (1992) *Fundamental Theory of Liquids; Method of Distribution Functions*, Adam Hilger, Bristol
10. Henderson, D (ed.) (1992) *Fundamentals of Inhomogeneous Fluids*, Marcel Dekker, New York.
11. Kjellander, R. and Mitchell, D. J. (1992) An Exact but Linear and Poisson–Boltzmann–like Theory for Electrolytes and Colloid Dispersions in the Primitive Model, *Chem. Phys. Letters* **200**, 76–82.
12. Attard, P. (1993) Asymptotic Analysis of Primitive Model Electrolytes and the Electrical Double Layer, *Phys. Rev. E* **48**, 3604–3621.
13. Kjellander, R. and Mitchell, D. J. (1994) Dressed–ion Theory for Electrolyte Solutions – a Debye–Hückel–like Reformulation of the Exact Theory for the Primitive Model, *J. Chem. Phys.* **101**, 603–626.
14. Leote de Carvalho, R. J. F. and Evans, R. (1994) The Decay of Correlations in Ionic Fluids, *Mol. Phys.* **83**, 619–654.
15. Ennis, J., Kjellander, R. and Mitchell, D. J. (1995) Dressed Ion Theory for Bulk Symmetric Electrolytes in the Restricted Primitive Model, *J. Chem. Phys.* **102**, 975–991.
16. Kjellander, R. (1995) Modified Debye–Hückel Approximation with Effective Charges: An Application of Dressed Ion Theory for Electrolyte Solutions, *J. Phys. Chem.* **99**, 10392–10407.
17. Attard, P. (1995) Ion Condensation in the Electric Double Layer and the Corresponding Poisson–Boltzmann Effective Surface Charge, *J. Phys. Chem.* **99**, 14174–14181.
18. Ennis, J., Marcelja, S and Kjellander, R. (1996) Effective Surface Charge for Symmetric Electrolytes in the Primitive Model Double Layer, *Electrochim. Acta* **41**, 2115–2124.

19. Kjellander, R. and Mitchell, D. J. (1997) Dressed Ion Theory for Electric Double Layer Structure and Interactions; An Exact Analysis, *Mol. Phys.* **91**, 173–188.
20. Kjellander, R. and Ulander, J. (1998) Effective Ionic Charges, Permittivity and Screening Length; Dressed Ion Theory Applied to 1:2 Electrolyte Solutions, *Mol. Phys.* **95**, 495–505.
21. Ulander, J. and Kjellander, R. (1998) Screening and Asymptotic Decay of Pair Distributions in Asymmetric Electrolytes, *J. Chem. Phys.* **109**, 9508–9522.
22. Leote de Carvalho, R. J. F., Evans, R. and Rosenfeld, Y. (1999) Decay of Correlations in Fluids: The One–Component Plasma from Debye–Hückel to the Asymptotic–High–Density Limit, *Phys. Rev. E* **59**, 1435–1451.
23. Kjellander, R. and Ulander, J. (2000) Charge Renormalization and Asymptotic Decay in Classical Coulomb Systems, *J. Phys. IV France* **10**, Pr5–431–436.
24. Kjellander, R. and Ulander, J. (2001) The Decay of Pair Correlation Functions in Ionic Fluids A Dressed Ion Theory Analysis of Monte Carlo Data, *J. Chem. Phys.* **114**, 4893–4904.
25. Stell, G. (1976) Correlation Functions and their Generating Functionals: General Relations with Applications to the Theory of Fluids, in C. Domb and M. S. Green (eds.), *Phase Transitions and Critical Phenomena*, Academic Press, London, **Vol. 5**, pp. 205–258.
26. Choquard, Ph. (1972) Equilibrium Field Theory of Classical Ionic Fluids, *Helv. Phys. Acta* **45**, 913915; Choquard, Ph. and Sari, R. R. (1972) Onset of Short Range Order in a One–Component Plasma, *Phys. Lett.* **40A**, 109–110.
27. Hill, T. L. (1959) Exact Definition of Quasi–Thermodynamic Point Functions in Statistical Mechanics, *J. Chem. Phys.* **30**, 1521–1523.
28. Morita, T. and Hiroike, K. (1960) A New Approach to the Theory of Classical Fluids I, *Progr. Theor. Phys.* **23**, 1003–1027.
29. Kjellander, R. and Marcelja, S (1985) Inhomogeneous Coulomb Fluids with Image Interactions between Planar Surfaces. I, *J. Chem. Phys.* **82**, 2122–2135.
30. Kjellander, R. (1988) Inhomogeneous Coulomb Fluids with Image Interactions between Planar Surfaces, II. On the Anisotropic Hypernetted Chain Approximation, *J. Chem. Phys.* **88**, 7129– 7137. (Erratum (1988) *ibid* **89**, 7649.)
31. Sokolowski, S. (1980) On the Solution of the Non–uniform Percus–Yevick equation, *J. Chem. Phys.* **73**, 3507–3508.
32. Sokolowski, S. (1983) Adsorption of Hard Spheres via the Non–uniform Percus–Yevick Equation, *Mol. Phys.* **49**, 1481–1488.
33. Plischke, M. and Henderson, D. (1986) Density Profiles and Pair Correlation Functions of Lennard–Jones Fluids near a Hard Wall, *J. Chem. Phys.* **84**, 2846–2852.
34. See p. 74 in [8].
35. Stillinger, F. H. and Lovett, R. (1968) Ion–Pair Theory of Concentrated Electrolytes. I. Basic Concepts, *J. Chem. Phys.* **48**, 3858–3868.
36. Stillinger, F. H. and Lovett, R. (1968) Ion–Pair Theory of Concentrated Electrolytes. II. Approximate Dielectric Response Calculation, *J. Chem. Phys.* **48**, 3869–3884.
37. Mitchell, D. J. and Ninham, B. W. (1968) Asymptotic Behaviour of the Pair Distribution Function of a Classical Electron Gas, *Phys. Rev.* **174**, 280–289.
38. Mitchell, D. J. and Ninham, B. W. (1978) Range of the Screened Coulomb Interaction in Electrolytes and Double Layer Problems, *Chem. Phys. Lett.* **53**, 397–399.
39. McBride, A., Kohonen, M. and Attard, P. (1998) The Screening Length of Charge–Asymmetric Electrolytes: A Hypernetted Chain Calculation, *J. Chem. Phys.* **109**, 2423–2428.

40. Stell, G. and Lebowitz, J. L. (1968) Equilibrium Properties of a System of Charged Particles, *J. Chem. Phys.* **48**, 3706–3717.
41. Hall, D. G. (1973) An Attempt to Define Generally Isothermal Differences in the Chemical Potential of Individual Ionic Species in Solution, *J.C.S. Faraday Trans. II* **69**, 1391–1401.
42. Stell, G. (1995) Criticality and Phase Transitions in Ionic Fluids, *J. Stat. Phys.* **78**, 197–238.
43. Shmidt, A. B. (1984) Poisson–Boltzmann Equation for Asymmetrically Charged Electrolytes, *Soviet Electrochem.* **20**, 767–770.
44. Outhwaite, C. W., Molero, M. and Bhuiyan, L. B. (1991) Symmetrical Poisson–Boltzmann and Modified Poisson–Boltzmann Theories, *J. C. S. Faraday Trans.* **87**, 3227–3230.
45. Alexander, S., Chaikin, P. M., Grant, P., Morales, G. J., Pincus, P. and Hone, D. (1984) Charge Renormalization, Osmotic Pressure and Bulk Modulus of Colloidal Crystals: Theory, *J. Chem. Phys.* **80**, 5776–5781.
46. Ulander, J. (1999) *PhD thesis*, Göteborg University, Göteborg, Sweden.
47. Grahame, D. C. (1953) Diffuse Double Layer Theory for Electrolytes of Unsymmetrical Valence Types, *J. Chem. Phys.* **21**, 1054–1060.
48. Kekicheff, P., Marcelja, S., Senden, T. J. and Shubin, V. E. (1993) Charge Reversal Seen in Electrical Double Layer Interaction of Surfaces Immersed in 2:1 Calcium Electrolyte, *J. Chem. Phys.* **99**, 6098–6113.
49. Ulander, J., Greberg, H. and Kjellander, R. (2001) Primary and Secondary Effective Charges for Electrical Double Layer Systems with Asymmetric Electrolytes, *J. Chem. Phys.*, in press

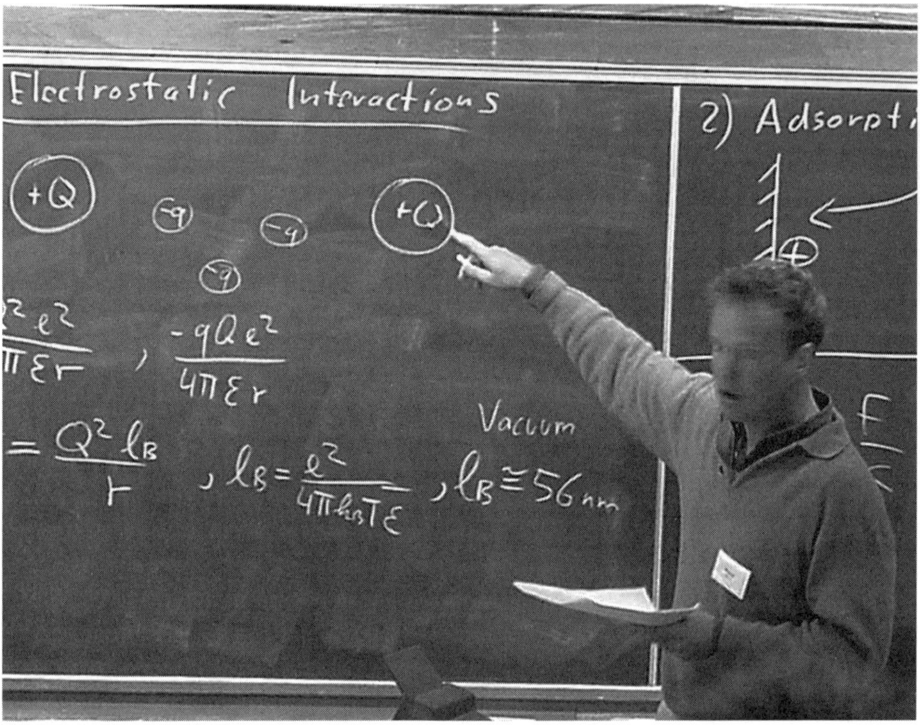

Electrostatic Interactions
Vacuum

FIELD-THEORETIC APPROACHES TO CLASSICAL CHARGED SYSTEMS

ANDRÉ G. MOREIRA and ROLAND R. NETZ
Max Planck Institute of Colloids and Interfaces
14424 Potsdam, Germany

1. Introduction

Electric charges and processes involving electrostatic interactions are abundant in biological, colloidal, and polymeric systems. Realizing that even the ubiquitous van-der-Waals or dispersion interactions are due to locally fluctuating electric fields (or, equivalently, polarization charges), one can justly say that interactions caused and mediated by permanent and induced charges constitute the most prominent factors determining the behavior and properties of all materials at mesoscopic scales[1].

Permanent charges on single molecules, surfaces, or interfaces in aqueous media arise via two routes: The substance can contain dissociable surface groups, which under suitable pH conditions may donate protons (in which case one speaks of *acidic* groups), thereby imparting negative charges to the surface, or accept protons (these are called *basic* groups) and thus produce positive charges on the surface. Secondly, small charged molecules, such as salt ions, can physically or chemically adsorb to a surface, thereby leading to an effective surface charge. In practice, one typically encounters a mixture of these mechanisms, such that the effective charge of a surface is controlled by the distribution of acidic and basic surface groups, solution pH, and bulk concentration of charged solutes. Surfaces or particles with permanent charges interact via Coulomb interactions, which can be much stronger than thermal energy.

[1] At length scales of elementary particles and atomic nuclei, the physics is dominated by the so-called weak and strong interactions. At galactic distances, the only sizable force is the gravitational one. Every-day-life processes are typically governed by a combination of gravitational and electrostatic/electrodynamic (which —according to the remark above— also include dispersion forces) interactions.

C. Holm et al. (eds.), Electrostatic Effects in Soft Matter and Biophysics, 367–408.
© 2001 *Kluwer Academic Publishers. Printed in the Netherlands.*

Induced charges arise via the polarization of atoms, molecules, and macroscopic bodies. For molecules that possess a permanent dipole moment (such as water), the macroscopic polarization contains a large contribution from the orientation of such molecular dipole moments. The interaction between induced charges gives rise to van-der-Waals forces, which act between all bodies and particles, regardless of whether they are charged, or whether they contain permanent dipole moments or not.

Three examples may suffice to demonstrate the diversity of phenomena induced by the presence of permanent and polarization charges:

— Human DNA, the storage medium of all genetic information, is a flexible biopolymer with a total length of roughly $2m$, bearing a total negative charge of about $10^{10}e$ (where e denotes an elementary charge), which is contained inside a nucleus with a diameter of less than $10\mu m$. In addition to the task of confining such a large, strongly charged object in a very small compartment, the DNA is incessantly replicated, repaired, and transcribed, which seems to pose an unsurmountable DNA-packaging problem[17]. Nature has solved this by an ingenious multi-hierarchical structure. On the lowest level, a short section of the DNA molecule, consisting of 146 base pairs (corresponding to a length of roughly $50nm$) is wrapped twice around a positively charged protein (the so-called histone). In experiments[55], it has been shown that a tightly wrapped state is only stable for intermediate, physiological salt concentrations. Theoretically, this is explained by the balance between electrostatic DNA–DNA repulsion (favoring a straight DNA conformation) and the electrostatic DNA–histone attraction, which sensitively depends on the salt concentration[25]. Similar complexes between charged spherical objects and oppositely charged polymers are also studied experimentally in the context of colloid-polymer and micelle-polymer interactions.

— Polymer science and technology have revolutionized the design, fabrication, and processing of modern materials and form an integral part of every-day life. Classical polymer synthesis is based on hydro-carbon chemistry and thus leads to polymers which are typically insoluble in water. In the quest for cheap, environmentally friendly, and non-toxic materials, attention has been shifted to charged polymers, so-called *polyelectrolytes*, since they are typically water-soluble[6, 12, 16]. For some polyelectrolytes, the affinity for water is so high that they are righteously called *super-adsorbing polymers*: They can bind amounts of water in multiple excess of their own weight. This property is put to good use in many practical applications such as diapers.

— Colloids[2] dispersed in aqueous solvents experience mutual attrac-

[2] The term *colloid* refers to any object that is larger than $1nm$ and smaller than a few microns and thus encompasses proteins, polymers, clusters, micelles, and so on.

tions due to van-der-Waals forces[52] or other, solvent-structure-induced forces[38, 28]. They thus tend to aggregate and form large agglomerates[53]. Large aggregates typically sediment, thereby destroying the dispersion. In colloidal science, this process is called coagulation or flocculation, depending on the strength and range of the inter-colloidal forces involved. In many industrial applications (for example dispersion paints, food emulsions such as mayonnaise or milk), stability of a dispersion is a desirable property, in other applications (such as sewage or waste-water treatment) it is not[36]. One way to stabilize a colloidal dispersion against coagulation is to impart permanent charges to the colloids: Typically, similarly charged particles repel each other (although we will study an exception to this rule in Section 4), such that van-der-Waals attraction (which is always stronger than electrostatic repulsion at small distances) cannot induce aggregation[3]. Many structures obtained with charged colloids bear resemblance with atomic structures, but occur on length and time scales that are much easier to observe experimentally. To some extent, colloidal systems can thus be used as models for ordering phenomena on the atomistic scale.

In these examples, electrostatic interactions provide the driving force for the salient features described, which therefore have to be included in any theoretical description. The reduced electrostatic interaction between two point-like charges (throughout this work, all energies are given in units of the thermal energy $k_B T$) can be written as $U(r) = Q_1 Q_2 v(r)$ where

$$v(r) = \frac{e^2}{4\pi\varepsilon_0 k_B T r} \tag{1}$$

is the Coulomb interaction between two elementary charges, Q_1 and Q_2 are the reduced charges in units of the elementary charge e, and ε_0 is the vacuum dielectric constant[4]. The interaction depends only on the distance r between the charges. Electrostatic interactions are additive, therefore the total electrostatic energy of a given distribution of charges results from adding up all pairwise interactions between charges according to Eq.(1). *In principle*, the equilibrium behavior of an ensemble of charged particles (e.g. a salt solution) follows from the partition function, i.e., the weighted sum over all different microscopic configurations, which —via the Boltzmann factor— depends on the electrostatic energy of each configuration. *In practice*, however, this route is complicated for several reasons:

[3] A second method of stabilizing a colloidal dispersion is to graft polymers to the surface of the colloids. If the polymers are under good-solvent conditions, they will swell and inhibit close contacts between two colloids. For this task, charged polymers are ideal, since they strongly swell in water.

[4] Note that the Systéme International (SI) is used, so that the factor 4π appears in the Coulomb law but not in the Poisson equation.

i) The Coulomb interaction, Eq.(1), is very long-ranged, such that many particles are coupled due to their simultaneous electrostatic interactions (as we will see, this is especially true for low densities). Electrostatic problems are therefore typically *many-body problems.* As is well known, even the problem of only three bodies interacting via gravitational potentials (which are analogous to Eq.(1) except that they are always attractive) defies closed-form solutions. To make things worse, even if we consider only two charged particles, the problem effectively becomes a many-body problem, as described below.

ii) In almost all cases, charged objects are dissolved in water. As all molecules and atoms, water is polarizable and thus reacts to the presence of a charge with polarization charges. In addition, and this is by far a more important effect, water molecules carry a permanent dipole moment, and are thus partially oriented in the vicinity of charged objects. The polarization effect of the solvent can to a good approximation[5] be taken into account by introducing an effective dielectric constant ε[35, 8, 13, 43]. Note that for water, $\varepsilon/\varepsilon_0 \approx 80$, so that electrostatic interactions are much weaker in water than in air (or some other low-dielectric solvent).

iii) In all biological and most industrial applications, water contains mobile salt ions. Salt ions of opposite charge are drawn to the charged objects and form a loosely bound counter-ion cloud and thus effectively reduce the charges of the objects; this effect is called *screening*. The effect of charge screening is dramatically different from the presence of a polarizable environment. As has been shown by Debye and Hückel some 70 years ago[14], screening modifies the electrostatic interaction such that it falls off exponentially with distance.

Starting point for the present treatment is the partition function of a system of mobile charges, interacting via Coulomb interactions. Using exact transformations, this problem is rendered in terms of a field theory. Quantum effects are always neglected, which is an acceptable approximation except maybe for processes involving protons. Formulation in terms of a field theory allows to use the vast body of knowledge and methods accumulated in the field of standard field theory, such as described in Refs. [1, 30, 44]. This is the topic of Section 2, where in a rather general fashion standard results of the statistical mechanics of interacting particles and field theoretic methods are reviewed. These techniques are then applied to the specific case of a single charged wall in Section 3 and to the case of two charged walls in Section 4. The field-theoretic methods described here are good for analyzing limiting behaviors, and we demonstrate that the mean-field or Poisson-Boltzmann approach is valid and in fact becomes

[5] Deviations from this approximation take the form of a momentum-dependent dielectric function $\tilde{\varepsilon}(k)$.

asymptotically exact for weakly charged objects and low-valent counter-ions, while the opposite limit of highly charged objects and high-valent counter-ions is accurately described by a virial approach. A lot can be learned from these two asymptotic methods. Experimentally, one is often in an intermediate situation, which is typically better described by integral-equation approaches, as is discussed in the contribution by R. Kjellander[24].

2. Field Theories for Interacting Particles

Introduction In this section we consider statistical ensembles of particles interacting via arbitrary potentials and show how to describe them using field-theoretic methods. Two different strategies discussed in detail are the saddle-point or mean-field method and the virial-expansion method. However, there is a great deal of freedom in deriving field theories, that means, there is not a unique field-theoretic description of such an ensemble. The appropriate descriptive framework has to be chosen with great care such as to fit to the system under consideration.

The statistical mechanics of ensembles of interacting particles starts (and very often also ends) with the calculation of the partition function, from which all relevant equilibrium observables can be calculated. The canonical partition function for a system of N particles interacting via a general pair-potential $v(\mathbf{r})$ is given by

$$Z_N = \frac{1}{N!}\prod_{j=1}^N \int \frac{d\mathbf{r}_j}{\lambda_t^3} \exp\left\{ -\sum_{j>k} v(\mathbf{r}_j - \mathbf{r}_k) - \sum_j \int d\mathbf{r}\sigma(\mathbf{r})v(\mathbf{r}-\mathbf{r}_j) \right. \\ \left. -\frac{1}{2}\int d\mathbf{r}d\mathbf{r}'\sigma(\mathbf{r})v(\mathbf{r}-\mathbf{r}')\sigma(\mathbf{r}') \right\} \tag{2}$$

where in addition we let all particles interact with some fixed distribution of particles $\sigma(\mathbf{r})$. This distribution might play the role of a bounding box, for charged systems this distribution corresponds to a fixed charge distribution.[6] The length λ_t is the thermal wavelength, which results from integrating out kinetic degrees of freedom, the precise value of which is unimportant. Defining a particle density operator $\hat{\rho}(\mathbf{r})$, which implicitly depends on all particle positions, as

$$\hat{\rho}(\mathbf{r}) \equiv \sum_{j=1}^N \delta(\mathbf{r}-\mathbf{r}_i), \tag{3}$$

[6] In principle, the interaction between the distribution $\sigma(\mathbf{r})$ and the particles could be different from the interaction between the particles themselves, a case which for simplicity we do not consider here.

the partition function (2) can be written in a more compact fashion as

$$
\begin{aligned}
Z_N[h] &= \frac{1}{N!}\prod_{j=1}^{N}\int\frac{d\mathbf{r}_j}{\lambda_t^3}\exp\left\{-\frac{1}{2}\int d\mathbf{r}d\mathbf{r}'[\hat{\rho}(\mathbf{r})+\sigma(\mathbf{r})]v(\mathbf{r}-\mathbf{r}')\right. \\
&\left.[\hat{\rho}(\mathbf{r}')+\sigma(\mathbf{r}')]+\frac{N}{2}v(0)+\int d\mathbf{r}h(\mathbf{r})\hat{\rho}(\mathbf{r})\right\}.
\end{aligned} \tag{4}
$$

The self-energy term in Eq.(4), proportional to $v(0)/2$, simply subtracts the diagonal term in the double integral. It will be later shown to renormalize the chemical potential (or fugacity) in a rather trivial way. The generating field $h(\mathbf{r})$ has been added so that the density expectation value can be calculated according to

$$
\langle\rho(\mathbf{r})\rangle = \left.\frac{\delta \ln Z_N[h]}{\delta h(\mathbf{r})}\right|_{h=0}. \tag{5}
$$

Higher-order correlation functions, so-called cumulant averages, follow by taking multiple functional derivatives of the logarithmic partition function (the so-called generating functional for connected correlation function, see problem at the end of this section) with respect to the generating field $h(\mathbf{r})$ according to

$$
\langle\rho(\mathbf{r})\rho(\mathbf{r}')\rangle_C = \langle\rho(\mathbf{r})\rho(\mathbf{r}')\rangle - \langle\rho(\mathbf{r})\rangle\langle\rho(\mathbf{r}')\rangle = \left.\frac{\delta^2 \ln Z_N[h]}{\delta h(\mathbf{r})\delta h(\mathbf{r}')}\right|_{h=0}. \tag{6}
$$

At this point, there is not much one can do with the partition function (4), mainly because the density operator $\hat{\rho}$ enters in a quadratic fashion. The main goal in this chapter is to show how to get rid of the configurational integral. This can be done at the cost of introducing a fluctuating field. The main step in deriving a field theory consists in introducing a unit operator which couples to the density operator $\hat{\rho}$,

$$
1 = \int \mathcal{D}\rho\, \delta(\rho-\hat{\rho}) = \int \mathcal{D}\rho\mathcal{D}\phi \exp\left\{\imath\int d\mathbf{r}\phi(\mathbf{r})[\rho(\mathbf{r})-\hat{\rho}(\mathbf{r})]\right\}, \tag{7}
$$

where we used the integral representation of the delta function. This unit operator can be introduced into the partition function at no cost. However, once introduced, it allows us to replace density operators $\hat{\rho}(\mathbf{r})$ by the corresponding fluctuating density fields $\rho(\mathbf{r})$ wherever we wish to do so. As a result, the partition function now reads[7]

$$
\begin{aligned}
Z_N[h] &= \frac{1}{N!}\prod_{j=1}^{N}\int\frac{d\mathbf{r}_j}{\lambda_t^3}\int\mathcal{D}\rho\mathcal{D}\phi\exp\Big\{+\frac{N}{2}v(0)+\imath\int d\mathbf{r}\phi(\mathbf{r})[\rho(\mathbf{r})-\hat{\rho}(\mathbf{r})] \\
&\quad -\frac{1}{2}\int d\mathbf{r}d\mathbf{r}'[\rho(\mathbf{r})+\sigma(\mathbf{r})]v(\mathbf{r}-\mathbf{r}')[\rho(\mathbf{r}')+\sigma(\mathbf{r}')] \\
&\quad +\int d\mathbf{r}h(\mathbf{r})\hat{\rho}(\mathbf{r})\Big\}. \qquad (8)
\end{aligned}
$$

The main achievement of this transformation is that now the particle coordinates only enter linearly and can be integrated out exactly, after which the partition function reads

$$
\begin{aligned}
Z_N[h] &= \int\mathcal{D}\rho\mathcal{D}\phi\exp\Big\{-\frac{1}{2}\int d\mathbf{r}d\mathbf{r}'[\rho(\mathbf{r})+\sigma(\mathbf{r})]v(\mathbf{r}-\mathbf{r}')[\rho(\mathbf{r}')+\sigma(\mathbf{r}')] \\
&\quad +\imath\int d\mathbf{r}\phi(\mathbf{r})\rho(\mathbf{r})\Big\}\frac{1}{N!}\left[e^{v(0)/2}\int\frac{d\mathbf{r}}{\lambda_t^3}e^{h(\mathbf{r})-\imath\phi(\mathbf{r})}\right]^N. \qquad (9)
\end{aligned}
$$

The partition function becomes even simpler upon transformation to the grand-canonical ensemble according to

$$
Z_\lambda[h]=\sum_{N=0}^{\infty}\tilde{\lambda}^N Z_N[h] \qquad (10)
$$

where $\tilde{\lambda}$ is the bare fugacity which is related to the particle chemical potential μ by $\tilde{\lambda}=e^{\mu}$. Using the definition of the exponential function, $e^x=\sum_{N=0}^{\infty}x^N/N!$, the grand-canonical partition function can be written as

$$
Z_\lambda[h]=\int\mathcal{D}\rho\mathcal{D}\phi\exp\{-H[\rho,\phi,h]\} \qquad (11)
$$

with

$$
\begin{aligned}
H_\lambda[\rho,\phi,h] &= \frac{1}{2}\int d\mathbf{r}d\mathbf{r}'[\rho(\mathbf{r})+\sigma(\mathbf{r})]v(\mathbf{r}-\mathbf{r}')[\rho(\mathbf{r}')+\sigma(\mathbf{r}')] \\
&\quad -\imath\int d\mathbf{r}\phi(\mathbf{r})\rho(\mathbf{r})-\lambda\int d\mathbf{r}e^{h(\mathbf{r})-\imath\phi(\mathbf{r})} \qquad (12)
\end{aligned}
$$

where we defined a rescaled fugacity as $\lambda=\tilde{\lambda}e^{v(0)/2}/\lambda_t^3$. As follows from the definition of the grand-canonical partition function, Eq.(10), the expec-

[7] Note that we replaced both density operators in the quadratic term by density fields, but not the operator coupled to the generating field h. This is somewhat arbitrary, but will lead to a form which will prove to be useful for us for the present purpose. In Section (2.4) we will come back to this point and take a slightly different route.

tation value of the particle number is given by

$$\langle N \rangle = \tilde{\lambda} \frac{\partial \ln Z_\lambda}{\partial \tilde{\lambda}} = \lambda \frac{\partial \ln Z_\lambda}{\partial \lambda} = \lambda \int d\mathbf{r} \langle e^{-\imath\phi(\mathbf{r})} \rangle. \tag{13}$$

As can be be seen, the expectation value of the particle number, $\langle N \rangle$, is independent of any multiplicative factor of the fugacity. The standard way of saying this is that the chemical potential is only defined up to a constant. The particle-density expectation value follows according to Eq.(5) as

$$\langle \rho(\mathbf{r}) \rangle = \lambda \langle e^{-\imath\phi(\mathbf{r})} \rangle \tag{14}$$

and of course satisfies the normalization condition $\int d\mathbf{r} \langle \rho(\mathbf{r}) \rangle = N$ as follows by comparison of Eqs.(13) and (14). The (disconnected) two-point-correlation function is given by

$$\langle \rho(\mathbf{r}) \rho(\mathbf{r}') \rangle = \lambda^2 \langle e^{-\imath\phi(\mathbf{r}) - \imath\phi(\mathbf{r}')} \rangle = \left. \frac{\delta^2 Z_\lambda[h]/Z_\lambda[0]}{\delta h(\mathbf{r}) \delta h(\mathbf{r}')} \right|_{h=0}. \tag{15}$$

Up to now all manipulations are exact, and for the general model, we exhausted the reservoir of possible exact transformations. We have succeeded in transforming the original problem, Eq.(2), where particle coordinates appear explicitly and in a basically intractable fashion, into a field-theory. We are not closer to obtaining the partition function exactly, since to perform the integrals over the fluctuating fields ρ and ϕ in Eq.(11) is at least as difficult as the integration over individual particle positions in Eq.(2). The advantage is that a field-theoretic formulation is more amenable to a systematic treatment, and a lot of knowledge about how to manipulate field theories has been accumulated over the past years, see standard text books[1, 30, 37, 44]. In the following, we will use a standard set of different approximations, which each become accurate and useful for different systems or different parameters.

Problem 1 In analogy to Eq.(6), calculate the four-point cumulant correlation function $\langle \rho(\mathbf{r}_1)\rho(\mathbf{r}_2)\rho(\mathbf{r}_3)\rho(\mathbf{r}_4) \rangle_C$ from the grand-canonical partition function (11) and compare it with the disconnected four-point correlation function which follows from a generalization of Eq.(15) .

Problem 2 In analogy to Eq.(13), calculate the quantity $\sqrt{\langle N^2 \rangle - \langle N \rangle^2} / \langle N \rangle$, which is a measure of particle-number fluctuations in the grand-canonical ensemble. Discuss under which conditions this quantity is zero in the thermodynamic limit and under which conditions it diverges. What does this mean?

2.1. DENSITY-FUNCTIONAL THEORY ON THE SADDLE-POINT LEVEL

The saddle-point method or mean-field-approximation (both names, although they do not strictly denote the same thing, are used interchangeably in literature) starts with introducing a dummy parameter, the loop parameter ℓ, into the partition function (11) in the following way

$$Z_\lambda[h] = \int \mathcal{D}\rho \mathcal{D}\phi \exp\{-\ell H_\lambda[\rho, \phi, h]\}\,. \tag{16}$$

The saddle-point method becomes accurate for large values of ℓ and consists in finding the value of the fluctuating fields ρ and ϕ which maximize the integrand. We will later see that this approximation scheme can be carried on in a systematic way where the inverse loop parameter ℓ^{-1} functions as an expansion parameter which counts the number of loops in a diagrammatic representation (which explains the name of ℓ). The loop parameter ℓ at this point has no physical meaning, but it will help us to structure and organize the perturbation scheme. At the end of our calculations we of course have to put $\ell = 1$. In Sections 3 and 4 we will encounter situations where a loop parameter appears which is a combination of physical parameters (such as temperature, ion valence, etc). The saddle point equations which determine the optimal values of the fluctuating fields, ρ_{SP} and ϕ_{SP}, which, as we will demonstrate soon, are in fact saddle points, are given by

$$\frac{\delta H_\lambda[\rho, \phi]}{\delta \rho(\mathbf{r})} = 0\,, \qquad \frac{\delta H_\lambda[\rho, \phi]}{\delta \phi(\mathbf{r})} = 0. \tag{17}$$

Note that we skip the dependence on the generating field h. Using H as defined in Eq.(12), the solution to the second equation is

$$\phi_{SP}(\mathbf{r}) = \imath \ln(\rho(\mathbf{r})/\lambda), \tag{18}$$

which, upon insertion into the Hamiltonian Eq.(12), leads to the saddle-point Hamiltonian

$$\begin{aligned} H_\lambda[\rho, \phi_{SP}] &= \frac{1}{2}\int \mathrm{d}\mathbf{r}\mathrm{d}\mathbf{r}'[\rho(\mathbf{r}) + \sigma(\mathbf{r})]v(\mathbf{r}-\mathbf{r}')[\rho(\mathbf{r}') + \sigma(\mathbf{r}')] \\ &\quad + \int \mathrm{d}\mathbf{r}\ \rho(\mathbf{r})\left(\ln[\rho(\mathbf{r})/\lambda] - 1\right). \end{aligned} \tag{19}$$

At this stage the density field ρ is still a fluctuating field which is to be integrated over. To complete the saddle-point analysis, we extremize Eq.(19) with respect to the remaining field ρ, which leads to the final saddle point equation

$$\int \mathrm{d}\mathbf{r}'\ v(\mathbf{r}-\mathbf{r}')[\rho_{SP}(\mathbf{r}') + \sigma(\mathbf{r}')] + \ln[\rho_{SP}(\mathbf{r})/\lambda] = 0. \tag{20}$$

At this saddle point, the grand-canonical Gibbs potential is given by

$$\mathcal{G}_{\lambda}^{SP} = H_{\lambda}[\rho_{SP}, \phi_{SP}] \tag{21}$$

and is thus obtained by inserting the saddle-point solution for the density ρ_{SP}, Eq.(20), into the Hamiltonian Eq.(19). Note that at this level the description is somewhat mixed, in that a density field ρ appears while the particle number N is not conserved but controlled by the fugacity λ.

Because the saddle-point density $\rho_{SP}(\mathbf{r})$ is a real function, the saddle-point value of ϕ is purely imaginary, as follows from Eq.(18). The functional integral over ϕ in Eq.(11) is a contour integral along the real axis. As the integrand contains no poles in the complex plane of ϕ, we can distort the integration contour so that it passes through this extremum. By virtue of the Cauchy-Riemann conditions, we know that the extremum is in fact a saddle point, which justifies the notation for this procedure, *saddle point approximation.* The functional $H_{\lambda}[\rho, \phi]$ shows a minimum at the saddle point as ϕ is varied along the real axis, but it is a maximum with respect to variation along the imaginary axis.

The normalization condition Eq.(13) leads to the expected result $N = \int \mathrm{d}\mathbf{r} \rho_{SP}(\mathbf{r})$. The canonical free energy follows from the Gibbs potential Eq.(21) by reverting the Legendre transform introduced in Eq.(10) according to

$$\mathcal{F}_{N}^{SP} = \mathcal{G}_{\lambda}^{SP} + N \ln \tilde{\lambda} \tag{22}$$

which leads to

$$\begin{aligned} \mathcal{F}_{N}^{SP} &= \frac{1}{2} \int \mathrm{d}\mathbf{r}\mathrm{d}\mathbf{r}' [\rho_{SP}(\mathbf{r}) + \sigma(\mathbf{r})] v(\mathbf{r} - \mathbf{r}') [\rho_{SP}(\mathbf{r}') + \sigma(\mathbf{r}')] \\ &\quad + \int \mathrm{d}\mathbf{r}\, \rho_{SP}(\mathbf{r}) \left(\ln[\lambda_t^3 \rho_{SP}(\mathbf{r})] - 1 \right). \end{aligned} \tag{23}$$

The saddle-point density distribution in the canonical ensemble ρ_{SP} is determined by the extremum of this functional given by the solution of the equation

$$\int \mathrm{d}\mathbf{r}'\, v(\mathbf{r} - \mathbf{r}') [\rho_{SP}(\mathbf{r}') + \sigma(\mathbf{r}')] + \ln[\lambda_t^3 \rho_{SP}(\mathbf{r})] = 0. \tag{24}$$

Equation (23) constitutes the mean-field free-energy functional for interacting particle systems.[8] Its main strength lies in the application to charged

[8] Strictly speaking, the free energy (23) is not a functional, since the saddle-point density distribution ρ_{SP} is not a free argument but fixed by the saddle-point equation Eq.(24). This point will be discussed in more detail in Section 2.2.1. Note also that we have skipped the dependence on the self energy $v(0)$, which is irrelevant.

systems, where it is known under the name of the Poisson-Boltzmann equation. Its main deficiencies are revealed when applied to particles which interact via hard-core interactions. In this case, the interparticle potential v is infinity for a certain range of distances, and the free-energy functional Eq.(23) is ill defined. We will in Section 2.4 see how to deal with such interactions as well. In the next section we will discuss how to systematically improve upon the saddle-point approximation by expanding the field-theoretic action around the saddle-point.

Problem 3 Check the stability of the saddle point $H_\lambda[\rho_{SP}, \phi_{SP}]$ of Eq.(19) by looking at the determinant of the Hessian (formed by second derivatives with respect to the two fields).

2.2. SADDLE-POINT AND BEYOND

For the systematic expansion in inverse powers of the loop parameter ℓ we start with the double-functional integral defined in Eq.(16) where the Hamiltonian H (which is by field-theorist usually called the *action*) is defined in Eq.(12). Since the integral over the field ρ is Gaussian, and we know how to do Gaussian functional integrals (in fact, Gaussian functional integrals are the only type of functional integral we know how to do), we can eliminate the field ρ exactly and obtain

$$Z_\lambda[h] = \int \frac{\mathcal{D}\phi}{Z_v} \exp\{-\ell H_\lambda[\phi, h]\} \tag{25}$$

with the action defined by

$$\begin{aligned} H_\lambda[\phi, h] &= \frac{1}{2}\int \mathbf{dr}\mathbf{dr}'\phi(\mathbf{r})v^{-1}(\mathbf{r}-\mathbf{r}')\phi(\mathbf{r}') + \imath\int \mathbf{dr}\phi(\mathbf{r})\sigma(\mathbf{r}) \\ &\quad -\lambda\int \mathbf{dr}e^{h(\mathbf{r})-\imath\phi(\mathbf{r})}. \end{aligned} \tag{26}$$

We now have a standard field theory (which in a slightly different form is known as the Sine-Gordon field theory) with only one fluctuating field. In deriving this exact result (we remind the reader that Eq.(25) in conjunction with Eq.(26) is an exact representation of the original problem of Eq.(2)) we introduced the inverse potential v^{-1} defined by

$$\int \mathbf{dr}\, v^{-1}(\mathbf{r}-\mathbf{r}')v(\mathbf{r}''-\mathbf{r}) = \delta(\mathbf{r}''-\mathbf{r}'), \tag{27}$$

the standard definition of a functional inverse (we will see that for the Coulomb interaction the functional inverse v^{-1} is well behaved and simple; this is obviously not the case for many other interaction potentials, which limits the applicability of the saddle-point approach to the case of charged systems). The symbol Z_v denotes the determinant of the Gaussian integral

$$Z_v = \int \mathcal{D}\phi \exp\left\{-\frac{1}{2}\int \mathrm{d}\mathbf{r}\mathrm{d}\mathbf{r}'\phi(\mathbf{r})v^{-1}(\mathbf{r}-\mathbf{r}')\phi(\mathbf{r}')\right\} \tag{28}$$

and is a measure of the free energy of vacuum fluctuations. It is one of the contributions to the van-der-Waals interaction, which is due to fluctuations of the electric field (or electrostatic potential). The saddle-point derived in the preceding section can also be obtained from the action in Eq.(26) and is defined by the saddle-point equation (in the remainder of this section we set $h = 0$)

$$\int \mathrm{d}\mathbf{r}'\, v^{-1}(\mathbf{r}-\mathbf{r}')\phi_{SP}(\mathbf{r}') + \imath\sigma(\mathbf{r}) + \imath\lambda \mathrm{e}^{-\imath\phi_{SP}(\mathbf{r})} = 0. \tag{29}$$

In a problem the reader will verify that Eqs.(20) and (29) are equivalent.

In order to go beyond the saddle-point or mean-field level, we expand the Hamiltonian (26) around the saddle point $\phi_{SP}(\mathbf{r})$. We write

$$\phi(\mathbf{r}) = \phi_{SP}(\mathbf{r}) + \chi(\mathbf{r})/\ell^{1/2} \tag{30}$$

and thus obtain the formal expansion for the action

$$H_\lambda[\phi] = H_\lambda[\phi_{SP}] + \sum_{j=2}^{\infty}\frac{1}{\ell^{j/2}j!}\int \mathcal{H}^{(j)}(\{\mathbf{r}_j\})\chi(\mathbf{r}_1)\cdots\chi(\mathbf{r}_j)\mathrm{d}\mathbf{r}_1\cdots\mathrm{d}\mathbf{r}_j. \tag{31}$$

The vertex functions $\mathcal{H}^{(j)}$ are defined by

$$\mathcal{H}^{(j)}(\{\mathbf{r}_j\}) \equiv \left.\frac{\delta^j H[\phi]}{\delta\phi(\mathbf{r}_1)\cdots\delta\phi(\mathbf{r}_j)}\right|_{\phi=\phi_{SP}} \tag{32}$$

and of course depend on the saddle-point function $\phi_{SP}(\mathbf{r})$ around which one expands. The first explicit vertex function are

$$\mathcal{H}^{(2)}(\mathbf{r}_1,\mathbf{r}_2) = v^{-1}(\mathbf{r}_1-\mathbf{r}_2) + \lambda\delta(\mathbf{r}_1-\mathbf{r}_2)\mathrm{e}^{-\imath\phi_{SP}(\mathbf{r}_1)} \tag{33}$$

and

$$\mathcal{H}^{(3)}(\mathbf{r}_1,\mathbf{r}_2,\mathbf{r}_3) = -\imath\lambda\delta(\mathbf{r}_1-\mathbf{r}_2)\delta(\mathbf{r}_1-\mathbf{r}_3)\mathrm{e}^{-\imath\phi_{SP}(\mathbf{r}_1)}. \tag{34}$$

The linear order in the expansion in Eq.(31) is absent by virtue of the saddle point equation

$$\mathcal{H}^{(1)}(\mathbf{r}) = \left.\frac{\delta H[\phi]}{\delta\phi(\mathbf{r})}\right|_{\phi=\phi_{SP}} = 0. \tag{35}$$

Using the definition of the two-point correlation function $G(\mathbf{r}, \mathbf{r}')$ (which also depends on the saddle point function ϕ_{SP})

$$\int \mathrm{d}\mathbf{r}\, G(\mathbf{r}, \mathbf{r}')\mathcal{H}^{(2)}(\mathbf{r}, \mathbf{r}'') = \delta(\mathbf{r}' - \mathbf{r}''), \tag{36}$$

the partition function Eq.(25) can be written as

$$\begin{aligned} Z_\lambda &= \mathrm{e}^{-\ell H_\lambda[\phi_{SP}]} \int \frac{\mathcal{D}\chi}{Z_v} \exp\left\{ -\frac{1}{2} \int \mathrm{d}\mathbf{r}\mathrm{d}\mathbf{r}' \chi(\mathbf{r}) G^{-1}(\mathbf{r}, \mathbf{r}')\chi(\mathbf{r}') \right. \\ &\quad \left. - \sum_{j=3}^{\infty} \frac{1}{\ell^{j/2-1} j!} \int \mathcal{H}^{(j)}(\{\mathbf{r}_j\})\chi(\mathbf{r}_1)\cdots\chi(\mathbf{r}_j)\mathrm{d}\mathbf{r}_1\cdots\mathrm{d}\mathbf{r}_j \right\}. \end{aligned} \tag{37}$$

The grand-canonical Gibbs potential $\mathcal{G}_\lambda = -\ln Z_\lambda$ can now be evaluated by standard saddle-point methods, treating ℓ^{-1} as the expansion parameter. We obtain including the one-loop order, ℓ^0,

$$\begin{aligned} \mathcal{G}_\lambda &= \ell H_\lambda[\phi_{SP}] - \ln \int \frac{\mathcal{D}\chi}{Z_v} \exp\left\{ -\frac{1}{2} \int \mathrm{d}\mathbf{r}\mathrm{d}\mathbf{r}' \chi(\mathbf{r}) G^{-1}(\mathbf{r}, \mathbf{r}')\chi(\mathbf{r}') \right\} \\ &\quad + \mathcal{O}(\ell^{-1}). \end{aligned} \tag{38}$$

The interested reader will be asked to obtain the next-leading order (containing all two-loop diagrams) in a problem.

Next we would like to calculate the counterion density to one-loop order. Note that obtaining the density from the free energy (38) according to Eq.(5) is rather complicated since the saddle-point field ϕ_{SP} depends implicitly on the generating field h. We therefore use the result for the grand-canonical density (14) in conjunction with the decomposition (30) and obtain

$$\begin{aligned} \langle \rho(\mathbf{r}) \rangle &= \lambda \mathrm{e}^{-\imath\phi_{SP}(\mathbf{r})} \left\langle \mathrm{e}^{-\imath\chi(\mathbf{r})/\sqrt{\ell}} \right\rangle \\ &= \lambda \mathrm{e}^{-\imath\phi_{SP}(\mathbf{r})} \left\langle 1 - \frac{\imath\chi(\mathbf{r})}{\ell^{1/2}} - \frac{\chi^2(\mathbf{r})}{2\ell} \right\rangle + \mathcal{O}(\ell^{-3/2}) \end{aligned} \tag{39}$$

where the expectation values are to be determined with respect to the full partition function (37). Since both the expectation value (39) and the action in Eq.(37) are expanded in inverse powers of ℓ, we can perform a full expansion of the expectation value and obtain

$$\begin{aligned} \langle \rho(\mathbf{r}) \rangle &= \lambda \mathrm{e}^{-\imath\phi_{SP}(\mathbf{r})} \left[1 + \frac{\imath}{6\ell} \int_{\mathbf{r}_1,\mathbf{r}_2,\mathbf{r}_3} \langle \chi(\mathbf{r})\chi(\mathbf{r}_1)\chi(\mathbf{r}_2)\chi(\mathbf{r}_3)\rangle_G \mathcal{H}^{(3)}(\mathbf{r}_1, \mathbf{r}_2, \mathbf{r}_3) \right. \\ &\quad \left. - \frac{1}{2\ell} \left\langle \chi^2(\mathbf{r}) \right\rangle_G \right] + \mathcal{O}(\ell^{-2}) \qquad (40) \\ &= \lambda \mathrm{e}^{-\imath\phi_{SP}(\mathbf{r})} \left[1 + \frac{\lambda}{6\ell} \int_{\mathbf{r}_1} \mathrm{e}^{-\imath\phi_{SP}(\mathbf{r}_1)} \left\langle \chi(\mathbf{r})\chi^3(\mathbf{r}_1) \right\rangle_G - \frac{1}{2\ell} \left\langle \chi^2(\mathbf{r}) \right\rangle_G \right]. \end{aligned}$$

In the last equation we used the explicit result for the cubic vertex function (34). The G-subscript denotes the expectation value with respect to the Gaussian, quadratic action (see next section for explicit examples). Using standard techniques for calculating Gaussian expectation values (see Section 2.3), we finally obtain

$$\begin{aligned}\langle \rho(\mathbf{r}) \rangle &= \lambda \mathrm{e}^{-\imath \phi_{SP}(\mathbf{r})} \left[1 + \frac{\lambda}{2\ell} \int_{\mathbf{r}_1} \mathrm{e}^{-\imath \phi_{SP}(\mathbf{r}_1)} G(\mathbf{r},\mathbf{r}_1) G(\mathbf{r}_1,\mathbf{r}_1) - \frac{1}{2\ell} G(\mathbf{r},\mathbf{r}) \right] \\ &\quad + \mathcal{O}(\ell^{-2}). \end{aligned} \tag{41}$$

As one can see, the particle density splits into the separate orders in powers of the loop parameter according to

$$\langle \rho(\mathbf{r}) \rangle = \rho_0(\mathbf{r}) + \ell^{-1} \rho_1(\mathbf{r}) + \mathcal{O}(\ell^{-2}) \tag{42}$$

where the leading order, equivalent to the mean-field result (18), is given by

$$\rho_0(\mathbf{r}) = \lambda \mathrm{e}^{-\imath \phi_{SP}(\mathbf{r})} \tag{43}$$

and the one-loop order by

$$\rho_1(\mathbf{r}) = \frac{\lambda \mathrm{e}^{-\imath \phi_{SP}(\mathbf{r})}}{2} \left[\lambda \int_{\mathbf{r}_1} \mathrm{e}^{-\imath \phi_{SP}(\mathbf{r}_1)} G(\mathbf{r},\mathbf{r}_1) G(\mathbf{r}_1,\mathbf{r}_1) - G(\mathbf{r},\mathbf{r}) \right]. \tag{44}$$

In the grand-canonical ensemble, the fugacity λ is fixed by the chemical potential of the reservoir, which is the appropriate ensemble for treating adsorption phenomena from bulk solutions. In the canonical ensemble, on the other hand, the particle number N is fixed and the fugacity λ is determined by the normalization (13) which can be conveniently solved by expanding the fugacity according to

$$\lambda = \lambda_0 + \ell^{-1} \lambda_1 + \mathcal{O}(\ell^{-2}). \tag{45}$$

Reexpanding the expression for the density (41) and the saddle-point equation (35), the contributions to the loop-wise expanded density in the canonical ensemble read

$$\rho_0(\mathbf{r}) = \lambda_0 \mathrm{e}^{-\imath \phi_{SP}(\mathbf{r})} \tag{46}$$

and

$$\begin{aligned}\rho_1(\mathbf{r}) &= \frac{\lambda_0 \mathrm{e}^{-\imath \phi_{SP}(\mathbf{r})}}{2} \left[\lambda_0 \int_{\mathbf{r}_1} \mathrm{e}^{-\imath \phi_{SP}(\mathbf{r}_1)} G(\mathbf{r},\mathbf{r}_1) G(\mathbf{r}_1,\mathbf{r}_1) - G(\mathbf{r},\mathbf{r}) \right] \\ &\quad + \lambda_1 \mathrm{e}^{-\imath \phi_{SP}(\mathbf{r})} \left[1 - \lambda_0 \int_{\mathbf{r}_1} \mathrm{e}^{-\imath \phi_{SP}(\mathbf{r}_1)} G(\mathbf{r},\mathbf{r}_1) \right] \end{aligned} \tag{47}$$

Note that in obtaining Eq.(47) we also expanded the saddle-point equation (35) in powers of ℓ^{-1}. Expanding the normalization condition (13) in powers of the loop parameter, the particle number N turns out to be determined by the zero-loop density, $N = \int \mathrm{d}\mathbf{r}\rho_0(\mathbf{r})$, leading to

$$\lambda_0 = \frac{N}{\int \mathrm{d}\mathbf{r}\ \mathrm{e}^{-\imath\phi_{SP}(\mathbf{r})}}. \tag{48}$$

At all higher orders, the normalization condition (13) leads to $\int \mathrm{d}\mathbf{r}\rho_k(\mathbf{r}) = 0$ (with $k > 0$), giving at first order

$$\lambda_1 = \frac{\lambda_0 \int_{\mathbf{r}} \mathrm{e}^{-\imath\phi_{SP}(\mathbf{r})} \left[G(\mathbf{r},\mathbf{r}) - \lambda_0 \int_{\mathbf{r}_1} \mathrm{e}^{-\imath\phi_{SP}(\mathbf{r}_1)} G(\mathbf{r},\mathbf{r}_1) G(\mathbf{r}_1,\mathbf{r}_1)\right]}{2\int_{\mathbf{r}} \mathrm{e}^{-\imath\phi_{SP}(\mathbf{r})} \left[1 - \lambda_0 \int_{\mathbf{r}_1} \mathrm{e}^{-\imath\phi_{SP}(\mathbf{r}_1)} G(\mathbf{r},\mathbf{r}_1)\right]}. \tag{49}$$

The one-loop correction to the particle density distribution (47) in conjunction with the fugacity expressions (48) and (49) constitute the main result of this section. We will in Sections 3 and 4 compare the one-loop prediction for the particle density with simulation results and see that the agreement is quite good if the loop parameter ℓ is large enough.

Problem 4 Verify that Eqs.(20) and (29) are equivalent.

Problem 5 Extend the loop expansion of the Gibbs potential (38) to second order in the inverse loop parameter ℓ^{-1}. Use the techniques introduced in Section 2.3. Drawing a vertex with n arms for each vertex function $\mathcal{H}^{(n)}$, connected by lines determined by the correlation functions, sketch the diagrams and convince yourself that the resulting diagrams contain indeed two loops. Note: There are three different diagrams. One of the two-loop diagrams is one-particle reducible, i.e. it can be decomposed into two disconnected parts by cutting one line. Such one-particle reducible diagrams play a certain role in establishing density-functional theories.

Problem 6 Calculate the connected or cumulant two-point correlation function to one-loop order.

2.2.1. *Density-functional theories within loop expansion*

In our loop expansion we have treated the generating field for density correlation functions, h, as the independent variable, see the exact partition function (25). It is often desirable to have an expression for a thermodynamic potential with the density expectation value $\langle\rho\rangle$ functioning as the independent variable, which is the central theme within density functional theories. In section 2.1 we have written down such an expression on the mean-field (saddle-point) level. However, the general derivation of a density functional within a loop-expansion is vastly different and involves a perturbative Legendre transform, using methods developed by Schwinger[48] and Brézin, Le Guillou, and Zinn-Justin[9]. To sketch the idea of these calculations, one notes that the density-dependent thermodynamic potential $\Gamma[\langle\rho\rangle]$ is related to the free energy $\mathcal{F}[h] = -\ln Z[h]$ by[9]

$$\Gamma[\langle\rho\rangle] = \mathcal{F}[h] + \int \mathrm{d}\mathbf{r}\; h(\mathbf{r})\langle\rho(\mathbf{r})\rangle. \tag{50}$$

The density expectation value follows from the free energy as

$$\langle\rho(\mathbf{r})\rangle = -\delta\mathcal{F}[h]/\delta h(\mathbf{r}). \tag{51}$$

This relation has to be inverted and inserted into Eq.(50) so that h is eliminated from the potential Γ. This program can for example be carried out within a loop expansion. In Ref.[41] this has been done for a solution of charged ions next to a charged wall, where, for simplicity, the expectation value of the electrostatic potential, $\langle\phi(\mathbf{r})\rangle$, has been treated as the independent variable. However, for charged systems, the Poisson equation gives an exact relation between the potential and the particle density, so that the potential is as good as the density.

2.3. GAUSSIAN EXPECTATION VALUES

As an explicit example, we calculate the four-point correlation function within a Gaussian field theory, which can be easily generalized by the reader to an arbitrary situation. To proceed, the quartic Gaussian correlation function is defined as

$$\langle\chi(\mathbf{r}_1)\chi(\mathbf{r}_2)\chi(\mathbf{r}_3)\chi(\mathbf{r}_4)\rangle_G = \frac{\int\mathcal{D}\chi\;\chi(\mathbf{r}_1)\chi(\mathbf{r}_2)\chi(\mathbf{r}_3)\chi(\mathbf{r}_4)\exp\left\{-\frac{1}{2}\int\mathrm{d}\mathbf{r}\mathrm{d}\mathbf{r}'\chi(\mathbf{r})G^{-1}(\mathbf{r},\mathbf{r}')\chi(\mathbf{r}')\right\}}{\int\mathcal{D}\chi\exp\left\{-\frac{1}{2}\int\mathrm{d}\mathbf{r}\mathrm{d}\mathbf{r}'\chi(\mathbf{r})G^{-1}(\mathbf{r},\mathbf{r}')\chi(\mathbf{r}')\right\}}. \tag{52}$$

In order to make progress, one introduces the Gaussian generating functional for disconnected correlation functions

$$Z_G[\omega] = \int\mathcal{D}\chi\exp\left\{-\frac{1}{2}\int\mathrm{d}\mathbf{r}\mathrm{d}\mathbf{r}'\chi(\mathbf{r})G^{-1}(\mathbf{r},\mathbf{r}')\chi(\mathbf{r}') + \int\mathrm{d}\mathbf{r}\omega(\mathbf{r})\chi(\mathbf{r})\right\} \tag{53}$$

As is quite easy to see, the expectation value (52) can be expressed as

$$\langle\chi(\mathbf{r}_1)\chi(\mathbf{r}_2)\chi(\mathbf{r}_3)\chi(\mathbf{r}_4)\rangle_G = \left.\frac{\delta^4 Z_G[\omega]/Z_G[\omega=0]}{\delta\omega(\mathbf{r}_1)\delta\omega(\mathbf{r}_2)\delta\omega(\mathbf{r}_3)\delta\omega(\mathbf{r}_4)}\right|_{\omega=0}. \tag{54}$$

[9] The potential Γ is the generating functional for one-particle irreducible diagrams, i.e., the Legendre transformation (50) removes all one-particle reducible diagrams, i.e., diagrams, that can be disconnected by cutting a single line.

By completing the square, we obtain

$$Z_G[\omega]/Z_G[\omega = 0] = \exp\left\{\frac{1}{2}\int d\mathbf{r}d\mathbf{r}'\omega(\mathbf{r})G(\mathbf{r},\mathbf{r}')\omega(\mathbf{r}')\right\}. \tag{55}$$

Each derivative in Eq.(54) pulls down a factor $G\omega$ from the exponential in the generating functional (55). Since the generating field is set to zero at the end, all field factors ω that are pulled down have to be canceled by derivatives, which explains why only even expectation values survive. The combinatorics of derivative-taking gives one contribution to the correlation function per possible pairing of fields (Wick's theorem). For the quartic correlation function one obtains

$$\begin{aligned}\langle\chi(\mathbf{r}_1)\chi(\mathbf{r}_2)\chi(\mathbf{r}_3)\chi(\mathbf{r}_4)\rangle_G &= G(\mathbf{r}_1,\mathbf{r}_2)G(\mathbf{r}_3,\mathbf{r}_4) + G(\mathbf{r}_1,\mathbf{r}_3)G(\mathbf{r}_2,\mathbf{r}_4)\\ &\quad + G(\mathbf{r}_1,\mathbf{r}_4)G(\mathbf{r}_2,\mathbf{r}_3). \end{aligned} \tag{56}$$

Problem 7 Convince yourself that the cubic correlation function $\langle\chi(\mathbf{r}_1)\chi(\mathbf{r}_2)\chi(\mathbf{r}_3)\rangle_G$ vanishes.

Problem 8 Calculate the connected Gaussian quartic correlation function from the generating functional for connected correlation function, $\ln Z[\omega]$, and compare your result with the result (56). Are the terms in Eq.(56) connected or disconnected (draw the corresponding diagrams where each external field point is denoted by a dot and G, the propagator, is denoted by a line)?

2.4. DENSITY-FUNCTIONAL-THEORY WITHIN VIRIAL EXPANSION

Let us reconsider the exact partition function of interacting particles as derived in the beginning of this section. In a slight deviation from the derivation of Eq.(8), we use the delta-operator-identity (7) for the density operator only on the quadratic part in (4) and obtain

$$\begin{aligned} Z_N[h] &= \frac{1}{N!}\prod_{j=1}^{N}\int\frac{d\mathbf{r}_j}{\lambda_t^3}\int\mathcal{D}\rho\mathcal{D}\phi\exp\left\{\frac{N}{2}v(0) + \int d\mathbf{r}h(\mathbf{r})\hat{\rho}(\mathbf{r})\right.\\ &\quad -\frac{1}{2}\int d\mathbf{r}d\mathbf{r}'v(\mathbf{r}-\mathbf{r}')\left[\rho(\mathbf{r})\rho(\mathbf{r}') + \sigma(\mathbf{r})\sigma(\mathbf{r}')\right]\\ &\quad \left. -\int d\mathbf{r}d\mathbf{r}'\hat{\rho}(\mathbf{r})v(\mathbf{r}-\mathbf{r}')\sigma(\mathbf{r}') + \imath\int d\mathbf{r}\phi(\mathbf{r})[\rho(\mathbf{r}) - \hat{\rho}(\mathbf{r})]\right\}. \end{aligned} \tag{57}$$

Performing the Legendre transformation to the grand-canonical ensemble, as defined in Eq.(10), and integrating out the field ρ exactly, similar to the derivation of Eqs.(25) and (26), we obtain the partition function

$$Z_\lambda[h] = e^{-\frac{1}{2}\int_{\mathbf{r},\mathbf{r}'}\sigma(\mathbf{r})v(\mathbf{r}-\mathbf{r}')\sigma(\mathbf{r}')} \int \frac{\mathcal{D}\phi}{Z_v} \exp\left\{-\frac{1}{2}\int_{\mathbf{r},\mathbf{r}'}\phi(\mathbf{r})v^{-1}(\mathbf{r}-\mathbf{r}')\phi(\mathbf{r}') + \lambda\int_{\mathbf{r}} e^{h(\mathbf{r})-\imath\phi(\mathbf{r})-\int_{\mathbf{r}'}v(\mathbf{r}-\mathbf{r}')\sigma(\mathbf{r}')}\right\}, \tag{58}$$

which is but a slight modification of the formulation in Eqs.(25) and (26). In the loop expansion one uses the fact that ℓ is large. We will in the following chapters see when this is a good approximation. A different approximation, the virial expansion, consists in treating the fugacity λ as the small expansion parameter (we have therefore set $\ell = 1$). We will in the following sections see that there are indeed situations, typically when the particle density is small, where this approximation scheme is the method of choice. In fact, for charged systems, loop and virial expansion correspond to the two oppositely limiting cases (reality usually lies somewhere in the middle).

To proceed, the fugacity or virial expansion of the partition function (58) can be written in a very compact form as

$$Z_\lambda[h] = e^{-\frac{1}{2}\int_{\mathbf{r},\mathbf{r}'}\sigma(\mathbf{r})v(\mathbf{r}-\mathbf{r}')\sigma(\mathbf{r}')} \cdot \sum_{j=0}^{\infty}\frac{\lambda^j}{j!}\left\langle\prod_{k=1}^{j}\int d\mathbf{r}_k e^{h(\mathbf{r}_k)-\imath\phi(\mathbf{r}_k)-\int_{\mathbf{r}'}v(\mathbf{r}_k-\mathbf{r}')\sigma(\mathbf{r}')}\right\rangle_v \tag{59}$$

where the expectation value is performed with respect to the partition function Z_v introduced in Eq.(28). All expectation values appearing in this expansion are simple Gaussians which can be trivially performed, even in cases when the functional inverse of the interaction, v^{-1}, is singular (as for example for the hard-core interaction). Up to second order, we obtain explicitly

$$Z_\lambda[h] = e^{-\frac{1}{2}\int_{\mathbf{r},\mathbf{r}'}\sigma(\mathbf{r})v(\mathbf{r}-\mathbf{r}')\sigma(\mathbf{r}')}\Big[1 + (\tilde\lambda/\lambda_t^3)\int d\mathbf{r}_1 e^{h(\mathbf{r}_1)-\int_{\mathbf{r}'}v(\mathbf{r}_1-\mathbf{r}')\sigma(\mathbf{r}')} + \frac{(\tilde\lambda/\lambda_t^3)^2}{2}\int d\mathbf{r}_1 d\mathbf{r}_2 e^{h(\mathbf{r}_1)+h(\mathbf{r}_2)-\int_{\mathbf{r}'}v(\mathbf{r}_1-\mathbf{r}')\sigma(\mathbf{r}')-\int_{\mathbf{r}'}v(\mathbf{r}_2-\mathbf{r}')\sigma(\mathbf{r}')-v(\mathbf{r}_1-\mathbf{r}_2)}\Big] + \mathcal{O}(\tilde\lambda^3). \tag{60}$$

The Gibbs potential $\mathcal{G}_\lambda[h] = -\ln Z_\lambda[h]$, is obtained as

$$\mathcal{G}_\lambda[h] = \frac{1}{2}\int_{\mathbf{r},\mathbf{r}'}\sigma(\mathbf{r})v(\mathbf{r}-\mathbf{r}')\sigma(\mathbf{r}') - (\tilde{\lambda}/\lambda_t^3)\int \mathrm{d}\mathbf{r}_1 \mathrm{e}^{h(\mathbf{r}_1)-\int_{\mathbf{r}'} v(\mathbf{r}_1-\mathbf{r}')\sigma(\mathbf{r}')}$$
$$+\frac{(\tilde{\lambda}/\lambda_t^3)^2}{2}\int \mathrm{d}\mathbf{r}_1\mathrm{d}\mathbf{r}_2 \mathrm{e}^{h(\mathbf{r}_1)+h(\mathbf{r}_2)-\int_{\mathbf{r}'} v(\mathbf{r}_1-\mathbf{r}')\sigma(\mathbf{r}')-\int_{\mathbf{r}'} v(\mathbf{r}_2-\mathbf{r}')\sigma(\mathbf{r}')}$$
$$\left[1-\mathrm{e}^{-v(\mathbf{r}_1-\mathbf{r}_2)}\right] + \mathcal{O}(\tilde{\lambda}^3). \tag{61}$$

The function in the square brackets is called the Mayer function and has proven very useful in the past because it is a finite, short-ranged and well-behaved function for hard-core interactions. In the virial expansion, it is much simpler to derive a density-functional theory according to the Legendre transform (50) as described in the preceding section for the loop expansion. The density distribution follows according to Eq.(51) from Eq.(61) as

$$\rho(\mathbf{r}) = (\tilde{\lambda}/\lambda_t^3)\mathrm{e}^{h(\mathbf{r})-\int_{\mathbf{r}'} v(\mathbf{r}-\mathbf{r}')\sigma(\mathbf{r}')} \tag{62}$$
$$-(\tilde{\lambda}/\lambda_t^3)^2\mathrm{e}^{h(\mathbf{r})-\int_{\mathbf{r}'} v(\mathbf{r}-\mathbf{r}')\sigma(\mathbf{r}')}\int \mathrm{d}\mathbf{r}_1 \mathrm{e}^{h(\mathbf{r}_1)-\int_{\mathbf{r}'} v(\mathbf{r}_1-\mathbf{r}')\sigma(\mathbf{r}')}\left[1-\mathrm{e}^{-v(\mathbf{r}_1-\mathbf{r})}\right].$$

To perform the inversion to the density as the independent variable, we invert Eq.(62) in powers of the density, leading to

$$(\tilde{\lambda}/\lambda_t^3)\mathrm{e}^{h(\mathbf{r})-\int_{\mathbf{r}'} v(\mathbf{r}-\mathbf{r}')\sigma(\mathbf{r}')} =$$
$$\rho(\mathbf{r}) + \rho(\mathbf{r})\int \mathrm{d}\mathbf{r}_1\rho(\mathbf{r}_1)\left[1-\mathrm{e}^{-v(\mathbf{r}_1-\mathbf{r})}\right] + \mathcal{O}(\rho^3). \tag{63}$$

Inserting the expansion (63) into the Gibbs potential (61), and performing the back-Legendre transformation with respect to the density distribution ρ, according to Eq.(50), and with respect to the particle number N, according to Eq.(22), the potential $\Gamma_N[\rho]$ is given by

$$\Gamma_N[\rho] = \int_{\mathbf{r}}\rho(\mathbf{r})\left(\ln\left[\lambda_t^3\rho(\mathbf{r})\right]-1\right) + \frac{1}{2}\int_{\mathbf{r},\mathbf{r}'}\sigma(\mathbf{r})v(\mathbf{r}-\mathbf{r}')\sigma(\mathbf{r}') \tag{64}$$
$$+\int_{\mathbf{r},\mathbf{r}'}\sigma(\mathbf{r})v(\mathbf{r}-\mathbf{r}')\rho(\mathbf{r}') + \frac{1}{2}\int_{\mathbf{r},\mathbf{r}'}\rho(\mathbf{r})\rho(\mathbf{r}')\left[1-\mathrm{e}^{-v(\mathbf{r}-\mathbf{r}')}\right] + \mathcal{O}(\rho^3).$$

This expression is the non-local density functional theory within the virial expansion. We note that expression (64) is very similar to the saddle-point density functional (23). In fact, the only difference is that in (64) the interaction v in the density-density coupling term appears in the exponential. One could be tempted to think that (23) might be obtained from (64) by expanding the exponential with respect to v. However, this completely misses the point, since the two approaches are fundamentally different and follow from systematic expansions of the the partition function with respect to the inverse loop parameter ℓ^{-1} and the fugacity λ, respectively. The similarity between the two approaches on the leading level is an interesting fact, however, on the next-leading levels, the resulting theories are vastly different.

Problem 9 Using the normalization condition (13), derive from Eq.(60) the free energy according to Eq.(22) in the absence of external fields, i.e., $h = 0$, where the average particle density is $\overline{\rho} = N/V$. What is the relation to the density functional (64)? For a hard-core potential with hard-core diameter a, calculate the coefficient in front of the quadratic density term $\overline{\rho}^2$, which is the so-called second virial coefficient.

3. Counter-Ion Distributions at Charged Planes

Introduction The Poisson-Boltzmann or mean-field approach gives asymptotically exact counter-ion density profiles around charged objects in the weak-coupling limit of low valence and high temperature. A theory is derived which becomes exact in the opposite limit of strong coupling. Formally, it corresponds to a standard virial expansion. Long-range divergences, which render the virial expansion intractable for homogeneous bulk systems, are shown to be renormalizable for the case of inhomogeneous distribution functions by a systematic expansion in inverse powers of the coupling parameter. For a planar charged wall, the analytical results compare quantitatively with extensive Monte-Carlo simulations.

The Poisson-Boltzmann (PB) approach is known to give reliable results only in the limit of low-valence ions or high temperatures[23, 4, 46, 41]. Corrections to PB have been attributed to fluctuations, or, in other words, correlations between ions, and, if present, additional non-electrostatic interactions between ions. These corrections are particularly important for the interaction between macroscopic similarly charged objects, where they can lead to attractions[23, 4, 46, 51, 27, 20, 18, 11]. In as much as the PB approach is accurate for weakly charged systems, no such theory was available for the distribution of counter ions around charged objects in the opposite limit of high-valence ion; moreover, it was not clear whether such a limit exists and whether it is physically meaningful. In this section we show, using field-theoretic methods, that while PB corresponds to the asymptotically exact theory in the *weak-coupling limit*, (corresponding to low-valence ions or high temperatures), the *strong-coupling theory* becomes asymptotically exact in the opposite limit of high-valence ions or low temperatures and constitutes a physically sound limit. For the case of a planar charged wall, we give explicit results for the asymptotic density profile in the strong-coupling limit. We also have performed extensive Monte-Carlo (MC) simulations of this system[33]. The resulting density profiles agree for weak and for strong coupling with predictions from PB theory and our strong-coupling theory, respectively. The strong-coupling limit is experimentally easily reached at room temperatures with highly charged walls and/or multivalent counter ions and thus relevant from the application point of view.

To proceed, consider the energy W (in reduced units of the thermal energy k_BT) of a system of N ions of valence q at an impenetrable and oppositely charged wall of surface charge density σ_s,

$$W = \sum_{j<k}^{N} \frac{\ell_B q^2}{|\mathbf{r}_j - \mathbf{r}_k|} + 2\pi q \ell_B \sigma_s \sum_{j=1}^{N} z_j \tag{65}$$

where

$$\ell_B = \frac{e^2}{4\pi\varepsilon k_B T} \tag{66}$$

is the Bjerrum length which measures the distance at which two unit charges interact with thermal energy; in water $\ell_B \approx 0.7nm$. The dielectric constant is assumed to be homogeneous throughout the system. The Gouy-Chapman length

$$\mu = \frac{1}{2\pi q \sigma_s \ell_B} \tag{67}$$

measures the distance from the wall at which the potential energy of an ion reaches the thermal energy. Rescaling all lengths by μ according to $\mathbf{r} = \mu\tilde{\mathbf{r}}$, the energy reads

$$W = \sum_{j<k}^{N} \frac{\Xi}{|\tilde{\mathbf{r}}_j + \tilde{\mathbf{r}}_k|} + \sum_{j=1}^{N} \tilde{z}_j \tag{68}$$

and thus only depends on the coupling parameter

$$\Xi = 2\pi q^3 \ell_B^2 \sigma_s. \tag{69}$$

Using the fact that the typical distance of an ion from the wall is in reduced units $\tilde{z} \sim 1$ (which holds both in the weak and strong-coupling limits, as we will show below) the confinement energy, the second term in Eq.(68), is of order unity. The typical distance between ions scales as $\tilde{r} \sim \Xi^{1/2}$ (for $\Xi > 1$ and assuming that the ions form a two-dimensional layer) or like $\tilde{r} \sim \Xi^{1/3}$ (for $\Xi < 1$ and assuming a liquid-like structure) and thus the repulsive energy, the first term in Eq.(68), is of order $\Xi^{1/2}$ for $\Xi > 1$ or $\Xi^{2/3}$

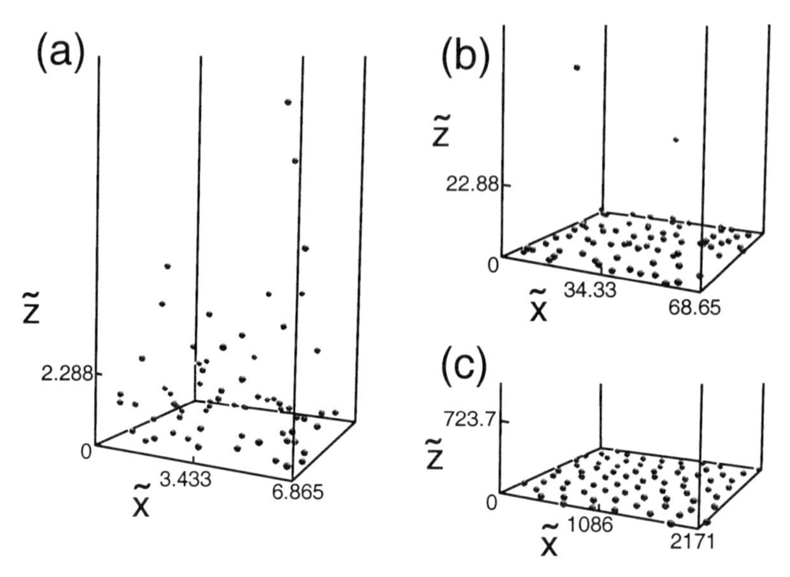

Figure 1. Snapshots of distributions of 75 ions for different values of the coupling constant Ξ from Monte-Carlo simulations. a) Weak-coupling limit where Poisson-Boltzmann theory is accurate, $\Xi = 0.1$, b) intermediate coupling regime, $\Xi = 10$, and c) strong-coupling limit, $\Xi = 10000$. Note that in a) 8 particles which are located far away from the charged surface are not shown.

for $\Xi < 1$. It thus follows on rather general grounds that for a large coupling parameter, $\Xi > 1$, the ionic structure is dominated by mutual repulsions, suggesting crystallization, while for weak coupling $\Xi < 1$ these repulsions should be rather unimportant. In Fig.1 we show ion-distribution snapshots from Monte-Carlo simulations for $\Xi = 0.1, 10, 10^4$. For small Ξ, Fig.1a, repulsive ion-ion interactions are indeed not playing a dominant role, the ion distribution is rather disordered and mean-field theory should work, since each ion moves in a diffuse cloud of neighboring ions. For large Ξ, on the other hand, ion-ion repulsions are strong and ion-ion distances are large compared to the distance from the wall. In Fig.1c the ions form a flat layer on the charged wall. A two-dimensional one-component plasma is known to crystallize for plasma parameter $\Gamma \approx 125$[7]. From the definition of the two-dimensional plasma parameter[7] we obtain the relation $\Xi = 2\Gamma^2$. This leads to a crystallization threshold (in units of our coupling parameter) of $\Xi \approx 31000$ meaning that all three systems in Fig.1 are below the crystallization threshold. However, mean-field theory is expected to break down, at least for the system shown in Fig.1c, because each ion moves, though confined by its immediate neighbors in the lateral directions, almost independently from the other ions along the vertical direction (which constitutes the soft mode). Let us now see how these notions can be made concrete in a field-theoretic formulation.

3.1. MAPPING ON A FIELD THEORY

The partition function of a system of N counter ions interacting via reduced Coulomb interactions $v_C(\mathbf{r}) = 1/r$ among themselves and with a charge distribution $\sigma_C(\mathbf{r})$ is given by

$$Z_N = \frac{1}{N!}\prod_{j=1}^{N}\int\frac{\mathrm{d}\mathbf{r}_j}{\lambda_t^3}\theta(z_j)\exp\left\{-\frac{1}{2}\int\mathrm{d}\mathbf{r}\mathrm{d}\mathbf{r}'\,\sigma_C(\mathbf{r})v_C(\mathbf{r}-\mathbf{r}')\sigma_C(\mathbf{r}')\right. \tag{70}$$

$$\left.-q^2\ell_B\sum_{j<k}v_C(\mathbf{r}_j-\mathbf{r}_k)+q\ell_B\int\mathrm{d}\mathbf{r}\sigma_C(\mathbf{r})\sum_j v_C(\mathbf{r}-\mathbf{r}_j)+\sum_j h(\mathbf{r}_j)\right\}$$

where the field h has been added to calculate density distributions later on. Note that we have defined the sign of the charge distribution, which is opposite from the sign of the charged particles, to be positive, i.e., $\sigma_C(\mathbf{r}) \geq 0$. The Heavyside function is defined by $\theta(z) = 1$ for $z > 0$ and zero otherwise and restricts the configurational integral to the upper half space ($z > 0$). At this point we employ a Hubbard-Stratonovitch transformation and perform a Legendre transformation to the grand-canonical ensemble, $Z_\lambda = \sum_N \tilde{\lambda}^N Z_N$, as explained in Section 2, which leads to[46, 39, 40, 41]

$$Z_\lambda = \int\frac{\mathcal{D}\phi}{Z_v}\exp\left\{-\frac{1}{2\ell_B q^2}\int\mathrm{d}\mathbf{r}\mathrm{d}\mathbf{r}'\phi(\mathbf{r})v_C^{-1}(\mathbf{r}-\mathbf{r}')\phi(\mathbf{r}')\right.$$

$$\left.+\frac{\imath}{q}\int\mathrm{d}\mathbf{r}\sigma_C(\mathbf{r})\phi(\mathbf{r})+\lambda\int\mathrm{d}\mathbf{r}\,\theta(z)\mathrm{e}^{h(\mathbf{r})-\imath\phi(\mathbf{r})}\right\} \tag{71}$$

where we introduced the notation $Z_v = \sqrt{\det v_C}$ in agreement with the definition (28) and the rescaled fugacity $\lambda = \tilde{\lambda}\mathrm{e}^{v_C(0)/2}/\lambda_t^3$. Next we rescale the action, similarly to our rescaling analysis of the Hamiltonian in the preceding section. All lengths are rescaled by the Gouy-Chapman length,

$$\mathbf{r} = \mu\tilde{\mathbf{r}}. \tag{72}$$

The grand-canonical partition function can be written as

$$Z_\lambda = \int\frac{\mathcal{D}\phi}{Z_v}\exp\left\{-\frac{1}{\Xi}H[\phi,h]\right\}. \tag{73}$$

Using the explicit form for the charge distribution

$$\sigma_C(\mathbf{r}) = \sigma_s \delta(z), \tag{74}$$

the Hamiltonian can be written as

$$\begin{aligned} H[\phi, h] &= \frac{1}{2} \int \mathrm{d}\tilde{\mathbf{r}} \mathrm{d}\tilde{\mathbf{r}}' \phi(\tilde{\mathbf{r}}) v_C^{-1}(\tilde{\mathbf{r}} - \tilde{\mathbf{r}}') \phi(\tilde{\mathbf{r}}') - \frac{\imath}{2\pi} \int \mathrm{d}\tilde{\mathbf{r}} \delta(\tilde{z}) \phi(\tilde{\mathbf{r}}) \\ &\quad - \frac{\Lambda}{2\pi} \int \mathrm{d}\tilde{\mathbf{r}} \theta(\tilde{z}) \mathrm{e}^{h(\tilde{\mathbf{r}}) - \imath\phi(\tilde{\mathbf{r}})} \end{aligned} \tag{75}$$

where the rescaled fugacity Λ is defined by

$$\Lambda = 2\pi\lambda\mu^3\Xi = \frac{\lambda}{2\pi\sigma_s^2 \ell_B}. \tag{76}$$

We have thus shown that the coupling parameter Ξ is the only parameter also in the field-theoretic formulation (as would be expected from the simple scaling argument advanced in the preceding section) and plays the role of the inverse loop parameter. The expectation value of the counter-ion density, $\langle \rho(\tilde{\mathbf{r}}) \rangle$, follows by taking a functional derivative with respect to the generating field h, $\langle \rho(\tilde{\mathbf{r}}) \rangle = \delta \ln Z_\lambda / \delta h(\tilde{\mathbf{r}}) \mu^3$, giving rise to the rescaled density distribution

$$\tilde{\rho}(\tilde{\mathbf{r}}) = \frac{\langle \rho(\tilde{\mathbf{r}}) \rangle}{2\pi \ell_B \sigma_s^2} = \Lambda \langle \mathrm{e}^{-\imath\phi(\tilde{z})} \rangle. \tag{77}$$

The normalization condition for the counter-ion distribution, $\mu \int \mathrm{d}\tilde{z} \rho(\tilde{z}) = \sigma_s / q$, which follows directly from the definition of the grand-canonical partition function, leads to

$$\int_0^\infty \mathrm{d}\tilde{z}\ \tilde{\rho}(\tilde{\mathbf{r}}) = \Lambda \int_0^\infty \mathrm{d}\tilde{z} \langle \mathrm{e}^{-\imath\phi(\tilde{z})} \rangle = 1 \tag{78}$$

and will be used to fix the fugacity Λ. This is an important equation since it shows that the expectation values of the fugacity term in Eq.(75) is bounded and of the order of unity.

3.2. MEAN-FIELD OR POISSON-BOLTZMANN APPROACH

Let us first repeat the saddle-point analysis, which, because of the structure of the functional integrand in Eq.(73), should be valid for $\Xi \ll 1$. First we note that the inverse Coulomb operator follows from Poisson's law, which in reduced units reads $\delta(\mathbf{r}) = -4\pi \nabla^2 v_C(\mathbf{r})$, as

$$v_C^{-1}(\mathbf{r}) = -\nabla^2 \delta(\mathbf{r}) / 4\pi. \tag{79}$$

The Hamiltonian (75) can therefore be rewritten as

$$H[\phi, h] = \frac{1}{4\pi} \int \mathrm{d}\tilde{\mathbf{r}} \left[\frac{1}{2} [\nabla \phi(\tilde{\mathbf{r}})]^2 - 2\imath\delta(\tilde{z})\phi(\tilde{\mathbf{r}}) - 2\Lambda\theta(\tilde{z})\mathrm{e}^{h(\tilde{\mathbf{r}}) - \imath\phi(\tilde{\mathbf{r}})} \right]. \quad (80)$$

The saddle-point equation, which follows from the equation $\delta H[\phi]/\delta\phi(\tilde{\mathbf{r}}) = 0$, reads

$$\frac{\mathrm{d}^2\phi(\tilde{z})}{\mathrm{d}\tilde{z}^2} = 2\imath\Lambda\mathrm{e}^{-\imath\phi(\tilde{z})} \quad (81)$$

with the boundary condition $\mathrm{d}\phi(\tilde{z})/\mathrm{d}\tilde{z} = -2\imath$. The solution of the saddle-point equation is

$$\imath\phi_{SP}(\tilde{z}) = 2\ln\left(1 + \Lambda^{1/2}\tilde{z}\right) \quad (82)$$

while the boundary condition leads to $\Lambda_0 = 1$, which shows that the saddle-point approximation is indeed valid in the limit $\Xi \ll 1$. Combining Eqs.(77) and (82), the density distribution of counter ions at the saddle-point level is given by the well-known Poisson-Boltzmann result

$$\tilde{\rho}_0(\tilde{z}) = \Lambda_0\mathrm{e}^{-\imath\phi_{SP}(\tilde{z})} = \frac{1}{(1+\tilde{z})^2} \quad (83)$$

which of course satisfies the normalization condition (78).

3.3. LOOP EXPANSION

The loop expansion proceeds along the line developed in Section 2.2 where the parameters introduced there are given in the present case by $\ell = 1/\Xi$ and $\lambda = \Lambda/2\pi$ and the normalization is determined by (78). The two-point propagator is in analogy with Eq.(33) given by

$$\mathcal{H}^{(2)}(\tilde{\mathbf{r}}_1, \tilde{\mathbf{r}}_2) = \frac{1}{4\pi}\left[-\nabla^2 + 2\Lambda\theta(\tilde{z}_1)\mathrm{e}^{-\imath\phi_{SP}(\tilde{\mathbf{r}}_1)}\right]\delta(\tilde{\mathbf{r}}_1 - \tilde{\mathbf{r}}_2). \quad (84)$$

Since the problem is translationally invariant parallel to the charged plate, we can introduce the two-dimensional Fourier-transform of the propagator

$$G(\tilde{z}, \tilde{z}', \tilde{p}) = \int \mathrm{d}\tilde{x}\mathrm{d}\tilde{y}\; \mathrm{e}^{\imath\tilde{p}_x(\tilde{x}-\tilde{x}') + \imath\tilde{p}_y(\tilde{y}-\tilde{y}')} G(\tilde{\mathbf{r}}, \tilde{\mathbf{r}}'). \quad (85)$$

Using that $\Lambda_0 = 1$ to leading order in the loop expansion, the propagator G, as defined in (36), is determined by the equation

$$\left[-\frac{\mathrm{d}^2}{\mathrm{d}\tilde{z}^2} + \tilde{p}^2 + 2\theta(\tilde{z})\mathrm{e}^{-\imath\phi_{SP}(\tilde{z})}\right] G(\tilde{z}, \tilde{z}', \tilde{p}) = \delta(\tilde{z} - \tilde{z}'). \quad (86)$$

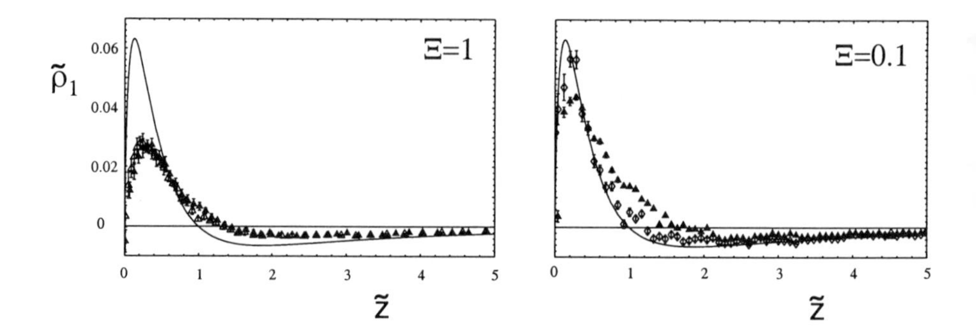

Figure 2. Field-theoretic prediction for the one-loop contribution to the rescaled density profile, $\tilde{\rho}_1$ (solid line), compared with numerical results for the density difference $(\tilde{\rho}-\tilde{\rho}_0)/\Xi$ for a) $\Xi = 1$, $N = 75$ particles and 10^7 Monte-Carlo steps (MCS) (filled triangles), $N = 200$ particles and 2×10^6 MCS (open triangles), and b) $\Xi = 0.1$, $N = 200$ particles and 5×10^6 MCS (filled triangles), $N = 600$ particles and 10^6 MCS (open diamonds). The theoretical one-loop prediction describes the numerical data very well for $\Xi = 0.1$, for $\Xi = 1$ higher-order corrections seem to be important.

The solution is

$$
\begin{aligned}
G(\tilde{z},\tilde{z}',\tilde{p}) &= \frac{2\pi}{\tilde{p}^3}\left(\tilde{p}+\frac{1}{\tilde{z}'+1}\right) \\
&\times \left[\left(\tilde{p}-\frac{1}{\tilde{z}+1}\right)e^{\tilde{p}(\tilde{z}-\tilde{z}')} + \left(\frac{\tilde{p}+\frac{1}{\tilde{z}+1}}{1+2\tilde{p}+2\tilde{p}^2}\right)e^{-\tilde{p}(\tilde{z}+\tilde{z}')}\right]
\end{aligned}
\tag{87}
$$

for $\tilde{z}' > \tilde{z} > 0$. The solution for $\tilde{z} > \tilde{z}' > 0$ follows from Eq.(87) by an interchange of the arguments $\tilde{z}$ and $\tilde{z}'$. Similarly to Section 2.2, the loop expansion is done by expanding the rescaled particle density $\tilde{\rho}$ and the rescaled fugacity Λ in inverse powers of the loop parameter, that means in powers of the coupling parameter Ξ, according to

$$
\tilde{\rho}(\tilde{z}) = \tilde{\rho}_0(\tilde{z}) + \Xi\tilde{\rho}_1(\tilde{z}) + \mathcal{O}(\Xi^2) \;\;,\;\; \Lambda = \Lambda_0 + \Xi\Lambda_1 + \mathcal{O}(\Xi^2) \tag{88}
$$

where $\tilde{\rho}_0$, equivalent to the mean-field result, is given in Eq.(83) and $\Lambda_0 = 1$. The one-loop contribution to the density distribution, $\tilde{\rho}_1$ is determined by Eqs.(47) and (49) where $\lambda_0 = \Lambda_0/2\pi = 1/2\pi$ has to be inserted. In Fig. 2 we compare the field-theoretic prediction for the one-loop contribution to the rescaled density profile, $\tilde{\rho}_1$ (solid line), with numerical results for the density difference $(\tilde{\rho}-\tilde{\rho}_0)/\Xi$ for a) $\Xi = 1$ and b) $\Xi = 0.1$ for different particle numbers. The theoretical one-loop prediction describes the numerical data very well for $\Xi = 0.1$, for $\Xi = 1$ the agreement is rather poor which means that in the latter case the higher-order loop corrections must be relevant.

3.4. STRONG-COUPLING OR VIRIAL APPROACH

Let us now consider the opposite limit, when the coupling constant Ξ is large. In this case, the saddle-point approximation breaks down, since the prefactor in front of the action in Eq.(73) becomes small. This can also be seen from our systematic loop-wise expansion around the saddle point, since the one-loop correction to the saddle-point solution is proportional to Ξ and thus leads to unphysical negative densities for $\Xi > 10$. However, from the field-theoretic partition function Eq.(73), it is self-evident what has to be done in this limit. Since the fugacity term is bounded, as evidenced by Eq.(78), one can expand the partition function (and also all expectation values) in powers of Λ/Ξ. This is nothing but a virial expansion, which has been thoroughly reviewed in Section 2.4. The normalization condition Eq.(78) can then be solved by an expansion of the fugacity in inverse powers of the coupling constant,

$$\Lambda = \Lambda_0 + \Lambda_1/\Xi + \mathcal{O}(1/\Xi^2), \tag{89}$$

and this indeed leads to an expansion of the density profile with the small parameter $1/\Xi$ (although the resulting perturbation series is not a strict power series in $1/\Xi$ as we will demonstrate shortly[42]). While the standard virial expansion fails for homogeneous bulk systems because of infra-red divergences[31] (see problem at the end of this section), these divergences are renormalized for the present case of inhomogeneous distribution functions via the normalization condition Eq.(78) as we will now demonstrate. In analogy to the treatment in Section 2.4 leading to Eq.(58), it is useful to formulate the field theory such that the source terms proportional to σ are contained in the fugacity term,

$$\begin{aligned} Z_\lambda[h] = \; & \mathrm{e}^{-\frac{\ell_B}{2}\int_{\mathbf{r},\mathbf{r}'}\sigma_C(\mathbf{r})v_C(\mathbf{r}-\mathbf{r}')\sigma_C(\mathbf{r}')} \\ & \times \int \frac{\mathcal{D}\phi}{Z_v} \exp\left\{ -\frac{1}{2\ell_B q^2}\int_{\mathbf{r},\mathbf{r}'} \phi(\mathbf{r}) v_C^{-1}(\mathbf{r}-\mathbf{r}')\phi(\mathbf{r}') \right. \\ & \left. -\lambda \int_{\mathbf{r}} \theta(z) \mathrm{e}^{h(\mathbf{r}) - \imath\phi(\mathbf{r}) + \ell_B q \int_{\mathbf{r}'} v_C(\mathbf{r}-\mathbf{r}')\sigma_C(\mathbf{r}')} \right\}, \end{aligned} \tag{90}$$

which is only a slight modification of the formulation in Eq.(71). Rescaling again all lengths by the Gouy-Chapman length, $\mathbf{r} = \mu\tilde{\mathbf{r}}$, the grand-canonical partition function can be written in the same form as in (73). Using the explicit form for the charge distribution (74) and the rescaled fugacity Λ as defined in (76), the Hamiltonian in (73) can be written as

$$\begin{aligned} H[\phi,h] = \; & \frac{1}{2}\int_{\tilde{\mathbf{r}},\tilde{\mathbf{r}}'} \phi(\tilde{\mathbf{r}}) v_C^{-1}(\tilde{\mathbf{r}}-\tilde{\mathbf{r}}')\phi(\tilde{\mathbf{r}}') + \frac{1}{8\pi^2}\int_{\tilde{\mathbf{r}},\tilde{\mathbf{r}}'} \delta(\tilde{z}) v_C(\tilde{\mathbf{r}}-\tilde{\mathbf{r}}')\delta(\tilde{z}') \\ & -\frac{\Lambda}{2\pi}\int_{\tilde{\mathbf{r}}} \theta(\tilde{z}) \mathrm{e}^{h(\tilde{\mathbf{r}}) - \imath\phi(\tilde{\mathbf{r}}) - \tilde{z}} \end{aligned} \tag{91}$$

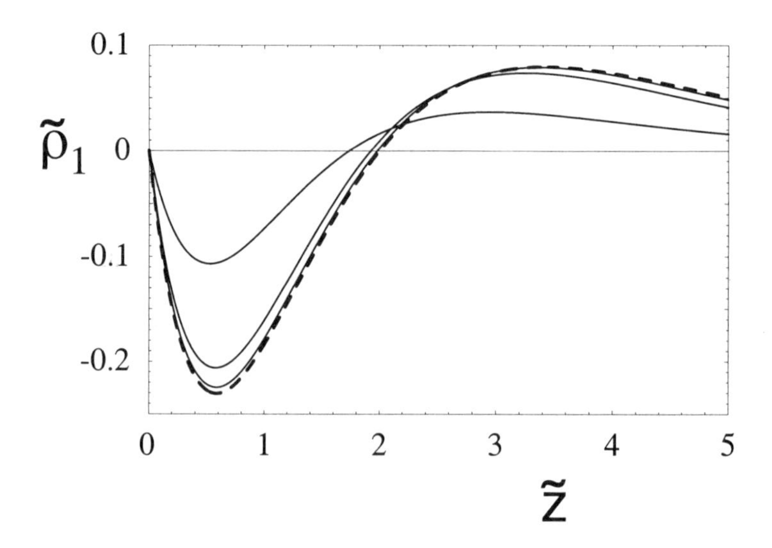

Figure 3. Theoretical prediction for the next-leading contribution to the rescaled density profile, $\tilde{\rho}_1$, within the virial expansion according to Eq.(96). Shown are data in the asymptotic regime $\Xi \to \infty$ (broken line), and for $\Xi = 1, 5, 10$ (solid lines) which gradually approach the asymptotic limit.

which is a slightly different form of Eq.(75). The Gibbs potential is calculated analogously to Eq.(61). The expectation value of the counter-ion density, $\langle \rho(\tilde{\mathbf{r}}) \rangle$, follows by taking a functional derivative with respect to the generating field h, $\langle \rho(\tilde{\mathbf{r}}) \rangle = -\delta G_\lambda / \delta h(\tilde{\mathbf{r}}) \mu^3$, and the rescaled density distribution is defined as $\tilde{\rho}(\tilde{\mathbf{r}}) = \langle \rho(\tilde{\mathbf{r}}) \rangle / (2\pi \ell_B \sigma_s^2)$ and is given in the virial expansion as

$$\tilde{\rho}(\tilde{\mathbf{r}}) = \Lambda \mathrm{e}^{-\tilde{z}} - \frac{\Lambda^2}{2\pi\Xi} \int \mathrm{d}\tilde{\mathbf{r}}' \, \theta(\tilde{z}') \mathrm{e}^{-\tilde{z}-\tilde{z}'} \left[1 - \mathrm{e}^{-\Xi v_C(\tilde{\mathbf{r}}-\tilde{\mathbf{r}}')}\right]. \tag{92}$$

Note that we have rescaled the fugacity Λ by an unimportant multiplicative constant. Expanding the rescaled density as

$$\tilde{\rho}(\tilde{z}) = \tilde{\rho}_0(\tilde{z}) + \tilde{\rho}_1(\tilde{z})/\Xi + \mathcal{O}(1/\Xi^2), \tag{93}$$

and solving the normalization condition Eq.(78) perturbatively, we obtain for the fugacity coefficients $\Lambda_0 = 1$ and

$$\Lambda_1 = \frac{1}{2\pi} \int \mathrm{d}\tilde{\mathbf{r}}' \mathrm{d}\tilde{z} \, \mathrm{e}^{-\tilde{z}-\tilde{z}'} \left[1 - \mathrm{e}^{-\Xi v_C(\tilde{\mathbf{r}}-\tilde{\mathbf{r}}')}\right]. \tag{94}$$

The density distribution is to leading order given by

$$\tilde{\rho}_0(\tilde{z}) = \mathrm{e}^{-\tilde{z}} \tag{95}$$

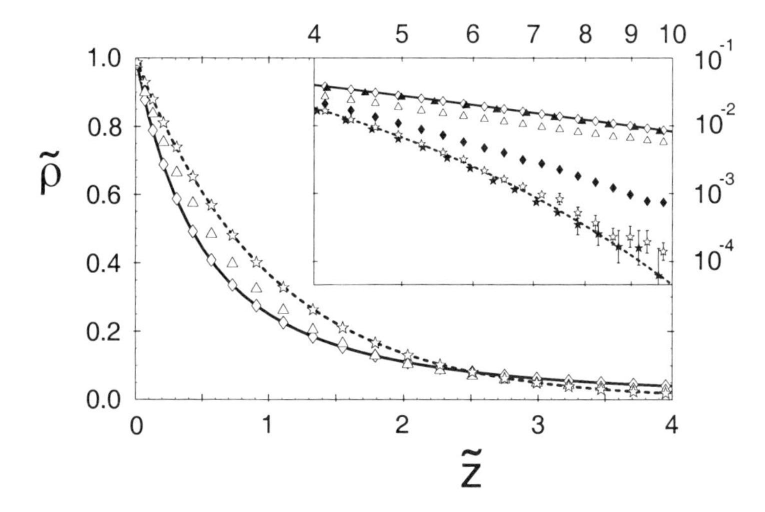

Figure 4. Rescaled counter-ion density distribution $\tilde{\rho} = \rho/2\pi\ell_B\sigma_s^2$ as a function of the rescaled distance from the wall $\tilde{z} = z/\mu$. The inset shows Monte Carlo data for coupling constants $\Xi = 10^5, 10^4, 100, 10, 1, 0.1$ (from bottom to top) in a double-logarithmic plot, the main figure shows data for $\Xi = 0.1$ (open diamonds), $\Xi = 10$ (open triangles) and $\Xi = 10^4$ (open stars). The solid and broken lines denote the Poisson-Boltzmann and strong-coupling predictions, Eqs.(83) and (95), respectively. All data were obtained with 75 particles and 10^6 Monte Carlo steps (MCS), except the data for $\Xi = 0.1$ where 600 particles were simulated. Error bars are smaller than the symbol size if not shown.

and to next-leading order by

$$\tilde{\rho}_1(\tilde{z}) = \Lambda_1 \mathrm{e}^{-\tilde{z}} - \frac{1}{2\pi}\int \mathrm{d}\tilde{\mathbf{r}}'\ \mathrm{e}^{-\tilde{z}-\tilde{z}'}\left[1 - \mathrm{e}^{-\Xi v_C(\tilde{\mathbf{r}}-\tilde{\mathbf{r}}')}\right]. \tag{96}$$

The result for $\tilde{\rho}_1(\tilde{z})$ is plotted in Fig.3 for different values of the coupling parameter Ξ. The fact that the first correction depends on the value of Ξ demonstrates that the strong-coupling theory is not an expansion in inverse powers of Ξ, but rather a fugacity perturbation. One notes that the functional form looks roughly opposite to the one-loop result shown in Fig.2. We note that for $\Xi < 2$ the resulting density shows an unphysical negative region. The density profile in the strong coupling limit $\Xi \to \infty$, Eq.(95), is thus given by a simple exponential, while the profile in the weak-coupling (Poisson-Boltzmann) limit $\Xi \to 0$, Eq.(83), is given by a power law. Just to avoid confusion at this stage, we stress that the exponential density profile Eq.(95) has nothing to do with the Debye-Hückel approximation. It is true that an exponential density profile (though with a different prefactor) also follows from linearizing the differential equation Eq.(81). However, the linearized solution can never be a more faithful representation of the true density profile than the full non-linear solution, and secondly, nothing

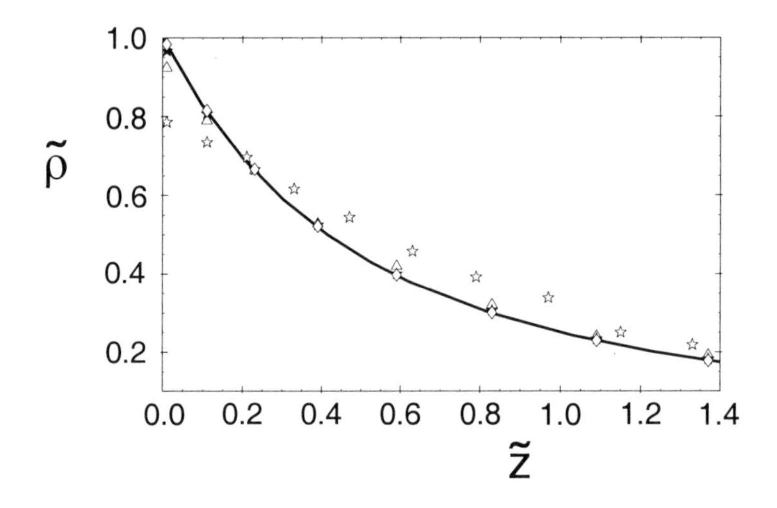

Figure 5. Results for the rescaled counter-ion density distribution $\tilde{\rho}$ as a function of the rescaled distance from the wall $\tilde{z}$ for $\Xi = 0.1$ and 5 particles and 10^8 MCS (open stars), 15 particles and 10^7 MCS (open triangles), 35 particles and 10^7 MCS (filled stars), and 600 particles and 10^6 MCS (open diamonds). Finite size are negligible except for the simulations using 5 particles and (to a lesser degree) 15 particles.

in the saddle-point equation indicates that linearization should be valid for large values of Ξ, because Eq.(81) does not depend on Ξ. It is the saddle-point approach itself (the framework within which the Debye-Hückel approximation can be formulated) which becomes invalid when Ξ becomes large.

In Fig.4 we show counter-ion density profiles obtained using Monte Carlo simulations for various values of the coupling parameter Ξ. As can be seen, the PB density profile Eq.(83) is only realized for $\Xi < 1$, while the strong-coupling profile Eq.(95) is indeed the asymptotic solution and agrees with simulation results for $\Xi > 10^4$. Experimentally, a coupling parameter $\Xi = 100$, which is already quite close to the strong-coupling limit (see Fig.4 inset), is reached with divalent ions for a surface charged density $\sigma_s \approx 3.6nm^{-2}$, which is feasible with compressed charged monolayers, and with trivalent counter ions for $\sigma_s \approx 1nm^{-2}$, which is a typical value. The strong-coupling limit is therefore experimentally accessible. In our simulations we employ periodic boundary conditions in the lateral directions,[10] while the ions are unconfined vertically. Since the number of particles in

[10] Forces and energies within the periodic boundary conditions are calculated following the approach by Lekner and Sperb described in J. Lekner, Physica A **176**, 524 (1991) and R. Sperb, Mol. Simul. **20**, 179 (1998), where the sum over the periodic images is transformed into a rapidly converging expansion in terms of Bessel functions.

the simulation, N, is related to the linear box size (in units of the Gouy-Chapman length μ) by $\tilde{L} = L/\mu = \sqrt{2\pi N \Xi}$, effects due to the finite number of particles and effects due to the finite lateral box size are connected. In Fig.5 we show MC density profiles for a fixed coupling constant $\Xi = 0.1$ and for various number of particles. For all but the smallest systems with 5 and 15 particles finite-size effects are negligible.

In summary, we have derived within a field-theoretic framework the strong-coupling theory for counter-ion distributions, valid in the limit of large valences and/or low temperatures. This theory becomes valid in the opposite limit where the Poisson-Boltzmann approach is accurate. It corresponds to a standard virial expansion of the field-theoretic action, which gives meaningful results because long-range divergences, which spoil the free-energy virial expansion for bulk systems and lead to non-analytic terms [40], are subtracted by a fugacity renormalization. We note that an exponential density profile (albeit with a different prefactor) has been derived in the strong-coupling limit by Shklovskii using a heuristic model where ions bound to the wall are in chemical equilibrium with free ions[49, 50].

Finally, we note that the strong-coupling limit not necessarily entails unrealistically small Gouy-Chapman lengths. Writing the Gouy Chapman length as $\mu = \ell_B q^2/\Xi$, it is clear that μ can be fixed at a moderate value for highly valent ions even when $\Xi \gg 1$. In this case, however, one would have to worry about the effect of discrete surface charges.

Problem 10 Calculate the second-virial coefficient in the bulk for a system of charged particles and show that the virial expansion breaks down.

Problem 11 Calculate moments of the density distribution in the weak-coupling and the strong-coupling limit. Discuss your results and compare them qualitatively with the snapshots in Fig.1.

Problem 12 Use the general result for the connected two-point correlation function at the one-loop level considered in Problem 6 and the specific form (87) to determine its asymptotic behavior for large lateral distances. For simplicity, choose $\tilde{z} = \tilde{z}' = 0$. What is the intuitive explanation for your result?

4. Binding of Similarly Charged Plates: A Global Analysis

Introduction Similarly and highly charged plates in the presence of multivalent counter ions attract each other, leading to electrostatically bound states. Using Monte-Carlo simulations we obtain the inter-plate pressure in the global parameter space. The equilibrium plate separation, where the pressure changes from attractive to repulsive, exhibits a novel unbinding transition. A systematic and asymptotically exact strong-coupling field-theory (which corresponds to the standard virial expansion) yields the bound state from a competition between counter-ion entropy and electrostatic attraction, in agreement with simple scaling arguments.

Experimentally, it has been known for a long time that highly charged planar surfaces attract each other in the presence of multivalent counter ions, inducing bound states. This electrostatic binding restricts the swelling of calcium clay particles[22] and leads to much reduced water uptake of charged lamellar membrane systems[54]. Attractive forces between charged surfaces have also been observed with the surface force apparatus[21]. Monte-Carlo simulations indeed confirmed that for a given surface charge density there exists a threshold counter-ion valence above which attraction can be observed over some range of plate separations[19]. Similar results have also been observed for charged cylinders[18, 20] and for charged colloidal particles[11, 26, 32].

Theoretically, these observations came as a surprise, since the mean-field or Poisson-Boltzmann (PB) theory, which works usually quite well for charged systems, predicts only repulsive forces between charged objects[2]. This contradiction between observation and PB prediction resulted in an immense theoretical activity, which aimed at understanding the simple model of two uniformly and similarly charged planar surfaces interacting across a gap of width d filled with point-like counter ions. Clearly, reality is much more complicated due to additional interactions[10], but even this simple model, which we will consider in the following, is quite challenging. A number of approaches were proposed which incorporate counter-ion correlations that are neglected within PB. The first were integral-equation theories[23], perturbative expansions around the PB theory[3, 46], and local density-functional theory[51, 15, 5], which compare well with simulation results and exhibit attraction. If the two plates are far apart from each other, the counter-ion clouds can be viewed as condensed on the plates, and the resulting simplified model can be solved within a Gaussian[3, 45, 29] or harmonic-plasmon approximation[27]. These approaches either involve numerics and do not provide much physical insight, or they are valid for large separations and cannot be used to characterize the bound state.

Of great significance is the fact that when the electrostatic force is attractive, the equilibrium distance d^* between the plates is smaller than the typical lateral distance a between counter ions (as we will demonstrate later on), thus rendering a quasi-two-dimensional layer of counter ions[47]. In the first part of this section, we will use this fact and present a scaling argument for the attraction between two plates, valid for $d \ll a$. We will then demonstrate that this scaling analysis is equivalent to the leading order of a systematic field-theory, valid in the strong-coupling (SC) limit (corresponding to large plate-charge density σ or large counter-ion valence q), where it agrees with extensive Monte-Carlo (MC) simulations[34]. Our MC results span the complete parameter space. Whenever attractive forces between the plates exist, they induce a bound equilibrium state, which exhibits a novel unbinding transition.

The simple scaling argument for attraction between charged plates starts with partitioning the system into isolated counterions sandwiched between two finite plate segments of area $A = q/2\sigma_s$. Neglecting ion-ion interactions should be valid for $d \ll \sqrt{A}$. Denoting the distance between the counterion and the plates as x and $d-x$, respectively, we obtain for the electrostatic interaction energies, in units of k_BT and for $d \ll \sqrt{A}$, the results $W_1 = 2\pi\ell_B q\sigma_s x$ and $W_2 = 2\pi\ell_B q\sigma_s(d-x)$, respectively, as follows from the potential at an infinite charged wall and omitting constant terms. The sum of the two interactions is $W_{1+2} = W_1 + W_2 = 2\pi\ell_B q\sigma_s d$ which shows that i) no pressure is acting on the counterion since the forces exerted by the two plates exactly cancel and ii) that the counterion mediates an effective attraction between the two plates. The interaction between the two plates is proportional to the total charge on one plate, $A\sigma_s$, and for $d \ll \sqrt{A}$ given by $W_{12} = -2\pi A\ell_B\sigma_s^2 d$. Since the system is electroneutral, $q = 2A\sigma_s$, the total electrostatic energy is $W_{el} = W_{12} + W_1 + W_2 = 2\pi A\ell_B\sigma_s^2 d$, leading to an electrostatic pressure $P_{el} = -\partial(W_{el}/A)/\partial d = -2\pi\ell_B\sigma_s^2$. The two plates attract each other. The entropic pressure due to counter-ion confinement is $P_{en} = 1/Ad = 2\sigma_s/qd$. The equilibrium plate separation is characterized by zero total pressure, $P_{tot} = P_{el} + P_{en} = 0$, leading to an equilibrium plate separation $d^* = 1/\pi\ell_B q\sigma_s$. By construction, the derivation for d^* is valid only for $d^* < a$ (where a is the average lateral distance between ions as defined by $\pi a^2 = q/2\sigma_s$), equivalent to the condition $\Xi = 2\pi\ell_B^2\sigma_s q^3 > 4$, i.e. for large values of the coupling constant Ξ. Surprisingly, these results for P_{tot} and d^* become exact in the SC limit $\Xi \to \infty$, as we will demonstrate in the following. In fact, the attraction between charged plates, as derived here for $d \ll a$, is conceptually simpler than the PB result of repulsion, because the latter case involves many-body effects.

4.1. MAPPING ON A FIELD THEORY

To proceed with our systematic field theory, consider the partition function for N counter ions confined between two parallel plates at distance d

$$Z_N = \frac{1}{N!}\prod_{j=1}^{N}\int \mathrm{d}\mathbf{r}_j\ \theta(z_j)\theta(d-z_j)\mathrm{e}^{-H} \tag{97}$$

where the Heavyside function is defined by $\theta(z) = 1$ for $z > 0$ and zero otherwise. Introducing the counter-ion density operator $\hat{\rho}(\mathbf{r}) = \sum_{j=1}^{N} \delta(\mathbf{r} - \mathbf{r}_j)$ the Hamiltonian can be written as

$$H = \frac{\ell_B}{2} \int \mathrm{d}\mathbf{r}\mathrm{d}\mathbf{r}'[q\hat{\rho}(\mathbf{r}) - \sigma_s\delta(z) - \sigma_s\delta(d-z)]v_C(\mathbf{r}-\mathbf{r}')$$
$$[q\hat{\rho}(\mathbf{r}') - \sigma_s\delta(z') - \sigma_s\delta(d-z')] - \int \mathrm{d}\mathbf{r}\; \hat{\rho}(\mathbf{r})h(\mathbf{r}) \tag{98}$$

where $v_C(\mathbf{r}) = 1/r$ is the Coulomb interaction and the field h has been added to calculate density distributions later on. The characteristic length scales are the Bjerrum length $\ell_B = e^2/4\pi\varepsilon k_B T$ and the Gouy-Chapman length $\mu = 1/2\pi\ell_B q\sigma_s$, which measure the distance at which the interaction between two unit charges and between a counter ion and a charged wall reach thermal energy, respectively. Rescaling all lengths by the Gouy-Chapman length according to $\mathbf{r} = \mu\tilde{\mathbf{r}}$ and $d = \mu\tilde{d}$, as done before, the Hamiltonian becomes

$$H = \frac{1}{8\pi^2\Xi} \int \mathrm{d}\tilde{\mathbf{r}}\mathrm{d}\tilde{\mathbf{r}}'[2\pi\Xi\hat{\rho}(\tilde{\mathbf{r}}) - \delta(\tilde{z}) - \delta(\tilde{d}-\tilde{z})]v_C(\tilde{\mathbf{r}}-\tilde{\mathbf{r}}')$$
$$[2\pi\Xi\hat{\rho}(\tilde{\mathbf{r}}') - \delta(\tilde{z}') - \delta(\tilde{d}-\tilde{z}')] - \int \mathrm{d}\tilde{\mathbf{r}}\; \hat{\rho}(\tilde{\mathbf{r}})h(\tilde{\mathbf{r}}) \tag{99}$$

and thus only depends on the coupling parameter $\Xi = 2\pi q^3\ell_B^2\sigma_s$. At this point we employ a Hubbard-Stratonovitch transformation, followed by a Legendre transformation to the grand-canonical ensemble, $Z_\lambda = \sum_N \tilde{\lambda}^N Z_N$, introducing the fugacity $\tilde{\lambda}$. The inverse Coulomb operator follows from Poisson's law as $v_C^{-1}(\mathbf{r}) = -\nabla^2\delta(\mathbf{r})/4\pi$, which leads to

$$Z_\lambda = \int \frac{\mathcal{D}\phi}{Z_v} \exp\left\{-\frac{1}{8\pi\Xi}\int \mathrm{d}\tilde{\mathbf{r}}\left[[\nabla\phi(\tilde{\mathbf{r}})]^2 - 4\imath\delta(\tilde{z})\phi(\tilde{\mathbf{r}}) - 4\imath\delta(\tilde{d}-\tilde{z})\phi(\tilde{\mathbf{r}})\right.\right.$$
$$\left.\left. -4\Lambda\theta(\tilde{z})\theta(\tilde{d}-\tilde{z})\mathrm{e}^{h(\tilde{\mathbf{r}})-\imath\phi(\tilde{\mathbf{r}})}\right]\right\} \tag{100}$$

where we used the notation $Z_v = \sqrt{\det v_C}$ and the rescaled fugacity Λ is defined by $\Lambda = 2\pi\lambda\mu^3\Xi = \lambda/(2\pi\sigma_s^2\ell_B)$, as in Eq.(76). The expectation value of the counter-ion density, $\langle\rho(\tilde{\mathbf{r}})\rangle$, follows by taking a functional derivative with respect to the generating field h, $\langle\rho(\tilde{\mathbf{r}})\rangle = \delta\ln Z_\lambda/\delta h(\tilde{\mathbf{r}})\mu^3$, giving rise to

$$\tilde{\rho}(\tilde{\mathbf{r}}) = \frac{\langle\rho(\tilde{\mathbf{r}})\rangle}{2\pi\ell_B\sigma_s^2} = \Lambda\langle\mathrm{e}^{-\imath\phi(\tilde{z})}\rangle. \tag{101}$$

The normalization condition for the counter-ion distribution, $\mu\int \mathrm{d}\tilde{z}\rho(\tilde{z}) = 2\sigma_s/q$, which follows directly from the definition of the grand-canonical partition function, leads to

$$\Lambda \int_0^{\tilde{d}} \mathrm{d}\tilde{z} \langle \mathrm{e}^{-\imath\phi(\tilde{z})} \rangle = 2, \tag{102}$$

which is the slightly modified version of the normalization condition of the case of a single charged wall, Eq.(78).

4.2. MEAN-FIELD OR POISSON-BOLTZMANN APPROACH

Let us first repeat the saddle-point analysis, which, because of the structure of the action in Eq.(100), should be valid for $\Xi \ll 1$. The saddle-point equation reads $\partial^2\phi(\tilde{z})/\partial\tilde{z}^2 = 2\imath\Lambda \mathrm{e}^{-\imath\phi(\tilde{z})}$, compare Eq.(81). The symmetric solution of this differential equation is

$$\imath\phi_{SP}(\tilde{z}) = 2 \ln \cos\left(\Lambda^{1/2}[\tilde{z} - \tilde{d}/2]\right). \tag{103}$$

The normalization condition (102) leads to the equation

$$\Lambda_0^{1/2} \tan[\tilde{d}\Lambda_0^{1/2}/2] = 1, \tag{104}$$

which is solved by $\Lambda_0 \simeq 2/\tilde{d}$ for $\tilde{d} \ll 1$ and $\Lambda_0 \simeq \pi^2/4\tilde{d}^2$ for $\tilde{d} \gg 1$. From Eqs.(101) and (103), the rescaled density distribution of counter ions is given by the well-known PB result

$$\tilde{\rho}_0(\tilde{z}) = 1/\cos^2\left(\Lambda_0^{1/2}[\tilde{z} - \tilde{d}/2]\right). \tag{105}$$

4.3. STRONG-COUPLING OR VIRIAL APPROACH

Let us now consider the opposite limit, when the coupling constant Ξ is large. In this case, the saddle-point approximation breaks down, since the prefactor in front of the action in Eq.(100) becomes small. Since the fugacity term is bounded, as evidenced by Eq.(102), one can expand the partition function (and also all expectation values) in powers of Λ/Ξ (which is equivalent to a virial expansion). For the expectation value determining the density Eq.(101) the leading two orders in the virial expansion are

$$\tilde{\rho}(\tilde{z}) = \Lambda + \frac{\Lambda^2}{2\pi\Xi} \int \mathrm{d}\tilde{\mathbf{r}}' \left(\mathrm{e}^{-\Xi v(\tilde{\mathbf{r}}-\tilde{\mathbf{r}}')} - 1\right). \tag{106}$$

The normalization condition Eq.(102) can then be solved by an expansion of the fugacity in inverse powers of the coupling constant, $\Lambda = \Lambda_0 + \Lambda_1/\Xi + \ldots$, in agreement with Eq.(89). To leading order, we obtain

$$\Lambda_0 = 2\mathrm{e}^{\Xi v(0)/2}/\tilde{d} \tag{107}$$

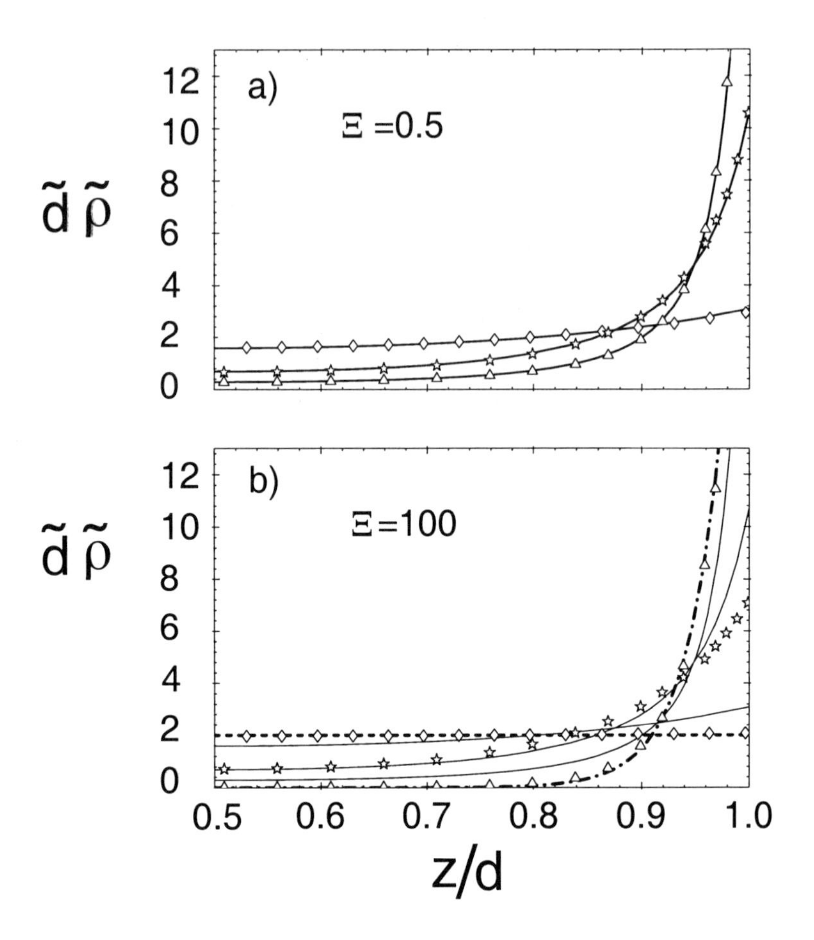

Figure 6. MC results for the rescaled counter-ion density $\tilde{d}\tilde{\rho}$ as a function of the rescaled distance from the wall z/d in the a) PB limit for $\Xi = 0.5$ and in the b) SC limit for $\Xi = 100$ for various plate separations $\tilde{d} = d/\mu = 1.5$ (open diamonds), $\tilde{d} = 10$ (open stars), and $\tilde{d} = 30$ (open triangles). In a) MC results agree well with the corresponding PB predictions (Eq.(105), solid lines), whereas in b) results for $\tilde{d} = 1.5$ agree with the asymptotic SC prediction, Eq.(108) (dashed line) and for $\tilde{d} = 30$ with a double-exponential curve (see text).

and thus the density distribution is (in agreement with our scaling analysis) to leading order indeed a constant given by

$$\tilde{\rho}_0(\tilde{z}) = 2/\tilde{d} + \mathcal{O}(\Xi^{-1}) \tag{108}$$

In Fig.6a we show counter-ion density profiles obtained using MC simulations[11] for small coupling parameter $\Xi = 0.5$ for various plate distances,

[11] Simulations were typically performed for 10^6 MC steps with 150 counter ions, in which case finite-size effects are negligible.

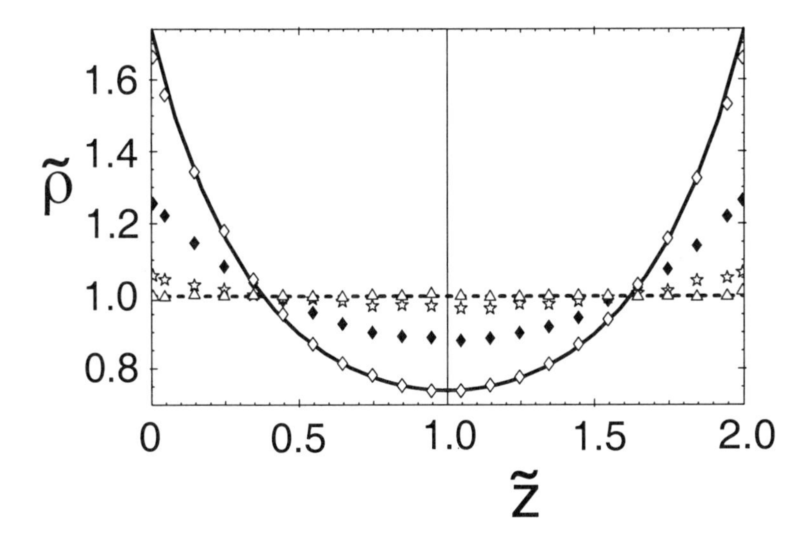

Figure 7. MC results for rescaled counter-ion density profiles $\tilde{\rho} = \rho/2\pi\ell_B\sigma_s^2$ for fixed plate separation $\tilde{d} = d/\mu = 2$ as a function of the rescaled distance $\tilde{z} = z/\mu$ from one wall. Symbols correspond to coupling parameters $\Xi = 0.5$ (open diamonds), $\Xi = 10$ (filled diamonds), $\Xi = 100$ (open stars), and $\Xi = 10^5$ (open triangles), exhibiting clearly the crossover from the PB prediction (solid line, Eq.(105)) to the SC prediction (broken line, Eq.(108)).

which are well described by the PB profiles Eq.(105) shown as solid lines. Fig.6b shows that for $\Xi = 100$ PB (thin solid lines) is inadequate. As suggested by our scaling analysis, the asymptotic SC result Eq.(108) should be valid for $d/a = \tilde{d}/\Xi^{1/2} < 1$ only, since otherwise ion-ion interactions become important. For $\tilde{d} = 3/2$ (open diamonds) we find $d/a = 0.15$, and indeed Eq.(108) is accurate. For $\tilde{d} = 10$ (open stars) we find $d/a = 1$, the density profile is neither described by Eq.(108) nor (105). Finally, for $\tilde{d} = 30$ (open stars) we find $d/a = 3$, the two layers are decoupled and the density profile is described by a double exponential $\tilde{\rho}(\tilde{z}) = (\mathrm{e}^{-\tilde{z}} + \mathrm{e}^{\tilde{z}-\tilde{d}})/(1 - \mathrm{e}^{-\tilde{d}})$ (dashed-dotted line), which is the superposition of the density profiles of two isolated charged surfaces in the SC limit (see preceding section). The crossover from PB to SC is demonstrated in Fig.7, where we plot density profiles for fixed plate separation $\tilde{d} = 2$ for various coupling parameters Ξ.

Using the contact value theorem, the pressure P between the two plates, which follows from the partition function via $P = \partial \ln Z_\lambda / A\mu\partial\tilde{d}$, is related to the counter ion density at a plate by[19, 41]

$$\tilde{P} = \frac{P}{2\pi\ell_B\sigma_s^2} = \tilde{\rho}(\tilde{d}) - 1. \tag{109}$$

The first term is the entropic pressure due to counter-ion confinement, the

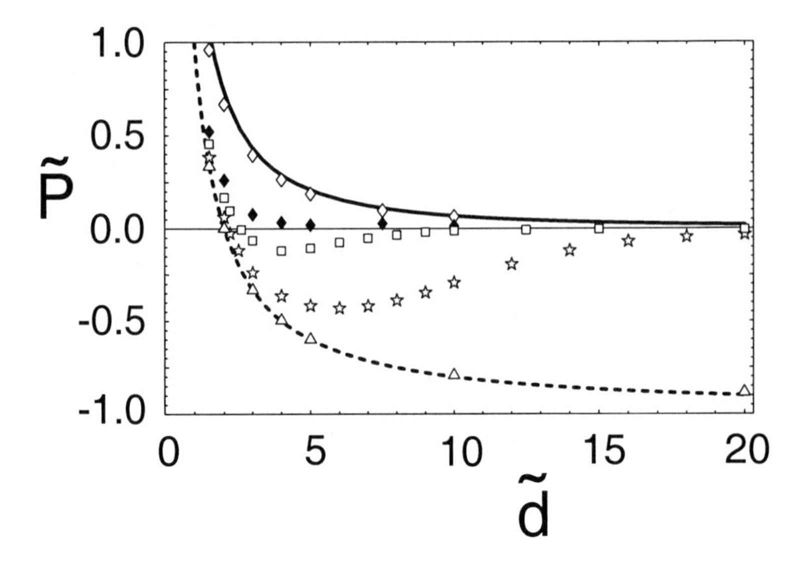

Figure 8. MC results for the rescaled pressure $\tilde{P} = P/2\pi\ell_B\sigma_s^2$ as a function of the rescaled plate separation $\tilde{d} = d/\mu$ for the same parameter values as in Fig.2 (and $\Xi = 20$, open squares), compared with the PB prediction $\tilde{P} = \Lambda_0$ (solid line) and the SC prediction $\tilde{P} = 2/\tilde{d} - 1$ (broken line).

second term is due to electrostatic interactions between the counterions and the charged plates. Numerically, the contact ion density $\tilde{\rho}(\tilde{d})$ is obtained from the density profiles by extrapolation. In Fig.8 we show numerical pressure data for various values of Ξ. Attraction (negative pressure) is obtained for $\Xi > 10$. The numerical pressure for $\Xi = 0.5$ (open diamonds) agrees well with the PB prediction (solid line), which from Eqs.(109) and (105) is given by $\tilde{P} = \Lambda_0$ with Λ_0 determined by $\Lambda_0^{1/2} \tan[\tilde{d}\Lambda_0^{1/2}/2] = 1$. The small distance range of most data, and the complete pressure data for $\Xi = 10^5$ (open triangles) are well described by the SC prediction (broken line). It results from combining Eqs.(109) and (108) and is given by $\tilde{P} = 2/\tilde{d} - 1$, from which the equilibrium separation, which corresponds to the minimum of the effective plate-plate interaction, is obtained as $\tilde{d}^* = 2$. Incidentally, this is exactly the scaling prediction for the pressure derived in the beginning of this section.

Finally, combining all pressure data, we obtain the global phase diagram shown in Fig.9, featuring two regions where the inter-plate pressure is attractive and repulsive. The dividing line between those regions, which corresponds to the equilibrium plate separation $\tilde{d}^*$ in the bound state, is determined over 4 decades of the coupling constant Ξ. In the limit of large coupling constants, the phase boundary saturates at $\tilde{d}^* = 2$, in agreement

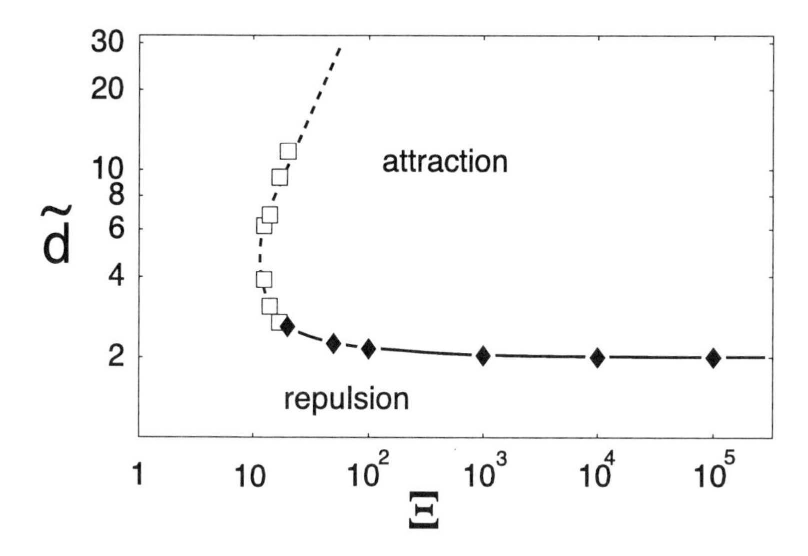

Figure 9. MC phase diagram showing regions of attractive and repulsive pressure as a function of plate separation $\tilde{d}$ and coupling strength Ξ. Attraction only occurs for intermediate distances and $\Xi > 12$. The equilibrium plate separation (determined by minimization of the free energy, solid line) exhibits a discontinuous unbinding to infinity at $\Xi^{**} \approx 17$ and saturates at $\tilde{d}^* = 2$ for $\Xi \to \infty$.

with our scaling argument and the leading result of our SC theory. The threshold coupling constant to observe attraction between the plates is $\Xi^* \simeq 12$. As Ξ decreases from large values, the equilibrium separation grows and exhibits a discontinuous transition [34] at $\Xi = \Xi^{**} \simeq 17$. This unbinding transition could experimentally be observed with charged lamellar or clay systems by raising the temperature.

As mentioned before, the ratio of plate separation and lateral ion-distance is $d/a = \tilde{d}/\Xi^{1/2}$, which crosses the equilibrium separation line in Fig.9 at $\tilde{d} \approx 3$. For the most part of the equilibrium line in Fig.9 the counter-ion distribution is indeed two-dimensional, and many-ion effects can be neglected except very close to Ξ^*. Correlations between counter ions, except the trivial lateral exclusion correlation which keeps ions apart from each other, are therefore mostly unimportant in the bound state. This explains why the simple single-ion scaling argument advanced in the beginning, which turns out to be exact in the limit $\Xi \to \infty$, works so well. A more formal explanation is given in Ref.[42] where the next-leading correction is calculated and compared with the leading term, leading to an equivalent conclusion. Recently, there has been an active discussion on the significance of Wigner crystallization (which occurs at $\Xi_{WC} \simeq 15600$ in the limit $\tilde{d} \to 0$) for the phenomena involving attraction between similarly charged plates.

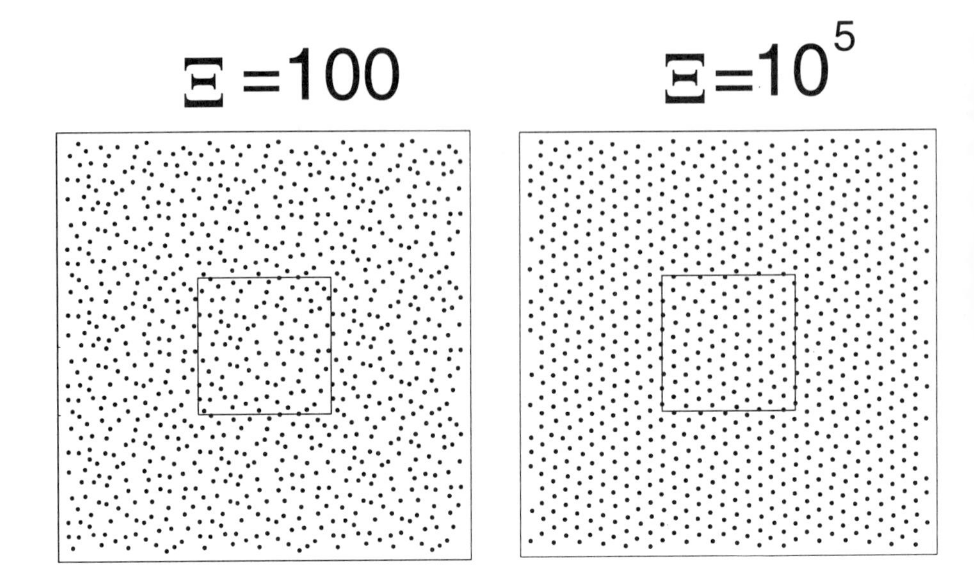

Figure 10. Top-view snapshots of Monte-Carlo simulations at a plate separation $\tilde{d} = 2$ for coupling parameter $\Xi = 100$, showing liquid-like behavior, and $\Xi = 10^5$, exhibiting crystalline order, indicative of Wigner crystallization.

In Fig.10 we show top-view snapshots of the Monte-Carlo simulations for $\Xi = 100$ and $\Xi = 10^5$ for $\tilde{d} = 2$. In agreement with the estimated Wigner crystallization threshold, $\Xi_{WC} \simeq 15600$, the snapshot for $\Xi = 100$ shows liquid behavior, while the snapshot for $\Xi = 10^5$ exhibits crystalline order. Since the experimentally relevant attraction occur for values $\Xi < 100$, it seems that Wigner crystallization is not connected or responsible for the attraction between similarly charged plates. On the other hand, treating the strongly correlated liquid layer of counter-ions like a Wigner crystal is in many cases a reasonable approximation[50].

Problem 13 Derive the contact value theorem Eq.(109) from the partition function.

References

1. D.J. Amit, *Field Theory, the Renormalization Group, and Critical Phenomena* (World Scientific, Singapore, 1984).
2. D. Andelman, in *Handbook of Biological Physics*, edited by R. Lipowsky and E. Sackmann, Volume I, p. 603 (Elsevier, 1995).
3. P. Attard, R. Kjellander, D.J. Mitchell, and B. Jönsson, J. Chem. Phys. **89**, 1664 (1988).

4. P. Attard, D.J. Mitchell, and B.W. Ninham, J. Chem. Phys. **88**, 4987 (1988); *ibid.* **89**, 4358 (1988).
5. M.C. Barbosa, M. Deserno, and C. Holm, Europhys. Lett. **52**, 80 (2000).
6. Barrat J.-L.; Joanny J.-F. *Adv. Chem. Phys.* **1996**, *94*, 1.
7. M. Baus and J.P. Hansen, Phys. Rep. **59**, 1 (1980).
8. C.J.F. Böttcher, *Theory of Electric Polarization* (Elsevier, Amsterdam, 1973).
9. E. Brézin, J.C. Le Guillou, and J. Zinn-Justin, in *Phase Transitions and Critical Phenomena*, Vol. VI, C. Domb and M.S. Green, eds. (Academic Press, N.Y., 1976).
10. Y. Burak and D. Andelman, Phys. Rev. E **62**, 5296 (2000).
11. J.C. Crocker and D.G. Grier, Phys. Rev. Lett. **77**, 1897 (1996).
12. H. Dautzenberg; W. Jäger; B.P.J. Kötz; C. Seidel; D. Stscherbina *Polyelectrolytes: Formation, characterization and application*, Hanser Publishers: Munich, Vienna, New York, 1994.
13. P.W. Debye, Phys. Z. **13**, 97 (1912).
14. P.W. Debye and E. Hückel, Phys. Z. **24**, 185 (1923).
15. A. Diehl, M.N. Tamashiro, M.C. Barbosa, and Y. Levin, Physica A **274**, 433 (1999).
16. S. Förster; M. Schmidt *Adv. Polym. Sci.* **1995**, *120*, 50.
17. W.M. Gelbart, contribution in this volume.
18. N. Grønbech-Jensen, R.J. Mashl, R.F. Bruinsma, and W.M. Gelbart, Phys. Rev. Lett. **78**, 2477 (1997).
19. L. Guldbrand, B. Jönsson, H. Wennerström, and P. Linse, J. Chem. Phys. **80**, 2221 (1984).
20. B.-Y. Ha and A.J. Liu, Phys. Rev. Lett. **79**, 1289 (1997).
21. P. Kékicheff, S. Marĉelja, T.J. Senden, and V.E. Shubin, J. Chem. Phys. **99**, 6098 (1993).
22. R. Kjellander, S. Marcelja, and J.P. Quirk, J. Colloid Interface Sci. **126**, 194 (1988).
23. R. Kjellander and S. Marĉelja, Chem. Phys. Lett. **112**, 49 (1984); J. Chem. Phys. **82**, 2122 (1985); Chem. Phys. Lett. **127**, 402 (1986).
24. R. Kjellander, contribution in this volume.
25. K.-K. Kunze and R.R. Netz, Phys. Rev. Lett. **85**, 4389 (2000).
26. A.E. Larsen and D.G. Grier, Nature **385**, 231 (1997).
27. A.W.C. Lau, D. Levine, and P. Pincus, Phys. Rev. Lett. **84**, 4116 (2000).
28. R. Lipowsky, Progr. Colloid Polym. Sci. **111**, 34 (1998).
29. D.B. Lukatsky and S.A. Safran, Phys. Rev. E **60**, 5848 (1999).
30. S.-K. Ma, *Modern Theory of Critical Phenomena* (Addison-Wesley, Redwood City, 1976).
31. Virial expansion methods are discussed in D.A. McQuarrie, *Statistical Mechanics* (Harper & Row, New York, 1976), Chap. 15.
32. R. Messina, C. Holm, and K. Kremer Phys. Rev. Lett. **85**, 872 (2000); Europhys. Lett. **51**, 461 (2000).
33. A.G. Moreira and R.R. Netz, Europhys. Lett. **52**, 705 (2000).
34. A.G. Moreira and R.R. Netz, Phys. Rev. Lett. **87**, 078301 (2001).
35. P.F. Mossotti, Bibl. Univ. Modena **6**, 193 (1847).
36. D.H. Napper, *Polymeric Stabilization of Colloidal Dispersions* (Academic Press, London, 1983).
37. J.W. Negele and H. Orland, *Quantum Many-Particle Systems* (Addison-Wesley, Reading, 1988).
38. R.R. Netz, Phys. Rev. Lett. **76**, 3646 (1996).
39. R.R. Netz and H. Orland, Europhys. Lett. **45**, 726 (1999).
40. R.R. Netz and H. Orland, Eur. Phys. J. E **1**, 67 (2000).

41. R.R. Netz and H. Orland, Eur. Phys. J. E **1**, 203 (2000).
42. R.R. Netz, to be published.
43. L. Onsager, J. Am. Chem. Soc. **58**, 1486 (1936).
44. G. Parisi, *Statistical Field Theory* (Addison-Wesley, Redwood City, 1988).
45. P.A. Pincus and S.A. Safran, Europhys. Lett. **42**, 103 (1998).
46. R. Podgornik and B. Zeks, J. Chem. Soc. Faraday Trans. 2 **84**, 611 (1988).
47. I. Rouzina and V.A. Bloomfield, J. Phys. Chem. **100**, 9977 (1996).
48. J. Schwinger, Proc. Natl. Acad. Sci. **37**, 452 (1951); **37**, 455 (1951); **44**, 956 (1958).
49. B.I. Shklovskii, Phys. Rev. E **60**, 5802 (1999).
50. B.I. Shklovskii, contribution in this volume.
51. M.J. Stevens and M.O. Robbins, Europhys. Lett. **12**, 81 (1990).
52. Verwey E.J.W.; Overbeek J.Th.G. *Theory of the Stability of Lyophobic Colloids*, Elsevier: Amsterdam, New York, 1948.
53. D.A. Weitz and J.S. Huang in *Kinetics of Aggregation and Gelation*, P. Family and D.P. Landau, eds. (Elsevier, Amsterdam, 1984).
54. H. Wennerström, A. Khan, and B. Lindman, Adv. Colloid Interface Sci. **34**, 433 (1991).
55. T.D. Yager; C.T. McMurray; K.E. van Holde *Biochem.* **1989**, *28*, 2271.

INTERACTIONS AND CONFORMATIONAL FLUCTUATIONS IN MACROMOLECULAR ARRAYS

RUDOLF PODGORNIK
LPSB - NICHD, National Institutes of Health, Bethesda, MD 20892-5626
Department of Physics, Faculty of Mathematics and Physics, University of Ljubljana, Jadranska 19, SI-1000 Ljubljana, and Department of Theoretical Physics, J. Stefan Institute, Jamova 39, SI-1000 Ljubljana, Slovenia

1. Introduction and outline

I will review the work on the equation of state in macromolecular arrays, meaning mostly ordered hexagonal DNA arrays and ordered multilamellar lipid membrane arrays. First I will discuss the interactions between charged macromolecules of different geometrical shape, such as DNA cylinders and planar lipid bilayers, and then develop a statistical mechanical theory for the equation of state that will include also conformational fluctuations of the macromolecules in the array. I will start with a very simple model and then refine it to the extent that the theoretically derived results become comparable with experiments. Further on I will analyze the way bare interactions couple with fluctuations and draw some general conclusions regarding their importance in ordered DNA arrays and ordered multilamellar lipid membrane arrays. The main difference in behaviors of the two systems stems as much from a different internal geometry (1D for polymers vs. 2D for membranes) as well as from the fact that in condensed DNA phases the positional order is short ranged (liquid like) while it is quasi long ranged (smectic) in membrane assemblies. I will conclude by examining the interaction renormalization of the bending moduli of DNA in hexagonal arrays and of lipid membranes in multilamellar arrays. From here I will show that in the low salt limit the effect of conformational fluctuations on the equation of state becomes negligible. In this case the osmotic pressure stemming only from the counterions can account for most of the available experimental data. The lecture will thus give a comprehensive introduction to the equation of state for both types of biological macromolecules under various solution conditions with monovalent salts and counterions.

C. Holm et al. (eds.), Electrostatic Effects in Soft Matter and Biophysics, 409–440.
© 2001 *Kluwer Academic Publishers. Printed in the Netherlands.*

In the phase diagram of DNA in aqueous solutions there is a range extending from the DNA crystal at densities corresponding to interaxial separations below ≈ 24 Å all the way to the nematic - isotropic transition at densities corresponding to interaxial separations ≈ 120 Å (for a bathing solution of 0.5 M NaCl) where DNA is orientationally ordered or even shows long range hexatic order perpendicular to the long axes of the molecules but is positionally still a liquid with only short range positional order [1]. Similarly there is a range of lipid concentrations well above the CMC where lipids exist in bilayer form making a lamellar L_α phase with quasi long range positional order of the lamellae [2] where fluctuations in the local positions depend only weakly (logarithmically) on the size of the sample.

In this part of the phase diagram of both systems, the equation of state, *i.e.* the dependence of the osmotic pressure on the macromolecular (DNA or lipid) concentration, has been studied very carefully. In this lecture I will make an attempt to explain in all necessary detail the theoretical underpinnings of these equations of state. It would suit my purposes ideally if I could compare the equation of state of a charged linear polymer such as DNA with the equation of state of charged lipid bilayers, but the experiments performed up to this time do not permit that quite yet. There are persistent salt equilibration problems in the multilamellar lipid case that have not yet been fully overcome. Nevertheless I believe that even the comparison of the behavior of DNA arrays with neutral lipid membrane arrays is definitively worth a try.

2. DLVO interactions

DLVO forces [3], refer to the (Poisson - Boltzmann or Debye-Hückel) mean-field screened electrostatic interactions and Lifshitz - van der Waals electromagnetic fluctuation forces in the context of pairwise interaction between charged cylindrical and planar macromolecules.

2.1. POISSON-BOLTZMANN THEORY

We start by writing down the expression for the non-equilibrium mean-field free energy density (f) of a gas of mobile charged particles *i.e.* counterions and salt ions. It can be written as a difference of the electrostatic field energy density (w) and the ideal entropy density of the mobile charge carriers (s)

$$\begin{aligned} f(\mathbf{E}(\mathbf{r}), \rho_i(\mathbf{r})) &= w - Ts \\ &= \tfrac{1}{2}\epsilon\epsilon_0 \mathbf{E}^2(\mathbf{r}) + kT \sum_i \left(\rho_i(\mathbf{r}) ln\left(\frac{\rho_i(\mathbf{r})}{\rho_i^0}\right) - (\rho_i(\mathbf{r}) - \rho_i^0) \right), \end{aligned} \tag{1}$$

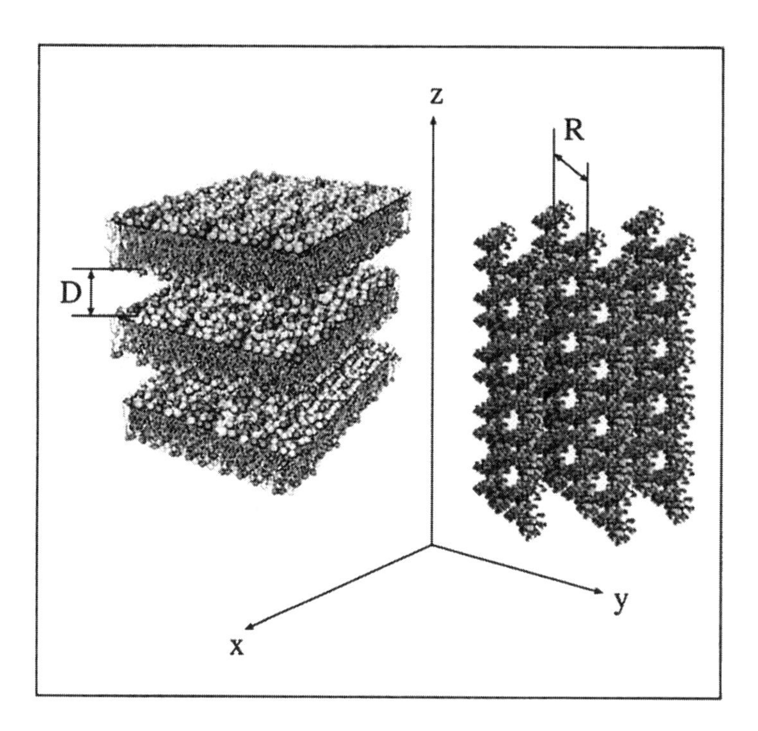

Figure 1. The schematic geometry of an array of DNA molecules with long range orientational order and short range positional order and an array of lipid membranes in the L_α with quasi-long range positional order. Both are drawn approximately to scale. Adapted in part from Rädler JO, Koltover I, Salditt T, Safinya CR *Science* **275** 78-81 (1997).

where $\mathbf{E}(\mathbf{r}) = -\nabla\phi(\mathbf{r})$ is the local electric field that satisfies the Poisson equation

$$\epsilon\epsilon_0 \ \mathrm{div}\mathbf{E}(\mathbf{r}) = \sum_i e_i\rho_i(\mathbf{r}). \tag{2}$$

Here ϵ is the static dielectric constant, kT is the thermal energy, $\rho_i(\mathbf{r})$ is the density of the mobile charge carrier i and ρ_i^0 is the charge density of the mobile charged species i in the reservoir. By minimizing the free energy given by the volume integral of Eq. 1 with respect to the electric field and taking into account the Poisson equation Eq. 2, one obtains the Poisson - Boltzmann equation in the form

$$\nabla^2\phi(\mathbf{r}) = -\frac{1}{\epsilon\epsilon_0}\sum_i \rho_i(\phi(\mathbf{r})) = -\frac{e_0 n_0}{\epsilon\epsilon_0}\sum_i z_i e^{-\beta z_i e_0\phi(\mathbf{r})}, \tag{3}$$

that gives the equilibrium profile of the electric field in the system. If the magnitude of the surface charges is not too large, one has $\beta e_0\phi(\mathbf{r}) \ll 1$. The

Poisson - Boltzmann equation can be linearized and reduced to the Debye - Hückel equation [4] that describes screened electrostatic interactions with a screening (Debye) length equal to $\lambda_D = \left(\sum_i \frac{e_0^2 n_0 z_i^2}{\epsilon\epsilon_0 kT}\right)^{-1/2}$.

The forces between macromolecules mediated by the equilibrium distribution of counterions and salt ions between them can be obtained by either of two approaches. In the first approach one evaluates the stress tensor, composed of the Maxwell and the osmotic part,

$$\sigma_{ij} = \epsilon\epsilon_0 \left(E_i E_j - \tfrac{1}{2}E^2\delta_{ij}\right) + kT\sum_i \rho_i(\phi(\mathbf{r}))\delta_{ij}, \tag{4}$$

at an appropriate plane of symmetry of the interacting molecules. In many cases it is simplest to take the stress tensor at the surface of the macromolecules, in which case the forces are obtained *via* a contact theorem [5], or at the symmetry plane in the middle of the space between the two macromolecules. If the line joining them is along the z axis and S is the plane perpendicular to it, the force between two charged macromolecules is obtained as

$$F = \int_S d^2\mathbf{r}\ (\sigma_{zz}(z=0) - \sigma_{zz}^{(0)}) = \int_S d^2\mathbf{r}\ (\sigma_{zz}(z=D/2) - \sigma_{zz}^{(0)}), \tag{5}$$

where $z = 0$ stands symbolically for the midplane position, $z = D/2$ for the surface of the macromolecule and $\sigma_{zz}^{(0)}$ is the stress tensor in the bulk.

Another approach would be to evaluate the complete equilibrium free energy $\mathcal{F}$ from Eq. 1 *via* a volume integral over all the space available to the mobile charges

$$\mathcal{F} = \int d^3\mathbf{r}\ f(\mathbf{E}(\mathbf{r}), \rho_i(\mathbf{r})) \tag{6}$$

where the electrostatic field and the densities of the mobile charges are obtained from the solution of the full or linearized Poisson - Boltzmann equation Eq. 3. The forces between charged bodies are then obtained as a derivative of the free energy with respect to the separation D between the bodies

$$F = \left(\frac{\partial\mathcal{F}}{\partial D}\right)_{T,\mu}, \tag{7}$$

at constant temperature and chemical potential of the mobile charged species.

2.2. INTERACTION BETWEEN CYLINDRICAL MACROMOLECULES

The mean-field electrostatic interaction can be explicitly evaluated on the Debye - Hückel level [6, 7], assuming that in an aggregate of long cylindrical molecules with local hexagonal symmetry the interactions are pairwise additive, resulting in an interaction potential or mean-field interaction free energy $\mathcal{F}(R)$ between two parallel charged rods of radius a, length L and at an interaxial separation R of the form [8]

$$\mathcal{F}(R) = kT\frac{\ell_B}{b^2}K_0(R/\lambda_D)\ L \tag{8}$$

where $\ell_B = e_0^2/4\pi\epsilon\epsilon_0 kT$ is the Bjerrum length, being 7.4 Å for water at room temperature. For monovalent salts with ionic strength $I = \sum_i n_0 z_i^2$ expressed in moles [M] per liter, $\lambda_D = 3.08\,\text{Å}\ /\sqrt{I[M]}$. Eq. 8 refers to two cylinders with surface charge density σ, modeled as two infinitely thin lines with a linear charge density described *via* an effective separation b between charges along the line $1/b = 2\pi\sigma\lambda_D/e_0K_1(a/\lambda_D)$ [9] (for DNA $b = l_{PO_4}$, where $l_{PO_4} \sim 1.7\,\text{Å}$ is the separation between the phosphate charges along DNA [10]). For large d/λ_D the Bessel function K_0 can be approximated by $K_0(d/\lambda_D) \approx \sqrt{\frac{\pi}{2}}\frac{e^{-R/\lambda_D}}{\sqrt{R/\lambda_D}}$.

In the case of very large surface charges the nonlinearities of the Poisson-Boltzmann equation effectively change the surface charge density entering the above equation Eq. 8 *via* $b \longrightarrow \ell_B$ while leaving the separation dependence largely unaltered. Sometimes these non-linearity effects, that become pronounced depending on whether the s.c. Manning parameter $\zeta = l_B/b$ is larger or smaller than one, would be referred to as Manning condensation [8]. Thus for two parallel molecules in the limit of large linear charge densities the interaction free energy would assume the form

$$\mathcal{F}(R) = \frac{kT}{\ell_B}K_0(R/\lambda_D)\ L. \tag{9}$$

Besides the mean-field forces due to the fixed charge distribution on the macromolecules one also has the forces due to the fluctuating electromagnetic fields permeating the space in and around the cylinders. These macroscopic van der Waals interactions can be computed in the framework of the Lifshitz theory [11] and are thus referred to as Lifshitz - van der Waals interactions. In this framework the fluctuation interactions are determined by the frequency (ω) dependence of the dielectric functions of the interacting macromolecules parallel ($\epsilon_{||}(\omega)$) and perpendicular ($\epsilon_{\perp}(\omega)$) to the long axis as well as the dielectric function ($\epsilon_m(\omega)$) of the bathing medium, *i.e.* an aqueous solution, and lead [12] to the interaction free energy between two parallel cylindrical macromolecules of the form

$$\mathcal{F}(R) = -\frac{3\ kT}{8\pi}\left(\pi a^2\right)^2 \mathcal{A}\frac{L}{R^5}, \tag{10}$$

where I introduced the Hamaker constant $\mathcal{A}$ as

$$\mathcal{A} = \sum_{n=0}^{\infty}{}' \left(\Delta_\perp^2(i\omega_n) + \tfrac{1}{4}\Delta_\perp(i\omega_n)\Gamma(i\omega_n) + \tfrac{3}{2^7}\Gamma^2(i\omega_n)\right), \tag{11}$$

as is usually done in the framework of the DLVO theory [13]. The primed summation means that the $n = 0$ term should be taken with a weight $\frac{1}{2}$. Also I have assumed that the magnetic properties of all the media involved are the same. I have used the following abbreviations $\Delta_{||}(i\omega_n) = (\epsilon_{||}(i\omega_n) - \epsilon_m(i\omega_n))/(\epsilon_m(i\omega_n))$, $\Delta_\perp(i\omega_n) = (\epsilon_\perp(i\omega_n) - \epsilon_m(i\omega_n))/(\epsilon_\perp(i\omega_n) + \epsilon_m(i\omega_n))$ and $\Gamma(i\omega_n) = \Delta_{||}(i\omega_n) - 2\Delta_\perp(i\omega_n)$. The dielectric response at imaginary frequencies $\epsilon(i\omega)$ can be shown *via* the Kramers - Kronig relations to be a monotonic function of ω. In the case that the bathing medium is an electrolyte solution the $n = 0$ term in the Hamaker constant is exponentially screened with half the Debye length [11].

Apart from the van der Waals forces one also has attractive counterion correlation forces, which account for all the thermal effects not taken into account on the Poisson-Boltzmann (mean-field) level and will be introduced in other lectures [4, 11]. They are usually negligible for univalent salts and/or counterions, the case addressed in these lecture notes.

It turns out that after we insert the appropriate numbers for $\epsilon_{||}(\omega)$ and $\epsilon_\perp(\omega)$ valid for DNA [10], the Lifshitz - van der Waals interaction in the whole regime of DNA densities investigated in this lecture is about two orders of magnitude smaller than the mean-field repulsive interactions and can thus be safely ignored.

2.3. INTERACTIONS IN PLANAR ARRAYS OF MACROMOLECULES

In the case of interacting planar macroions of surface area S in a multilamellar aggregate the interaction free energy is again straightforward to calculate. Andelman [14] has defined several regions in the parameter space where explicit forms for the interaction free energy can be written down. In the Debye-Hückel region [15] corresponding to $\lambda_{DH} \ll D$, where D is the surface to surface separation between the neighboring lamellae, with small surface charge

$$\mathcal{F}(D) = \frac{kT\lambda_D}{\pi\ell_B b^2}\left(\mathrm{cth}(\kappa D) - 1\right) S, \tag{12}$$

where the separation of charges on the surface is given by the Gouy-Chapman length $b = e_0/8\pi^2|\sigma|\ell_B$. In the same range of D but with large surface charge density , one can use the superposition approximation, at the midplane obtaining

$$\mathcal{F}(D) = \frac{8\ kT}{\pi \lambda_D l_B}\ e^{-\kappa D} S. \tag{13}$$

Just as in the case of cylindrical macromolecules one also has the forces due to fluctuating electromagnetic fields permeating space in and around the macromolecules. If the dielectric function of the macromolecule, *i.e.* the lipid bilayer, is $\epsilon_l(\omega)$ and that of the intervening aqueous solution $\epsilon_m(\omega)$ then the unretarded van der Waals interaction between two semi-infinite regions of dielectric material m over a slab of thickness D is [13]

$$\mathcal{F}(D) = -\frac{\mathcal{A}}{12\pi D^2}\ S, \tag{14}$$

where S is the surface area and the Hamaker constant in planar geometry has been defined as

$$\mathcal{A} = \tfrac{3kT}{2} {\sum_{n=0}^{\infty}}' \int_0^{\infty} u\ du\ \ln\Big(1 - \big(\tfrac{\epsilon_m(i\omega_n) - \epsilon_l(i\omega_n)}{\epsilon_m(i\omega_n) + \epsilon_l(i\omega_n)}\big)^2 e^{-2u}\Big). \tag{15}$$

Again I have assumed that the magnetic properties of all the media involved are the same. If the two macromolecular regions are of finite width then the dependence of the interaction energy on D is modified [11].

Here we have omitted the effects of retardation assuming that the separation between the macromolecules is small enough. As in the case of cylindrical macromolecules I will not address the counterion correlation forces since they form the subject matter of other lectures [4].

Contrary to the case of DNA, the Lifshitz - van der Waals interactions in lipid membrane arrays immersed in an aqueous solution are comparable in magnitude to the mean-field electrostatic repulsions and can in general not be ignored. For most lipid membranes interacting across an aqueous solution a typical value of the Hamaker constant is estimated to lie in the range $10^{-13} - 10^{-14}$ ergs [3, 16]. If the aqueous solution is in fact an electrolyte solution the $n = 0$ term of the Hamaker constant is again exponentially screened with half the Debye length of the electrolyte solution.

3. Non-DLVO forces

One of the (many) important assumptions in the foundations of the Poisson - Boltzmann and Lifshitz - van der Waals interactions is that water is a structureless medium that can be characterized solely through its frequency dependent dielectric response and that the interacting macromolecules can be characterized solely through their continuous charge distribution on their surface. Theoretical work as well as numerous experiments suggest that at small separations between interacting molecules progressively more and more of the molecular details of the bathing solvent as well as interacting surfaces come into play when one tries to understand the nature of the interactions in the last few Angstroms. We still have no clear and widely accepted theoretical picture of the state of affairs in this regime of intersurface separations.

There is however no need to stay in this limbo since the experiments suggest [16] that these short range (SR) interactions whatever they might be due to, can in many cases be described *via* a semi-empirical force law of the form

$$\begin{aligned} \mathcal{F}_{SR}(D) &= F_0 \; e^{-D/\xi} S \\ \mathcal{F}_{SR}(R) &= F_0 \; K_0(R/\xi) L. \end{aligned} \tag{16}$$

where the first equation is valid for the short range interactions between planar macromolecules and the second one between cylindrical macromolecules. The numerical value of ξ is between $1 - 3$ Å[16, 17]. I will use this empirical form $\mathcal{F}_{SR}$ of the interaction free energy together with the mean-field electrostatic $\mathcal{F}_{ES}$ and Lifshitz - van der Waals interactions $\mathcal{F}_{vdW}$ in order to have an interaction free energy expression valid in the whole regime of intersurface separations.

Without going too much off the track let me just state that it seems quite probable that most of the effects described by Eqs. 16 are due to the structure and structuring effects of water close to molecular surfaces. Therefore one often refers to this short range interaction as hydration interaction.

4. Equation of state for hexagonal and multilamellar assemblies

An equation of state in general means a connection between the osmotic pressure Π and the macromolecular density ρ of the assembly

$$\Pi = \Pi(\rho) \tag{17}$$

and is thus obtained *via* a complete statistical mechanical treatment of a solution composed of cylindrical or planar macromolecules and the bathing medium. While the intermolecular degrees of freedom are taken into account through interaction potentials described in the previous two sections, the intramolecular degrees of freedom are usually treated within a mesoscopic elastic model that substitutes macroscopic elasticity for the complicated short-range intramolecular potentials acting between different segments of the macromolecules [18]. Though it would seem off hand a bit far fetched to expect that macroscopic elasticity can satisfactorily describe the complex short ranged intramolecular interactions it turns out that this indeed is so.

Elastic models for cylindrical and planar macromolecules have been well worked out [19]. The general idea is that the trace over all the microscopic degrees of freedom is assumed to result exactly in the mesoscopic Hamiltonian itself that one now uses in the partition function to evaluate the trace over mesoscopic degrees of freedom.

A most naive approach to the equation of state would be to simply forget the intramolecular degrees of freedom, assume that the macromolecules are ideally rigid and assemble into a crystal of either lamellar or hexagonal symmetry with perfect positional order. In this case the osmotic pressure of such a system can be obtained *via* Ewald force summation (in the case of short range interactions even this is dispensable) involving the intermolecular potentials leading directly to the equation of state. For the multilamellar array the equation of state would thus read

$$\Pi(\rho) = -\left(\frac{\partial(\mathcal{F}(D)/S)}{\partial D}\right)_{T,\mu} \qquad \mathcal{F} = \mathcal{F}_{SR} + \mathcal{F}_{vdW}, \quad \rho = \frac{1}{D}, \tag{18}$$

where S is the area of the lamellae, D is the interlamellar separation and ρ is the one-dimensional density. The free energy contains only the short range (hydration) part and the van der Waals part. As I already noted in the introduction I shall limit myself mostly to the case of uncharged lipid bilayers where there are no pertinent salt equilibration problems. Note that all the contributions to the free energy scale linearly with area S.

For the hexagonal array of DNA molecules the corresponding equation of state can be obtained from

$$\Pi(\rho) = -\left(\frac{\partial(\mathcal{F}(D)/L)}{\sqrt{3}\ R\partial R}\right)_{T} \qquad \mathcal{F} = \mathcal{F}_{SR} + \mathcal{F}_{ES}, \quad \rho = \frac{2}{\sqrt{3}\ R^2}. \tag{19}$$

L is the length of the molecules in the array and ρ is now the two-dimensional density. Note that there is no van der Waals contribution (see above) to the free energy in this case and that all the parts of the total free energy scale linearly with length L of the molecules.

We can now take these equations of state and compare them to experiments performed on DNA [20] and lipid membrane arrays [21] composed of neutral lipids DPPC and DMPC. Performing this type of exercise one is immediately convinced that something crucial is missing as there is practically no correspondence (except at very high densities) between the experiment and this type of simpleminded theory, see Fig. 2. For the multilamellar case I have taken $H = 4 \times 10^{-14}$ ergs [3, 16] and the magnitude of the short range interactions is obtained from the fit to the data at high macromolecular densities as $F_0 = 2.1 \times 10^1 erg/cm^2$ for lipids and $F_0 = 1.3 \times 10^{-7} erg/cm$

for DNA. In the case of DNA the magnitude of the surface charge is taken at the Manning value [8]. Obviously this equation of state underestimates the energetics of the system.

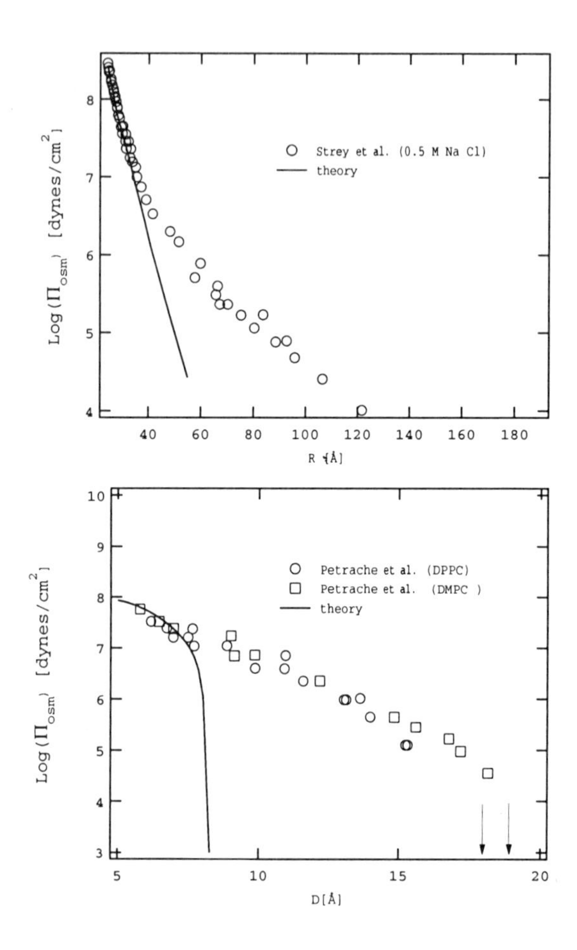

Figure 2. A fit of the equations of state Eqs. 18 and 19 to the DNA [20] and lipid membranes, composed of neutral lipids DPPC and DMPC, [21] experiments. Obviously there are large discrepancies on this level between the theoretical predictions and actual data. The arrows indicate the stable equilibrium separation at zero osmotic pressure.

5. The effect of thermal fluctuations

I now relax the constraint that the macromolecules in the array are ideally stiff and describe them as deformable elastic bodies *via* mesoscopic elasticity theory [18, 19]. Since I am dealing with an array I might as well construct a mesoscopic elastic Hamiltonian for the whole array. Because in the regime of relevant densities the hexagonal array is either in the line hexatic or the cholesteric phases [20] and the multilamellar array is in the smectic L_α phase [2] I can use the mesoscopic elastic Hamiltonians pertaining to them. The goal here will be to combine the effects of thermally driven elastic fluctuations of the macromolecules with the interactions between them and calculate their combined effect on the equation of state.

This approach is based on ideas first introduced by Helfrich in the '70's and later worked out in detail by Lipowsky, Leibler and others in the '80's [22], who showed in the context of membranes that thermal motion can have a profound effect on interactions between flexible macromolecules. In this lecture we will limit ourselves to the regime of macromolecular densities higher than the cholesteric-isotropic transition of DNA or the unbinding transition for lipid membranes, deeply within the ordered phase. Thus: no phase transitions.

5.1. A MACROSCOPIC THEORY OF THE EQUATION OF STATE IN A HEXAGONAL ARRAY

Let us first consider the elastic free energy of a nematic: a three dimensional liquid with long-range orientational order with an average director **n** along the z-axis. We forget about the hexatic or even the cholesteric order because it is on a much larger scale then the nematic order. There are three kinds of deformations to quadratic order in **n** with symmetry $C_{\infty h}$ [18]: splay, twist and bending. The corresponding elastic constants for these deformations are the Frank constants K_1, K_2 and K_3 [23]. For small deviations of the director field $\mathbf{n}(\mathbf{r})$ around its average orientation along the z axis $\mathbf{n}(\mathbf{r}) \approx (\delta n_x(\mathbf{r}), \delta n_y(\mathbf{r}), 1)$, with $\mathbf{r} = (z, \mathbf{r}_\perp)$ the free energy assumes the form

$$\mathcal{F}_N = \tfrac{1}{2} \int d^2\mathbf{r}_\perp dz \left[K_1 \left(\nabla_\perp . \delta\mathbf{n}\right)^2 + K_2 \left(\nabla_\perp \times \delta\mathbf{n}\right)^2 + K_3 \left(\partial_z \delta\mathbf{n}\right)^2 \right] \quad (20)$$

For polymer nematics we have to consider the fact that the director field and the density of polymers $\rho = \rho_0 + \delta\rho$ are coupled [24]. If the polymers were infinitely long and stiff the coupling would be given by the continuity equation for the director field:

$$\partial_z \delta\rho + \rho_0 \nabla_\perp . \delta\mathbf{n} = 0. \quad (21)$$

This constraint, however, is softened if the polymer has a finite length ℓ or a finite elastic modulus K_c. On length scales larger than ℓ or K_c/kT the polymer can either fill the voids with its own ends or fold back on itself and can splay without any density change [20]. Following [25, 26] this can be expressed by introducing a Lagrange multiplier G, a measure of how effectively the constraint Eq. 21 is enforced. Density changes are then expanded to second order in density deviations $\delta\rho(\mathbf{r}_\perp, z) = \rho(\mathbf{r}_\perp, z) - \rho_0$. $\mathcal{B}$ is the polymer solution bulk compressibility modulus. The total mesoscopic free energy, or equivalently the mesoscopic Hamiltonian $\mathcal{H}$ is now obtained in the form [27]

$$\mathcal{H} = \mathcal{F}_0(\rho_0) + \tfrac{1}{2}\int d^2\mathbf{r}_\perp dz \left[\mathcal{B}\left(\tfrac{\delta\rho}{\rho_0}\right)^2 + G\left(\partial_z\delta\rho + \rho_0\nabla_\perp.\delta\mathbf{n}\right)^2\right] + \mathcal{F}_N \quad (22)$$

where G is given by

$$G = \frac{k_B T \ell}{2\rho_0}, \quad (23)$$

with ℓ never to exceed K_c/kT [26]. In the limit of finite polymer length G is also finite and can be obtained from the observation that $\partial_z\delta\rho + \rho_0\nabla_\perp.\delta\mathbf{n}$ equals the difference between the number of chain heads and tails [26].

In order to calculate the contribution to the free energy due to thermal fluctuations on the mesoscopic scale one has to take a trace of $\exp(-\beta\mathcal{H})$ with respect to density and nematic director fluctuations. The macroscopic free energy that now contains also the contributions of the thermal fluctuations on the mesoscopic scales described by the Hamiltonian Eq. 22 can be obtained as

$$\mathcal{F} = -kT\,\ln\langle e^{-\beta\mathcal{H}}\rangle = \tfrac{1}{2}\,k_B T \iint \tfrac{d^2q_\perp dq_z}{(2\pi)^3}\,\ln\left(K_1 q_\perp^2 q_z^2 + K_3 q_z^4 + \mathcal{B}q_\perp^2\right). \quad (24)$$

The problem now is that the above integral requires a cutoff meaning that the higher order terms in $q_\perp$ are more important than the ones we have kept here. For a moment let us assume that this free energy is nevertheless valid and we may calculate

$$\begin{aligned}\frac{\partial\mathcal{F}}{\partial\mathcal{B}} &= \tfrac{1}{2}\,k_B T V \iint \tfrac{q_\perp dq_\perp dq_z\ q_\perp^2}{(2\pi)^2\ K_1 q_\perp^2 q_z^2 + K_3 q_z^4 + \mathcal{B}q_\perp^2} \\ &= \tfrac{k_B T\ V}{16\pi} \int \frac{q_\perp^3\,dq_\perp}{\sqrt{\mathcal{B}q_\perp^2}\sqrt{K_1 q_\perp^2 + 2\sqrt{\mathcal{B}K_3 q_\perp^2}}}. \end{aligned} \quad (25)$$

This integral depends essentially on the upper cutoff for $q_\perp = q_{\perp max}$ and we obtain

$$\frac{\partial\mathcal{F}}{\partial\mathcal{B}} = k_B T\frac{V}{4\pi}\frac{\mathcal{B}K_3}{K_1^2\sqrt{\mathcal{B}K_1}}\ F\left(\frac{q_{\perp max}}{2\sqrt{\frac{\mathcal{B}K_3}{K_1^2}}}\right), \quad (26)$$

where the function $F(x)$ has been defined as $F(x) = \int_0^x \frac{u^{3/2}du}{\sqrt{1+u}}$. From here we obtain the two limiting forms of the free energy as

$$\mathcal{F} \simeq \frac{k_B T\ V}{5 \times 2^{3/2}\pi} \sqrt[4]{\frac{\mathcal{B}}{K_3}}\, q_{\perp max}^{5/2} + \cdots \qquad ; \ q_{\perp max} \ll 2\sqrt{\frac{\mathcal{B}K_3}{K_1^2}} \tag{27}$$

$$\mathcal{F} \simeq \frac{k_B T\ V}{16\pi} \sqrt{\frac{\mathcal{B}}{K_1}}\, q_{\perp max}^{2} + \cdots \qquad ; \ q_{\perp max} \gg 2\sqrt{\frac{\mathcal{B}K_3}{K_1^2}}. \tag{28}$$

Obviously the long-wavelength physics is very complicated and depends crucially on the values of typical polymer length and the ratios of elastic constants. However it is also dependent on the $q_\perp$ cutoff. We have to either eliminate the cutoff by including higher order terms in the original Hamiltonian or choose a meaningful cutoff.

One can show [20] that a consistent value of the cutoff has to be proportional to the "Brillouin zone" radius $q_{\perp max} \simeq \frac{\pi}{R}$, where R is the effective separation between the polymers in the nematic phase. This is a physically meaningful and appropriate cutoff because the underlying macroscopic elastic model has, by definition, to break down at wavelengths comparable to the distance between molecules.

Putting in the numbers valid for DNA arrays one realizes that in the regime of densities considered here we are always in the Eq. 27 limit.

We would now have to derive the mesoscopic elastic moduli from the microscopic interactions described *via* a pair potential $U(\mathbf{r})$ (the interaction free energy) between the segments of the macromolecules. At present this program is too ambitious and we simply exploit the standard *ansatz* for the different elastic moduli [23]

$$\begin{aligned} K_1 &= K_2 \simeq U(R)/R \\ K_3 &\simeq \rho_0 K_c + U(R)/R \\ \mathcal{B} &\simeq V\frac{\partial^2 \mathcal{F}_0(V)}{\partial V^2} = \frac{\sqrt{3}}{4}\left(\frac{\partial^2(\mathcal{F}/L)}{\partial^2 R} - \frac{1}{R}\frac{\partial(\mathcal{F}/L)}{\partial R}.\right), \end{aligned} \tag{29}$$

where ρ_0 is the 2D density of the macromolecules perpendicular to their long axes, K_c is the elastic rigidity modulus of a single polymer molecule and we assumed that the polymers have a local hexagonal packing symmetry with an average separation R between first neighbors [20].

Using now the form of the bare interaction free energy $\mathcal{F}(R) = \mathcal{F}_{SR}(R) + \mathcal{F}_{ES}(R)$ appropriate for a DNA array (see above) including the short range and the electrostatic part of the interaction as well as the fact that at all relevant densities $K_3 \sim eq\rho_0 K_c$, we are able to fit the calculated equation of state obtained from Eq. 27 to the experimental equation of state [20], see Fig. 3. The values for the DNA bending rigidity and the Debye length obtained from such a fit are comfortably within the expected range [20]. It is a bit surprising that the effective charge is about half the amount expected on the basis of the Manning condensation theory.

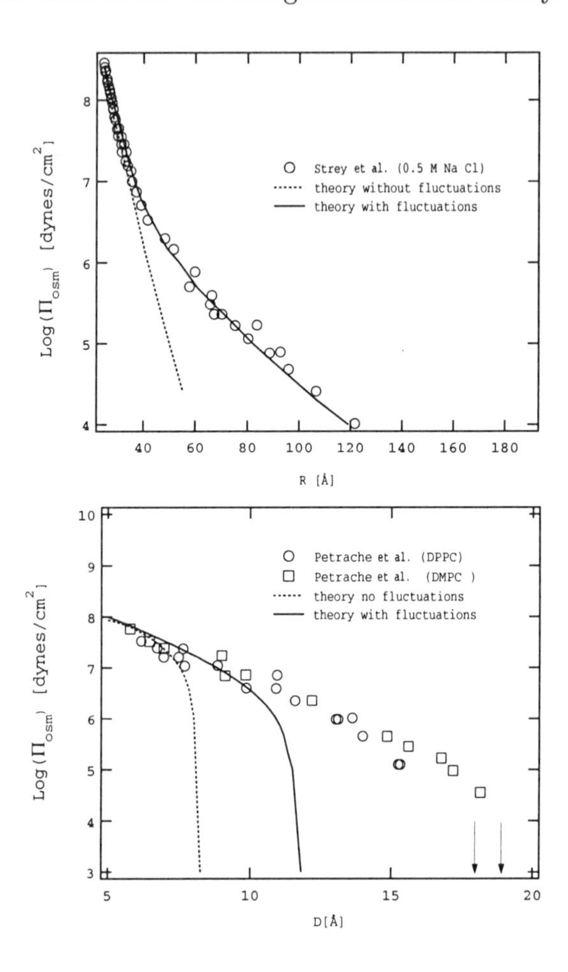

Figure 3. A fit of the lowest order fluctuation equations of state, Eqs. 27 and 34, to DNA [20] and lipid membrane [21] data. While for DNA theory and experiment essentially overlap, there is as yet no correspondence between them in the case of lipid membranes.

5.2. A MACROSCOPIC THEORY OF THE EQUATION OF STATE IN A LAMELLAR ARRAY

For lipid membranes the evaluation of the equation of state from the mesoscopic Hamiltonian is simpler because the director field and the density of the membranes along the (z)-axis are not coupled [23]. The free energy is thus composed of the elastic free energy of the membranes and the compressional free energy perpendicular to the planes of the membranes [18].

For small deviations of the director field $\mathbf{n}(\mathbf{r})$ (that is taken in the direction of the local normal to the bilayer surfaces) from its average direction taken along the z axis, $\mathbf{n}(\mathbf{r}) \approx (\delta n_x(\mathbf{r}), \delta n_y(\mathbf{r}), 1)$, the nematic free energy takes the form [23]

$$\mathcal{H}_N = \tfrac{1}{2} \int d^2\mathbf{r}_\perp dz \left[K_1 \left(\nabla_\perp . \delta\mathbf{n}\right)^2 + K_2 \left(\nabla_\perp \times \delta\mathbf{n}\right)^2 + K_3 \left(\partial_z \delta\mathbf{n}\right)^2 \right]. \quad (30)$$

Furthermore since one assumes that the deviations of the membranes from the planar shape are small, we have to the lowest order $\delta\mathbf{n}(\mathbf{r}) \simeq (1, \nabla_\perp u(\mathbf{r}_\perp, z))$, where $u(\mathbf{r}_\perp)$ is the local deviation of the membranes from the plane in the z direction. Since the density and the nematic deformation in smectic arrays are not coupled, one has $\delta\rho/\rho \simeq \frac{\partial u(\mathbf{r}_\perp, z)}{\partial z}$. One obtains for the mesoscopic Hamiltonian to the quadratic order [23]

$$\begin{aligned} \mathcal{H} &= \tfrac{1}{2}\mathcal{B} \int d^2\mathbf{r}_\perp dz \left(\tfrac{\delta\rho}{\rho_0}\right)^2 + \mathcal{H}_N = \\ &= \tfrac{1}{2}\mathcal{B} \int d^2\mathbf{r}_\perp dz \left(\tfrac{\partial u(\mathbf{r}_\perp, z)}{\partial z}\right)^2 + \tfrac{1}{2} K_1 \int d^2\mathbf{r}_\perp dz \left(\nabla_\perp^2 u(\mathbf{r}_\perp, z)\right)^2. \end{aligned} \quad (31)$$

Summing now over all the displacement modes in Fourier space just as in the case of the hexagonal array, yields the following form of the macroscopic free energy

$$\mathcal{F} = -kT \; ln \langle e^{-\beta\mathcal{H}} \rangle = \tfrac{1}{2}\, k_B T \iint \tfrac{d^2 q_\perp dq_z}{(2\pi)^3} \; ln \left(K_1 q_\perp^4 + \mathcal{B} q_z^2\right). \quad (32)$$

To evaluate the fluctuation part of the free energy I now proceed as follows. First I calculate

$$\frac{\partial \mathcal{F}}{\partial \mathcal{B}} = \tfrac{1}{2}\, k_B T V \; \iint \tfrac{q_\perp dq_\perp dq_z}{(2\pi)^2} \; \tfrac{q_z^2}{K_1 q_\perp^4 + \mathcal{B} q_z^2} \; = \tfrac{\pi k_B T V}{8\sqrt{K_1 \mathcal{B}}} \; \int q_z dq_z. \quad (33)$$

This integral depends again on the upper cutoff for $q_z = q_{z\ max}$. From here I obtain the free energy in the form

$$\mathcal{F} \simeq \frac{\pi k_B T V}{8} \sqrt{\frac{\mathcal{B}}{K_1}}\; q_{z\ max}^2. \quad (34)$$

As before a meaningful cutoff for $q_{z\ max}$ would be proportional to the Brillouin zone radius $q_{z\ max} \simeq \frac{\pi}{D}$ where D is the average separation between the membranes in the stack. Again the mesoscopic elastic Hamiltonian has to break down below the scales set by $q_{z\ max}$.

In this case the osmotic compressibility modulus is taken in the form involving only the bare pair interactions between the molecules $U(\mathbf{r})$

$$\mathcal{B} \simeq V \frac{\partial^2 \mathcal{F}(D)}{\partial V^2} = D \frac{\partial^2 \mathcal{F}(D)/S}{\partial^2 D}, \quad (35)$$

where in this case $\mathcal{F}(D) = \mathcal{F}_{SR}(D) + \mathcal{F}_{vdW}(D)$ with D being the average first neighbor separation between the membranes in the array. Again we limit ourselves to the uncharged lipid membranes where the electrostatic term in the interaction is missing. For the bilayer bending rigidities we use the values obtained by Petrache *et al.* [21].

Surprisingly for a lipid membrane array the difference between the experiment and the theory on this level of approximation is still substantial, see Fig. 3. I interpret this difference as signifying that the compressibility modulus itself, and not just the osmotic pressure, has to be calculated by taking into account the thermal fluctuations. This is a much more ambitious project than the simple macroscopic elastic theory that appears to work for DNA. Let us see what we can do about it.

6. Fluctuation contribution to the compressibility modulus

At this point one can use any of the advanced theories that take into account the thermal fluctuations at a deeper level than the macroscopic theory of the previous section and that allow also for the thermal fluctuation effects on the value of the compressibility modulus itself [28]. Such theories are well worked out also in other areas of physics such as e.g. magnetic vortex arrays in type II superconductors. There are different approaches that one can follow. One could either formulate the problem in the language of the functional renormalization group [28] or on the level of a variational calculation of the compressibility modulus [29]. To remain as close as possible to the approach outlined before, we chose the second, *i.e.* variational, approach. It usually fares quite well even when compared to the more powerful renormalization group approach.

6.1. VARIATIONAL CALCULATION OF THE OSMOTIC PRESSURE: MULTILAMELLAR SYSTEM

I now improve on our estimate of the equation of state by evaluating the change in the osmotic compressibility not just osmotic pressure consistently for a fluctuating 2D multilamellar system. This was done first by Podgornik and Parsegian [30] and later by Nagle and coworkers [21]. In a multilamellar stack I assume that the interaction Hamiltonian can be written as:

$$\begin{aligned} \mathcal{H} &= \tfrac{1}{2}K_C \sum_{n=1}^{\infty} \int d^2\mathbf{r}_\perp \left(\boldsymbol{\nabla}_\perp^2 z_n(\mathbf{r}_\perp)\right)^2 \\ &\quad + \tfrac{1}{2}\sum_{n=1}^{\infty} \int d^2\mathbf{r}_\perp \; V\left(z_{n+1}(\mathbf{r}_\perp) - z_n(\mathbf{r}_\perp)\right), \end{aligned} \tag{36}$$

where $z_n(\mathbf{r}_\perp)$ describes the local displacement of the n-th membrane in a stack in the z direction and $V(z)$ is the interaction potential between two neighboring membranes in a stack. I assume that the interactions are sufficiently short ranged that the next-to-first-neighbor interactions are negligible. Introducing the following parametrization of the local displacement

$$z_n(\mathbf{r}_\perp) = nD + u_n(\mathbf{r}_\perp) \qquad u_n(\mathbf{r}_\perp) = \sum_{\mathbf{Q},q_z} u\left(\mathbf{Q},q\right) e^{i\mathbf{Q}\mathbf{r}_\perp} e^{inDq_z}, \tag{37}$$

I define a harmonic reference Hamiltonian, the parameters of which will be evaluated variationally, in the form

$$\mathcal{H}_0 = \tfrac{1}{2}\sum_{\mathbf{Q},q_z}\left(K_C Q^4 + 4\mathcal{B}\sin^2\tfrac{q_z D}{2}\right)|u\left(\mathbf{Q},q_z\right)|^2 + V_0(D). \tag{38}$$

Evaluating the reference free energy with the help of $\mathcal{F}_0 = -kT\,\ln\left\langle e^{-\beta\mathcal{H}_0}\right\rangle$ I obtain

$$\mathcal{F}_0 = V_0(D) + \tfrac{kT}{2}\sum_{\mathbf{Q},q_z} ln\left(1 + \frac{4\mathcal{B}\sin^2\frac{q_z D}{2}}{K_C Q^4}\right) = V_0(D) + \frac{kT}{2\pi D}\sqrt{\frac{\mathcal{B}}{K_C}}. \tag{39}$$

If I now compare the free energy of the exact Hamiltonian with the free energy of the trial Hamiltonian I have the Bogolyubov inequality $\mathcal{F} \leq \mathcal{F}_0 + \langle \mathcal{H} - \mathcal{H}_0, \rangle_0$, where the subscript 0 in the last term means that the average should be taken with respect to $\mathcal{H}_0$. The idea now is (Feynman - Kleinert [31]) to obtain an optimal upper bound for the free energy. This is obtained by treating $V_0(D)$ and $\mathcal{B}$ as variational parameters. Variation of the free energy in the Bogolyubov inequality with respect to these two parameters produces an optimal upper bound for the free energy. After defining

$$\sigma^2(\mathbf{Q},q_z) = \left\langle |u\left(\mathbf{Q},q_z\right)|^2\right\rangle = \frac{kT}{K_C Q^4 + 4\mathcal{B}\sin^2\frac{q_z D}{2}}, \tag{40}$$

and

$$V_{\sigma^2}(D) = \int dz\; V(z)\sum_{\mathbf{Q},q}\exp\left(iq(z-D) - \beta q_z^2\Delta^2\right), \tag{41}$$

where $V(\mathbf{Q},q)$ is the Fourier transform (wave-vector $\mathbf{k} = (\mathbf{Q},q)$) of the interaction potential Eq. 36, with

$$\Delta^2 = \sum_{\mathbf{Q},q}(1-\cos qD)\sigma^2(\mathbf{Q},q) = \frac{kT}{16K_C D\sqrt{K_C\mathcal{B}}}, \tag{42}$$

it is now straightforward to show that the variational equations for $V_0(D)$ and $\mathcal{B}$ can be obtained in the form

$$\begin{aligned} \mathcal{B}(1-\cos qD) &= \tfrac{1}{2}\frac{\partial}{\partial\sigma^2(\mathbf{Q},q)}V_{\sigma^2}(D) \\ V_0(D) &= \tfrac{1}{2}V_{\sigma^2}(D)-\mathcal{B}\Delta^2. \end{aligned} \tag{43}$$

The optimized free energy is given by its value in the reference system Eq. 39 with $V_0(D)$ and $\mathcal{B}$ calculated from Eq. 43.

When I compare the equation of state obtained from the free energy Eq. 39 I see a much better quantitative agreement with experiment, see Fig. 4. Obviously most of the thermal fluctuation effects have thus been taken into account. The same type of fit has also been carried out by Petrache *et al.* [21].

6.2. VARIATIONAL CALCULATION OF THE OSMOTIC PRESSURE: HEXAGONAL COLUMNAR SYSTEM

For a system with a hexagonal local symmetry I follow closely the calculation of Volmer and Schwartz [29] derived for the system of magnetic vortex line in type II superconductors. Apart from the difference between elastic energies of a vortex line vs. a flexible polymer the two cases are analogous. The interaction Hamiltonian of an oriented polymer array with hexagonal local symmetry can be written in the form

$$\begin{aligned} \mathcal{H} &= \tfrac{1}{2}K_C\sum_{n,m=1}^{\infty}\int dz\left(\frac{\partial^2\mathbf{r}_\perp{}^{(n,m)}(z)}{\partial z^2}\right)^2 \\ &+\tfrac{1}{2}\sum_{n,m\neq n'm'}^{\infty}\int dz\, V\left(\mathbf{r}_\perp{}^{(n,m)}(z)-\mathbf{r}_\perp{}^{(n',m')}(z)\right), \end{aligned} \tag{44}$$

where $\mathbf{r}_\perp{}^{(n,m)}(z)$ is the local displacement of a polymer chain at the (n,m) lattice position perpendicular to the z axis. In principle the indices n,m would run through all the positions of the polymers at a certain planar cross section through the nematic but because of the short range nature of the interaction and computational convenience I restrict them to nearest neighbors [29]. Again instead of using this non-harmonic Hamiltonian I will take a simpler reference Hamiltonian of a general harmonic form. In the parameterization

$$\begin{aligned} \mathbf{r}_\perp{}^{(n,m)}(z) &= \mathbf{R}_{nm}+\mathbf{u}^{(n,m)}(z), \\ \mathbf{u}^{(n,m)}(z) &= \sum_{\mathbf{Q}_\perp,q_z}\mathbf{u}(\mathbf{Q}_\perp,q_z)e^{iq_z z+i\mathbf{Q}_\perp\mathbf{R}_{nm}} \end{aligned} \tag{45}$$

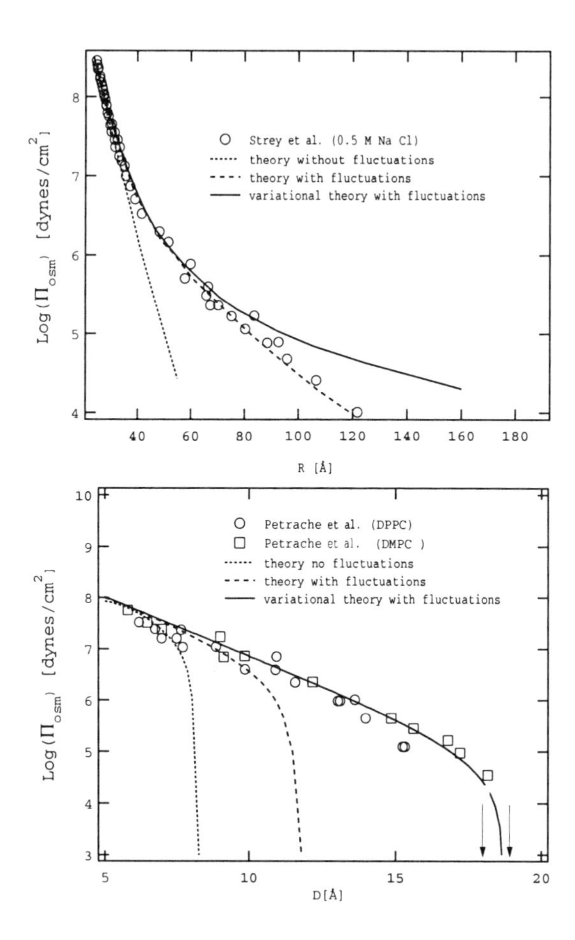

Figure 4. A fit of the variational equation of state obtained from Eq. 39 to the measured neutral lipid membrane array equation of state data [21], lower graph. The theory is now completely consistent with experiments. Surprisingly the same type of fit with a variational equation of state for DNA, upper graph, obtained from Eq. 50 fares much worse than the macroscopic fluctuation theory, see Fig. 3.

where $\mathbf{R}_{nm} = n\mathbf{a}_1 + m\mathbf{a}_2$, with $\mathbf{a}_1$ and $\mathbf{a}_2$ the two base vectors of the hexagonal lattice perpendicular to the long axis z of the molecules. I take the reference Hamiltonian in the harmonic form

$$\mathcal{H}_0 = \tfrac{1}{2} \sum_{\mathbf{Q}_\perp, q_z} \left(K_C q_z^4 \delta_{ik} + \mathcal{B}_{ik}(\mathbf{Q}_\perp)\right) u_i(\mathbf{Q}_\perp, q_z) u_k(-\mathbf{Q}_\perp, -q_z) + V_0(\mathbf{a}), \quad (46)$$

where $\mathcal{B}_{ik}(\mathbf{Q}_\perp) = \mathcal{B}_{ik} \sum_{\mathbf{a}} 4\sin^2 \frac{\mathbf{Q}_\perp \mathbf{a}}{2}$ and the sum over $\mathbf{a}$ refers to summation over the positions of nearest neighbors. By exactly the same arguments as in the case of a multilamellar system I now derive an upper bound for the free energy corresponding to Eq. 44 *via* a variational estimate for the parameters in the reference Hamiltonian Eq. 46. Defining thus

$$\sigma_{ik}(\mathbf{Q}_\perp, q_z) = \langle u_i(\mathbf{Q}_\perp, q_z) u_k(-\mathbf{Q}_\perp, -q_z)\rangle = \frac{kT}{K_C q_z^4 \delta_{ik} + \mathcal{B}_{ik}(\mathbf{Q}_\perp)} \qquad (47)$$

with

$$V_{\sigma_{ij}(\mathbf{Q}_\perp, q_z)}(\mathbf{a}) = \int d^2\mathbf{r}_\perp\, V(\mathbf{r}_\perp) \sum_{\mathbf{a}} \sum_{\mathbf{k}} e^{i\mathbf{k}(\mathbf{a}-\mathbf{r}_\perp) - 2\beta k_i k_j \sum_{\mathbf{Q}_\perp, q_z} \sin^2 \frac{\mathbf{Q}_\perp \mathbf{a}}{2} \sigma_{ij}(\mathbf{Q}_\perp, q_z)}, \qquad (48)$$

the variational equations can be written in the form

$$\begin{aligned} 2\mathcal{B}_{ij} \sum_{\mathbf{a}} \sin^2 \frac{\mathbf{Q}_\perp \mathbf{a}}{2} &= \frac{1}{2} \frac{\partial V_{\sigma_{ij}(\mathbf{Q}_\perp, q_z)}(\mathbf{a})}{\partial \sigma_{ij}(\mathbf{Q}_\perp, q_z)} \\ V_0(\mathbf{a}) &= \tfrac{1}{2} V_{\sigma_{ij}(\mathbf{Q}_\perp, q_z)}(\mathbf{a}) - \tfrac{1}{2} \sum_{\mathbf{Q}_\perp, q_z} \mathcal{B}_{ik}(\mathbf{Q}_\perp)\, \sigma_{ij}(\mathbf{Q}_\perp, q_z) \end{aligned} \qquad (49)$$

and the free energy can be obtained from

$$\mathcal{F}_0 = -kTln\left\langle e^{-\beta \mathcal{H}_0} \right\rangle = V_0(\mathbf{a}) + \frac{kT}{2} \mathrm{Tr} \sum_{\mathbf{Q}_\perp, q_z} ln\left(\delta_{ik} + \frac{\mathcal{B}_{ik}(\mathbf{Q}_\perp)}{K_C q_z^4} \right). \qquad (50)$$

Very similar equations have already been derived in the case of magnetic vortex arrays [29]. Though these equations look more complicated than in the multilamellar case, the only real difference between the two is the local symmetry of the assembly: lamellar vs. hexagonal columnar order.

However, when we compare the equation of state obtained from Eq. 50 with the one obtained from Eq. 27 we see that it fares much worse when compared with experiments or compared with the macroscopic fluctuation equation of state developed in section 4.1, see Fig. 4. This is surprising and demands an explanation.

The reason for this is found out to be quite simple. The variational *ansatz* with the bounded positional correlation function $\sigma_{ik}(\mathbf{Q}_\perp, q_z)$ fits best the description of a solid, with long range positional order. For a fluid, with only short range positional correlations, which is what the DNA array is at these densities, the positional correlations (as opposed to density correlations which of course remain finite) diverge in the thermodynamic limit and can not be described with a finite correlation function. Therefore the variational theory, strictly applicable only to a hexagonal crystal, fares much worse than the mesoscopic fluctuation theory of Section 4.1, if applied to a DNA array at densities between the crystalline and isotropic phases.

A variational theory similar in spirit to what I developed above was also put forth by de Vries [32] basing his analysis on previous work by Odijk [33]. Numerical results of this approach are indistinguishable from those presented above.

7. Interaction renormalization of the bending moduli

Until now we have studied the modification in the interactions between macromolecules wrought by their finite flexibility. The equation of state that we obtained for lamellar and hexagonal arrays contains also the material properties of the interacting macromolecules such as their mesoscopic elastic rigidities. In order to be consistent one also has to evaluate the effect of the interactions on those. We shall only consider the effect of long range intermolecular electrostatic interactions on the renormalization of the bending moduli.

7.1. LAMELLAR GEOMETRY

Let us do that first for the lamellar case that has been extensively reviewed by Andelman [14] and Fogden and Ninham [34]. The idea of these calculations is to solve the PB equation with non-planar boundaries and expand the corresponding electrostatic free energy in terms of the curvature of the boundaries.

Without going into formal details I will now list some of the main results dealing with the effect of the interactions on the bending rigidities in the case of a lamellar system. For a very thin membrane in the Debye-Hückel limit [14] one derives the change of the bending rigidity due to long range intramolecular interactions as [35]

$$K_C \longrightarrow K_C + \frac{3\,\sigma^2\lambda_D^3}{8\epsilon\epsilon_0} = K_C + \frac{3kT\lambda_D^3}{8\pi\ell_B b^2}, \tag{51}$$

where ℓ_B and the Gouy-Chapman length b have already been introduced in Sections 1.2 and 1.3. Note that electrostatic interactions on this level of approximation stiffen up the lipid membrane surface. This expression however works only for small surface charges *i.e.* in the linearized Debye-Hückel approximation. For higher charges, corresponding to the intermediate regime in Andelman's sense [14], the renormalized bending rigidity can be obtained in the form [36]

$$K_C \longrightarrow K_C + \frac{kT\lambda_D}{2\pi\ell_B}, \tag{52}$$

which can be derived heuristically from Eq. 51 by the substitution $b \longrightarrow \lambda_D$.

In the case of salt free, counterion-only limit the renormalization of the bending modulus is also dependent on the amount of space available for the counterions, *i.e.* to the interlamellar spacing, since the screening length scales as the amount of space available to counterions [3]. In the case of large surface charge density one is left with [37]

$$K_C \longrightarrow K_C + const. \frac{kTD}{\ell_B}, \tag{53}$$

where $const. = \mathcal{O}(1)$. On the other hand in the limit of low surface charge when the counterions behave almost as an ideal gas one derives [38]

$$K_C \longrightarrow K_C + const. \frac{kTD^3}{b^2 \ell_B}. \tag{54}$$

If the surface charge density is high enough there is an additional effect that one has to realize when evaluating the electrostatic contribution to the bending modulus of a membrane. It stems from the ionic density correlations that can not be described on the level of the mean-field Poisson - Boltzmann theory.

Shklovskii and coworkers [39] have formulated a simple scheme, conceptually based on the Wigner crystal picture of a strongly correlated electron gas, to take these correlations into account for highly charged macromolecules where one can envision the counterions of valency Z as tightly associated with the macro molecular surface. In this case the electrostatic interactions decrease the bending modulus according to

$$K_C \longrightarrow K_C - const.\ kT\ Z^2 \left(\frac{\ell_B}{b}\right) \left(\frac{h}{b}\right)^2, \tag{55}$$

where h is the thickness of the membrane and $b = \left(2Ze/\sqrt{3}\sigma\right)^{1/2}$ is the average separation between the counterions of charge Ze_0 on the surface. Note again that the renormalized bending rigidity on this level of approximation gets smaller than the bare value.

This is however not the whole story. In a series of papers de Vries, see [40] and references therein, has explored the effect of thermal fluctuations in multilamellar systems in the salt free case for a much broader range of conditions than addressed here. Depending on the value of the Debye screening length, the separation between the bilayers and their surface charge he derived different regimes in the equation of state differing quite substantially in the role of the fluctuations. In the rather limited case treated here, corresponding to very low salt and relatively small interlamellar separations, the fluctuation and electrostatics in deed couple only through the renormalized bending rigidity and harmonic fluctuation theory.

A similar roster of formulas can be assembled for the electrostatic renormalization of the bending modulus or equivalently the persistence length of a cylindrical macromolecule such as DNA in a hexagonal array [8]. In the case of low linear charge density of the macromolecule one has the famous Odijk-Skolnick-Fixmann [41, 42] result

$$K_C \longrightarrow K_C + \frac{kT\ell_B}{4(b\kappa)^2}. \tag{56}$$

In the case of high linear charge density, surpassing the Manning limit $b < \ell_B$, it is much more difficult if not impossible to obtain a closed formula for the persistence length renormalization. Different numerical estimates [43, 44] nevertheless allow for certain conjectures. For high enough salt and linear charge density above the Manning limit one can write

$$K_C \longrightarrow K_C + \frac{kT}{4\ell_B\kappa^2}. \tag{57}$$

It is generally believed that this formula captures the bending rigidity renormalization of DNA satisfactorily [45]. Recent experiments on stretching of single DNA molecules certainly corroborate this point of view [46, 47].

The above renormalized values for the persistence length are basically valid if the chain is very stiff to start with. In this case the chain stiffens up because upon bending the segments of the chain that were initially far apart get closer and since the interaction potential is repulsive it tends to stiffen up the chain. This mechanism of chain stiffening is usually referred to as "kinematic" stiffening.

If the limit of very stiff chain is no longer applicable, the renormalization of the persistence length is much more difficult to obtain and there is as yet no consensus emerging on what to expect. There have been several attempts ([48] and references therein) at evaluating the renormalized persistence length in different regimes defined by the elastic constant of the chain and the linear charge density of the chain, and what does transpire from these calculations is that the only parameter governing the behavior of the renormalized persistence length is $\mathcal{G}_0 = K_C\ell_B/kTb^2$. There still is disagreement on what exactly the different regimes are.

The most difficult regime to analyze is the small coupling regime defined by $\mathcal{G}_0 \ll 1$. For small enough $\mathcal{G}_0$ we argued [48] that one is first to encounter a regime of effective rigidities pertaining to very small bare elastic constants, *i.e.* the chain is originally very flexible, where the bending rigidity renormalization scales as

$$K_C \longrightarrow K_C + \left(\frac{K_C\ell_B}{kTb^2}\right)^{1/6}\left(\frac{kT}{K_C\kappa}\right)^{7/6}. \tag{58}$$

The persistence length renormalization in this case is not "kinematic" as is the case for high $\mathcal{G}_0$. Rather it is due to close proximity of different segments of the chain that are far apart along the chain. The mechanism of elastic modulus renormalization is thus completely different then in the Odijk-Skolnick-Fixman case. For even smaller values of $\mathcal{G}_0$ there is finally a regime described by

$$K_C \longrightarrow K_C + \left(\frac{K_C \ell_B}{kTb^2}\right)\left(\frac{kT}{K_C \kappa}\right)^7. \tag{59}$$

I should stress that these two regimes are obtained only at extremely (possibly, at least for DNA, unphysical) low values of the coupling parameter $\mathcal{G}_0$. Though there is wide consensus on the form of the persistence length renormalization in the OSF regime there is no similar consensus regarding the expressions for K_C in the low coupling regime.

In the salt free case there has been precious little work and reliable estimates are hard to obtain. Based on the numerical analysis of the solutions of the PB equation with counterions only [42, 43] one can conclude that the renormalized bending rigidity would in this case behave as

$$K_C \longrightarrow K_C + const. \frac{kT\ell_B R^2}{b^2}, \tag{60}$$

which for charges that surpass the Manning limit $b < \ell_B$ becomes

$$K_C \longrightarrow K_C + const. \frac{kTR^2}{\ell_B}. \tag{61}$$

Again as is usual for the salt free systems these equations can be interpreted as if the screening length would become of the order of the average separation R between the cylindrical molecules in a hexagonal array.

As in the case of planar macromolecules the effects of the counterion correlations show up also in the renormalization of the persistence length of cylindrical macromolecules. Shklovskii and coworkers [39] have evaluated the renormalization of the persistence length for tightly associated counterions of valency Z when they form a strongly correlated 2D liquid on and along the surface of the macromolecule. Their result is

$$K_C \longrightarrow K_C - const.\ kT\ Z^{7/2} \ell_B \left(\frac{a}{b}\right)^3, \tag{62}$$

where $b = \sqrt{2Ze/\sqrt{3}\sigma}$ is the average separation between the counterions on the surface and a is the radius of the cylinder. A limiting result for the renormalization of the persistence length was derived also by Hansen *et al.* [49] when the correlation contribution can be considered as a small perturbation to the mean field result. In this case one obtains

$$K_C \longrightarrow K_C + \frac{kT\ell_B}{4(b\kappa)^2} - \frac{kT}{12}\left(\frac{\lambda_D^3}{a^2}\right). \tag{63}$$

The second term on the r.h.s. is the OSF result and the next one describes the effect of correlations. Clearly in this limit the OSF and the correlation contributions are additive.

The form of the bending modulus renormalization thus obviously depends critically on the regime of counterion charge magnitude and salt concentration. Parts of the phase space of these variables, especially for cylindrical macromolecules have still remained unexplored especially those where the non-linear effects stemming from the Poisson - Boltzmann equation would presumably start to play a role [8].

8. Equation of state for salt free systems

In the case of salt free systems we have two different effects conspiring to rule out the fluctuations as a significant contributor to the total osmotic pressure. First of all it is the direct interactions between the macromolecules which are long ranged since for a salt free system the effective screening length is on the order of the average separation between the macromolecules, see Sections 1.2 and 1.3. Also the long range nature of the interactions makes the bending moduli depend on the average separation too, making them quite large (in the case that the correlation effects are negligible, *i.e.* if the counterions are monovalent) at sufficiently low density of macromolecules where the fluctuation effects should be highest, see Section 6.1 and 6.2. We will show how these two facts conspire to quench the effect of fluctuations on the equation of state. Without conformational fluctuations the equation of state is much easier to obtain than in the case of added salt where fluctuations play a prominent role.

The equation of state for both systems, hexagonal DNA arrays and multilamellar lipid membrane arrays, has been measured in nominally salt free conditions. For DNA the recent work by Raspaud *et al.* [50] and the old data by Auer and Alexandrowicz [51] cover a range of DNA densities spanning many orders of magnitude. The equation of state for charged lipids such as PG and PI at salt free conditions has been studies by Cowley *et al.* [53] and also provides a reliable set of data for gauging the accuracy of the theory, which is what we shall do in the next two sections.

8.1. LIPID MEMBRANE ARRAY

If we have only counterions with no additional salt the Poisson - Boltzmann equation can be solved analytically in a cell composed of an inner charged surface and an outer neutral surface where by symmetry the transverse component of the electrostatic field has to be zero [3]. A cell is thus composed of a charged surface with a fixed surface charge density σ and a symmetry surface at $z = D/2$ where the z-component of the electric field vanishes. This surface simulates the effects of the first neighbor at $z = D$. Solving the PB equation in a planar cell geometry one obtains for the osmotic pressure [3]

$$\Pi(D) = kT\ n(D/2) = \frac{kT}{2\pi l_B}\kappa^2(D), \tag{64}$$

where $\kappa(D)$ is the D dependent screening length, being the solution of

$$\kappa(D)D\ \tan\frac{\kappa(D)D}{2} = \frac{D}{b}, \tag{65}$$

and $n(D/2)$ is the concentration of counterions at the midpoint between the membranes, see Eq. 5. Eq. 65 can not be solved analytically but in two limiting cases as discussed in detail by Andelman [14]. In the ideal gas limit corresponding to $b \gg D$ and $\lambda_{DH} \gg D$, the limiting expression for the interaction free energy can be derived as

$$\mathcal{F}(D) = \frac{kT}{\pi \ell_B b}\ln DS. \tag{66}$$

In the Gouy-Chapman region however, which corresponds to $\lambda_{DH} \gg D$ and $b \ll D$, one remains with

$$\mathcal{F}(D) = \frac{\pi\ kT}{2 l_B D}S. \tag{67}$$

Using now the form of the free energy for the salt free case (counterions only) Eqs. 66,67 and the renormalized values of the bending rigidities Eq. 54 (where I assume that in this case the largest contribution to the total bending rigidity comes from the electrostatic part which is evaluated in the low coupling limit) I can derive the magnitude of the fluctuation contribution, based on the theory developed in Sections 4.1 and 4.2. The following scaling forms valid in the lamellar geometry describe the fluctuation $\mathcal{F}_1$ and the bare contribution $\mathcal{F}_0$ to the total free energy of the system

$$\mathcal{F}_1 \sim \sqrt{\frac{\mathcal{B}}{K_C}} \sim D^{-5/2} \quad vs. \quad \mathcal{F}_0 \sim lnD \qquad \text{(small } \sigma\text{)},$$

$$\mathcal{F}_1 \sim \sqrt{\frac{\mathcal{B}}{K_C}} \sim D^{-2} \quad vs. \quad \mathcal{F}_0 \sim D^{-1} \qquad \text{(large } \sigma\text{)}. \tag{68}$$

The l.h.s. of the above comparisons corresponds to the magnitude of the fluctuation effects and the r.h.s. to the bare interactions, where the macromolecules are considered to be ideally rigid. Clearly for all values of the surface charge σ and sufficiently large D, the bare interactions are larger than the fluctuation modified interactions, in contrast to the case of screened electrostatic interactions. In the salt free limit the equation of state derived for an ideally rigid lattice should thus suffice.

One should again keep in mind however [40] that the above statements do not have universal validity and that the effect of the thermal fluctuations can have rather different consequences for a broader range of interlamellar separations, salt and surface charge then treated here. Especially in the case of low density surfactant systems the effects of thermal fluctuations are more complicated than alluded to above.

Numerical evaluation of the equation of state Eq. 64 compares well with experiments [53] performed on multilamellar arrays of salt free charged lipids PI and PG, if one adds a short range, hydration contribution Eq. 16 on top of the salt free electrostatics, see Fig. 5. We obtain a good fit if we assume there is one charge per 1400 Å^2 of the lipid surface. This is consistent with the amount of structural charge in the bilayers composed of egg PC, PG and PI. Thus the salt free equation of state Eq. 64 suffices for description of charged lipid data for most of the accessible macromolecular densities except close to molecular contact.

8.2. DNA ARRAY

If we have a salt free, *i.e.* counterion only, case for cylinders in an array with hexagonal symmetry the Poisson - Boltzmann equation can be solved analytically within the framework of the s.c. cell model [54]. A cell is composed of a single charged macromolecular cylinder of radius a and an outer cylindrical wall of radius $R/2$, simulating the effect of nearest neighbors at separation R, where by symmetry the radial component of the electrostatic field at the cell wall has to be zero. The Fuoss - Katchalsky - Lifson [54] solution leads to the following form of the osmotic pressure in the array [56]

$$\Pi(R) = kT\ n(R/2) = \frac{(1 + z^2(R))}{2\pi l_B\ R^2}, \tag{69}$$

where $n(R/2)$ is the concentration of counterions at the outer wall of the cell, where again by symmetry the electric field vanishes, and $z(R)$ is the solution of

$$ln(\frac{a}{R}) = \frac{\arctan\left(\frac{1-Q}{z(R)}\right) - \arctan\left(\frac{1}{z(R)}\right)}{z(R)}, \tag{70}$$

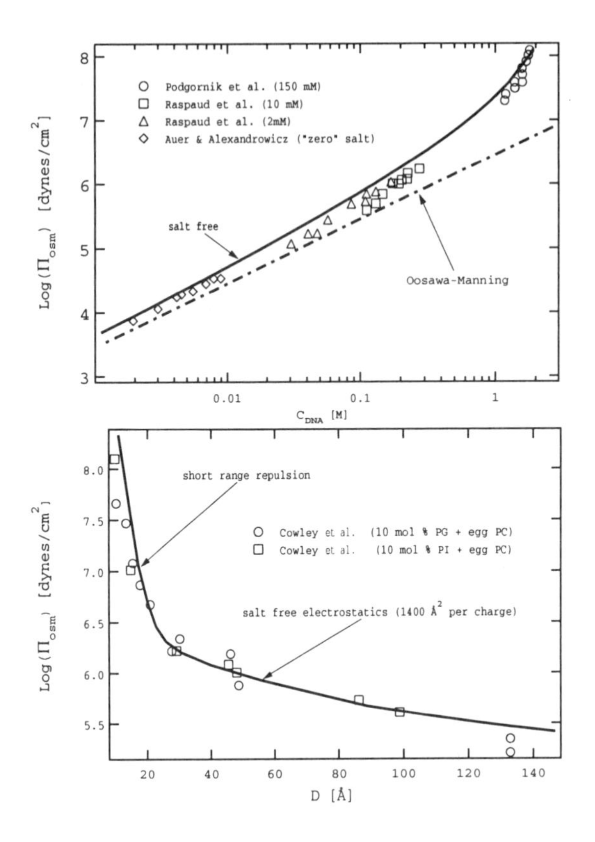

Figure 5. Equation of state for salt free DNA and lipid membrane array. Experimental data for DNA arrays in the medium, low and high densities are taken from [50], [51] and [52]. I have added the standard Manning - Onsager limiting law [55] for comparison. The salt free lipid data are taken from [53]. In both cases the simple equation of state for salt free systems, Eqs. 64 and 69, does a very good job in explaining the data for all but extremely large concentrations close to macromolecular contact.

where $Q = \ell_B/\ell_{PO_4}$ is the dimensionless charge on the surface of the cylinder. a is of course the radius of the cylinder and ℓ_{PO_4} is the separation between the phosphate charges along the contour of DNA. Eq, 70 can not be solved analytically for general R.

One can however derive explicit dependence of the interaction free energy on the interaxial spacing between nearest neighbors R, in the limit of small and large R. It can be derived in the form

$$\mathcal{F}(R \longrightarrow \infty) \simeq \frac{kT\ L}{4\sqrt{3}\pi\ell_B} \ln \pi R^2, \tag{71}$$

and

$$\mathcal{F}(R \longrightarrow a) \simeq \frac{kT\ L}{4\sqrt{3}\pi l_{PO_4}} \ln\pi(R^2 - a^2). \tag{72}$$

The two free energies obviously have the ideal gas form (the ln of the space available to the counterions) with a different number of effective free particles per unit length of the macromolecules (note the change of $\ell_B \longrightarrow b = l_{PO_4}$, where $l_{PO_4} \sim 1.7\,\text{Å}$ is the separation between the phosphate charges along DNA [10, 49]), in the limit of large and small macroion concentrations.

Using these explicit limiting results we can now estimate the effect of fluctuations on the equation of state in a hexagonal array. If the interaction free energy is given by Eqs. 71, 72 then the renormalized bending rigidity is obtained from Eqs. 60, 61 leading to the following estimates

$$\mathcal{F}_1 \sim \sqrt[4]{\frac{\mathcal{B}}{K_C}} \sim R^{-1} \qquad vs. \qquad \mathcal{F}_0 \sim ln\pi R^2 \quad (\text{small } \sigma),$$

$$\mathcal{F}_1 \sim \sqrt[4]{\frac{\mathcal{B}}{K_C}} \sim (R-a)^{-1} \qquad vs. \qquad \mathcal{F}_0 \sim ln\pi(R^2 - a^2) \quad (\text{large } \sigma), \tag{73}$$

and again the fluctuation terms are seen to scale with a larger negative power of R then the bare interactions, making them essentially irrelevant in the calculation of the equation of state. Again this runs in the opposite direction from the case of screened electrostatics where the fluctuation enhanced interactions are essential for understanding of the equation of state.

Evaluating $\Pi(R)$ for DNA one can quite successfully fit the measured equation of state [50, 51] in a very broad range of DNA densities with the structural parameters appropriate for the distribution of charge along DNA. Only at very high densities is there any discrepancy between the salt free calculated equation of state and the measurements, where the measurements decay faster with interaxial spacing than the electrostatic theory alone would predict. Again this is most probably due to the short range structural force kicking in at these high densities.

Comparing the measured and theoretical equations of state derived from Eq. 64 and 69 we see that they really describe the experiments performed in nominally salt free conditions quite well for all but extremely large macromolecular densities, see Fig. 5. Especially impressive is the case of DNA arrays where the theory and experiment match in a range of DNA densities spanning several orders of magnitude. Only at very large macromolecular concentrations corresponding to separations of a few Ångstroms before the contact do we see any noticeable deviation from theory.

Acknowledgements

I would like to thank Adrian Parsegian, Per Lyngs Hansen, Stephanie Tristram-Nagle and John Nagle for their helpful comments on the MS.

References

1. R. Podgornik, H.H. Strey and V.A. Parsegian, *Curr. Op. in Colloid and Interf. Sci.* **3** (1998) 534.
2. J.F. Nagle and S. Tristram-Nagle, *Biochim. Biophys. Acta* **1469** 159 (2000)
3. J. Israelachvili, *Intermolecular and Surface Forces* Hardcover 2nd edition Academic Press (1998)
4. R. Kjellander, these lecture notes. R.R. Netz, these lecture notes.
5. W.M. Gelbart, R.F. Bruinsma, P.A. Pincus and V.A. Parsegian, *Physics Today* **53** (2000) 38.
6. S.L. Brenner and D.A. McQuarrie, *J.Coll.Interf.Sci* **44** 298 (1973)
7. A.E. James and D.J.A. Williams, *J. Coll. Interf. Sci.* **79** 33 (1981)
8. K.S. Schmitz, *Macroions in Solution and Colloidal Suspension*, VCH Publishers, New York (1993).
9. S. L. Brenner and V. A. Parsegian, Biophys. J. 14, 327-334 (1974).
10. V.A. Bloomfield, D.M. Crothers and I. Tinoco, *Nucleic Acids: structures, properties and functions*, University Science Book, Sausalito (2000).
11. J. Mahanty and B.W. Ninham, *Dispersion Forces* (Academic Press, London, 1976)
12. V.A. Parsegian, *J Chem. Phys.* **56** 4393 (1972).
13. B.V. Derjaguin, N.V. Churaev and V.M. Muller and , *Surface Forces*, Consultants Bureau, New York (1987).
14. D. Andelman, in *Structure and Dynamics of Membranes*, Eds. R. Lipowsky and E. Sackmann, North Holland, New York (1995) 603.
15. E.G. Verwey and J.T.G. Overbeek, *The Theory of the Stability of Lyophilic Colloids* (Elsevier, Amsterdam, 1948)
16. R.P. Rand and V.A. Parsegian, *Biochim. Biophys. Acta* **988** 351 (1989)
17. T.J. McIntosh, *Curr. Op. Struct. Biol.* **10** 481 (2000)
18. P. M. Chaikin, T. C. Lubensky, *Principles of Condensed Matter Physics* (Cambridge Univ. Pr., 2000)
19. A.G.. Petrov, *The Lyotropic State of Matter: Molecular Physics and Living Matter Physics* (Gordon and Breach Science Publishers, 1999).
20. H.H. Strey, V.A. Parsegian and R. Podgornik: Equation of state of polymeric liquid crystals, *Phys. Rev. E* **59** (1999) 999-1008. H.H. Strey, R. Podgornik, D.C. Rau and V.A. Parsegian, *Curr. Opin. Struc. Biol.* **8** (1998) 309-313. R. Podgornik, H.H. Strey and V.A. Parsegian, *Curr. Opin. Coll. Interf. Sci.* **3** (1998) 534-539.
21. H.I. Petrache, N. Gouliaev, S. Tristram - Nagle, R. Zhang, R.M. Suter and J.F. Nagle, *Phys. Rev. E* **57** 7014 (1998)
22. R. Lipowsky, in *Structure and Dynamics of Membranes*, Eds. R. Lipowsky and E. Sackmann, North Holland, New York (1995) 521.
23. P. De Gennes and J. Prost, *The Physics of Liquid Crystals*, 2nd ed. (Oxford University Press, Oxford, 1993).
24. R. Meyer, in *Polymer Liquid Crystals*, edited by A. Ciferri, W. Krigbaum, and R. Meyer (Academic, New York, 1982), p. 133.

25. P. Le Doussal and D. R. Nelson, Europhys. Lett. **15**, 161 (1991).
26. R. Kamien, P. Doussal, and D. Nelson, Phys. Rev. A **45**, 8727 (1992).
27. J. Selinger and R. Bruinsma, Phys. Rev. A **43**, 2910 (1991).
28. R. Lipowsky, in *Structure and Dynamics of Membranes*, Eds. R. Lipowsky and E. Sackmann, North Holland, New York (1995) 603.
29. A. Volmer, M. Schwartz, *Eur. Phys. J* **B 7** 211 (1999)
30. R. Podgornik and V.A. Parsegian, *Langmuir* **8** 557 (1992)
31. H. Kleinert, *Path Integrals in Quantum Mechanics, Statistics, and Polymer Physics* (World Scientific Pub Co, 1995)
32. R. de Vries, *J. Phys. II France* **4** 1541 (1994)
33. T. Odijk, *Langmuir* **8** 1690 (1992). T. Odijk, *Europhys. Lett.* **24** 177 (1993) .
34. A, Fogden adn B.W. Ninham, *Adv. in Coll. Interf. Sci.* **83** 85 (1999)
35. M. Winterhalter and W, Helfrich, *J.Phys.Chem.* **92** 6865 (1989)
36. H. Lekkerkerker, *Physica* **A 159** 319 (1989)
37. P.G. Higgs and J.-F. Joanny, *J.Phys. (France)* **51** 2307 (1990)
38. J.L. Harden, C. Marques, J.-F. Joanny and D. Andelman, *Langmuir* **8** 1170 (1992)
39. T.T. Nguyen, I. Rouzina, B.I. Shklovskii, *Phys. Rev. E* **60** 7032 (1999)
40. H. von Berlepsch and R. de Vries, *European Phys. J. E* **1** 141 (2000)
41. T. Odijk, *J. Polym. Sci.* **15**, 477 (1977).
42. J. Skolnick and M. Fixman, *Macromolecules* **10**, 944 (1977).
43. M. Fixman, *J. Chem. Phys.* **76** 6346 (1982).
44. M. Le Bret, *J.Chem.Phys.* **76** 243 (1982)
45. Various authors in issue 10 of *J.Phys.Chem.* **96** (1992)
46. C. Bustamante, S.B. Smith, J. Liphardt and D. Smith, *Curr. Opinion in Structural Biology* **10** 279 (2000) .
47. C.G. Baumann, S.B. Smith, V.A. Bloomfield and C. Bustamante, *Proc. Natl. Acad. Sci.* **94** 6185 (1997).
48. P.L. Hansen and R. Podgornik, *J.Chem.Phys.* (2000) in press.
49. Hansen PL, Svenšek D, Parsegian VA, Podgornik R, Phys. Rev. E **60** 1956 (1999).
50. E. Raspaud, M. da Conceicao and F. Livolant, *Phys. Rev. Lett.* **84** 2533 (2000).
51. H.E. Auer and Z. Alexandrowicz, *Biopolymers* **8** 1 (1969).
52. Podgornik, R., Rau, D. C. & Parsegian, V. A. (1989) *Macromolecules* **22**, 1780; Strey, H. H., Parsegian, V. A. & Podgornik, R. (1999) *Phys. Rev. E* **59**, 999.
53. A.C. Cowley, N.L. Fuller, R.P. Rand and V.A. Parsegian, *Biochemistry 17* 3163 (1978)
54. R. Fuoss, A. Katchalsky, and S. Lifson, Proc. Natl. Acad. Sci USA 37 (1951) 579; S.Lifson, and Katchalsky A., J. Polymer Sci. **XIII** (1954) 43.
55. F. Oosawa, *Polyelectrolytes* (Marcel Dekker, Inc., New York, 1971).
56. P.L. Hansen, R. Podgornik and V.A. Parsegian, *Phys. Rev. Lett.* (2000) submitted.

STRUCTURE AND PHASEBEHAVIOR OF CATIONIC-LIPID DNA COMPLEXES

JOACHIM O. RÄDLER
Max-Planck-Institute for Polymer Research
Ackermannweg 10, D-55128 Mainz, Germany

1. Introduction

In recent years aggregates made of nucleic acid and cationic lipid received much attention as novel gene delivery systems for human gene therapy [Behr, 1994; Crystal, 1995; Miller, 1998]. The interest in cationic lipid-DNA (CL/DNA) complexes arose from the discovery by Felgner and his group in 1987 showing that cationic liposomes enhance the transfer of exogeneous DNA into eucaryotic cells [Felgner, 1987]. This finding was the result of a long arduous attempt to package DNA into liposomes in order to facilitate its transfer across the plasma membrane of cells. The guiding idea was to encapsulate nucleic acid into liposomes, that are capable to fuse with the plasma membrane of the target cell. Viral infection serves as a shining example that this task can be achieved with paramount efficiency. Synthetic gene carriers, however, still don't reach the efficiency of viruses by almost three orders of magnitudes. Clearly viruses have evolved over many millions of years into perfect gene packing and transferring vehicles. In a separate article by Gelbart in this volume the physics of compaction of nucleic acid in viruses is described in detail. For double stranded DNA of several kilo-base pair length the conformational entropy, the elastic energy as well as the electrostatic repulsion of the negatively charged phosphate groups work against the compression of DNA into capsules that are smaller than the DNA radius of gyration. Viruses make use of suitable condensing agents and a rigid capsule. The viral capsid is endowed with targeting, endosomolytic and nuclear translocation proteins that serve specific functions to make it possible for the virus to reach the nucleus of the target cell. In synthetic gene delivery systems these viral properties should be mimicked. It is surprising that there are cationic agent that alone are able to condense

C. Holm et al. (eds.), *Electrostatic Effects in Soft Matter and Biophysics*, 441–460.
© 2001 *Kluwer Academic Publishers. Printed in the Netherlands.*

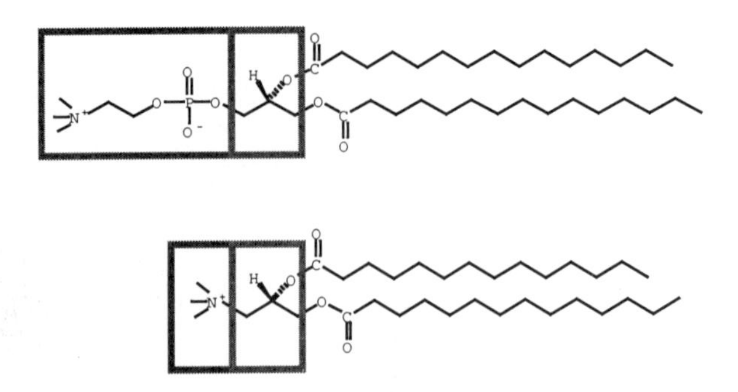

Figure 1. Schematic drawing of a lecithin (DMPC) and its cationic lipid derivative (DMTAP). Both molecules have alkyl chains attached to a glycerol backbone. The phosphocholin (PC) headgroup is zwitterionic while the trimethyl ammonium group (TAP) is cationic.

DNA in such a way that a detectable amount of gene transfer (transfection) takes place. Nowadays cationic lipids and some cationic polymers are the most prominent transfection agents on the market. The resulting complexes are referred to as *lipoplexes* and *polyplexes* [Felgner, 1997]. However, it is still not well understood how these complexes mediate gene transfer. It becomes more and more evident that the physico-chemical properties of the DNA complexes are intimately related to transfer efficiency [Safinya, 1999]. In particular the structure and structural transitions of cationic lipid/DNA (CL/DNA) complexes have only recently been elucidated. In this articles we will introduce the formation of CL/DNA complexes, the molecular structure and discuss possible mechanisms of cell entry.

2. Structural Characterization of Lipoplex Mesophases

2.1. CATIONIC LIPIDS

Cationic liposomes are usually composed of two components : a cationic amphiphile and a natural helper lipid. The helper lipid is needed, since not all cationic lipids form lipid bilayers by themselves. The nature of the co-lipid and the molar ratio of cationic-to-co-lipid strongly influence the bilayer forming and fusogenic properties of the mixed cationic liposomes deal. In principle, of course, we could use more than two components or even imagine a magic lipid cocktail for the use of gene delivery. For practical purposes, however, it is wise to restrict the basic studies to two-component lipid mixtures. In fact we will find that even in this case the phase behavior is rather complex. Cationic lipids are synthetic molecules that are generally not found in nature. Reviews on the chemistry of cationic lipids are found in reference [Miller 1998; Lasic, 1996]. In Figure 1 we show the chemical struc-

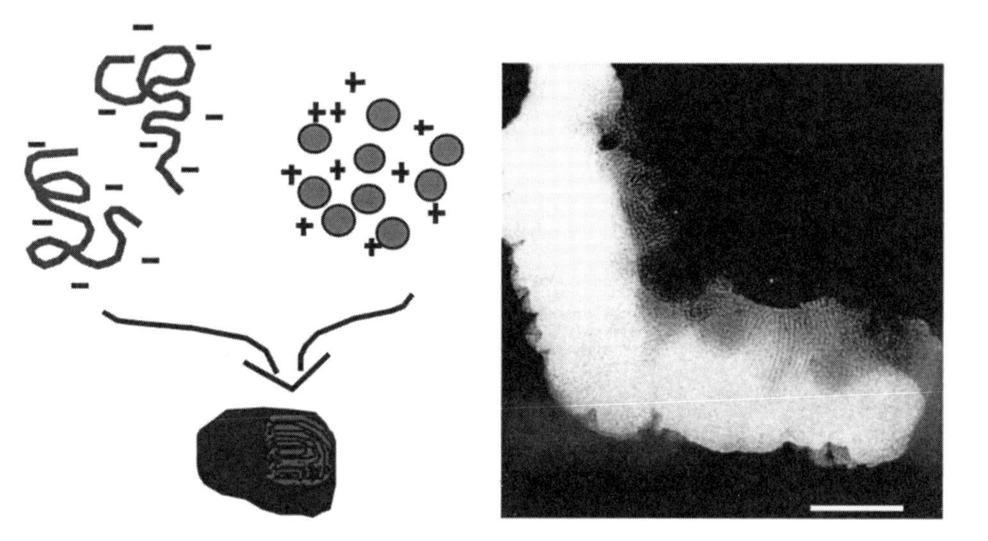

Figure 2. (a) Schematic representation of DNA molecules and cationic liposomes condensing into cationic lipid-DNA (CL/DNA) complexes.
(b) Electron microscopy image of a CL/DNA aggregate showing an amorphous shape and hints of a lamellar fine structure.

ture of a particular cationic lipid : dioleoyl-trimethyl-ammonium propane (DOTAP) [Silvius 1991]. This lipid can be regarded as a cationic analog to the natural zwitterionic lecithin dioleoylphosphatidyl-choline. Both amphiphiles have a glycerol backbone with two alkyl chains attached. The cationic head group consists of a trimethyl ammonium group while differ in the cationic group TAP which replaces the zwitterionic cholin group of DOPC. Cationic lipids self-assemble in aqueous solution into higher order aggregates like bilayer sheets, micelles or tubular structures. The prevailant structure in dilute solutions are bilayer, that close up into single walled liposomes. Liposome suspensions can be processed further by sonication or extrusion in order to achieve monodisperse size distributions of unilamellar vesicles.

2.2. CATIONIC LIPID – DNA COMPLEXES

The complex formation of cationic liposomes with nucleic acid occurs readily upon mixing cationic lipid and DNA suspensions. Nucleic acid binds to the cationic liposomes due to electrostatic attraction. In many cases the cationic liposomes collapse and fuse with the nucleic acid into compact aggregates as sketched in Figure 2a. The system readily precipitates into submicron size lipoplexes. First evidence of lipoplexes and their structure was obtained by cryo-TEM or freeze fracture electron microscopy work [Ghirlan-

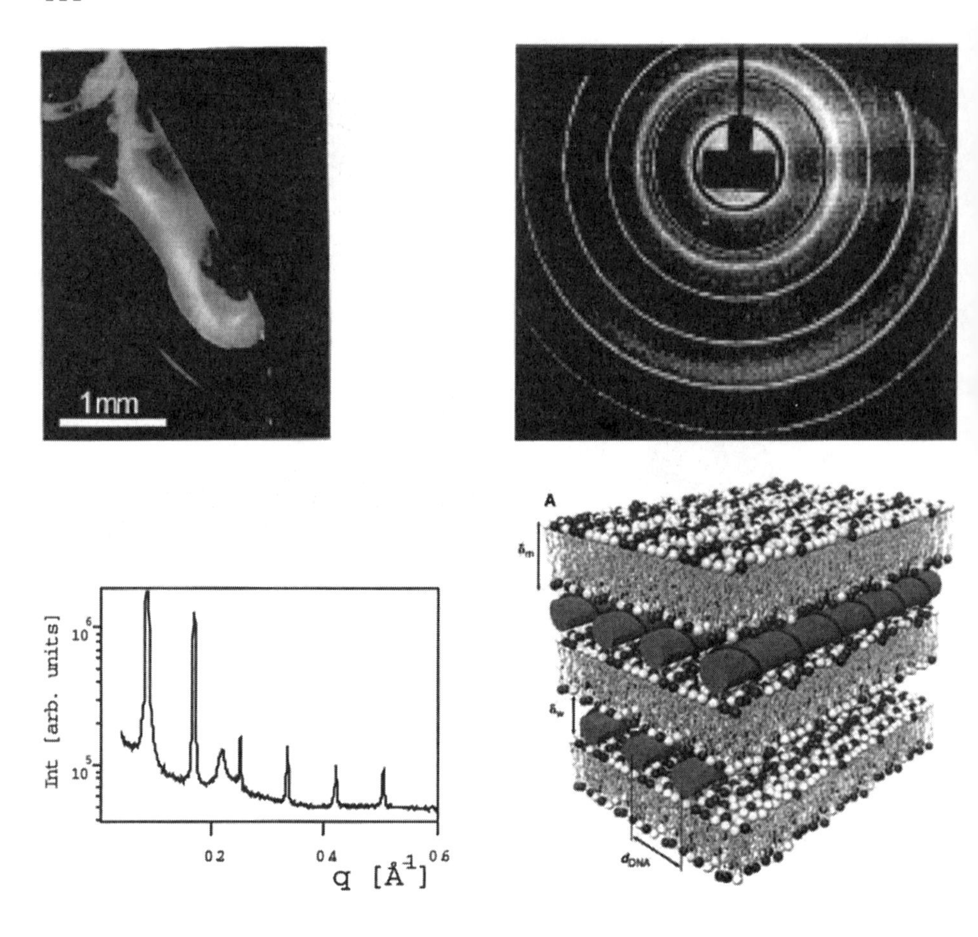

Figure 3. (a) Dense preparation of lipid-DNA complexes in a X-ray glass capillary.
(b) Typical Debye Scherrer rings obtained by small angle X-ray scattering (SAXS).
(c) Radially integrated scattering intensity.
(d) Schematic drawing of the molecular order in lamellar CL/DNA complexes.

do, 1992; Gustaffson, 1995; Sternberg ,1994; Zabner, 1995; Lasic, 1997]. In Figure 2b we give an example of a complex of DOPC/DOTAP/λ−DNA imaged by electron microscopy in a negative staining procedure. The complexes have amorphous irregular shapes and appear highly condensed with sometimes regular fine textures. Surprisingly the CL/DNA complexes are far more ordered on the molecular level, when probed by X-ray diffraction.

2.3. SMALL ANGLE X-RAY SCATTERING

The bulk structure of CL/DNA complexes is best studied by small angle X-ray scattering (SAXS). To this end concentrated samples are prepared in quartz capillaries by injecting 25mg/ml cationic liposome suspension into a 5mg/ml DNA solution (Figure 3a). After a few days of equilibration period such samples exhibit a reproducible small angle X-ray diffraction pattern that is characteristic for an isotropically distributed powder of crystallites. Figures 3b shows the diffraction micrograph of a lamellar CL/DNA phase. Sharp equally spaced diffraction rings are measured. In Figure 3c the corresponding radially averaged intensity corrected for the polarization factor is shown as a function of the scattering vector, $q = 4\pi \sin\vartheta/\lambda$, whereby 2ϑ denotes the scattering angle. The diffraction maxima fulfill the Bragg condition.

$$2d_{\mathrm{L}} \sin\vartheta = n\lambda \tag{1}$$

where d_{L} is the lamellar repeat distance of the lipid-DNA stacking and n the order of the diffraction peak. The molecular arrangement for this particular case is shown in Figure 3d. Monolayer of DNA are intercalated between cationic lipid bilayer. The lipid membrane repeat distance $d_{\mathrm{L}} = \delta_m + \delta_w$ is given by the sum of the membrane thickness, δ_m and the thickness of the water phase, δ_w , that surrounds the DNA strands. The second characteristic length scale involved is the DNA-DNA packing distance $d_{\mathrm{DNA}} = 2\pi/q_{\mathrm{DNA}}$, i.e. the interhelical spacing in plans. This length appears as a diffuse reflection at $q_{\mathrm{DNA}} \approx 0.2\text{Å}^{-1}$ in Figure 3c. The reason for diffuse scattering is the dominance of thermal fluctuations in a low dimensional lattice. Salditt et al. showed that the shape of the diffuse scattering of the intercalated DNA is in agreement with a harmonic model of a two-dimensional smectic phase [Salditt, 1997; Salditt, 1998]. Theoretical work that followed this work showed that the fluctuations in a system of stacked smectic lattice exhibit a rich scenario depending on the coupling between adjacent lattices [OHern 1998; Golubovic 1998; OHern, 1999; Golubovic, 2000]. In particular a novel phase was described, which was termed "sliding phase", where orientational order is preserved, while positional coherence between neighbouring layers is lost. In the other extreme, if the coupling between layers is strong long range positional order of the DNA from layer to layer can result in a two-dimensional columnar order of the DNA rods. Recent experiments by Artzner et al. demonstrated that in this case the columnar lattice has centered rectangular symmetry [Artzner et al.]. Figure 4 shows a schematic drawing, the X-ray scattering data and an electron density map calculated from the diffraction peaks. The peaks indexed (2n,0) are due to the lamellar stacking, while the reflections [(1,1);(3,1);(5,1) etc.] arise from the DNA superlattice. Indication for interlayer coupling of the DNA lattices was also found in the fine texture of cryo-EM micrographs of onionlike CL/DNA complexes [Battersby 1998].

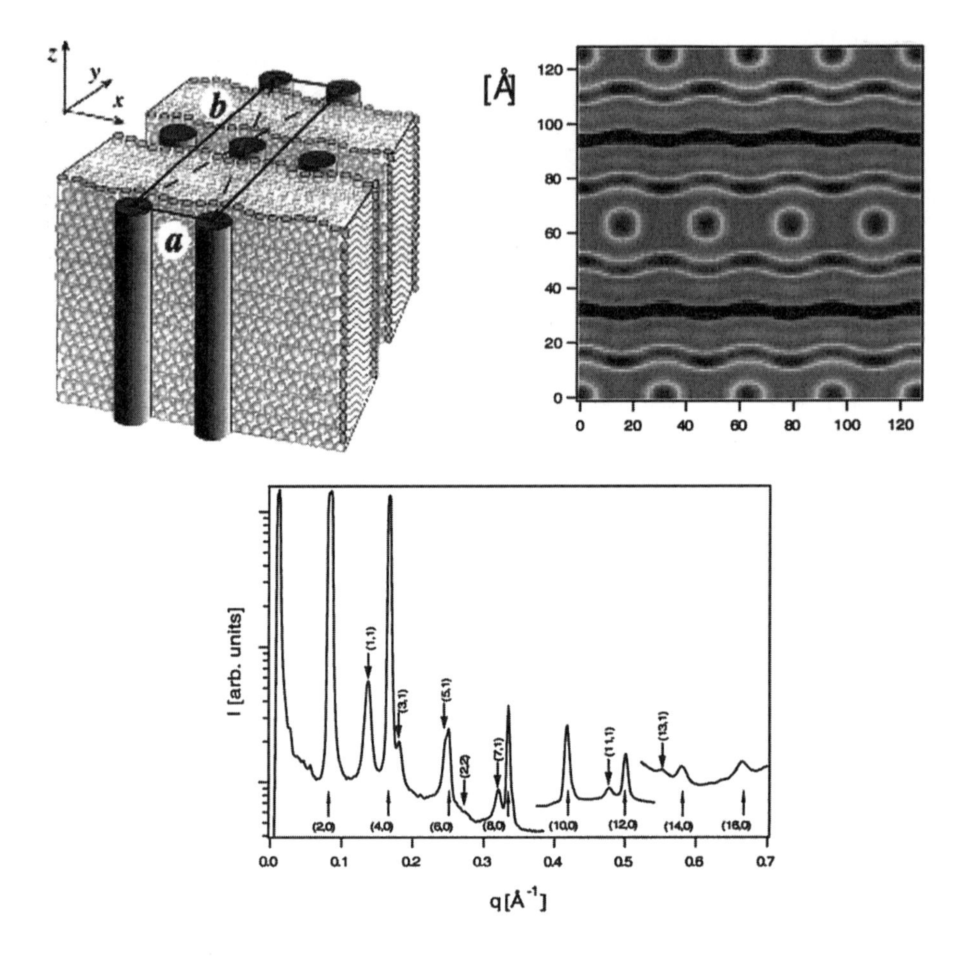

Figure 4. (a) Schematic drawing of a rectangular-columnar DNA super-lattice embedded in a lamellar lipid bilayer stack
(b) Electron density map of the rectangular-columnar structure derived from the SAXS data shown in (c)
(c) Small angle X-ray scattering intensity of a DMPC/DMTAP-DNA mixture [Artzner 2001].

2.4. MESOMORPHIC POLYMORPHISM

Over the past five years the bulk structure of numerous cationic lipid-DNA phases was studied [Artzner 1998, Battersby 1998, Boukhnikachvili, Rädler 1997, Sternberg 1994, Koltover 1998, Pitard,1997]. Most of the structures found are related to the lyotropic mesophases of pure amphiphilic system without DNA. For lipid mesophases Luzatti introduced a nomenclature[Tardieu]. In this article will adopted this nomenclature for the CL/DNA complexes, but add the index "c" in order to indicate that the

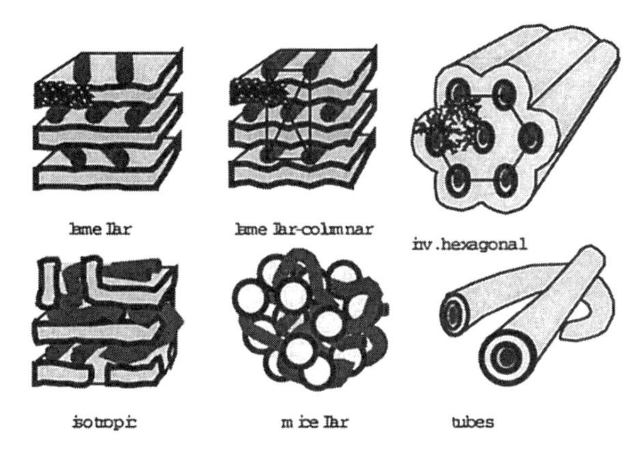

Figure 5. Overview on the different mesomorphic structures found for CL/DNA composite phases. (from reference [Artzner 2000])

phases differ from classical amphiphilic mesophases by the fact that they are complexed with DNA. Figure 2 presents an overview on possible lipid-DNA morphologies. The lamellar phase, L_α^c, was introduced in the previous chapter and is the predominant structure for CL/DNA aggregates. In some cases the lamellar phase evolves into a rectangular columnar structure at low-temperature [Artzner 1998, Battersby 1998]. Beside the lamellar stacking several other arrangements are possible. In particular an inverted hexagonal phase, denoted H_{II^c}, was found by X-ray scattering, where the D-NA strands coated by a lipid monolayer are arranged in hexagonal columnar order [Ghirlando, 1992; Koltover, 1998]. A micellar lipid-DNA phase was described by Boukhnikachvili et al. [Boukhnikachvili, 1997] and evidence for tubular structures stems from electron microscopy studies [Sternberg, 1994]. The lower row in Figure 5 shows furthermore a hypothetical isotropic phase, which should lack long range order in the lipid as well as the DNA moiety. In general the structure of CL/DNA composite phases depends on the "shapes" of the amphiphilic molecules. Amphiphiles with a small headgroup area compared to the hydrophobic tail region tend to form inverted structures with negative curvature [Israelachvili 1992]. On the other hand amphiphiles with large headgroup-to-tail ratio hold a positive natural curvature and tend to form micellar or cylindrical structures. The bilayer forming lipids have almost zero intrinsic curvature. Knowing the mesophase of the cationic lipid gives some indication for the expected structure of a CL/DNA complex. However, in the following we demonstrate experiments, where the observed structure allows quantitative comparison with theory based on a rigorous calculation of the total free energy of CL/DNA aggregates. The later includes electrostatic, elastic and entropic terms. Moreover,

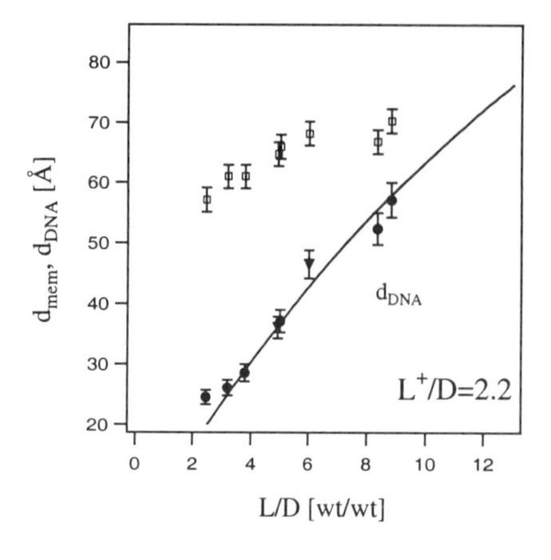

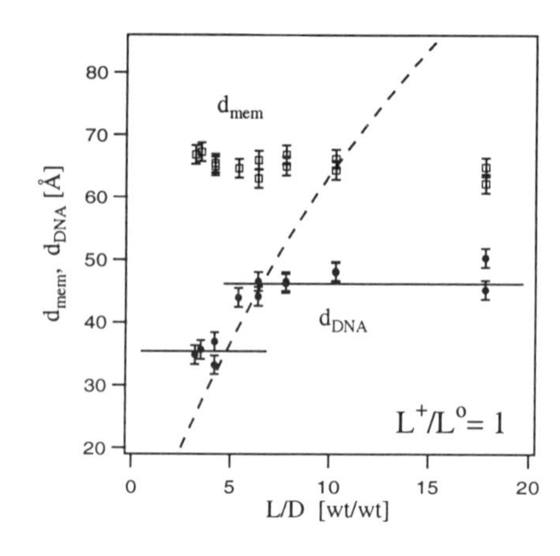

Figure 6. (a) DOPC/DOTAP/λ-DNA-complexes : The DNA-DNA separation distance is plotted as a function of the lipid-to-DNA ratio, L/D at constant isoelectric charge ratio
(b) The DNA-DNA separation distance as a function of the lipid-to-DNA ratio at fixed membrane composition $L^+/L^0 = 1$ (from reference [Rädler 1998]).

the theory explains the surprising structural transitions found by small angle X-ray scattering.

From a wider perspective the classification of CL/DNA mesophases can be extended also to any amphiphile-polyelectrolyte systems that self-assembles by opposite charge of the two components. In fact lamellar composites of cationic surfactants and oppositely charged polyelectrolytes were reported in material science (Antonietti, 1994; Ponomarenko, 1996; Dautzenberg, 1997), and also biologically relevant mixtures of charged lipids with polypeptides and polysugars were studied (de Kruiff *et al.*, 1985; Zschörnig *et al.*, 1992, Subramanian, 2000).

3. Phase transitions in lipid-DNA complexes

3.1. LYOTROPIC PHASE BEHAVIOR

The neutral helper lipid plays an important role in the properties of CL/DNA complexes. Technically we can describe dependence on composition as lyotropic phase behavior. For the model system DOPC/DOTAP/DNA the lyotropic behavior as a function of cationic mol fraction and lipid-to-DNA ratio is well documented [Rädler 1997, Rädler 1998, Koltover et al. 1999]. Let us denote the three components of the mixture: D [mg], L^0 [mg], and L^+ [mg], corresponding to the weights of DNA, neutral lipid, and cationic lipid, respectively. The spacings in CL/DNA complexes depend on the lipid composition and the lipid-to-DNA ratio L/D, where $L = L^+ + L^0$

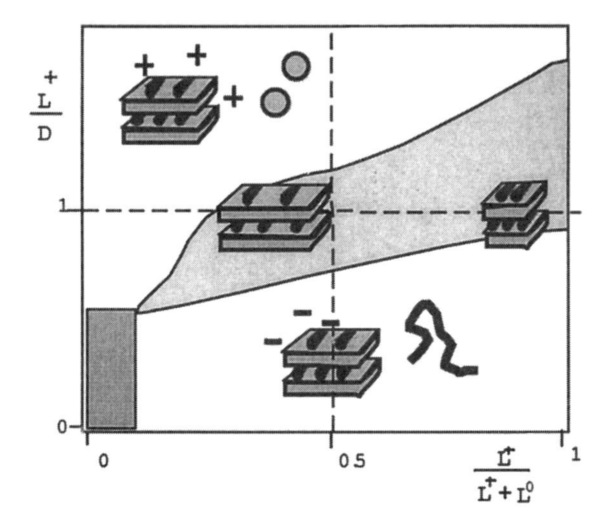

Figure 7. Schematic phase diagram. The one-phase region corresponds to isoelectric lamellar complexes. In the two neighboring regimes complexes coexist with either DNA or lipid. (drawing after ref.[Harries 1998]).

denotes the total amount of lipid. The first question that can be addressed, is the structure as a function of the lipid-to-DNA mass ratio, L/D, when the composition of the cationic liposomes is held constant. It turns out that a lamellar composite structure forms over a wide range. However, a single lamellar condensed phase with only excess water exists only in a narrow regime around the isoelectric condition, i.e. when charge neutral, complexes with equal number of cationic lipids and DNA phosphate groups, are formed. Away from the isoelectric condition complexes coexist with either excess lipid or excess DNA. X-ray data provide the membrane repeat spacing and the DNA packing distance in all cases (see Figure 6). In the coexistence regime the internal interhelical spacing becomes independent from the lipid-to-DNA ratio. Surprisingly the equilibrium value of the packing distance of complexes in the presence of excess DNA or excess cationic liposomes differ. This phenomenon is known as overcharging of the complexes and is described in more detail in the article be Gelbart in this volume. A calculation of the free energy of the lamellar phase, based on non-linear Poisson Boltzmann equation, elastic contributions and entropy of the counterions and lipids, predicts indeed the observed discontinuity [Bruinsma, 1998; Harries, 1998, Dan 1997, May, 2000]. Theory also predicts the distribution of counter-ions in CL/DNA complexes. For free DNA in aqueous solution most of its counter-ions are condensed on the strand. In contrast, in CL/DNA the cationic lipid molecules compensates for the anionic phosphate groups and act like counter-lipids whose mobility is re-

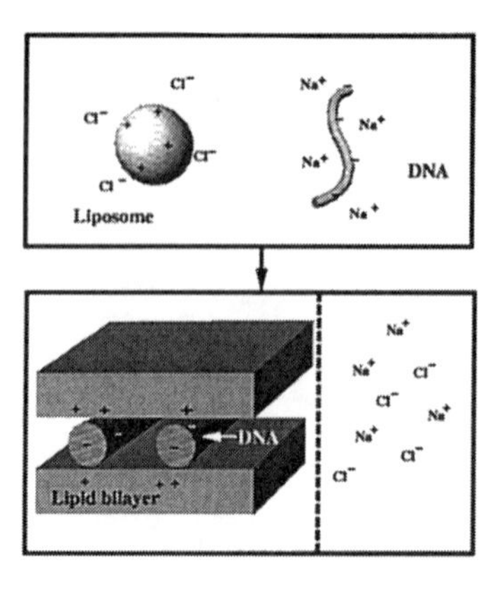

Figure 8. In the condensed lipid-DNA complexes the cationic lipids serve as counter-lipids to the anionic phosphate groups and Nä+ and Cl- ions are released to the excess water phase (from ref. [Wagner 1999]).

stricted to the plane of the membrane. As a consequence the counter ions of the DNA as well as the counter ions of the cationic membrane are released from the complexes (see Figure 8). The release of counter ions can in fact be measured by an relative increase of the ionic conductivity of the aqueous supernatant [Wagner 1999].

In a second kind of experiment the charge ratio of cationic lipid-to-DNA, L^+/D, is kept constant, while the charge density of the membrane, or the molar composition, L^+/L^0, of the membrane is varied. It can be shown that the DNA-DNA repeat distance is set by the charge density of the membrane and hence by the molar fraction of cationic lipid, if the composition of the lipid-DNA mixtures is systematically varied at a the isoelectric charge ratio. Assuming that the aggregates form one homogeneous phase conservation of the local volume fractions predicts [Rädler 1997]

$$d_{\mathrm{DNA}} = \frac{A_D \rho_D}{\delta_m \rho_L}(L/D) \tag{2}$$

where A_D and ρ_D denote the area cross-section and volume density of DNA and δ_{mem} and ρ_L the thickness and density of the lipid bilayer. Eq.(2) describes the interhelical spacing accurately in the regime between 25Å and 60Å (or Φ =0.25-0.6) in case of DOPC/DOTAP/DNA complexes using $A_{\mathrm{DNA}} = 186\text{Å}^2$, $\rho_{\mathrm{DNA}} = 1.5$g/ccm and $\rho_{mem} = 1.07$g/ccm. Note that the later constants are literature values and not fitting parameters and that hence the absolute DNA-DNA spacing can be calculated from the lipid-to-

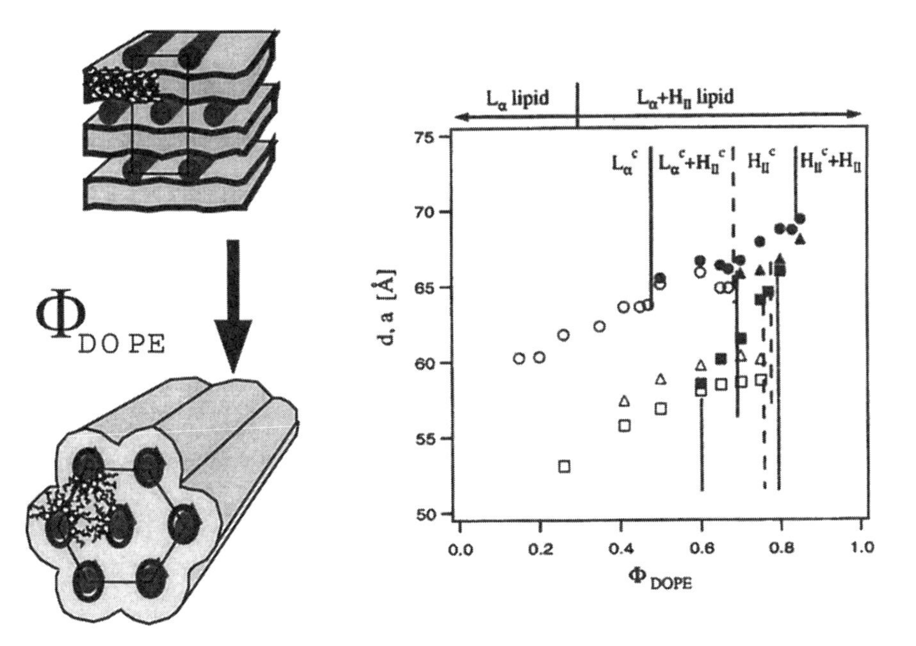

Figure 9. The lamellar-hexagonal transition as a function of the mol-percentage of the ethanolamine lipid DOPE. The plot depicts the evolution of the DNA lattice constant (after ref. [Koltover 1998]).

DNA mass ratio L/D without adjustable parameter. The variation in the DNA interaxial distance as a function of the charged-to-neutral lipid ratio unambiguously substantiates that (1) we directly probe the DNA behavior in multilayer assemblies and (2) the linear DNA chains confined between bilayers form a 2D smectic phase.

3.2. LAMELLAR-HEXAGONAL TRANSITION

Complexes of the cationic lipid DOTAP mixed with the helper lipid DOPE, which favors inverted structures were expected to form hexagonal phase [Farhood, 1995]. Experiments, however, revealed coexistence of lamellar and hexagonal phase for most mixtures. As shown in Figure 9 a pure inverted hexagonal phase exists only in a small distinct region around volume fraction $\Phi_{\mathrm{DOPE}} = 0.72$ [Koltover, 1998], while a pure lamellar phase exists at low molar fraction DOPE. The fact that coexistence occurs over a wide range is predicted by theoretical work [May, 1997; May, 2000]. The model takes a composition dependent spontaneous curvature, entropic mixing of the lipid and the electrostatic interactions into account. Theory predicts and experiment confirms that a transition from lamellar to hexagonal can be

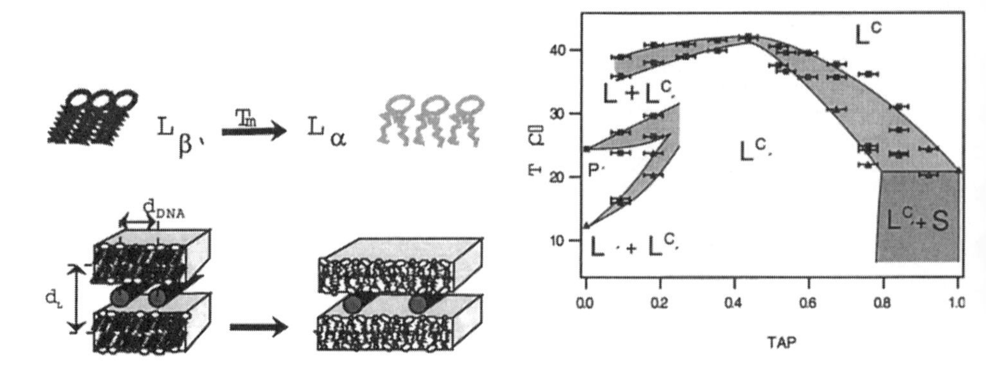

Figure 10. (a) Gel-liquid-crystalline phase transition of the lipid chains. (b) Thermotropic phase diagramm of isoelectric DMPC/DMTAP-DNA mixtures (from ref. [Zantl 1999])

induced by either adding curvature loving lipid or by adding co-surfactant that lower the bending modulus.

3.3. THERMOTROPIC PHASEBEHAVIOR

So far we have considered complexes, where the chains of the cationic lipid molecules were in the fluid phase. As studied by Luzatti and coworkers many years ago lipid mesophases undergo several transitions as a function of decreasing temperature into low-temperature phases with different kind of short ranged chain ordering [Tardieu 1973]. Many lipid bilayer systems undergo a transition from a fluid L_α state over the rippled P_β phase to the planar gel-phase with tilted chains, L'_β. The thermotropic phase diagram of binary mixtures of cationic lipids and helper lipids was studied for a few lipids [Silvius, 1991, Zantl 1999, Lewis, 2001] with saturated alkyl chains. Typically non-ideal mixing behavior is observed with a maximum of the transition temperature at equimolar lipid ratios. The question arises, how the thermotropic phase behavior of cationic lipids is affected upon condensation into CL/DNA phases. It appeared that lamellar intercalated structure persists independent of the lipid chain conformation [Zantl, 1999]. In lipid mixtures with saturated hydrocarbon chains the chain melting transition is accompanied by a relative extension of the lamellar arrangement due to the discontinuity of the lipid head group area with temperature. A thermotropic phase diagram of CL/DNA was established by using a simultaneously measuring the SAX and WAX scattering (Figure 10) [Rappolt]. The data on DMPC/DMTAP/DNA follow the same Eq.(2) within 1% accuracy for $\Phi = 0.5$. Most importantly we find that the equation also holds in the $L_{\beta'}$ gel-phase, if the $L_{\beta'}$ membrane thickness and the $L_{\beta'}$ lipid

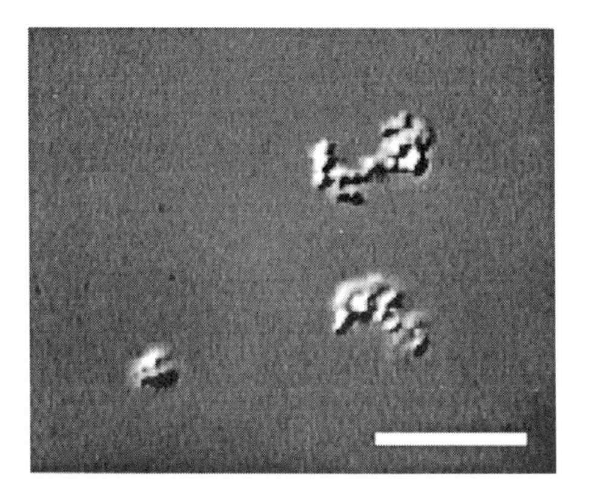

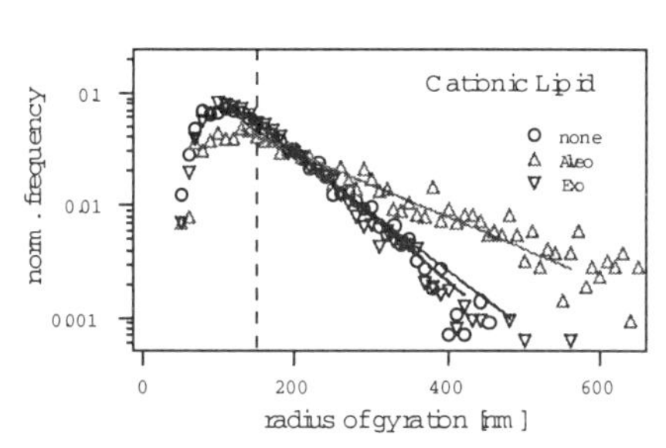

Figure 11. (a) Dendritic aggregated lipid-DNA complexes close to the isoelectric ratio (scale bar=10m).
(b) The size distribution can be determined by quantitative fluorescence microscopy. The distribution function follows an exponential law

mass density are used. The latter increases by 5% from the fluid to gel-like state. The same relative change is found in the measured volume change of the membrane $\Delta_V = \Delta_{\text{DNA}} + \Delta_{mem} \approx 5\%$ (table1). The structural changes at the $L_{\alpha^c} - L_{\beta'^c}$ transition are hence governed by the volumetric changes of the lipid bilayer.

4. Colloidal Aspects

4.1. SHAPE AND SIZE DISTRIBUTION

The shape and size distribution of lipoplexes depends strongly on concentration and buffer conditions. The size distribution of DNA-lipid aggregates can be measured by dynamic light scattering [Rädler 1998]. The average size of the DNA-lipid globules as a function of L/D exhibits a maximum at the point of charge neutrality L/D. This finding is in agreement with zeta-potential measurements which showed a negative surface potential for overcharged complexes and a positive surface potential for undercharged complexes. Optical studies confirm the aggregation behavior. The late stage of aggregation can be imaged and reveals fractal like objects composed of sub-micrometer globules Figure 11a. The kinetics of the complex formation involves three time scales: A rapid liposome-DNA *condensation* yielding globular complexes and a slower colloidal *aggregation* of globules that is strongly dependent on the surface charge of the globules. On an even longer time scale the diffusion-aggregated complexes undergo further compaction by fusion. Using fluorescence microscopy the size distribution of late stage complexes adsorbed to a solid substrate was evaluated by image processing

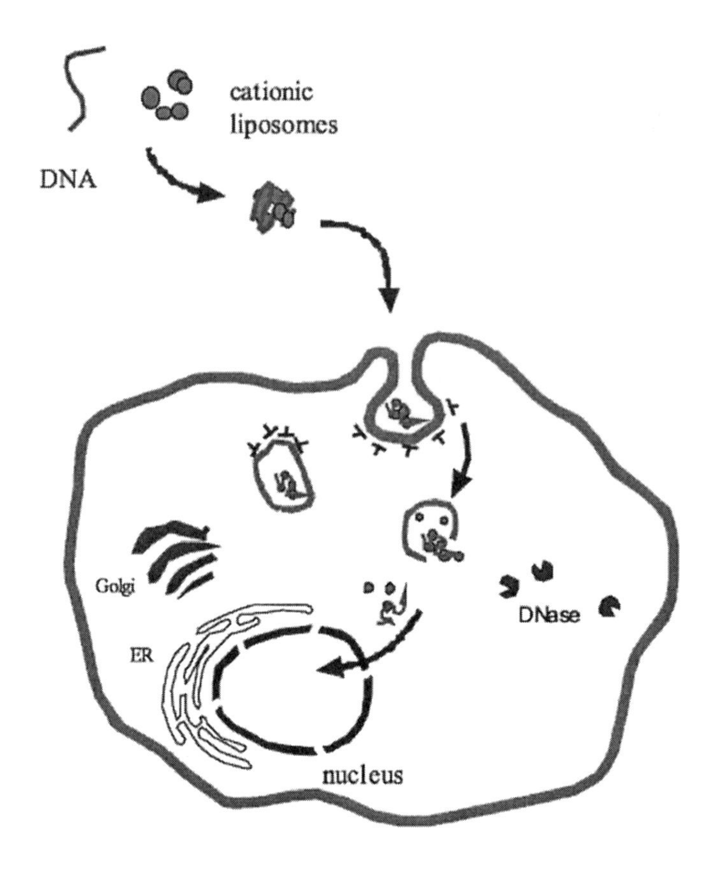

Figure 12. Path of entry of cationic lipid- DNA gene delivery complexes. Complexes are taken up by endocytosis, fuse with the endosome and are released into the cytosol. The transfer mechanism of DNA into the cell nucleus is unknown.

(Figure 11b) [Galneder, 2001]. The size distribution shows an exponential decay in the asymptotic regime indicative for growth by diffusion limited aggregation. The aggregation behavior can dramatically change in the presence of serum or protein rich buffers. This infers a severe limitation for the use of CL/DNA for *in vivo* applications. As an example, lung surfactant were shown to directly interfere with the transfection efficiency in cell culture [Ernst, 1999].

4.2. IMPLICATIONS OF STRUCTURAL RESEARCH FOR GENE THERAPY

The goal of structural research on CL/DNA complexes was to elucidate the physics of self-assembly for these systems. It is remarkable that in fact many cationic lipids assemble in a way that can be quantitatively. On a long term one wishes to use a rational design approach in order to improve on the transfer efficiency of synthetic CL/DNA vectors. Unfortunately the mechanism of gene transfer is yet poorly understood. Clearly many different barriers and hurdles have to be crossed by the complex. For this reason no simple relation between CL/DNA complex structure and the overall transfection efficiency is to be expected. However, different structures may have properties that have relevance in the gene transfer process. The hexagonal structure for example is more fusogenic than the lamellar phase. This was demonstrated by fusion experiments of CL/DNA with anionic giant vesicles [Koltover 1999]. On the other hand, compaction of plasmid DNA into lamellar CL/DNA aggregates protects DNA from being cut by restriction enzymes [Rädler 1998]. The pathway of CL/DNA gene delivery complexes is shown in Figure 12. Complexes attach to the cell surface and are taken up by endocytosis [Wrobel, Zabner]. As the pH decreases inside the endosome the CL/DNA fuses with the endosomal membrane and plasmid DNA is released into the cytosol. From there the DNA has to pass through the nuclear pores, which might be supported by an active transport mechanism. Since gene transfer occurs over many barriers, an ideal vector should be equipped with several components such that each component is designed to help to overcome one barrier. To this end Wagner et al. constructed transferin-polylysine-DNA complexes for specific targeting [Wagner, 1992] and Lee and Huang designed a folate targeted vector using polylysine-condensed DNA with an anionic lipid membrane coating [Lee, 1996]. Yet, the self-assembly of well defined multi-component vectors is still in its infancy and remains a formidable task for the future. The investigation of structure and phase behavior of binary and ternary complexes is a prerequisite on this long way.

References

1. Antonietti, M., J. Conrad, and A. Thünemann. (1994). Polyelectrolyte-Surfactant Complexes: A New Type of Solid, Mesomorphous Material. *Macromolecules.* **27**, 6007-6011.
2. Artzner, F., R. Zantl, G. Rapp, and J. O. Rädler. (1998). Evidence of a Rectangular Columnar Phase in Condensed Lamellar Cationic Lipid-DNA Complexes. *Phys. Rev. Lett.* **81**, 5015-5018.
3. Artzner, F., R. Zantl, J.O. Rädler (2000) Lipid-DNA and Lipid-Polyelectrolyte Mesophases : Structure and Exchange Kinetics in : Cellular and Molecular Biology (editor R. Wegmann, Paris) Vol. 46 (5), 967-978.
4. Artzner, F., R. Zantl, and J. O. Rädler. (2001). unpublished results.
5. Battersby, B. J., R. Grimm, S. Huebner, and G. Cevc. (1998). Evidence for Three-Dimensional Interlayer Correlations in Cationic-Lipid-DNA Complexes as Observed by Cryo-Electron Microscopy. *Biophys. Biochem. Acta.* **1372**, 379.
6. Behr, J.-P. (1994). Gene Transfer with Synthetic Cationic Amphiphiles: Prospects for Gene Therapy. *Bioconjugate Chemistry.* **5**, 382-389.

7. Boukhnikachvili, T., O. Aguerre-Chariol, M. Airiau, S. Lesieur, M. Ollivon, and J. Vacus. (1997). Structure of in-serum transfecting DNA-cationic lipid complexes. *FEBS Letters*. **409**, 188-194.
8. Bruinsma, R. (1998). Electrostatics of DNA-Cationic Lipid Complexes: Isoelectric Instability. *Eur. Phys. J. B*. **4**, 75.
9. Chaikin, P.M. and T. Lubensky (1995). Principles of condensed matter physics. Cambridge University Press.
10. Crystal, R. G. (1995). Transfer of Genes to Humans: Early Lessons and Obstacles to Success. *Science*. **270**, 404-410.
11. Dan, N. (1997). Multilamellar Structures of DNA Complexes with Cationic Liposomes. *Biophys. J.* **73**, 1842-1846.
12. Dautzenberg, H. (1997). Polyelectrolyte Complex Formation in Highly Aggregating Systems. *Macromolecules*. **30**, 7810-7815.
13. Ernst, N., W. Schmalix, J. O. Rdler, R. Galneder, E. Mayer, C. Planck, D. Reinhardt, and J. Rosenecker. Interaction of Liposomal and Polycationic Transfection Complexes with Pulmonary Surfactant (1999) *J. Gene Med.* **1**, 331-340.
14. Fang, Y., and J. Yang. (1997). Two-Dimensional Condensation of DNA Molecules on Cationic Lipid Membranes. *J.Phys.Chem. B*. **101**, 441-449.
15. Farhood, H., N. Serbina, and L. Huang. (1995). The role of dioleoyl phosphatidylethanolamine in cationic liposome mediated gene transfer. *Biochimica et Biophysica Acta*. **1235**, 289-295.
16. Felgner, P. L., T. R. Gadek, M. Holm, R. Roman, H. W. Chan, M. Wenz, J. P. Northrop, G. M. Ringold, and M. Danielson. (1987). Lipofection : A highly efficient, lipid-mediated DNA-transfection procedure. *Proc. Natl. Acad. Sci. USA*. **84**, 7413.
17. Felgner PL, Barenholz Y, Behr JP, et al., Nomenclature for synthetic gene delivery systems, HUM GENE THER 8 (5): 511-512 MAR 20 1997
18. Galneder, R., S. Gersting, J. Rosenecker, and J. O. Rädler. (2001). Size distribution of lipid and polymer based gene delivery complexes determined by quantitative fluorescence microscopy. unpublished.
19. Ghirlando, R., E. J. Wachtel, T. Arad, and A. Minsky. (1992). DNA Packaging Induced by Micellar Aggregates: A Novel in Vitro DNA Condensation System. *Biochemistry*. **31**, 7110-7119.
20. Golubovic, L., and M. Golubovic. (1998). Fluctuations of Quasi-Two-Dimensional Smectics Intercalated between Membranes in Multilamellar Phases of DNA-Cationic Lipid Complexes. *Phys. Rev. Lett.* **80**, 4341-4344.
21. Golubovic, L., Lubensky TC, O'Hern CS. (2000). Structural properties of the sliding columnar phase in layered liquid crystalline systems. *Phys. Rev. E*. **62 (1)**,: 1069-1094.
22. Gustafsson, J., G. Arvidson, G. Karlsson, and M. Almgren. (1995). Complexes between cationic liposomes and DNA visualized by cryo-TEM. *BBA*. **1235**, 305-312.
23. Harries, D., S. May, W. M. Gelbart, and A. Ben-Shaul. (1998). Structure, Stability and Thermodynamics of Lamellar DNA-Lipid Complexes. *Biophysical Journal*. **75**, 159.
24. Israelachvili, J.N.. (1992). Intermolecular and surface forces (2nd Ed.) London: Academic Press.
25. Koltover, I., T. Salditt, J. O. Rädler, and C. R. Safinya. (1998). The Inverted Hexagonal Phase of DNA-Cationic Liposome Complexes: Structure to Gene Release Mechanism Correlations. *Science*. **281**, 78-81.
26. Koltover, I, Salditt, T. and Safinya, C.R. (1999) Phase Diagram, Stability and

Overcharging of Lamellar Cationic Lipid-DNA Self-Assembled Complexes. *Biophys. J.* Vol.**77**, 915-924.
27. Lasic, D. D., H. Strey, M. C. A. Stuart, R. Podgornik, and P. M. Frederik. (1997). The Structure of DNA-Liposome Complexes. *J.Am.Chem.Soc.* **119**, 832-833.
28. Lasic, D. D., Templeton, N.S. (1996). Liposomes in Gene Therapy. *Adv. Drug Del. Rev.* **20**, 221-226.
29. Lee, R. J., and L. Huang. (1996). Folate-targeted, Anionic Liposome-entrapped Polylysine-condensed DNA for Tumor Cell-specific Gene Transfer. *Journal of Biological Chemistry.* **271**, 8481-8487.
30. Lewis,R.N.A.H., S. Tristram-Nagle, J. F. Nagle, and R. N. McElhaney (2001) The thermotropic phase behavior of cationic lipids: *BBA* **1510**, 70-82.
31. May, S., and A. Ben-Shaul. (1997). DNA-Lipid Complexes: Stability of Honeycomb-Like and Spaghetti-Like Structures. *Biophysical Journal.* **73**, 2427-2440.
32. May S, Harries D, Ben-Shaul (2000) A The phase behavior of cationic lipid-DNA complexes, *Biophys. J* **78 (4)**, 1681-1697.
33. Miller, A. D. (1998). Kationische Liposomen für die Gentherapie. *Angew. Chem.* **110**, 1862-1880.
34. Netz, R., and J.-F. Joanny. (1999). Adsorption of semiflexible polyelectrolytes on charged surfaces: charge compensation, charge reversal, and multi-layer formation. *Macromolecules.* in press.
35. O'Hern, C. S., and T. C. Lubensky. (1998). Sliding Columnar Phase of DNA-Lipid Complexes. *Phys. Rev. Lett.* **80**, 4345-4348.
36. O'Hern, C. S., T. Lubensky, and J. Toner. (1999). Thermodynamic Stability of Decoupled Lamellar Phases. *Phys. Rev. Lett* 83 (14): 2745-2748.
37. Pitard, B., O. Aguerre, M. L. Airiau, A.-M., T. Boukhnikachvili, G. Byk, C. Dubertret, C. Herviou, D. Scherman, J.-F. Mayaux, and J. Crouzet. (1997). Virus-sized self-assembling lamellar complexes between plasmid DNA and cationic micelles promote gene transfer. *Proc. Natl. Acad. Sci. USA.* **94**, 14412-14417.
38. Rädler, J. O., I. Koltover, T. Salditt, and C. R. Safinya. (1997). Structure of DNA-Cationic Liposome Complexes: DNA Intercalation in Multilamellar Membranes in Distinct Interhelical Packing Regimes. *Science.* **275**, 810-814.
39. Rädler, J. O., I. Koltover, A. Jamieson, T. Salditt, and C. R. Safinya. (1998). Structure and Interfacial Aspects of Self-Assembled Cationic Lipid-DNA Gene Carrier Complexes. *Langmuir.* **14**, 4272-4283.
40. Rappolt, M., and G. Rapp. (1996). Simultaneous Small- and Wide-Angle X-ray Diffraction During the Main Transition of Dimyristoylphosphatidylethanolamine. *Ber. Bunsenges. Phys. Chem.* **100**, 1153-1162.
41. Safinya, C.R. and Koltover I. (1999) Self-Assembled Structures of Lipid/DNA Nonviral Gene Delivery Systems from Synchrotron X-Ray Diffraction; in *Nonviral Vectors for Gene Therapy*, Academic Press, 91-117.
42. Salditt, T., I. Koltover, J. O. Rädler, and C. R. Safinya. (1997). Two Dimensional Smectic Ordering of Linear DNA Chains in Self-Assembled DNA-Cationic Liposome Mixtures. *Phys. Rev. Lett.* **79**, 2582-2585.
43. Salditt, T., I. Koltover, J. O. Rädler, and C. R. Safinya. (1998). Self-assembled DNA-Cationic Lipid Complexes: Two-Dimensional Smectic Ordering, Correlations and Interactions. *Phys. Rev. E.* **58**, 889-904.
44. Safinya,C.R. and I. Koltover (1998).) Self-Assembled Structures of Lipid/DNA Nonviral Gene Delivery Systems from Synchrotron X-Ray Diffraction; in *Nonviral Vectors for Gene Therapy*, Academic Press, 91-117.

45. Silvius, J.R. Anomalous mixing of zwitterionic and anionic phospholipids with double-chain cationic amphiphiles in lipid bilayers (1991) BBA 1070, 51.
46. Sternberg, B., F. L. Sorgi, and L. Huang. (1994). New structures in complex formation between DNA and cationic liposomes visualized by freeze-fracture electron microscopy. *FEBS letters.* **356**, 361-366.
47. Subramanian G, Hjelm RP, Deming TJ, et al., Structure of complexes of cationic lipids and poly(glutamic acid) polypeptides: A pinched lamellar phase, J AM CHEM SOC 122 (1): 26-34 JAN 12 2000
48. Tardieu, A., V. Luzzati, and F. C. Reman. (1973). Structure and Polymorphism of the Hydrocarbon Chains of Lipids: A Study of Lecithin-Water Phases. *J. Mol. Biol.* **75**, 711.
49. Träuble, H., and H. Eibl. (1974). Electrostatic effects on lipid phase transitions: membrane structure and ionic environment. *Proc. Natl. Acad. Sci. USA.* **71**, 214-219.
50. Wagner, E., C. Plank, K. Zatloukal, M. Cotten, and M. L. Birnstiel. (1992). Influenza virus hemagglutinin HA-2 N-terminal fusogenic peptides augment gene transfer by transferrin-polylysine-DNA complexes: Toward a synthetic virus-like gene-transfer vehicle. *Proc. Natl. Acad. Sci. USA.* **89**, 7934-7938.
51. Wagner K, Harries D, May S, et al., Direct evidence for counterion release upon cationic lipid-DNA condensation, LANGMUIR 16 (2): 303-306 JAN 25 2000
52. Wong GCL, Tang JX, Lin A, et al., Hierarchical self-assembly of F-actin and cationic lipid complexes: Stacked three-layer tubule networks, SCIENCE 288 (5473): 2035-+ JUN 16 2000
53. Wrobel, I., and D. Collins. (1995). Fusion of cationic liposomes with mammalian cells occurs after endocytosis. *BBA.* **1235**, 296-304.
54. Zabner, J., A. J. Fasbender, T. Moninger, K. A. Poellinger, and M. J. Welsh. (1995). Cellular and Molecular Barriers to Gene Transfer by a Cationic Lipid. *Journal of Biological Chemistry.* 270, 18997-19007.
55. Zantl, R., F. Artzner, G. Rapp, and J. Rädler. (1999). Chain melting transition in Cationic Lipid - DNA complexes. *Euro. Phys. Lett.* **45**, 97-103.
56. Zantl, R., L. Baicu, F. Artzner, I. Sprenger, G. Rapp, and J. Rädler. (1999). Thermotropic Phase Behavior of Cationic Lipid-DNA Complexes Compared to Binary Lipid Mixtures (1999) *J. Phys. Chem.* **103**: 10300-10310.

Pearl-Necklaces
Trinkets made out of Polyelectrolytes

SMALL ANGLE SCATTERING METHODS APPLIED TO POLYELECTROLYTE SOLUTIONS

MICHEL RAWISO
Institut Charles Sadron, CNRS-ULP, 6 rue Boussingault, 67083 Strasbourg cedex, France

Summary of the lectures

The series of two lectures that I gave in Les Houches (October 2000) was meant to be an introduction to small angle scattering methods[1-15] applied to polymer [16-19] and polyelectrolyte solutions[20-30]. They were restricted to coherent static scattering and only solutions at thermodynamic equilibrium were considered. They were divided into three main sections: an introduction to light neutron and X-ray scattering techniques, a discussion of the main results dealing with linear polyelectrolytes and a presentation of more recent neutron and X-ray scattering experiments with regularly branched polyelectrolytes.

After a short introduction into ion distributions [31-33], all the relevant structure or scattering functions [34-45] that yield information about the molecular and dielectric structures of polyelectrolyte solutions [46-54] were reviewed in the first section. This covers: partial (polyion-counterion) structure functions; density and charge structure functions; intra- (form factor) and intermolecular structure functions of polyions [55-58]. I took care to explain how and with which accuracy they could be measured from small angle neutron and X-ray scattering experiments. The main models for the dispersion state and the mean conformation of polyions as well as the distribution of counterions around polyions that are usually compared to scattering data were also presented[59-88].

In the second section, I attempted to sort out the significant parameters of the overall structure of linear polyelectrolytes that can be collected from the analysis of the scattering curves reported in the literature[89-150]. They were compared with those obtained from other techniques and with theoretical predictions as well as Monte Carlo and molecular dynamics

C. Holm et al. (eds.), Electrostatic Effects in Soft Matter and Biophysics, 461–468.
© 2001 *Kluwer Academic Publishers. Printed in the Netherlands.*

simulations, when necessary. I focused on hydrophilic synthetic polyelectrolytes with monovalent counterions. Natural polyelectrolytes were only mentioned as typical examples of rigid polyelectrolytes, and the specific case of hydrophobic synthetic polyelectrolytes [151-156] was discussed in C. Williams' lectures.

In the third section, I reported recent small angle neutron and X-ray scattering experiments on highly charged star-branched[157-166], comb-shaped[167-174] and ring[175-178] polyelectrolytes. Three regular architectures for sodium-sulphonated polystyrene (NaPSS) polyelectrolytes were considered. The aim was to show how geometrical parameters characterizing the average conformation of polyions can be measured more easily and how further structural aspects could be gained. I also emphasized the differences in the dispersion state of polyions for the various architectures. Finally, in the particular case of star-branched polyelectrolytes, the effect of an additional hydrophobic interaction on the structure of polyions was tackled. This part complemented C. Williams' lecture on hydrophobic polyelectrolytes, see her chapter in this volume.

References

1. A. Guinier, G. Fournet, *Small Angle Scattering of X-rays*, Wiley Interscience, New York (1955)
2. B. Jacrot, Rep. Prog. Phys. **39** (1976) 911
3. B. Farnoux, G. Jannink, in *Scattering Techniques Applied to Supramolecular and Non-Equilibrium Systems*, S.H. Chen, B. Chu and R. Nossal Ed., Plenum Press, New York (1981)
4. R.G. Kirste, R.C. Oberthür, in *Small Angle X-rays Scattering*, chp. 12 p. 387, O. Glatter and O. Kratky Ed., Academic Press, New York (1982)
5. D. McIntyre, F. Gornick (Ed.), *Light Scattering From Dilute Polymer Solutions*, Gordon & Breach, New York (1964)
6. M. Huglin (Ed.), *Light Scattering From Polymer Solutions*, Academic Press, New York (1972)
7. B. Cabane, in *Surfactant Solutions : Novel Techniques of Investigation*, chp. 2 p. 57, R. Zana Ed., Marcel Dekker, New York (1986); in *Colloïdes et Interfaces*, p. 101, M. Veyssié and A.M. Cazabat Ed., les Editions de Physique, Les Ulis (1984)
8. G. Jannink, Makromol. Chem. Makromol. Symp., **1** (1986) 67; in *The Physics and Chemistry of Aqueous Ionic Solutions*, p. 311, M-C. Bellissent-Funel and G.W. Neilson Ed., D. Reidel Publishing Company, Dordrecht (1986)
9. G.D. Wignall, in *Encyclopedia of Polymer Science and Engineering*, J.I. Kroschwitz Ed., Wiley, New York (1987)
10. B. Chu, *Laser Light Scattering*, Academic Press, New York (1991)
11. P. Lindner, T. Zemb (Ed.), *Neutron, X-ray and Light Scattering : Introduction to an Investigative Tool for Colloidal and Polymeric Systems*, North Holland Delta Series, Amsterdam (1991)
12. J.S. Higgins, H.C. Benoît, *Polymers and Neutron Scattering*, Oxford University Press, New York (1996)

13. J-P. Cotton, Adv.Colloid Interface Sci. **69** (1996) 1
14. J-P. Cotton, F. Nallet (Ed.), *Diffusion de Neutrons aux Petits Angles*, J. Phys. IV (France) **9** (1999)
15. R-J. Roe, *Methods of X-Ray and Neutron Scattering in Polymer Science*, Oxford University Press, New York (2000)
16. H. Yamakawa, *Modern Theory of Polymer Solutions*, Harper & Row, New York (1971)
17. P-G. de Gennes, *Scaling Concepts in Polymer Physics*, Cornell University Press, New York (1979)
18. M. Doi, S.F. Edwards, *The Theory of Polymer Dynamics*, Clarendon Press, Oxford (1986)
19. J. des Cloizeaux, G. Jannink, *Polymers in Solution: their Modelling and Structure*, Clarendon Press, Oxford (1990); *Les Polymères en Solution: leur Modélisation et leur Structure*, Les Editions de Physique, Les Ulis (1987)
20. C. Tanford, *Physical Chemistry of Macromolecules*, Wiley, New York (1961)
21. A. Katchalsky, Z. Alexandrovicz, O. Kedem, in *Chemical Physics of Ionic Solutions*, B.E. Conway and R.G. Barradas Ed., Wiley, New York (1966)
22. A Katchalsky, Pure and Appl. Chem. **26** (1971) 327
23. F. Oosawa, *Polyelectrolytes*, Marcel Dekker, New York (1970)
24. G.S. Manning, Quart. Rev. Biophys., **11** (1978) 179
25. T. Odijk, in *Ionic Liquids, Molten Salts, Polyelectrolytes*, K.H. Bennemann Ed., Springer Verlag, Berlin (1982)
26. M. Mandel, in *Encyclopedia of Polymer Science and Engineering*, Herman F. Mark et al Ed., Wiley, New York (1990)
27. K.S. Schmitz, *Macroions in Solution and Colloidal Suspensions*, VCH, New York (1993)
28. M. Hara (Ed.), *Polyelectrolytes: Science and Technology*, Marcel Dekker, New York (1993)
29. S. Förster, M. Schmidt, Adv. Polym. Sci. **120** (1995) 51
30. J-L. Barrat, J-F. Joanny, Adv. Chem. Phys. **100** (1996) 909
31. P. Debye, E. Hückel, Z. Phys. **24** (1923) 185
32. R.M. Fuoss, A. Katchalsky, S. Lifson, Proc. Natl. Acad. Sci. USA **37** (1951) 579
33. G.S. Manning, J. Chem. Phys. **51** (1969) 924 and 934 ; G.S. Manning, A. Holtzer, J. Chem. Phys. **77** (1973) 2066
34. P. Debye, Ann. Physik **46** (1915) 809
35. L. Van Hove, Phys. Rev. **95** (1954) 249
36. J-P. Cotton, H. Benoît, J. Phys. France **36** (1975) 905
37. J. des Cloizeaux, G. Jannink, Physica **102**A (1980) 120
38. G. Jannink, M. Nierlich, C. Williams, C.R.Acad. Sc. Paris **290**B (1980) 83
39. J-P. Cotton, J. Phys. IV (France) **9** (1999) Pr1-1
40. M. Rawiso, J. Phys. IV (France) **9** (1999) Pr1-147
41. G.D. Sears, Neutron News **3** (1992) 26
42. F.J. Millero, Chem. Rev. **71** (1971) 147; in Water and Aqueous Solutions, chp. 13 p. 519 R.A. Horne Ed., Wiley Interscience (1972)
43. J. F. Hinton, E.S. Amis, Chem. Rev. **71** (1971) 627
44. C. Tondre, R. Zana, J. Phys. Chem. **76** (1972) 3451; R. Zana, E.B. Yeager, *Colloques Internationaux du CNRS* , CNRS Ed., n° **246** (1976) 155
45. Y. Marcus, Chem. Rev. **88** (1988) 1475
46. J-P. Hansen, I.R. Mc Donald, *Theory of Simple Liquids*, chp. 10 p. 364, Academic Press, London (1986)

47. G. Jannink, J.R.C. van der Maarel, Biophysical Chemistry **41** (1991) 15
48. F.H. Stillinger, R. Lovett, J. Chem. Phys. **49** (1968) 1991
49. L. Blum, *Theoretical Chemistry: Advances and Prospectives*, vol 5, D. Henderson Ed., Academic Press (1980)
50. J.L. Lebowitz, B. Jancovici, Phys. Rev. Lett. **27** (1989) 1491; B. Jancovici, Journal of Statistical Physics 80 (1995) 445
51. F. Nallet, G. Jannink, J.B. Hayter, R. Oberthür, C. Picot, J. Phys. France **44** (1983) 87; F. Nallet, Thesis Université Paris VI (1983)
52. P.J. Derian, L. Belloni, M. Drifford, Europhys. Lett. **7** (1988) 243
53. J.R.C. van der Maarel, L.C.A. Groot, M. Mandel, W. Jesse, G. Jannink, V. Rodriguez, J. Phys. II (France) **2** (1992) 109
54. W. Kuntz, P. Calmettes, G. Jannink, L. Belloni, T. Cartailler, P. Turq, J. Chem. Phys. **96** (1992) 7034 (erratum J. Chem. Phys. **98** (1993) 1755)
55. C.E. Williams, M. Nierlich, J-P. Cotton, G. Jannink, F. Boué, M. Daoud, B. Farnoux, C. Picot, P-G. de Gennes, M. Rinaudo, M. Moan, C. Wolff, J. Polym. Sci. Polym. Lett. Ed. **17** (1979) 379
56. M. Nierlich, F. Boué, A. Lapp, R.C. Oberthür, J. Phys. (France) **46** (1985) 649; Colloid Polym. Sci. **263** (1985) 955
57. T. Csiba, G. Jannink, D. Durand, R. Papoular, A. Lapp, L. Auvray, F. Boué, J-P. Cotton, R. Borsali, J. Phys. II France **1** (1991) 381
58. F. Boué, J-P. Cotton, A. Lapp, G. Jannink, J. Chem. Phys. **101** (1994) 2562
59. G. Porod, Monatsh. Chem. **80** (1949) 251 ; O. Kratky, G. Porod, Rec. Trav. Chim. Pays Bas **68** (1949) 1106
60. L.D. Landau, E.M. Lifshitz, *Statistical Physics*, chp. XII, Addison-Wesley, Reading, Mass. (1969); *Physique Statistique*, chp. XII p. 433, MIR (1984)
61. H. Benoît, P. Doty, J. Phys. Chem. **57** (1953) 958
62. R.C. Oberthür, Makromol. Chem. **179** (1978) 2693
63. P. Sharp, V.A. Bloomfield, Biopolymers **6** (1968) 1201
64. J. des Cloizeaux, Macromolecules **6** (1973), 403
65. H. Yamakawa, J. Chem. Phys. **59** (1973) 3811 ; H. Yamakawa, M. Fujii, Macromolecules **7** (1974) 649
66. T. Yoshizaki, H. Yamakawa, Macromolecules **13**, (1980), 1518
67. J.S. Pedersen, P. Schurtenberger, Macromolecules **29** (1996) 7602
68. D. Pötschke, P. Hickl, M. Ballauff, P-O. Astrand, J.S. Pedersen, Macromol. Theory Simul. **9** (2000) 345
69. H. Hayashi, P.J. Flory, G.D. Wignall, Macromolecules **16** (1983) 1328
70. M. Rawiso, R. Duplessix, C. Picot, Macromolecules **20** (1987) 630
71. M. Rawiso, J-P. Aimé, J-L. Fave, M. Schott, M.A. Müller, M. Schmidt, H. Baumgartl, G. Wegner, J. Phys. (France) **49** (1988) 861
72. R.G. Kirste, Makromol. Chem. **101** (1967) 91; J. Polym. Sci. C **16** (1967) 2039; Kolloïd-Z.u. Z. Polymere **244** (1971) 290
73. H. Benoît, J. Polym. Sci. **11** (1953) 507
74. M. Daoud, J-P. Cotton, J. Phys. France **43** (1982) 531
75. T.A. Witten, P.A. Pincus, M.E. Cates, Europhys. Lett. **2** (1986) 137
76. W. Dozier, J.S. Huang, L.J. Fetters, Macromolecules **24** (1991) 2810
77. M. Adam, D. Lairez, Fractals **1** (1993) 149
78. D. Richter, O. Jucknischke, L. Willner, L.J. Fetters, M. Lin, J.S. Huang, J. Roovers, P.M. Toporowski, L.L. Zhou, J. Phys. IV (France) Suppl. **3** (1993) 3
79. G.S. Grest, L.J. Fetters , J.S. Huang , D. Richter, Adv. Chem. Phys. **100** (1996) 67

80. C.M. Marques, T. Charibat , D. Izzo, E. Mendes, Eur. Phys. J. B **3** (1998) 353
81. G. Jannink, P-G. de Gennes, J. Chem. Phys. **48** (1968) 2260
82. P-G. de Gennes, J. Phys. France **31** (1970) 235
83. M. Daoud, J-P. Cotton, B. Farnoux, G. Jannink, G. Sarma, H. Benoît, R. Duplessix, C. Picot, P-G. de Gennes, Macromolecules **8** (1975) 804
84. V. Borue, I. Erukhimovitch, Macromolecules **21** (1988) 3240
85. J-F. Joanny, L. Leibler, J. Phys. France **51** (1990) 545
86. T.A. Vilgis, R. Borsali, Phys. Rev. A **43** (1991) 6857
87. J.C. Brown, P.N. Pusey, J.W. Goodwin, R.H. Ottewill, J. Phys. A **8** (1975) 664
88. C.G. de Kruif, W.J. Briels, R.P. May, A. Vrij, Langmuir **4** (1988) 668
89. J-P. Cotton, M. Moan, J. Phys. (France) Lett. **37** (1976) L-75
90. P-G. de Gennes, P. Pincus, R.M. Velasco, F. Brochard, J. Phys. (France) **37** (1976) 1461
91. T. Odijk, Macromolecules **12** (1979) 688
92. M. Nierlich, C.E. Williams, F. Boué, J-P. Cotton, M. Daoud, B. Farnoux, G. Jannink, C. Picot, M. Moan, C. Wolf, M. Rinaudo, P-G. de Gennes, J. Phys. (France) **40** (1979) 701
93. J.B. Hayter, G. Jannink, F. Brochard-Wyart, P-G. de Gennes, J. Phys. (France) Lett. **41** (1980) L-451
94. N. Ise, T. Okubo, S. Kunugi, H. Matsuoka, K.I. Yamamoto, Y. Ishii, J. Chem. Phys. **81** (1984) 3294
95. G. Weill, Polym. Commun. **25** (1984) 147
96. M. Drifford, J-P. Dalbiez, J. Phys. Chem. **88** (1984) 5368
97. M. Drifford, J-P. Dalbiez, Biopolymers **24** (1985) 1501
98. K. Kaji, H. Urakawa, T. Kanaya, R. Kitamaru, J. Phys. (France) **49** (1988) 993
99. G. Weill, J. Phys. (France) **49** (1988) 1049
100. E.E. Maier, S.F. Schulz, R. Weber, Macromolecules **21** (1988) 1544
101. J-P. Martenot, J-C. Galin, C. Picot, G. Weill, J. Phys. (France) **50** (1989) 493
102. R. Krause, E.E. Maier, M. Deggelmann, M. Hagenbüchle, S.F. Schulz, R. Weber, Physica A **160** (1989) 135
103. M. Schmidt, Makromol. Chem. Rapid commun. **10** (1989) 89
104. S. Förster, M. Schmidt, M. Antonietti, Polymer **31** (1990) 781
105. J. Yamanaka, H. Matsuoka, H. Kitano, N. Ise, T. Yamaguchi, S. Saeki, M. Tsubokawa, Langmuir **7** (1991) 1928
106. G. Weill, Biophysical Chemistry **41** (1991) 1
107. A.V. Dobrynin, R.H. Colby, M. Rubinstein, Macromolecules **28** (1995) 1859
108. M. Milas, M. Rinaudo, R. Duplessix, R. Borsali, P. Lindner, Macromolecules **28** (1995) 3119
109. W. Essafi, Thesis Université Paris VI (1996)
110. S. Lifson, A. Katchalsky, J. Polym. Sci. **13** (1954) 43
111. M-N. Spiteri, Thesis Université Paris XI (1997)
112. K. Kassapidou, W. Jesse, M.E. Kuil, A. Lapp, S. Egelhaaf, J.R.C. van der Maarel, Macromolecules **30** (1997) 2671
113. J.R.C. van der Maarel, K. Kassapidou, Macromolecules **31** (1998) 5734
114. M. Dymitrowska, L. Belloni, J. Chem. Phys. **109** (1998) 4569; **111** (1999) 6633
115. R. Rulkens, G. Wegner, T. Thurn-Albrecht, Langmuir **15** (1999) 4022
116. X. Li, W.F. Reed, J. Chem. Phys. **94** (1991) 4568
117. W.F. Reed, S. Ghosh, G. Medjahdi, J. François, Macromolecules **24** (1991) 6189
118. M. Sedlak, Macromolecules **26** (1993) 1158
119. W.F. Reed, Macromolecules **27** (1994) 873

120. B.D. Ermi, E.J. Amis, Macromolecules **31** (1998) 7378
121. E. Buhler, M. Rinaudo, Macromolecules **33** (2000) 2098
122. R. Klucker, J-P. Munch, F. Schosseler, Macromolecules **30** (1997) 3839
123. P. Pfeuty, J. Phys. (France) Colloque C**2** (1978) C2-149
124. A.R. Khokhlov, J. Phys. A: Math. Gen. **13** (1980) 979; A.R. Khokhlov, K.A. Khachaturian, Polymer **23** (1982) 1742
125. E. Raphaël, J-F. Joanny, Europhys. Lett. **13** (1990) 623
126. K. El Brahmi, Thesis Université Louis Pasteur, Strasbourg (1991)
127. F. Schosseler, F. Ilmain, S.J. Candau, Macromolecules **24** (1991) 225; F. Schosseler, A. Moussaid, J-P. Munch, S.J. Candau, J. Phys. II (France) **1** (1991) 1197
128. A. Moussaid, F. Schosseler, J-P. Munch, S.J. Candau, J. Phys. II (France) **3** (1993) 573; A. Moussaid, Thesis Université Louis Pasteur, Strasbourg (1994)
129. A. Katchalsky, J. Polym. Sci. **7** (1951) 393
130. J.C. Leyte, M. Mandel, J. Polym. Sci A **2** (1964) 1879.
131. C. Heitz, M. Rawiso, J. François, Polymer **40** (1999) 1637
132. J. Plestil, Y.M. Ostanevich, V.Y. Bezzabotnow, D. Hlavata, J. Labsky, Polymer **27** (1986) 1241
133. M. Ragnetti, R.C. Oberthür, Colloid Polym. Sci. **264** (1986) 32
134. M.N. Spiteri, F. Boué, A. Lapp, J-P. Cotton, Phys. Rev. Lett. **77** (1996) 5218
135. T. Odijk, J. Polym. Sci. Polym. Phys. Ed. **15** (1977) 477
136. J. Skolnick, M. Fixman, Macromolecules **10** (1977) 944; M. Fixman, J. Chem. Phys. **76** (1982) 6346
137. M. Le Bret, J. Chem. Phys. **76** (1982) 6243
138. M.J. Stevens, K. Kremer, Phys. Rev. Lett. **71** (1993) 2228
139. J.L. Barrat, J.F. Joanny, Europhys. Lett. **3** (1993) 343; J. Phys II (France) **4** (1994) 1089
140. G. Weill, G. Maret, Polymer **23** (1982) 1990; G. Maret, G. Weill, Biopolymers**22** (1983) 2727; G. Weill, G. Maret, T. Odijk, Polymer Commun. **25** (1984) 147
141. M. Tricot Macromolecules **17** (1984) 1698
142. V. Degiorgo, F. Mantegazza, R. Piazza, Europhys. Lett. **15** (1991) 75
143. E. Nordmeier, W. Dauwe, Polymer J. **24** (1992) 229
144. J.R.C. van der Maarel, L.C.A. Groot, J.G. Hollander, W. Jesse, M.E. Kuil, J.C. Leyte, L.H. Leyte-Zuiderweg, M. Mandel, J-P. Cotton, G. Jannink, A. Lapp, B. Farago, Macromolecules **26** (1993) 7295
145. S.S. Zakharova, S.U. Egelhaaf, L.B. Bhuiyan, C.W. Outhwaite, D. Bratko, J.R.C. van der Maarel, J. Chem. Phys. **111** (1999) 10706
146. R.M. Nyquist, B-Y. Ha, A.J. Liu, Macromolecules **32** (1999), 3481
147. J. Ray, G.S. Manning, Macromolecules **32** (1999) 4588
148. M. Deserno, C. Holm, S. May, Macromolecules **33** (2000) 199
149. B. Guilleaume, J. Blaul, M. Wittemann, M. Rehahn, M. Ballauff, J. Phys.: Condens. Matter **12** (2000) A245; J. Blaul, M. Wittemann, M. Ballauff, M. Rehahn, J. Phys. Chem. B **104** (2000) 7077; B. Guilleaume, M. Ballauff, G. Goerigk, M. Wittemann, M. Rehahn, to be published
150. M. Deserno, C. Holm, J. Blaul, M. Ballauff, M. Rehahn, Eur. Phys. J. E **5** (2001) 97
151. W. Essafi, F. Lafuma, C.E. Williams, ACS Symposium Series 548, Macro-ion characterization from dilute solutions to complex fluids, chp. 21 p. 278, K.S. Schmitz Ed. (1994)
152. W. Essafi, F. Lafuma, C.E. Williams, J. Phys. II (France) **5** (1995) 1269 ; Eur. Phys. J. B **9** (1999) 261

153. Y. Kantor, M. Kardar, Europhys. Lett. **27** (1994) 643; Phys. Rev. E **51** (1995) 1299
154. A.V. Dobrynin, M. Rubinstein, S.P. Obukhov, Macromolecules **29** (1996) 2974 ; A.V. Dobrynin, M. Rubinstein, Macromolecules **32** (1999) 915
155. A.V. Lyulin, B. Dünweg, O.V. Borisov, A.A. Darinskii, Macromolecules **32** (1999) 3264
156. U. Micka, C. Holm, K. Kremer, Langmuir **15** (1999) 4033
157. M. Heinrich, Thesis Université Louis Pasteur, Strasbourg (1998)
158. M. Heinrich, M. Rawiso, J-G. Zilliox, P. Lesieur, J-P. Simon, Eur. Phys. J. E **4** (2001) 131; M. Heinrich, M. Rawiso, J-G. Zilliox, O. Diat, to be published
159. P. Pincus, Macromolecules **24** (1991) 2912
160. S. Misra, W.L. Mattice, D.H. Napper, Macromolecules **27** (1994) 7090
161. E.B. Zhulina, O.V. Borisov, T.M. Birshtein, J. Phys II (France) **2** (1992) 63; Macromolecules **28** (1995) 1491
162. O.V. Borisov, J. Phys. II France **6** (1996) 1; O.V. Borisov, E.B. Zhulina Eur. Phys. J. B **4** (1998) 205
163. C. Biver, R. Hahiharan, J. Mays, W.B. Russel, Macromolecules **30** (1997) 1787; R. Hahiharan, C. Biver, J. Mays, W.B. Russel, Macromolecules **31** (1998) 7506
164. P. Guenoun, S. Lipsky, J.W. Mays, M. Tirrel, Langmuir **12** (1996) 1425 ; P. Guenoun, H.T. Davis, M. Tirrel, J.W. Mays, Macromolecules **29** (1996) 3965 ; P. Guenoun, M. Delsanti, D. Gazeau, J.W. Mays, D.C. Cook, M. Tirrel, L. Auvray, Eur. Phys. J. B **1** (1998) 77; P. Guenoun, F. Muller, M. Delsanti, L. Auvray, Y.J. Chen, J.W. Mays, M. Tirrel, Phys. Rev. Lett. **81** (1998) 3872
165. X. Guo, M. Ballauff, Langmuir **16** (2000) 8719
166. Q. de Robillard, X. Guo, M. Ballauff, T. Narayanan, Macromolecules **33** (2000) 9109
167. M. Rawiso, J.M. Catala, F. Schnell, to be published
168. T.M. Birshtein, O.V. Borisov, Y.B. Zhulina, A.R. Khokhlov, T.A. Yurasova, Polymer Science USSR **29** (1987) 1293
169. C.M. Marques, Thesis Université Louis Pasteur, Strasbourg (1989)
170. G.H. Fredrickson, Macromolecules **26** (1993) 2825
171. M. Wintermantel, K. Fischer, M. Gerle, R. Ries, M. Schmidt, K. Kajiwara, H. Urakawa, I. Wataoka, Angew. Chem. Int. Ed. Engl. **34** (1995) 1472; M. Wintermantel, M. Gerle, K. Fischer, K. Kajiwara, I. Wataoka, H. Urakawa, K. Kajiwara, Y. Tsukahara, Macromolecules **29** (1996) 978; S.S. Seiko, M. Gerle, K. Fischer, M. Schmidt, M. Möller, Langmuir **13** (1997) 5368; I. Wataoka, H. Urakawa, K. Kajiwara, M. Schmidt, M. Wintermantel, Polymer International **44** (1997) 365
172. V. Castelletto, L.Q. Amaral, Macromolecules **32** (1999) 3469
173. M. Saariaho, A. Subbotin, O. Ikkala, G. ten Brinke, Macromol. Rapid Commun. **21** (2000) 110
174. D. Vlassopoulos, G. Fytas, B. Loppinet, F. Isel, P. Lutz, H. Benoît, Macromolecules **33** (2000) 5960
175. M. Rawiso, F. Boué, F. Isel, P. Lutz, to be published
176. D.F. Hodgson, E.J. Amis, J. Chem. Phys. **95** (1991) 7653
177. G.B. McKenna, G. Hadziioannou, P. Lutz, G. Hild, C. Strazielle, C. Straupe, P. Rempp, A.J. Kovacs, Macromolecules **20** (1987) 498; G. Hadziioannou, P.M. Cotts, G. ten Brinke, C.C. Han, P. Lutz, G. Hild, C. Strazielle, C. Straupe, P. Rempp, A.J. Kovacs, Macromolecules **20** (1987) 493
178. M. Müller, J.P. Wittmer, M.E. Cates, Phys. Rev. E **53** (1996) 5063

LATERAL CORRELATION OF MULTIVALENT COUNTERIONS IS THE UNIVERSAL MECHANISM OF CHARGE INVERSION

TOAN T. NGUYEN, ALEXANDER YU. GROSBERG and BORIS I. SHKLOVSKII
Department of Physics, University of Minnesota
116 Church St. Southeast, Minneapolis, Minnesota 55455

1. Introduction

Charge inversion is a counterintuitive phenomenon in which a strongly charged particle (a macroion) binds so many counterions that its net charge changes sign. As shown below the binding energy of a counterion with large charge Z is larger than k_BT, so that this net charge is easily observable: it is the net charge that determines a particle drift in a weak field electrophoresis. Charge inversion is possible for a variety of macroions, ranging from the charged surface of mica to charged lipid membranes, colloids, DNA or actin. Multivalent metal ions, small colloidal particles, charged micelles, short or long polyelectrolytes including DNA can play the role of multivalent counterions. Recently charge inversion has attracted significant attention [1-22].

Theoretically, charge inversion can be also thought of as an over-screening. Indeed, the simplest screening atmosphere, familiar from the linear Debye-Hückel (DH) theory, compensates at any finite distance only a part of the macroion charge. It can be proven that this is true also in the non-linear Poisson-Boltzmann (PB) theory. The statement that the net charge preserves sign of the bare charge agrees with the common sense. One can think that this statement is even more universal than results of PB equation. However, this presumption of common sense fails for screening by Z-valent counterions with large Z (Z-ions). In this case, most of counterions are localized at the very surface of macroion. The energy of their lateral Coulomb interaction may exceed k_BT by an order of magnitude or more. As a result [3, 4], at the macroion surface Z-ions form a two-dimensional

C. Holm et al. (eds.), Electrostatic Effects in Soft Matter and Biophysics, 469–486.
© 2001 *Kluwer Academic Publishers. Printed in the Netherlands.*

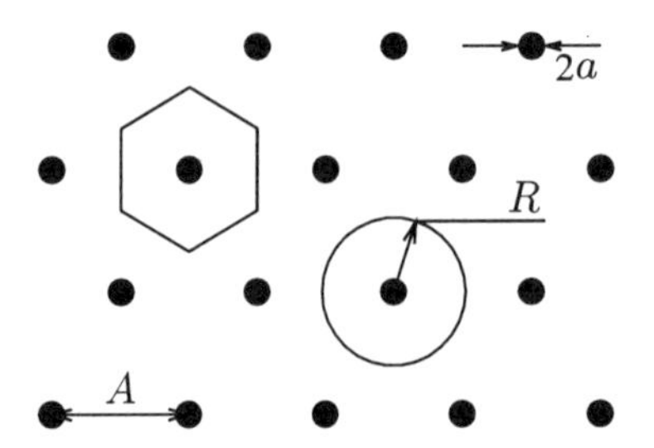

Figure 1. The Wigner crystal of positive Z-ions on the negative uniform background of surface charge. A hexagonal Wigner-Seitz cell and its simplified version as a disk with radius R are shown.

(2D) strongly correlated liquid (SCL) with properties resembling a Wigner crystal (WC), which is shown in Fig. 1.

The negative chemical potential of this liquid leads to a correlation induced attraction of Z-ions to the surface, which brings more of them to the surface than necessary to neutralize it. Role of correlations of Z-ions in another physical phenomenon attraction of two likely-charged surfaces was recognized even earlier [23]. This paper consists of two parts. In the first part, we review recent works on correlation induced charge inversion. Sec. II deals with small Z-ions, while in Sec. III the role of Z-ions is played by polyelectrolytes. In the second part (Sec. IV), we discuss proposed alternative mechanisms of charge inversion: metalization approach (mental smearing of Z-ions on the macroion surface) and counterion release. It is shown there that they are also based on the correlation physics, so that some of the apparent difference is purely semantic.

2. Correlations of small size multivalent counterions

Let us consider screening a macroion surface with a negative charge density $-\sigma$ by $Z:1$ salt with positive Z-ions in the case of large enough Z and σ. If such a surface is neutralized by Z-ions, the average distance between them in directions parallel to the plane equals $R_0 = (\pi\sigma/Ze)^{-1/2}$. If this distance is much larger than the Gouy-Chapman length $\lambda = Dk_BT/(2\pi Ze\sigma)$, which is the typical distance to the surface mean field PB approach obviously fail. The ratio of these two lengths $R_0/\lambda = 2\Gamma$, where $\Gamma = Z^2e^2/(R_0Dk_BT)$ is the dimensionless inverse temperature measured in units of the typical interaction energy and D is the dielectric constant of water. Here we are concerned with large enough Z and σ at which parameter $\Gamma \gg 1$. For example, at $Z = 3$ and $\sigma = 1.0$ e/nm^2 we get $\Gamma = 6.4$, $\lambda \simeq 0.08$ nm and $R_0 \simeq 1.0$ nm. In such conditions Gouy-Chapman solution of PB equation can not be valid any more. Indeed, if distance of a Z-ion to the surface x is in the interval $\lambda \ll x \ll R_0$, it does not feel other Z-ions but interacts only with the macroion surface. This interaction leads to an exponential decay of concentration of Z-ions as function of x in the range $x \ll R_0$ instead of $1/x^2$ Gouy-Chapman law. A new theory based on inequality $\Gamma \gg 1$ was suggested in Ref. [3, 4]. The main idea of this theory is that at $\Gamma \gg 1$ the screening atmosphere can be divided in two distinct phases: 2D SCL, which should be treated exactly, and the gas phase which at large distances x can be described by the PB equation and is in equilibrium with 2D SCL.

Thermodynamic properties of 2D liquid of classical charged particles on the neutralizing background, so called one-component plasma (OCP) are well known (see bibliography in Ref. [3, 4]). The chemical potential of OCP can be written as $\mu = \mu_{id} + \mu_c$, where μ_{id} and μ_c are its ideal and correlation parts. At $T = 0$ OCP forms WC. To calculate μ_c, we start from the energy of WC per Z-ion,

$$\varepsilon(n) \simeq -1.11Z^2e^2/RD = -1.96n^{1/2}Z^2e^2/D, \tag{1}$$

where $R = (\pi n)^{-1/2}$ is the radius of its Wigner-Seitz (WS) cell (See Fig. 1) and n is 2D concentration of Z-ions. This gives $\mu_c = \mu_{WC}$, where

$$\mu_{WC} = \frac{\partial[n\varepsilon(n)]}{\partial n} = -1.65\Gamma k_B T = -1.65Z^2e^2/DR. \tag{2}$$

In the range of temperatures where $1 \ll \Gamma \ll 130$ WC is melted and for SCL $\mu_c = \mu_{WC} + \delta\mu$, where $\delta\mu$ is a small positive correction [3, 4].

Physics of a large and negative μ_{WC} can be understood as follows. Let us imagine for a moment that an insulating macroion is replaced by a neutral metallic particle. In this case, each Z-ion creates an image charge of opposite sign inside the metal. Energy of attraction to the image is $U(x) = -(Ze)^2/4Dx$, where x is the distance from Z-ion to the surface. At the metal surface energy of the interaction with the image equals $U(a) = -(Ze)^2/4Da$, where a is radius of the counterion. The chemical potential μ_{WC} for a charged surface of an insulating macroion can be interpreted in a similar language of images. Consider bringing a new Z-ion to the macroion surface already covered by an adsorbed layer of Z-ions (See Fig. 2). This layer plays the role of a metal surface. Indeed, the new Z-ion repels nearest adsorbed ones, creating a correlation hole for himself. In other words, it creates a negative image. Calculation of the energy of attraction to the image in this case is, however, less trivial than in the case of a metal. The problem is that minimal size of the image in the adsorbed layer is equal to the WS cell radius R. (The adsorbed layer is a good metal only at larger scales.) Thus, for WC we arrive at $\mu_{WC} \sim -(Ze)^2/DR$. Eq. (2) provides the numerical coefficient in this expression. It is clear now that contrary to the case of a metallic surface, charge inversion for insulating macroion

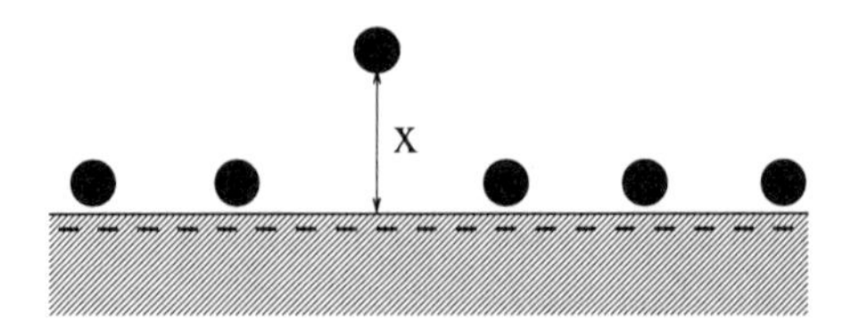

Figure 2. The origin of attraction of a new positive Z-ion to the already neutralized surface. Z-ions are shown by solid circles. The new Z-ion creates its negative correlation hole.

requires a finite density of an adsorbed layer, and, correspondingly, a finite bare charge density σ.

Correlation induced attraction of Z-ion to a neutral SCL has two interesting analogies in the solid state and atomic physics. The energy $|\mu_{WC}|$ plays the same role for Z-ions as the work function plays for electrons of a metal or the ionization energy plays in a many-electron atom. It is known that in the Thomas-Fermi approximation which neglects exchange and correlation holes, both the work function of a metal [24] and the ionization energy of a many-electron atom [25] vanish. For screening by Z-ions the PB approximation is an analog of the Thomas-Fermi approximation and it results in $\mu_c = 0$. Only correlations lead to a finite work function or ionization energy and to a finite $\mu_c \simeq \mu_{WC}$.

Let us now discuss behavior of the concentration of Z-ions, $N(x)$ near the surface. For this purpose let us extract a Z-ion from SCL and move it along the x axis. As it is mentioned above, this Z-ion leaves behind its correlation hole. In the range of distances $x \ll R$, the correlation hole is a disc of the surface charge with radius R (-WS cell) and the Z-ion is attracted to the surface by its uniform electric field $E = 2\pi\sigma/D$. We can say that our Z-ion does not see other Z-ions until $x \ll R$. Therefore, $N(x) = N_s \exp(-x/\lambda)$ at $x \ll R$. Here $N_s = n/\lambda$ is the three-dimensional concentration of Z-ions at the plane. At large enough x the concentration $N(x)$ should saturate at the level of

$$N_0 = N_s \exp(-|\mu_{WC}|/k_BT) = N_s \exp(-1.65\Gamma). \tag{3}$$

Let us find out how and where this saturation happens. At $x \sim R = 2\Gamma\lambda$ the ratio x/λ reaches roughly speaking a half of $|\mu_{WC}|/k_BT$. At $x \gg R$ the Z-ion creates the image with relatively small radius R which attracts Z-ion back and provides the decaying Coulomb correction to the activation energy of $N(x)$: $N(x) = N_s \exp\left(-[|\mu_{WC}| - Z^2e^2/4Dx]/k_BT\right)$. This correction is similar to the well known "image" correction to the work function of a metal [24]. At $x = Z^2e^2/4Dk_BT = \lambda\Gamma^2/2$ this correction reaches k_BT, so that $N(x)$ saturates at the value N_0.

The dramatic difference between the described above exponential decay of $N(x)$ and the Gouy-Chapman $1/(\lambda + x)^2$-law is obviously related to the correlation effects. Recently such an exponential decay of $N(x)$ with the following tendency to saturation was rederived in more formal way and confirmed by Monte-Carlo simulations [26]. At distances $x \gg \lambda\Gamma^2/2$, interaction of the removed ion with its correlation hole in SCL is not important and the correlation between ions of the gas phase are even weaker because $N(x)$ is exponentially small. Therefore, at larger distances one can describe $N(x)$ by PB equation [3, 4].

Studying charge inversion, we want to calculate the net charge density of the macroion surface $\sigma^* = -\sigma + Zen > 0$. Even though the system is not neutral, we can still use the chemical potential of a neutral OCP given by Eq. (2) after the following exact transformation. Indeed, let us add uniform charge densities $-\sigma^*$ and σ^* to the macroion surface. The first addition results in a neutral OCP. The second addition creates only uniform potential $e\psi(0)$ on the macroion surface. For example, if the macroion is a sphere with radius r and screening radius of solution is larger than r we get $\psi(0) = Q^*/Dr$, where $Q^* = 4\pi r^2\sigma^* = -Q + 4\pi r^2 neZ$ is the net charge of the sphere and $-Q = 4\pi r^2\sigma$ is its bare charge. It is important to emphasize that macroscopic net charge Q^* does not interact with OCP, because potential of the uniformly charged sphere at the neutral OCP is constant.

The condition of balance of the electro-chemical potential at the surface of macroion, μ, and in the bulk of solution, μ_b now can be written as $Ze\psi(0) = -\mu_{WC} + (\mu_b - \mu_{id}) = -\mu_{WC} + k_BT\ln(N_s/N) = k_BT\ln(N/N_0)$, where N is the concentration of Z-ions in the bulk of solution. It is clear that when $N > N_0$, the net charge density σ^* is indeed positive, i.e. has the sign opposite to the bare charge density $-\sigma$. The concentration N_0 is very small because $|\mu_{WC}|/k_BT = 1.65\Gamma \gg 1$. Therefore, it is easy to achieve charge inversion increasing N. At large enough N we have $|\mu_b - \mu_{id}| \ll |\mu_{WC}|$. This gives a simple equation for calculation of the maximal inverted charge density

$$\psi(0) = |\mu_{WC}|/Ze. \tag{4}$$

Let us consider a sphere with charge $-Q$ screened by $Z:1$ salt with a large concentration, N. In this case $\psi(0) = Q^*/Dr$ and Eq. (4) has a simple meaning: $|\mu_{WC}|/Ze$ is the "correlation" voltage which charges a spherical capacitor. Expressing R and $|\mu_{WC}|$ through Q and Z we arrive at the simple prediction [4] for the maximum possible inverted charge:

$$Q^* = 0.83\sqrt{QZe}. \tag{5}$$

This charge is much larger than Ze, but still is smaller than Q because of limitations imposed by the large charging energy of the macroscopic net charge. For example, for $Q = 100e, Z = 4$, we get $Q^* = 17e$. Eq. (5) was recently confirmed by numerical simulations [15, 22].

It was also shown in Ref. [3, 4], that in the case of a cylinder, the conventional picture of nonlinear screening known as the Onsager-Manning (OM) condensation should be strongly modified when dealing with multivalent ions. Consider a cylinder with a negative linear charge density $-\eta$ and assume that $\eta > \eta_z$, where $\eta_z = k_B T D/Ze$. OM theory [27], shows that such a strongly charged cylinder is partially screened by counterions residing at its surface, so that net linear charge density of the cylinder, $\eta^* = -\eta_z$. The rest of the charge is screened at much larger distances according to the linear DH theory. The OM theory uses PB mean field approach and, therefore, does not take into account lateral correlations of counterions. It is shown in Ref. [3, 4] that the correlation induced negative chemical potential μ_{WC} leads to inversion of the sign of η^* at $N > N_0$ in the case of cylinder, too. At large enough N inverted charge density η^*, can reach $k_B T D/e$.

Even stronger charge inversion of a spherical or cylindrical macroion can be obtained in the presence of substantial concentration of monovalent salt, such as NaCl, in solution. Monovalent salt screens long range Coulomb interactions stronger than short range lateral correlations between adsorbed Z-ions. Therefore, screening by monovalent salt diminishes the charging energy of the macroion much stronger than the correlation energy of Z-ions. As a results, the inverted charge Q^* becomes larger than that predicted by Eq. (5). Since, in the presence of a sufficient concentration of salt, the macroion is screened at the distance smaller than its size, the macroion can be thought of as an over-screened surface, with inverted charge Q^* proportional to the surface area. In this sense, overall shape of the macroion and its surface is irrelevant, at least to a first approximation. Therefore, we consider here a simple case: screening of a planar macroion surface with a negative surface charge density $-\sigma$ by solution with concentration N of $Z:1$ salt and a large concentration N_1 of a monovalent salt. Correspondingly, we assume that all weak interactions are screened with DH screening length $r_s = (8\pi l_B N_1)^{-1/2}$. For simplicity, we discuss here only a charged surface of insulating macroion with the same dielectric constant as water. Complications related to the difference between dielectric constants are discussed in Ref. [13].

The dependence of the charge inversion ratio, σ^*/σ, on r_s was calculated analytically [13] in two limiting cases $r_s \gg R_0$ and $r_s \ll R_0$, where $R_0 = (\pi\sigma/Ze)^{-1/2}$ is the radius a WS cell at the neutral point $n = \sigma/Ze$. At $r_s \gg R_0$ calculation starts from Eq. (4). Electrostatic potential can be calculated as potential of the plane with the charge density σ^* screened at the distance r_s. This gives $\psi(0) = 4\pi\sigma^* r_s$. At $r_s \gg R_0$ screening by monovalent ions does not change Eq. (2) substantially so that we still can use it in Eq. (4) which now describes charging of a plane capacitor by voltage $|\mu_{WC}|/Ze$. This gives [13]

$$\sigma^*/\sigma = 0.41(R_0/r_s) \ll 1 \quad (r_s \gg R_0). \tag{6}$$

Thus, at $r_s \gg R_0$ inverted charge density grows with decreasing r_s. Now we switch to the case of strong screening, $r_s \ll R_0$. It seems that in this case σ^* should decrease with decreasing r_s, because screening reduces the energy of SCL and leads to its evaporation. In fact, this is what eventually happens. However, there is a range where $r_s \ll R_0$, but the energy of SCL is still large. In this range, as r_s decreases, the repulsion between Z-ions becomes weaker and makes it easier to pack more Z-ions on the plane. Therefore, σ^* continues to grow with decreasing r_s. At $r_s \ll R_0$ it is convenient to minimize directly the free energy of WC of Z-ions at the macroion surface with respect of n. This free energy consists of nearest neighbor repulsion energies of Z-ions and the attraction energy of Z-ions to the charge surface. All interactions are screened according to DH theory, so that interaction of non-nearest neighbors can be neglected. This gives [13]

$$F = -4\pi\sigma r_s Zen/D + (3nZ^2e^2/DA)\exp(-A/r_s), \tag{7}$$

where $A = (2/\sqrt{3})^{1/2}n^{-1/2}$ is the lattice constant of the hexagonal WC. Minimizing this free energy with respect to n one arrives at

$$\sigma^*/\sigma = (\pi/2\sqrt{3})[R_0/r_s \ln(R_0/r_s)]^2 \quad (r_s \ll R_0). \tag{8}$$

Thus σ^*/σ grows with decreasing r_s and can become larger than 100%. At $r_s \sim R_0$ Eq. (8) and Eq. (6) match each other.

3. Correlations of adsorbed polyelectrolyte molecules

3.1. ADSORPTION OF ROD-LIKE POLYELECTROLYTES ON THE CHARGED PLANE

A practically important class of multivalent ions are polyelectrolyte (PE) molecules. In this section we discuss charge inversion caused by adsorption of long rod-like Z-ions. To make signs consistent with the case of DNA, we assume that PE charge is negative, $-\eta_0$ per unit length, while the macroion surface is a plane with positive charge density σ. The basic idea is once again the strong lateral correlations: due to the strong lateral repulsion, charged rods adsorbed at the surface tend to be parallel to each other and have a short range order of an one-dimensional WC in the perpendicular to rods direction (Fig. 3). It is well known that the bare charge density

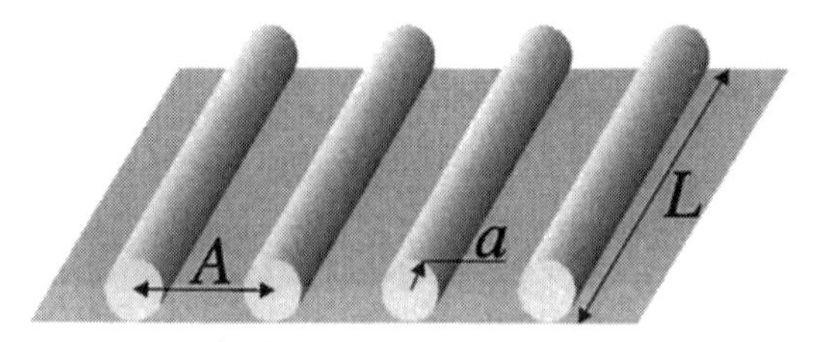

Figure 3. Rod-like negative Z-ions such as double helix DNA are adsorbed on a positive uniformly charged plane. Strong Coulomb repulsion of rods leads to one-dimensional crystallization with lattice constant A.

of DNA, $-\eta_0$, is about four times the critical density $-\eta_c = k_BT/e$ of OM condensation. Therefore, about three quarters of the bare charge of an isolated DNA is compensated by positive monovalent ions residing at its surface so that the net charge of DNA in the bulk solution is $\eta^* = -\eta_c$. This is generally not true for adsorbed DNA. Indeed, some OM-condensed ions, as they are repelled from the charged surface, can be pushed away from DNA and released into the solution when DNA is adsorbed [6]. This effect is particularly important when monovalent salt concentration is low, otherwise counterions do not gain enough entropy to justify the release.

It was shown [13] that the criterion of counterion release is governed by the comparison of screening length, r_s, of monovalent salt and the spacing between critically charged rods, $A_0 = \eta_c/\sigma$, when they neutralize the plane. If screening is strong and $r_s \ll A_0$, the potential of the surface is so weak that counterions are not released. In other words, charge density for each adsorbed DNA helix remains $\eta^* = -\eta_c$. Simultaneously, at $r_s \ll A_0$ the DH approximation can be used to describe screening of the charged surface by monovalent salt. Using these simplifications one can directly minimize the free energy of one-dimensional crystal of DNA rods on the positive surface written similarly to Eq. (7). Then the competition between attraction of DNA helices to the surface and the repulsion of the neighboring helices results in the negative net surface charge density $-\sigma^*$. The charge inversion ratio reads [9, 13] :

$$\sigma^*/\sigma = (\eta_c/\sigma r_s)/\ln(\eta_c/\sigma r_s) \quad (\eta_c/\sigma r_s \gg 1, r_s \ll A_0). \tag{9}$$

Thus the inversion ratio grows with decreasing r_s as in the case of spherical Z-ions. At small enough r_s and σ, the inversion ratio can reach 200% before DNA molecules are released from the surface. It is larger than for spherical ions, because in this case, due to the large length of DNA helix, the correlation energy remains large and WC-like short range order is preserved at smaller values of σ and/or r_s. In the works [13, 14], we called this phenomenon "giant charge inversion."

Let us switch now to the opposite extreme of weak screening by a monovalent salt, $r_s \gg A_0$. In this case, screening of the overcharged plane by monovalent salt becomes strongly nonlinear, with the Gouy-Chapman screening length $\lambda^* = Dk_BT/(2\pi e\sigma^*)$ much smaller than r_s. Furthermore, some counterions are released from DNA upon adsorption. As a result the absolute value of the net linear charge density of each adsorbed DNA, η^*, becomes larger than η_c. Two nonlinear equations for σ^* and η^* are derived in Ref. [13] and are discussed below in Sec. IV. Their solution at $r_s \gg A_0$ reads:

$$\sigma^*/\sigma = (\eta_c/2\pi a\sigma)\exp\left(-\sqrt{\ln(r_s/a)\ln(A_0/2\pi a)}\right), \qquad (10)$$

$$\eta^* = \eta_c\sqrt{\ln(r_s/a)/\ln(A_0/2\pi a)} \quad . \qquad (11)$$

At $r_s \simeq A_0$ we get $\eta^* \simeq \eta_c$, $\lambda^* \simeq r_s$ and $\sigma^*/\sigma \simeq \eta_c/(2\pi r_s\sigma)$ so that Eq. (10) crosses over smoothly to the strong screening result of Eq. (9). Since η^* can not be smaller than η_c, the fact that $\eta^* \simeq \eta_c$ already at $r_s \simeq A_0$ proves that at $r_s \ll A_0$, indeed, $\eta^* \simeq \eta_c$. Simultaneously, the fact that at this point $\lambda^* \simeq r_s$ means that $r_s \ll \lambda^*$ at $r_s \ll A_0$, so that screening in this regime becomes linear and is described by DH theory.

3.2. FLEXIBLE POLYELECTROLYTES WRAPPING AROUND THE CHARGED SPHERES

Until now we talked only about adsorption of a rod-like PE. If the persistence length L_p is finite, PE is released from the charged surface when r_s is very small. This happens when interaction of a PE segment of the length L_p with the surface becomes smaller than k_BT. On the other hand, at larger r_s a flexible PE lays flat at the surface. Due to its electrostatic rigidity neighboring molecules form WC-like SCL and behave similarly to that of a rod-like PE. The correlation induced adsorption of a flexible weakly charged PE on an oppositely charged surface was comprehensively studied in Ref. [19]. When surface is covered by less than one complete layer of PE, results are indeed close to presented above.

Let us also mention the charge inversion in the problem of complexation of a positive sphere and a flexible negative PE (Fig. 4). Refs. [5, 6, 8, 10] predicted substantial charge inversion in this case: more PE is wound around a sphere than necessary to neutralize it. Role of correlations in such charge inversion was recently emphasized in Ref. [16]. It was shown that neighboring turns repel each other and form almost equidistant solenoid, which locally resembles WC. The tail of PE repels PE adsorbed at the surface and creates a correlation hole, which attracts the tail back to the surface.

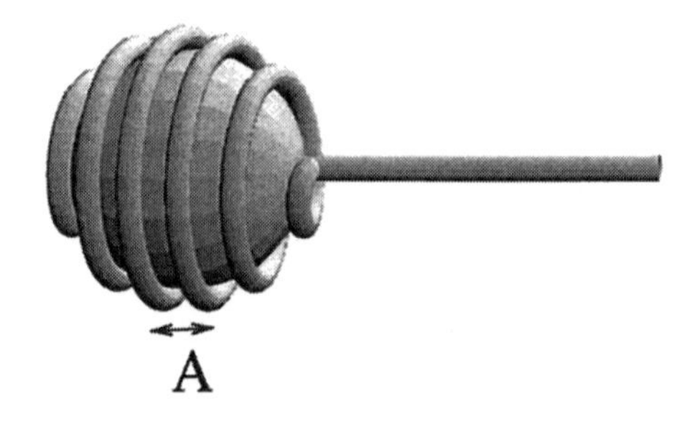

Figure 4. A PE molecule winding around a spherical macroion. Due to Coulomb repulsion, neighboring turns lie parallel to each other. Locally, they resemble a one-dimensional Wigner crystal with the lattice constant A.

Figure 5. Complexation of a negative PE molecule and many positive spheres in a necklace-like structure. On the surface of a sphere neighboring PE turns form WC similar to Fig. 4. At larger scale, charged spheres repel each other and form one-dimensional Wigner crystal along the PE molecule.

The last example of the correlation driven charge inversion we want to mention here is the complexation in a solution with given concentrations of a long flexible negative PE and positive spherical particles such as colloids, micelles, globular proteins or dendrimers [18]. PE binds spheres winding around them. If the total charge of PE in the solution is larger than the total charge of spheres, repulsive correlation of PE turns on a sphere surface lead to inversion of the net charge of each sphere. Negative spheres repel each other and form on the PE molecule periodic beads-on-a-string structure, which resembles 10 nm fiber of chromatin (Fig. 5). If the total charge of PE is smaller than the total charge of spheres, the latter are under-screened by PE and their net charges are positive. Bound spheres once more repel each other and form a periodic necklace. Because a segment of PE wound around a sphere interacts almost exclusively with this sphere, it plays the role of WS cell. These WC-like correlations lead to inversion of charge of a PE molecule: spheres bind to PE in such a great number that the net charge of the PE molecule becomes positive. It is shown in Ref. [18] that inverted charge of PE by the absolute value can be larger than the bare charge of PE even in the limit of very weak screening by monovalent salt. When a large concentration of a monovalent salt is added charge inversion can reach giant proportions. This theory is in a qualitative agreement with recent experiments on micelles-PE systems [11].

4. Are there other mechanisms of charge inversion?

However briefly, we completed the review of works on correlation induced charge inversion. It is now time to ask: Are there alternative, correlation-independent, electrostatic mechanisms leading to this phenomenon? Our answer is no, and we argue that correlations-based mechanism is the universal one. Specifically, we consider below two theories suggested in literature. One, pioneered by [5, 7], considers adsorbed Z-ions as a smeared continuous medium similar to a metal. We argue that this *metalization* approach is an approximation for correlation theory; we discuss when this Ansatz is accurate, and when it fails. The other theory, put forward by [6], views *counterion release* as the driving force behind charge inversion. To this end, we show that while counterion release is obviously favorable for charge inversion, it is itself driven by correlations.

4.1. METALIZATION APPROACH

This approach was clearly formulated for the cases of complexation of a PE with a sphere in [5] and for adsorption of flexible polymers on a charged plane in [7]. It considers the adsorbed Z-ions, at the macroion surface, as a smeared continuum, while still treating bulk solution as consisting of discrete charges. Why is smearing so much favorable as to cause adsorption of Z-ions to the macroion which is already neutralized or even overcharged?

The answer is that smearing amounts to *complete* neglect of self-energy $\varepsilon_{self} = Z^2e^2/2Da$ of adsorbed Z-ions; better to say, it neglects the energy of each Z-ion's electric field in the range of distances between the ion radius a and macroion size r. On the other hand, correlations can be viewed as physical mechanism eliminating *a part* of self-energy of each Z-ion corresponding to the field in the range of distances between correlation hole size R (which plays the role of screening radius of SCL) and macroion size r. In this sense, metalization overestimates the role of correlations and the charge inversion ratio, particularly when $R \gg a$.

To appreciate the difference between correlated and smeared set of Z-ions, it is worth comparing both to the random distribution of the same ions. Correlated configuration is more favorable than random: correlations happen because ions reconfigure themselves non-randomly to gain some free energy. Smeared continuum is also more favorable than random configuration of discrete ions, the difference being precisely the self-energy of ions.

Let us illustrate these ideas in the simplest geometry - a sphere with charge $-Q$ and radius r screened by spherical Z-ions of radius $a \ll r$ in the absence of monovalent salt. Correlation theory, as we have seen, results in Eq. (5) for this problem. What does metalization approach predicts for the same problem? Assuming $a \ll r$, it is easy to write down all relevant energies: $E_{met} = Q^{*2}/2Dr$, which corresponds to the uniformly smeared net charge Q^*; $E_{rand} = E_{met} + M\varepsilon_{self}$, which is the energy of the sphere with $M = 4\pi r^2 n$ randomly distributed Z-ions on its surface; and $E_{WC} = E_{rand} + M\varepsilon(n)$, which is the energy of SCL of Z-ions. Here $\varepsilon(n)$ is the energy of WC per Z-ion given by Eq. (1). The last term of E_{WC} is the correlation energy [28], it is the negative (favorable) contribution due to the difference between correlated and random configurations of Z-ions. As expected, $E_{met} < E_{WC} < E_{rand}$.

To address equilibrium charge inversion, let us now balance the chemical potential of Z-ions at the surface and in the bulk. Apart from translational entropy contribution negligible at high enough concentrations of Z-ions, we have: for the bulk part chemical potential is equal to the self-energy ε_{self}; and for the surface part $\mu = \partial E/\partial M$, where E is either E_{met}, or E_{rand}, or E_{WC}. Therefore, depending on the approximation we want to use, the equilibrium condition reads

$$\begin{aligned} \text{metal}: \quad \varepsilon_{self} &= Q^*_{met}Ze/Dr \\ \text{random}: \quad \varepsilon_{self} &= Q^*_{rand}Ze/Dr + \varepsilon_{self} \\ \text{SCL}: \quad \varepsilon_{self} &= Q^*_{SCL}Ze/Dr + \varepsilon_{self} - 1.65Z^2e^2/RD. \end{aligned} \tag{12}$$

Looking at the result starting from the correlation theory, we see that self-energies cancel and $Q^*_{SCL} = 1.65Zer/R$, producing Eq. (5). By contrast, random distribution leads to $Q^*_{rand} = 0$, i.e., charge inversion is impossible without correlations. Finally, metalization approach yields $Q^*_{met} = Zer/2a$. Comparing with result for SCL, metalization approach is off by a numerical factor only when $R \sim 2a$, while it substantially overestimates charge inversion at $R \gg a$.

We can repeat all the above arguments for adsorption of PE of length $L < r$, radius a, and charge density below OM threshold, $\eta < \eta_c$, on a sphere with radius r and charge $-Q$. At the surface of sphere PE molecules form one of several possible anisotropic WC-like phases. For example, it can be nematic SCL with distance A between almost parallel neighboring PE molecules or a solenoid with distance A between turns, where PE molecules continue each other simulating a long PE which winds around the sphere. In any case, balancing chemical potentials leads to the equations similar to Eqs. (12):

$$\begin{aligned} \text{metal}: \quad \varepsilon_{self} &= Q^*\eta L/Dr \\ \text{random}: \quad \varepsilon_{self} &= Q^*\eta L/Dr + \varepsilon_{self} \\ \text{SCL}: \quad \varepsilon_{self} &= Q^*\eta L/Dr + \varepsilon_{self} - (\eta^2/DL)\ln(L/A) \ , \end{aligned} \tag{13}$$

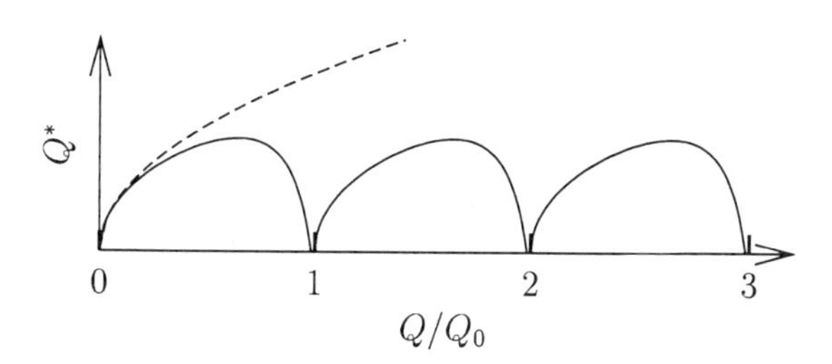

Figure 6. Schematic plot of the oscillating inverted charge of the sphere Q^* as a function of the absolute value Q of its bare charge. Here Q_0 is the charge of the full layer of Z-ions. Dashed line obeys Eq. (5), which is derived for Z-ions of vanishing radius a.

where $\varepsilon_{self} = \eta^2 L D^{-1} \ln(L/a)$. Thus, in full correlation theory, we get $Q^*_{WC} = (r\eta^2 L/Ze^2)\ln(L/A)$ because of cancellation of self-energies. Random distribution of positions and orientations of PE molecules does not lead to charge inversion. For the metalized state, we have self-energy only in the bulk, which gives $Q^*_{met} = (r\eta^2 L/Ze^2)\ln(L/a)$. In complete analogy with the previous case, we see that metalization approach generally overestimates charge inversion, particularly at $A \gg a$, i.e., when few PE molecules are adsorbed. With growing coverage the metalization approximation becomes increasingly accurate.

Qualitatively new effect, however, becomes important when macroion is so strongly charged that its neutralization requires almost full layer of Z-ions. Excluded volume effect of hard cores of Z-ions adds a positive term to the chemical potential of SCL, which is proportional to $k_B T$ and diverges at the full coverage simultaneously with pressure. Close to the full layer this term compensates and then over-compensates the negative Coulomb term μ_{WC}, so that charge inversion disappears. In the language of images, this happens because a full layer is incompressible and does not allow for correlation hole and image. For even larger macroion charge, the second layer starts to form, launching a new wave of charge inversion. In the beginning, charge inversion is small because all the attraction of a new Z-ion approaching the surface is provided by a weak interaction with an inflated image in the emerging second layer, where once again $A \gg a$. Continuing, we arrive at the prediction of oscillating inverted charge Q^* as a function of Q (see Fig. 6), where charge inversion vanishes every time when the top layer of Z-ions is full [29]. Metalization approximation fails to capture the physics of these oscillations.

The oscillations of charge inversion we arrived at are similar to the oscillations of the compressibility and magneto-capacitance in the quantum Hall effect, which are related to consecutive filling of Landau levels [30]. In this sense, we deal with a classical analog of the quantum Hall effect.

We would like to conclude this subsection emphasizing the difference between metalization approach and the PB one. PB approximation effectively smears Z-ions everywhere, both at the macroion surface and in the bulk of solution, while the metalization approach keeps Z-ions of the bulk discrete. Therefore, it is not surprising that the metalization approach somewhat overestimates charge inversion, while there is no charge inversion in PB approximation.

4.2. COUNTERION RELEASE

Consider now adsorption of DNA, or other polymer charged above the OM threshold η_c. The result of correlation theory for this case is presented in Sec. III. Another, seemingly independent reason for charge inversion, was suggested in [6], arguing that the entropy gain of counterions released by DNA *drives* charge inversion. Of course, this theory does not pretend for universality, as it applies to highly charged polymers only. Nevertheless, we want to understand the relations of counterion release and correlations theories.

Let us start with a simple qualitative argument. Imagine that neutralizing DNA molecules, along with their OM-condensed small ions, are adsorbing on the macroion (possibly releasing some of the ions), and currently the neutralization condition is achieved. Assume further that DNA rods are distributed randomly, uncorrelated in both positions and orientations. In this case, next arriving DNA molecule feels no average field, so that it has no reason to release its counterions. The situation is completely different if DNA molecules are correlated on the surface, where locally each molecule is surrounded by a correlation hole - positive stripe of the background charge (WS cell). The corresponding field, or positive potential of the WS cell, may cause release of counterions not only at the neutrality point, but even if the surface overall is overcharged ($\psi(0) < 0$). In other words, correlation hole, or adjustment of DNA molecules to each other, or image charge, or correlations (all synonyms!) is a necessary condition for *both* counterion release *and* charge inversion.

To understand this better, let us re-examine the fundamental physical conditions governing the adsorption equilibrium, namely that chemical potentials of both DNA and small ions should be the same at the surface of macroion and in the bulk of the solution. The equations expressing these conditions were derived in [14] (Eqs. (57) and (60) of that work). We rederive them below, as Eq. (15) and Eq. (16). We deal below only with the case of weak screening, when $r_s \gg A$.

Consider first chemical potential of small ions. It must be the same for ions in the bulk solution, ions condensed on the surface of dissolved DNA rods, and ions on those DNA rods which, in turn, are adsorbed on the macroion. The two corresponding equations read:

$$\begin{aligned} k_B T \ln(N_{1,sb}/N_1) &= (2e\eta_c/D)\ln(r_s/a) \\ k_B T \ln(N_{1,ss}/N_1) &= -e\psi(0) + (2e\eta^*/D)\ln(A/2\pi a) \ . \end{aligned} \tag{14}$$

The left-hand sides of Eqs. (14) is obviously the entropy loss of monovalent ions when they move from the solution to the surface of DNA and the right-hand side is the potential energy gain per ion. In the entropic part, N_1, $N_{1,sb}$ and $N_{1,ss}$ are the concentrations of counterions in the bulk and at the surface of the dissolved and adsorbed DNA, respectively. In the energetic part, $\psi(0)$ is the average potential of the uniformly over-charged plane with charge density σ^*, while logarithmic terms correspond to the potential of DNA having charge density $-\eta_c$ in the bulk, where it is screened by monovalent ions at the distance r_s, and DNA with charge density $-\eta^*$ at the surface, where it is screened by other DNA molecules at the distance $A/2\pi$. As regards $\psi(0)$, it is given by the Gouy-Chapman formula $\psi(0) \simeq -(2k_BT/e)\ln((r_s+\lambda^*)/\lambda^*)$, where $\lambda^* = Dk_BT/(2\pi e\sigma^*)$, because plane as a whole is non-linearly screened. Concentrations $N_{1,ss}$ and $N_{1,sb}$ are related to the corresponding net charges: $N_{1,ss}/N_{1,sb} = (\eta_0 - \eta^*)/(\eta_0 - \eta_c)$. According to Eq. (11) η^* differs only logarithmically from η_c. Therefore, we can neglect $\ln(N_{1,ss}/N_{1,sb})$ and use the same notation $N_{1,s}$ for both concentrations. Taking into account that $\eta_c = k_BTD/e$, we get

$$\eta_c \ln(\lambda^*/a) = \eta^* \ln(A/2\pi a). \tag{15}$$

Already at this stage we can look at what happens if we neglect correlations between adsorbed DNA rods. In this case, they are screened by small ions only, and the corresponding screening length is λ^* instead of $A/2\pi$. Eq. (15) then gives $\eta^* = \eta_c$. Thus, if adsorbed DNA molecules were uncorrelated, positioned and oriented randomly, they would not have released counterions upon adsorption.

Let us return to the strongly correlated distribution of DNA. To complete the picture, we concentrate on the chemical potential of DNA rods. Corresponding equilibrium condition is similar to Eq. (4):

$$\begin{aligned} L\eta^*\psi(0) = |\mu_c| + L(\eta^* - \eta_c)(k_BT/e)\ln(N_{1,s}/N_1) \\ -(L\,{\eta^*}^2/D)\ln(\lambda^*/a) + (L\eta_c^2/D)\ln(r_s/a) \ , \end{aligned} \tag{16}$$

where L is DNA length. In the right hand side, the first term, $\mu_c \simeq \mu_{WC} = -(L\,{\eta^*}^2/D)\ln(2\pi\lambda^*/A)$, is the correlation chemical potential of DNA at the macroion surface. This is the energy of interaction of one DNA rod charged to $-\eta^*$ with its correlation hole which is a positive background stripe of width A. The second term is precisely the entropy gain due to the release of counterions. The third and fourth terms together represent the increase of the DNA self-energy when it moves from the bulk solution where it is charged to $-\eta_c$ and screened at r_s to the surface where it is charged to $-\eta^*$ and screened at λ^*.

Fundamental equations (15), (16) are identical with Eqs. (57), (60) of Ref [14], which are derived there in a more rigorous way. The solution of these equations is given by Eqs. (10), (11), which simultaneously describe the correlation induced charge inversion and counterion release.

Let us now continue to explore what happens if we imagine that DNA is randomly distributed on the macroion surface with respect to both position and orientation of rods, i. e. if we artificially drop all correlation effects. As we already mentioned, Eq. (15) yields $\eta^* = \eta_c$ in this approximation. Furthermore, $\mu_c = 0$ in the absence of correlations, so that Eq. (16) yields $\psi(0) = 0$ and $\sigma^* = 0$. Thus, assuming adsorbed DNA randomly distributed on the macroion surface with respect to both positions and orientations of rods, we arrive at the conclusion that there is neither counterion release nor charge inversion. By contrast, lateral correlations leading to parallel alignment of DNA molecules, can drive charge inversion even without counterion release, as this happens, e.g., for weakly charged PE, for which $\eta_0 < \eta_c$. In this sense, we conclude that correlations are the driving force of both counterion release and charge inversion.

Returning to the correlated distribution of DNA let us consider also a limit of very large r_s, which is not very realistic, but interesting from a theoretical standpoint. According to the Eq. (11), DNA net charge η^* increases with increasing r_s, which physically means higher proportion of released counterions. To address this regime, we should make one step back including the $\ln(\eta_0 - \eta^*)/(\eta_0 - \eta_c)$ term in Eq. (15). For our purposes here it is sufficient just to approximate η^* by η_0 in the Eq. (16). First and third term in the right-hand side combine then to $(L\eta_0^2/D)\ln(A/2\pi a)$, which vanishes at $A = 2\pi a$ and is negligible around this point. Therefore, the averaged surface potential in this case

$$\eta_0\psi(0) = (\eta_0 - \eta_c)(k_BT/e)\ln(N_{1,s}/N_1) + (\eta_c^2/D)\ln(r_s/a), \tag{17}$$

looks like the sum of counterion release and metalization terms. Interestingly, both of them are driven by correlations!

We conclude repeating that the underlying physics of charge inversion is always determined by correlations. An explicit treatment of correlations provides regular and universal description of this phenomenon. Apart from charge inversion, the correlations manifest themselves also in a number of other ways, including metal-like properties of the macroion surface over certain range of length scales and including the release of counterions for the case of highly charged Z-ions, such as DNA.

T. T. N. and B. I. S. are supported by NSF DMR-9985785.

References

1. J. Ennis, S. Marcelja and R. Kjellander, Electrochim. Acta, **41**, 2115 (1996).
2. T. Wallin, P. Linse, J. Phys. Chem. **100**, 17873 (1996); **101**, 5506 (1997).
3. V. I. Perel and B. I. Shklovskii, Physica A 274, 446 (1999).
4. B. I. Shklovskii, Phys. Rev. E **60**, 5802 (1999).
5. E. M. Mateescu, C. Jeppersen and P. Pincus, Europhys. Lett. **46**, 454 (1999);
6. S. Y. Park, R. F. Bruinsma, and W. M. Gelbart, Europhys. Lett. **46**, 493 (1999);
7. J. F. Joanny, Europ. J. Phys. B **9** 117 (1999).
8. P. Sens, E. Gurovitch, Phys. Rev. Lett. **82**, 339 (1999).
9. R. R. Netz, J. F. Joanny, Macromolecules, **32**, 9013 (1999).
10. R. R. Netz, J. F. Joanny, Macromolecules, **32**, 9026 (1999).
11. Y. Wang, K. Kimura, Q. Huang, P. L. Dubin, W. Jaeger, Macromolecules, **32** (21), 7128 (1999).
12. M. Lozada-Cassou, E. Gonzalez-Tovar, and W. Olivares, Phys. Rev. E **60**, R17 (1999).
13. T. T. Nguyen, A. Yu. Grosberg, B. I. Shklovskii, Phys. Rev. Lett. **85**, 1568 (2000);
14. T. T. Nguyen, A. Yu. Grosberg, B. I. Shklovskii, J. Chem. Phys. **113**, 1110 (2000).
15. R. Messina, C. Holm, K. Kremer, Phys. Rev. Lett. **85**, 872 (2000); Euro. Phys. Lett., **51**, 461 (2000).
16. T. T. Nguyen, B. I. Shklovskii, Physica A, **293**, 324 (2001).
17. P. Chodanowski and S. Stoll, Macromolecules, **34**, 2320 (2001).
18. T. T. Nguyen, B. I. Shklovskii, J. Chem. Phys. **114**, 5905 (2001).
19. A. V. Dobrynin, A. Deshkovski and M. Rubinstein, Macromolecules **34**, 3421 (2001).
20. D. Andelman, J. F. Joanny, cond-mat/0011072
21. K. B. Zeldovich, A. R. Khokhlov, Eur.Phys. J. E. (2001).
22. M. Tanaka, A. Yu. Grosberg, J. Chem. Phys. (2001).
23. L. G. Gulbrand, Bo Jonsson, H. Wennerström, and P. Linse, J. Chem. Phys. **80**, 2221 (1984); R. Kjellander and S. Marcelja, Chem. Phys. Lett. **114**, 124(E) (1985); I. Rouzina and V. A. Bloomfield, J. Phys. Chem. **100**, 9977, (1996); N. Gronbech-Jensen, R. J. Mashl, R. F. Bruinsma, and W. M. Gelbart, Phys. Rev. Lett. **78**, 2477 (1997); Y. Levin, J. J. Arenzon, and J. F. Stilck, Phys. Rev. Lett. **83**, 2680 (1999); B. I. Shklovskii, Phys. Rev. Lett. **82**, 3268 (1999).
24. N. D. Lang, *Solid state physics*, Vol. 28, Edited by H. Ehrenreich, F. Seitz, and D. Turnbull, Academic Press, New York (1973).
25. L. D. Landau and E. M. Lifshitz, Quantum Mechanics (nonrelativistic theory), Chapter X, Oxford, Pergamon Press (1980).
26. A. G. Moreira, R. R. Netz, Europhys. Lett. **52**, 705 (2000).
27. G. S. Manning, J. Chem. Phys. **51**, 924 (1969).
28. L. D. Landau and E. M. Lifshitz, Statistical Physics, Part 1, Chapter VII, Oxford, Pergamon Press (1977).
29. T. T. Nguyen, B. I. Shklovskii, Phys. Rev. E (2001).
30. T. P. Smith, W. I. Wang and P. J. Stiles, Phys. Rev. B**34**, 2995 (1986); A. L. Efros, Solid State Comm. **65**, 1281 (1988); S. V. Kravchenko, D. A. Rinberg, S. G. Semenchinsky, and Pudalov, Phys. Rev. B**42**, 3741 (1990); J. P. Eisenstein, L. N. Pfeifer and K. W. West, Phys. Rev. Lett. **68**, 674 (1992).

HIGHLY CHARGED POLYELECTROLYTES : EXPERIMENTAL ASPECTS

CLAUDINE E. WILLIAMS
Laboratoire des Fluides Organisés (CNRS URA 792)
Physique de la Matière Condensée
Collège de France
Paris, France

Abstract. This chapter is an overview of the properties of linear, flexible and highly charged synthetic polyelectrolytes as seen by an experimentalist. Firstly, two characteristic properties of the single chain, i.e. extended conformation and charge renormalisation by counterion condensation, are examined; their implications to the behaviour of dilute solutions is considered. Then the structural properties of semi-dilute solutions of hydrophilic polyelectrolytes are introduced and the results of selected x-ray and neutron scattering experiments are compared to the predictions of the isotropic model of de Gennes *et al.* Most of these data corroborate the current models but some raise puzzling questions. Finally, the effects of the interplay between electrostatic interactions and hydrophobic ones (bad solvent conditions) on chain conformation are described.

1. Introduction

Polyelectrolytes are polymer chains containing a variable (usually large) amount of ionizable monomers. Once dissolved in an appropriate polar solvent such as water, the ion pairs dissociate. The electrostatic charges of one sign are localized on the chain whereas the large number of oppositely charged counterions are scattered in the solution. Polyelectrolytes are everywhere around us and in us. Most biopolymers, including DNA, RNA and proteins, are polyelectrolytes. The water solubility of synthetic polyelectrolytes permits a wide range of non-toxic, environment friendly and inexpensive formulations. Super-absorbing diapers, low fat food, washing detergents and their additives, waste water purifiers are amongst the most common examples. Thus phenomena specific to polyelectrolytes have strong implications in molecular and cell biology as well as technology. Despite more than fifty years of active experimental and theoretical efforts, many properties of charged polymers are still poorly understood, in con-

C. Holm et al. (eds.), Electrostatic Effects in Soft Matter and Biophysics, 487–506.
© 2001 *Kluwer Academic Publishers. Printed in the Netherlands.*

trast to their neutral counterparts. The complexity stems primarily from the simultaneous presence of long range (possibly screened) electrostatic interactions and short range excluded volume interactions and also to the crucial role of the counterions. From a theoretical point of view, polyelectrolytes combine the fields of the physics of polymers and the statistical mechanics of charged systems. The coarse-grained models that have been so successful in describing "universal" properties of neutral polymers are much less suitable for polyelectrolytes because some properties are not independent of the microscopic characteristics of the charged chain and of the type of counterions. Such is the case with counterion condensation a rather local phenomenon which, in turn, influences the inter-chain interactions (see Joanny's and Netz' chapters in these Proceedings). On the strictly experimental side, the difficulties are also numerous. Polyelectrolyte chains are strongly stretched and thus the concentration where they begin to overlap, c^*, is extremely low; furthermore because of the long-range nature of the electrostatic interaction, inter-chain interactions appear well below c^*; thus most techniques aimed at determining the chain configuration, such as x-ray and neutron scattering, are not sensitive enough to analyze the single chain behavior. A polyelectrolyte solution has at least three components and it is not straightforward to separate the contribution of the charged chains from that of the counterions to a given technique. There are also problems associated with the complex structure of water, its auto-dissociation and difficulties in removing foreign particles; consequently there is no such thing as a "salt free" polyelectrolyte (except in numerical simulations) and all samples contain some amount of added salt which is usually unspecified.

Despite these shortcomings the physical chemistry and thermodynamics of dilute solution (single chain behavior) have been thoroughly investigated and rather well understood in the fifties. However the question of how an assembly of intrinsically flexible charged chains is organized was still debated. A second era started when de Gennes and his collaborators introduced the isotropic model for semi-dilute solutions, inspired by the scaling description of overlapping neutral chains. One characteristic feature of charged polymers is the competition between the connectivity of the chains and the electrostatic repulsion between charged monomers which results in a broad maximum in the scattering function observed by X-ray and neutron scattering. This broad peak is often considered as a "signature" of polyelectrolyte solutions. During the last decade many new theoretical approaches, both analytical and computational, have appeared and a large amount of experimental information has been collected which have led to a deeper understanding of these complex systems. At the same time new questions have been raised, puzzles and surprises have appeared.

In this chapter I want to give the reader a flavor of some of the interesting problems of the physics of polyelectrolytes, as seen by an experimentalist. This is by no means a review article. I shall limit myself to the *static properties of linear, flexible and highly charged synthetic polyelectrolytes.* Among the large amount of experimental information which has been collected in the last decade, I shall select and discuss in more detail some well defined problems which either corroborate the theoretical predictions, in particular concerning the structure of the semi-dilute solutions of hydrophilic polyelectrolytes, or alternatively raise new, interesting and puzzling questions such as those related to the electrostatic persistence length of charged chains or counterion condensation in semi-dilute solutions. Many experimental results that will be discussed here rely on scattering techniques; for details on those the reader is referred to M. Rawiso's chapter. Some simple aspects of the theory are sometimes recalled. Repetitions are unavoidable. They might even be useful as each author has his/her own view and perception of a problem and one presentation can be better suited to a reader than another one.

After this discussion centered on the more "classic" hydrophilic polyelectrolytes I shall show how short range hydrophobic interactions may affect drastically the configuration and effective charge of polyelectrolyte chains where hydrophobic monomers have been introduced. Light, neutron, and x-ray scattering studies as well as osmometry and fluorescence measurements will be discussed in the light of theoretical predictions and numerical simulations. It should be stressed that the problem of hydrophobic interactions has much practical importance since most synthetic polymers are strongly hydrophobic and it is the addition of charged groups to an otherwise hydrophobic backbone that improve their solubility in water. It is also worthwhile mentioning that proteins are basically hydrophobic polyelectrolytes. Although their tertiary structure depends on complex site specific interactions, the Rayleigh type instability resulting in a compact pearl necklace structure could provide some insight into the onset of chain folding.

The chapter is organized as follows. After recalling the pertinent characteristic lengths of polyelectrolyte solutions, some properties of the single chain will be described. The next section will be devoted to a description of the structure of semi-dilute solutions of hydrophilic polyelectrolytes, i.e. when only electrostatic interactions are taken into account. Finally the last section will discuss the case of polyelectrolytes solutions with hydrophobic interactions.

2. Characteristic lengths

Polyelectrolyte solutions are characterized by a wide range of interactions varying from molecular distances (counterion-chain interaction) to almost micrometer lengths (long-range electrostatic forces). Some orders of magnitude are worth keeping in mind.

- Most flexible synthetic polyelectrolytes have a vinylic backbone. The *monomer length* a is about 2.5Å. For a chain with $N = 10^4$ monomers, the contour length L is 25,000Å.
- The solvent is characterized by the *Bjerrum length* l_B; it is the distance over which the electrostatic energy between two elementary charges e in a solvent of dielectric constant ε is compensated by the thermal energy k_BT.

$$l_B = e^2/\varepsilon k_B T \tag{1}$$

  l_B is equal to 7.12Å in water at 300°K.
- The *Debye-Hückel screening length* κ^{-1} is defined as

$$\kappa^2 = 4\pi l_B I \tag{2}$$

  where the ionic strength I is $\sum_i Z_i^2 c_i$ where Z is the valence and c the concentration of each ionic species i. Typically, κ^{-1} is of the order of 100Å for a 10^{-3} M solution. However it is important to remember that the screened Coulomb potential, $U(r) = (e^2/\varepsilon r)\exp(-r\kappa)$ introduced in the Debye-Hückel theory is valid only if the electrostatic effects are sufficiently weak.

3. Some properties of the single chain

3.1. A MATTER OF DEFINITIONS

Polyelectrolytes are said to be *weakly charged* when a small fraction of the monomers are charged and a large fraction of them are neutral; Coulomb interactions interplay with usual Van der Waals interaction. They are *highly charged* when a large fraction of monomers are charged; in this case Coulomb interactions dominate the solution behavior. Except when specifically stated we will be dealing with highly charged polyelectrolytes throughout this chapter.

The latter definitions should not be confused with the notion of *weak* and *strong* polyelectrolytes. In the weak case, the charged monomer units are derived from a weak acid, e.g., monomers with COOH groups. In solution, not all groups are dissociated and the degree of dissociation depends on the pH of the solution; each chain can be viewed as a random copolymer of monomers with COO- and COOH groups which fluctuate; the charges are said to be annealed [1]. This can have important implications since the charge distribution can adapt to local conditions [2]. For strong polyelectrolytes, e.g. with SO_3H units derived from a strong acid, all monomers are dissociated and their position along the chain is fixed; the charges are said to be quenched.

As with neutral polymers, the chain molecular architecture can be diverse. Stars, dendritic or branched polyelectrolytes have been synthesized. The distribution of charges along the chain, from blocky to completely random, introduces another level of complexity. Polyelectrolyte gels, consisting of polyelectrolytes molecules covalently cross-linked to form a tri-dimensional network have unique swelling properties which have led to the manufacture of super-absorbing material. Typically, one gram of dry gel can absorb one hundred times its weight of water. Such fascinating behavior is related to the presence of a large number of counterions which are confined into the network in order to insure macroscopic electro-neutrality, at the expense of a loss of translational entropy, and which create a large osmotic pressure responsible for the swelling. The mechanism by which ionic networks swell was studied and understood in the fifties [3]. It has been a very active field since then and new developments have appeared. However it is outside the scope of this course and it will not be discussed here.

3.2. CHAIN EXTENSION AT INFINITE DILUTION

It was recognized very early on that polyelectrolyte chains are very large objects. The large monotonous increase in reduced viscosity ($\eta_{\text{red}} = \eta_{\text{sp}}/c$ with c the volume fraction and η_{sp} the specific viscosity) as concentration decreases, observed already in the thirties [4], was soon interpreted as evidence of chain stretching for highly dilute and *salt-free* solutions. Based on a large body of experimental data, [5] Fuoss [6] developed in the fifties an empirical extrapolation law to zero concentration of the reduced viscosity, based on the assumption of a coil-to-rod transition upon dilution, for salt-free solutions. Consequently, highly charged polyelectrolytes were commonly pictured as rigid rods in dilute concentrations. As it happens more sophisticated experiments in later years [7] have shown that there is actually a maximum of the reduced viscosity at very low concentration thus casting doubts on the validity of Fuoss extrapolation. Indeed the situation is more complex and subtle and, although chain stretching is indeed dramatic, the chain retains some flexibility as we shall see in what follows. However despite its shortcomings the rod picture catches well some of the physics of polyelectrolytes.

The effect of the repulsive interaction between like charges on the chain conformation can be understood by a Flory-type calculation, due to Kuhn, Künzle and Katchalsky in 1948 [8] (before Flory published his own calculation for neutral chains with short-range excluded-volume interactions!). It relies on simplifying assumptions but gives a simple physical picture. Let us consider a chain with N monomers and assume that a fraction f of those are ionizable. Thus, in solution, the chain contains fN charged monomers and $(1-f)N$ neutral monomers, all randomly distributed. In a mean-field approach, the Flory-type energy for a chain of size R is

$$E_F = k_B T \frac{R^2}{Na^2} + k_B T \frac{(Nf)^2 l_B}{R} \tag{3}$$

The first term is the *elastic energy* where we assume that the chain has a Gaussian configuration when all electrostatic interactions are switched off, i.e. when the mean squared average end-to-end distance is $R_0^2 = Na^2$. The second term is the *electrostatic energy* due to the Nf monomers of charge e. Minimization with respect to R leads to

$$R \approx N f^{2/3} (l_B a^2)^{1/3} \tag{4}$$

This simple calculation leads to the important result that the size of a polyelectrolyte is proportional to the number of monomers and to the two third power of the charge fraction. At this point it should be stressed that the linear dependency of R with N *does not* imply that the chain is fully extended; it may retain some local molecular flexibility and still R would scale as N. The flexibility at short distances of stretched charged chains is most clearly seen in Monte-Carlo simulations [9].

A ≪blob≫ description is useful to get a better image of the chain conformation. [10], [11] (Details on scaling theory can be found in Joanny's chapter in these Proceedings.) Let us assume here that the chain is weakly charged and that the backbone (chain without charges) is in a θ-solvent [12]. We now look at the spatial monomer-monomer correlations and find that there is one important length which we call D, the *electrostatic blob* size. On length scales smaller than D, the electrostatic interaction is only a weak perturbation, the chain statistics are determined by the solvent quality and thus remain Gaussian in our case; if g_e monomers are involved, then $D^2 = g_e a^2$. On length scales larger than D, the electrostatic repulsion between blobs dominates and the chain has the conformation of a rod of N/g_e blobs of size D. The total length is

$$L = (N/g_e) D \tag{5}$$

with a linear dependence on N. The size of the electrostatic blob and the number of monomers involved depend on the linear charge density of the chain but not on its size. Indeed, using the fact that on a length scale D the electrostatic interaction is of the same order as the polymer fluctuations and that the sub-chain has a Gaussian configuration, one finds that

$$D \approx a \left(\frac{f^2 l_B}{a} \right)^{1/3} \qquad g_e \approx \left(\frac{f^2 l_B}{a} \right)^{-2/3} \tag{6}$$

Small angle x-ray and neutron scattering (SAXS and SANS) have the capability of measuring the monomer-monomer pair correlations and thus completely describe the single chain conformation. One would expect to observe the characteristic q^{-2} dependence of the scattered intensity at large scattering angles (short length scales), a signature of a Gaussian chain conformation and the q^{-1} dependence of a rod at small angles (large length scales). The crossover between the two regimes would correspond to D^{-1}. However single chain behavior is only observed at concentrations much too low to give a measurable signal in the best contrast conditions (see Rawiso's chapter for details). At practical concentrations the long range electrostatic interactions between the chains give rise to a chain-chain correlation peak which combines with the single chain contribution, making data analysis complex if not impossible.

3.3. COUNTERION CONDENSATION AND CHARGE RENORMALIZATION

When we looked at the chain conformation, we implicitly assumed that the entropy of mixing was driving the counterions to distribute uniformly in the solution. However when the chain is highly charged, the electrostatic interactions attract the counterions to the oppositely charged polymer chain. The potential close to the chain can be so high that for some counterions the gain in entropy of mixing is small compared to the electrostatic interaction; these counterions remain bound to the chain, so reducing the effective charge of the chain compared to the nominal (or chemical) charge. In other words, seen from a large distance the chain appears less charged than its chemical composition would predict. This phenomenon is known as *counterion condensation.* For most practical purposes one can simply consider that the chain has an effective charge f_{eff} which is lower than the bare or chemical charge f. Consequently counterion condensation will modify the inter-chain and intra-chain electrostatic interactions and influence the conformation of chains, their structure in solution and even their dynamic behavior. This exemplifies the complex coupling between short and large length scales which makes understanding polyelectrolytes difficult and challenging.

The distribution of charges around a single infinitely long *rod* has been first derived using Poisson-Boltzmann theory [13]. The cell model, where the rod-like macroions are confined into a periodic array of parallel cylindrical Wigner-Seitz cells, provides an analytical expression for the distribution of counterions for salt free solutions. In an alternate approach, due to Manning [14] and Oosawa [15], [16], which captures well the physics of the phenomenon, the counterions are assumed to be divided in two species, either free in the solution or condensed in a sheath around the chain. There is chemical equilibrium between the two species.

Imagine the chain as a rod of length L ($L >> a$) and of linear charge density $f = a/A$, where A is the distance between charges along the chain. The density of counterions at a distance r from the chain ($r << L$) is $n(r) = n_0 \exp -\psi(r)$ where $\psi(r)$ is the electrostatic potential calculated by Gauss theorem as $\psi(r) = 2l_B/A \ln(r)$. The total number of counterions $m(r)$ per unit length of rod within a cylinder r becomes

$$m(r) = \int_0^r n(r') 2\pi r' \, dr' = \int_0^r 2\pi n_0 r'^{1-2l_B/A} \, dr' \tag{7}$$

The behavior of this integral depends very much on the value of the coupling parameter $u = l_B/A$. When u is small, i.e. when f is small, the integral is dominated by its upper bound and the total number of counterions decreases as r decreases. On the other hand, when u is larger than 1 (highly charged chain), the integral diverges at its lower bound. There is a condensation of counterions until the value of u reaches 1, at which point the average distance between charges is equal to the Bjerrum length.

What does this imply in practical terms? For a vinyl polymer in water at room temperature, a is 2.5Å and l_B is 7.12Å; the onset of counterion condensation corresponds to a critical f equal to 0.35, i.e. when about every third monomer is charged. Let us imagine that chemistry allows us to gradually introduce more and more ionizable monomers in a neutral polymer. The effective (or net) charge fraction of the chain will increase as the chemical (or bare) charge fraction up to 0.35, then it will remain constant as more and more counterions are condensed. Thus in this regime the condensed counterions partially neutralize the bare charge density to an uniform effective charge density. The results of various techniques (osmometry, electrophoresis, conductivity...) which are sensitive to the number of *free* counterions in the solution have validated, at least qualitatively, the existence of counterion condensation even for intrinsically flexible chains [17]. They are discussed in details by Manning [18] in a recent paper where many references to experimental data can be found. Conductivity is by far the most commonly used technique; measurements of the counterion activity, of the electrophoretic mobility or of the osmotic pressure are also performed. For our part, we have used the latter technique to measure the number of osmotically active counterions in dilute solutions of acrylamide-*co*-sodium-2-acrylamide-2-methyl propane sulphonate (AMAMPS). Each sample of this flexible vinylic polyelectrolyte has a different number of randomly distributed ionizable sulphonate groups. At the precision of the technique (about 10%) the number of free counterions, and consequently the effective charge of the chains is found constant between f = 0.4 and 1. [19] A non-uniform distribution of counterions is also clearly seen in neutron scattering experiments [20]. Although the results are interpreted in terms of a continuous radial distribution in terms Poisson-Boltzmann equation, they clearly show that the concentration of counterions is much larger close to the chain.

Nowadays, the concept of counterion condensation is generally accepted. The idea of a renormalized charge above "Manning threshold" is currently assumed in scaling theories of polyelectrolytes, even at high concentrations. However the quantitative agreement between experiments and theory is still puzzling since the physical conditions of the supporting experiments are far from the basic assumptions of the theory. For instance the Manning-Oosawa (MO) theory is strictly valid for an infinite, uniformly charged rod at zero concentration and without added salt. Various attempts to take into account the effect of finite chains at nonzero concentration and/or with added salt have all predicted deviations to MO theory. The coupling between counterion condensation and chain conformation (flexible chain) has also been considered theoretically [21], [22], [23] and in Monte Carlo simulations [24], [25]. A collapse of the stretched chain is even predicted. [26], [27] Because these considerations have not found any definitive experimental corroboration I shall not discuss them further.

The picture that emerges is the following: the condensed counterions are confined into a sheath around the chain but they still retain some mobility *along* it [28]; they do not in general combine again with the charge on the chain. Ion pairing takes place only if the dielectric constant is too low for the charges to be dissociated (ionomer effect). These points are discussed extensively and very clearly in Oosawa's book. The existence of two distinct populations of counterions, i.e. free and condensed, is still very controversial. NMR experiments on the binding of [23]Na^+ on DNA show that the cation is territorially bound and exchanges rapidly with the ones in the rest of the solution [29]. However this behavior does not seem to be universal and depends on the nature of the cation. One experimental problem, as noted by Schmitz [30], is the lack of "operational definition" of the condensed counterions and the fact that different techniques do not "count" counterions in the same manner. Anyway there are two limiting cases when the notion of two states for the counterions is not valid. In an infinitely dilute solution of finite chains, there is no condensed state as the entropic cost would be infinitely large. In very concentrated solutions, the chains are so close to each other than the electrostatic potential and the distribution of counterions have to be uniform. In all practical cases, one could distinguish two populations if one defines as "condensed" those counterions that are entropically bound to the chain with an electrostatic energy larger than kT. [31]

Anticipating on the next section, it is worth noting that there appears to be a renormalization of all electrostatic interactions also in semidilute solutions, in the range of charge fraction where MO condensation should appear (i.e. for AMAMPS between f = 0.4 and 1). This implies that the effective charge is also constant although the combined SAXS and SANS experiments could not provide any exact value for it.[19]

4. Semi-dilute solutions of flexible, hydrophilic polyelectrolytes

Most realistic experiments with polyelectrolytes are done at concentrations where the chains are interacting and the single chain behavior is no longer relevant. In the fifties, when polyelectrolytes were the focus of intensive studies, it was firmly believed that the molecules were retaining their rod-like conformation and as concentration increases, they would form a lattice of rods. [32] No evidence for such a structure was found except for a s-ingle broad peak in scattering experiments whose position varies as $c^{1/2}$ as expected for a 2d (short range) order of rods. A breakthrough came in the seventies when de Gennes applied to polyelectrolytes the techniques of statistical physics that had been successful for neutral polymers and introduced the idea that a semi-dilute solution of polyelectrolytes remains *isotropic* at any concentration, the chains forming an entangled network. In what follows I shall first describe the main results of the scaling theory in its simplest form. More details about the theory of polyelectrolytes can be found in Joanny's and in Netz's chapters of this book and in a review article of Barrat and Joanny [33]. This theoretical framework will help me to discuss some of the most important and illustrative experimental data that have been accumulated in the last two decades. The choice of these is evidently be very personal. Some of the experimental breakthroughs are based on the technique of small angle neutron scattering and, to a lesser extent, of small angle x-ray scattering. For details on these, the reader is referred to Rawiso's chapter.

4.1. THE OVERLAP CONCENTRATION C*

In dilute solution, the chain are elongated (size L given by equation 5) and their average separation is d; as c increases one reaches a situation where d is still larger than L but becomes smaller than κ^{-1}, the Debye screening length. A peak appears in the scattering profile, characteristic of a liquid-like order; its position q^* is of the order of $2\pi/d \propto (c/N)^{1/3}$, as observed experimentally for sufficiently short polyelectrolytes [34], [76] and in simulations [35].

c^* can be defined as the concentration when d = L , i.e. at the onset of steric hindrance. Then

$$c^* \approx \frac{N}{L^3} = \frac{N}{((N/g_e)D)^3} \propto N^{-2} \tag{8}$$

Boris has verified this scaling law by compiling various literature results on poly(styrene sulphonate) of different molecular weights. However the absolute values of c^* are found to be higher than expected. This can be understood in terms of Monte-Carlo simulations which show clearly that well before c^* the chains begin to be less extended, shifting c^* away as it is approached. The experimental determination of c^* can indeed be tricky. The only meaningful criterion is to set it as the concentration where the q^* *versus* c exponent changes from 1/3 to 1/2 (semidilute behavior, see next paragraph); however the dilute/semi-dilute crossover is observed only with short chains. For longer chains it is rather arbitrarily set as the concentration when the solution viscosity is twice the solvent viscosity [37].

4.2. ISOTROPIC MODEL FOR SEMIDILUTE SOLUTIONS

We give here briefly the static scaling picture of de Gennes *et al.* [10] as revised by Rubinstein *et al.* [38], [39]. A more complete treatment will be found in Joanny's chapter. Let us consider an ensemble of chains as defined in section 3.1. We will introduce a parameter $B = Na/L$ which depends on solvent quality (i.e. good or theta solvent). Once again we will focus on salt-free solutions and look at the monomer-monomer correlations. The model postulates that the chains overlap and form a transient network above c^*. In this network, each chain is surrounded by an excluded volume or *correlation hole* from which other chains are excluded. There is an important correlation length ξ such that, for distances $r < \xi$, the electrostatic forces are dominant and the section of the chain has the same extended configuration as in dilute solution; for distances $r > \xi$, both electrostatic and excluded volume interactions are screened and the chain follows random walk statistics since both electrostatic and excluded volume interactions are screened. If we assume that ξ depends on c as a power law, that it should be independent of N and that $\xi = L$ at $c = c^* \cong N/L^3 \cong (B/a)^3/N^2$, then we find that

$$\xi \approx L(c^*/c)^{1/2} \approx (B/ca)^{1/2} \propto c^{-1/2}N^0 \tag{9}$$

Keep in mind that ξ scales as $c^{-1/2}$ and is thus proportional to κ^{-1} the Debye screening length (but not necessarily equal to it) [40]. Each correlation volume (or blob) $\xi^3 b$ contains g monomers and one chain amongst others is a random walk of N/g monomers and has a size

$$R \approx (a/cB)^{1/4} N^{1/2} \tag{10}$$

The concentration dependence ($c^{-1/4}$) is much stronger than for neutral polymers ($c^{-1/8}$).

Adding salt screens the electrostatic interactions. Eventually the polyelectrolyte reverts to neutral chain behavior when the salt screening length is smaller than the mesh size. For the sake of brevity, I will not discuss this here but refer the reader to the paper of Dobrynin *et al.* [38] (The original paper of Pfeuty [41] may also be consulted.) Briefly, any property X of a polyelectrolyte solution with added (monovalent) salt c_s can be expressed in terms of the same property without salt X_0 as $X \approx X_0(1 + 2Ac_s/c)^\alpha$ with the appropriate scaling exponent α.

Basically the theory provides the monomer-monomer correlations (counterions are not taken into account explicitely) at different length scales (statics) and time scales (dynamics, not considered here). Thus the validity of this model can be best tested with experiments which probe the pair correlations in Fourier space, i.e. using the techniques of small angle neutron scattering (SANS), small angle x-ray scattering (SAXS) and also static light scattering for probing very large distances. But before describing the experiments, first let us summarize the scaling predictions:

- In the absence of salt, the structure factor at small wave vectors, which is related to the osmotic compressibility and is dominated by the small ions, is very small and given by $S(q = 0) \tilde{=} 1/f$. [43]
- There is a broad peak in the salt-free structure factor at a finite wave vector $q^* \approx \xi^{-1} \propto c^{1/2}$. At this value, it can be shown that $S(q)$ is larger than $S(0)$ and therefore the profile is an increasing function of q at $q < \xi^{-1}$. Let us stress that the peak in the structure factor is not due to some order in the solution but it is related to the very small value of $S(q = 0)$ due to the constraint of electro-neutrality in the solution. Another picture interprets the peak as that of a liquid of correlation blobs of size ξ.
- At large q-values, corresponding to distances smaller than the mesh size ξ, the chain has a rod-like behavior and $S(q)$ should decrease as $1/q$.
- At high salt concentrations such that the counterions are localized in a sheath of size $\kappa^{-1} < \xi$, the solution behaves similarly to a neutral solution, the peak disappears and $S(q)$ decays monotonically from $S(q = 0) = 2c_s/f^2c$

There exists a large body of scattering data on highly charged polyelectrolytes. Most of them are concerned with poly(styrene sulphonate) considered as a flexible polymer and relevant here, and DNA with a large rigidity (generally viewed as a rod). Only what I believe are the most representative experimental results are listed in what follows. (For a more general review see [44] and Rawiso's chapter).

- For *all intrinsically flexible systems* a maximum in the scattering intensity profile has been observed by neutron [45], [46] and x-ray scattering [47], [48]. The peak is broad and no higher order peak has ever been observed with flexible chains at any concentration which rules out the presence of well ordered rodlike chains.
- For most systems the position of the maximum q^* scales as $c^{1/2}$. For PSS it has been verified over more than four orders of magnitude in concentration by combining SAXS, SANS and LS data. (However remember that this scaling law does not allow one to discriminate between the isotropic and the parallel rod models)
- The broad peak fades away as salt is added; the intensity of the low-q side of the peak increases, as expected since the osmotic compressibility, proportional to S(0), increases; a peak is no longer visible when the salt concentration is roughly equal to the number of free counterions in the solution ; as for the peak position, there is no definite answer in the literature: it either moves slightly to lower q's or does not move at all (this appears to be related to the difficulty to determine its precise position when the peak is skewed because of the higher intensity at $q<q^*$). [46], [19]
- The invariance of q^* with the linear charge density f above about 0.4 for vinylic polymers has usually been interpreted in terms of counterion condensation in semi-dilute solutions [19], [49] although it is fair to say that other interpretations have been suggested [50]. For smaller f's in the weakly charged regime, q^* varies as $f^{1/3}$.
- As noted earlier polyelectrolytes solutions are multicomponent systems and the observed scattered intensity has contributions from all components, monomers, counterions, coions, solvent. Elaborate and clever techniques of contrast variation in neutron scattering [42] have given access to all partial structure factors. The Zero Average Contrast (ZAC) method has been the most promising since it allows to get directly the chain form factor without need for delicate and time consuming extrapolations. Then an effective structure factor can be derived by dividing the scattered intensity at full contrast by the form factor. [51], [52] A peak is found in the structure factor, at variance with the idea [53] that there is a constrain for the chains to stay at a finite distance from each other without any particular accident in

the structure factor (correlation hole) but confirming that a soft order may exist in the solution.

- The chain rigidity seen as a q^{-1} dependence of the scattered intensity has been observed in the form factor determined in ZAC condition for well chosen systems (see paragraph 4.3). It is important to stress that it *cannot be detected* in SAXS or full contrast SANS experiments when all components, including counterions contribute to the scattering. Any q^{-1} dependence in these conditions is only fortuitous for intrinsically flexible polyelectrolytes.
- Determination of S(0) and its scaling dependence with various parameters and comparison with direct measurements of the osmotic pressure have been hindered by the intriguing presence of an excess intensity at very small angles (see next section) which does not allow a meaningful extrapolation of the scattering profile to q=0.

In conclusion, most of the experimental results support the scaling predictions of the isotropic model of de Gennes *et al.* for semi-dilute solutions of polyelectrolytes. In this respect scattering techniques, especially contrast variation methods in neutron scattering, have been priceless to determine the structural properties of model polyelectrolytes. Monte-Carlo simulations are also in qualitative agreement with this model. However it is fair to mention that there are still some indefectible tenants of an ordered model for flexible polyelectrolytes, in analogy to the situation of charged rods and globules.

4.3. PUZZLES AND RIDDLES

Although the isotropic model for semi-dilute solutions has been found adequate to describe the physical properties of a large number of flexible polyelectrolytes, many unsolved problems still exist and they are often at the heart of heated controversies. I shall just briefly mention two of those. The first one is concerned with the long range structure of semidilute solutions and the second one, with the rigidity of a polyelectrolyte chain due to electrostatic interactions.

It has been repeatedly suggested that "large aggregates" exist in semi-dilute solutions *without added salt.* This concept has been introduced to rationalize two unexpected and puzzling experimental features, namely, an upturn in the static x-ray and neutron scattering intensity profiles at very small angles associated with "objects" larger than a single chain (close to 100nm) and a slow mode in dynamic light scattering concomitant with the usual fast mode due to single chain movements in the semi-dilute solution. These observations have been reported for many different systems. Experimentally, the phenomena are difficult to catch and although many articles are published every year on the subject, it is difficult to get a definite picture of these "aggregates" [54]. Recent experiments by Brett and Amis [55] using coupled SANS and SLS in order to get structural data over a very broad range of distances give the most quantitative static picture of the aggregates to date and definitely rule out the possibility of artefacts. A sensible description of the experimental scene is available in [56] and enlightened comments can be found in [57]. Still the physical origin of the attractive forces that would produce the aggregates is still difficult to ascertain although it is clear from the majority of dynamic experiments that the basic mechanism is a coupling between polyions and counterions. There have been many theoretical approaches predicting attractive forces between like charged chains but these are concerned with ideal chains and often limited to very diluted systems. Indeed a quantitative treatment of the interactions in these highly charged systems, taking into account the counterions explicitely requires to use simplified models. These models give directions for further research but they are still far from real systems. For this particular problem the gap between theory and experiment is still wide open although striking breakthrough have been made recently in the understanding of attractive forces between like charge rods.

What is the effect of electrostatic interactions on the rigidity of intrinsically flexible polyelectrolytes? This is another simple question with no simple answer either theoretical or experimental. Odijk [58], [59] and, independently, Skolnick and Fixman [60] were the first to tackle the theory of the electrostatic persistence length. They used perturbation methods, starting from a rod-like chain. They predict that the total persistence length L_T is the sum of a bare persistence length L_0 (chain without charges) and of an electrostatic one L_{OSF} that scales as the *square of the Debye screening length* as has been rather well verified for stiff polyelectrolytes such as DNA. This finding was in direct opposition to scaling theories which assumed that the chain is flexible at all scales and predicted that L_{el} scales *linearly* with κ^{-1}. Further theoretical predictions, using various methods fall randomly into two categories, according to the κ^{-1} [61] or κ^{-2} [62] dependence of L_{el} on the screening length. Simulations do not give a clear answer either and a recent investigation indicate an exponent smaller than 1. [63] There is a possibility that the conflicting results are due to limitations of using Debye-Huckel screened interaction to describe polyelectrolyte systems. Clearly experiments are necessary. Unfortunately direct measurement of a persistence length at finite concentrations are difficult to perform. A recent SANS determination of the electrostatic rigidity of a fully charged PSS, using sophisticated contrast variation techniques showed that, *for the conditions of the experiment*, [64] $L_p \propto (c_p + 2c_s)^{-1/3}$, indicating that all ions contribute to the chain stiffness, including the condensed ones, and showing still another scaling law!

These are but a few of the complex problems that anyone working on polyelectrolytes is bound to encounter. It is these questions that make the field challenging and exciting.

5. Highly charged hydrophobic polyelectrolytes

The description of the statistical physics of polyelectrolyte chains that we have outlined in the preceding section assumes implicitly that the chain backbone is in a θ-solvent and that its properties are entirely determined by the electrostatic interactions (strong coupling limit). Numerous experiments have validated this assumption, using very often fully charged poly (styrene sulphonate), PSS, as a model compound. In all these studies, the fact that the uncharged polymer is highly hydrophobic seemed irrelevant in agreement with the theoretical predictions. The question we might ask at this point is the following: what happens if the hydrophobic character of the chain is increased by introducing some styrene monomers in the chain, still keeping the charge fraction in the strong coupling regime? According to what we have learned earlier, the effective charge of the chain should be renormalized by Manning condensation and the structure of the solution should remain unchanged since the electrostatic interactions are assumed to be dominant.[19] As you can guess this is not the case with these polymers and hydrophobic effects have drastic effects on the chain properties. Note that in this case the experiments have been leading theory.

The following discussion will be based on an experimental investigation using small angle X-ray and neutron scattering, light scattering, osmometry and fluorescence emission spectroscopy. [52], [65], [66], [67], [68], [69] The polyelectrolyte considered is a random copolymer of various compositions of styrene, a hydrophobic monomer, and styrene sulphonate, an ionizable water soluble monomer. The charge fraction was varied between 0.3 (limit of solubility in water) and 1 (fully charged) i.e. in a charge range where the average linear spacing between charges b is smaller than the Bjerrum length. The experiments have shown that:

- the expected renormalized effective charge is found only at f=1 and is continuously reduced as f decreases; at f=0.38 the reduction is about 75% of Manning's value as measured at f=1 (as shown by osmometry and confirmed by SAXS and SANS). This indicates that CC is not the only process involved and that some charges are trapped in the solution where they are no longer osmotically active.
- there are some diffuse hydrophobic regions in the solution (as shown by fluorescence emission of a pyrene probe) a fact which invalidates the

current description of the chain embedded in a continuous medium of dielectric constant ε

- the chain conformation gets more and more compact as f decreases. Around f=0.3, SAXS data can be interpreted as due to interacting weakly charged hard spheres, the effective charge being obtained by osmometry. This is also confirmed by SANS data (using contrast variation techniques) which show that the single chain form factor in semidilute solution is that of a wormlike chain when fully charged whereas it evolves towards a more spherical dense object as f decreases.
- addition of a good solvent for the backbone to the aqueous solution reverts to "normal" behavior
- all properties are continuous as f is varied and there is no evidence for any sharp transition.

The following qualitative picture has emerged from these studies. Close to longer sequences of hydrophobic monomers, the local ε is much smaller than the solvent value and the neighboring ionic monomers are not dissociated and remain as ion pairs (recall that PSS is an *ionomer* at low charge contents) . This has two consequences: a proportion of the counterions are bound on site and are not osmotically active, reducing even more the counterion contribution to the osmotic pressure; dipolar attractive forces produce aggregation in the diffuse hydrophobic regions, stabilizing the effect. It is important to note that a gel is not formed as concentration increases, indicating that these interactions are within a chain and not interchain.

At the same time and not quite independently, theorists [36] have revived the problem of a single (weakly charged) chain in a poor solvent where the overall shape is determined by the balance of the electrostatic repulsion and the surface tension. It was first suggested by Khokhlov [70] that the chain would minimize its energy if it took the shape of an elongated cylinder. However Dobrynin and Rubinstein [71] have shown that a more favorable shape would be that of a necklace with compact beads joined by narrow strings. This configuration results from a Rayleigh charge instability, similar to the breaking up of a charged drop into smaller ones; it results from a balance between interfacial energy and electrostatic energy. Changing the charge on the chain results in a cascade of abrupt transitions between necklaces with different numbers of beads. Above c^*, the overlap concentration for the pearl necklaces, there are two important regimes [72], [73] labelled "string controlled" and "bead controlled", depending on whether the concentration is sufficient for the inter-bead repulsion to become important. It is predicted that ξ scales as $c^{-1/2}$ in the first regime but as $c^{-1/3}$ in the second regime, a departure to the seemingly universal law for the polyelectrolyte peak! This finding is certainly comforting the experimentalists who were trying desperately to scale their data with an exponent 0.5... There is clearly qualitative agreement between the pearl necklace model of Dobrynin and Rubinstein. Although the theory cannot handle the complexities of real systems it provides trends and effects to look for. For instance recently the string-controlled and the bead- controlled regimes have been clearly identified in a model system of a polyelectrolyte in various non aqueous solvents of varying dielectric constant and quality [74]. Globules have also been seen in various simulations, the most extensive being the molecular dynamics simulations of Micka *et al* at high concentrations and using discrete ions [75].

If there is no doubt that a structure akin to a pearl necklace exists in hydrophobic polyelectrolytes, a detailed picture is still lacking. In particular, the chain organization in semi-dilute and concentrated solutions is still difficult to ascertain. Different model systems should be investigated where the hydrophobic character and the electrostatic interactions should be varied independently. Still, at this point, my message will be that polyelectrolytes still hide many secrets and that there are still new discoveries to be made.

References

1. Raphael E., Joanny J.-F., *Europhys. Lett.* **13**, 623, (1990).
2. Borukhov I., Andelman D., Borrega R., Cloitre M., Leibler L., Orland H., *J. Phys. Chem. B* **104**, 11027, (2000).
3. Flory P. J., *Principles of Polymer Chemistry*, Cornell University Press (Ithaca) 1953.
4. Staudinger H., *Die hochmolekülaren organischen Verbindungen Kautschuk und Cellulose*, Springer Verlag (Berlin) 1932 ; Kern W. Z., *Physic Chem A* **181**, 283, (1938).
5. Fuoss R.M., Strauss U.P., *J. Polym. Sci.* **3**, 246, (1948); Fuoss R.M., *J. Polym. Sci.* **3**, 603, (1948); correction *ibid.* **3**, 246, (1949); Katchalsky A., Eisenberg H., *J. Polym. Sci.* **6**, 145, (1951); Oth A., Doty P., *J. Phys. Chem.* **56**, 43, (1952).
6. R. M. Fuoss, *Disc.Faraday Soc.* **11**, 125, (1952).
7. The existence of a maximum has been reported in many experiments but was not easily accepted because it was thought to be due to artefacts due to uncontrolled remaining impurities in the solution, interaction with the wall of the viscosimeter, etc; for definite experiments see Cohen J., Priel Z., Rabin Y., *J. Chem. Phys.* **88**, 7111, (1988) or Yamanaka J., Matsuoka H., Kitano H. Hasegawa M., Ise N., *J. Am. Chem. Soc.* **112**, 587-592, (1990); a critical discussion on the complex rheology of polyelectrolytes can be found in Boris C.B., Colby R.H., *Macromolecules* **31**, 5746-5755, (1998).
8. Kuhn W., Kunzle O., Katchalsky A., *Helv. Chem. Acta* **31**, 1994, (1948).
9. Stevens M.J., Kremer K., *J. Chem. Phys.* **103**, 1669, (1995).
10. de Gennes P.G., Pincus P., Velasco R.M., Brochard F., *J. Phys. (Paris)* **37**, 1461, (1976).
11. de Gennes P.G., *Scaling Concepts in Polymer Physics*, Cornell Univ. Press, Ithaca, NY 1979.

12. The same reasoning can be applied to a chain in good solvent. The case of a bad solvent is more subtle and a globule/solvent surface tension contribution has to be included in the energy; this will be briefly treated in the section dealing with hydrophobic PE.

13. Fuoss R.M., Katchalsky A., Lifson S., *Proc. Natl. Acad. Sci. USA* **37**, 579 (1951); Alfrey T., Berg P., Morawetz H., *J. Polym. Sci* **7**, 543, (1951); Fixman M. J., *J. Chem. Phys.* **70**, 4995, (1979).
14. Manning G.S., *J. Chem. Phys.* **51**, 924 and 934, (1969).
15. Osawa F., *J. Polym. Sci* **13**, 93, (1954).
16. Oosawa F., *Polyelectrolytes* 1971 M. Dekker, New York.

17. It is important to stress that , unless specifically stated, we are *not* discussing here the case of polyelectrolytes with a long persistence length such as DNA, where the chain configuration is closer to that of a rod and where Manning condensation is also present

18. Manning G.S., *Ber. Bunsenges. Phys. Chem.* **100**, 909-922, (1966).
19. Essafi W., Lafuma F., Williams C.E., *Eur. Phys. J. B* **9**, 261-66, (1999).

20. Most quantitative determinations of the distribution of counterions by contrast matching SANS and their fit with Poisson-Boltzmann equation have dealt with rod-like polyelectrolytes (DNA and synthetic polymers). For an attempt with poly(styrenesulfonate), see Kassapidou K., Jesse W., Kuil M.E., Lapp A., Egelhaaf S., van der Maarel J.R.C., *Macromolecules* **30**, 2671-2684, (1997).

21. Ramanathan G.V., Woodbury C.P., *J. Chem. Phys.* **77**, 4133, (1982).
22. Gonzales-Mozuelos P., Olvera de la Cruz M., *J. Chem. Phys.* **103**, 3145, (1995).
23. Nyquist R.M., Ha B.Y., Liu A., *Macromolecules* **32**, 3481, (1999).
24. Stevens M. J., Kremer K., *J. Chem. Phys.* **103**, 1669, (1995).
25. Micka U., Holm C., Kremer K., *Langmuir* **15**, 4033, (1999).
26. Brilliantov N.V., Kuznetsov D.V., Klein R., *Phys. Rev. Lett.* **81**, 1433-1436, (1998).
27. Schiessel H., Pincus P., *Macromolecules* **31**, 7953-7959, (1998).

28. The longitudinal movement and the subsequent escape of a counterion is clearly seen in the simulations of M. Stevens (unpublished results).

29. Bleam M.L., Anderson C.F., Record Jr M.T., *Proc. Natl. Acad. Sci. USA* **77**, 3085, (1980).
30. Schmitz K.S. in *Macro-ion Characterization*, chapter 1, ACS, Washington 1994.
31. Schmitz K. S., *Langmuir* **13**, 5849, (1997).
32. Katchalsky A., Alexandrowicz Z., Kedem O. in *Chemical Physics of Ionic Solutions*, Wiley, New York 1966.
33. Barrat J.L., Joanny J.-F., *Adv. Chem. Phys.* **100**, 909-922, (1996).
34. Kaji K., Urakawa H., Kanaya T. and Kitamaru R., *J. Phys. France* **49**, 993-1000, (1988).
35. M.J. Stevens, K. Kremer, in *Macro-ion Characterization*, K.S. Schmitz, ed., ACS, Washington 1994.
36. Boris D.C., PhD thesis, Univ. of Rochester, New York 1996.
37. Boris D.C., Colby R.H., *Macromolecules* **31**, 5746-5755, (1998).
38. Dobrynin A., Colby R. H., Rubinstein M., *Macromolecules* **28**, 1859, (1995).
39. Rubinstein M, Colby R. H., Dobrynin A., *Phys. Rev. Lett.* **73**, 2776, (1994).

40. We have assumed here that there is only one characteristic length in the problem, i.e. that the electrostatic persistence length of an intrinsically flexible polyelectrolyte is proportional to the Debye screening length. This is still a disputed fact. See section 4.3 and, for instance, ref. 11.

41. Pfeuty P.J., *J. Phys. France Coll C2* **39**, 149, (1978).

42. Higgins J.S., Benoit H.C., *Polymers and Neutron Scattering*, Oxford University Press 1996.
43. f is actually an effective f which takes into account only those counterions which are free (osmotically active) in the solution.
44. Förster S., Schmidt M., *Adv. Polym. Sci.* **120**, 51, (1995).
45. Cotton J.P., Moan M., *J. Phys. Lettres (Paris)* **37**, 75-77, (1976).
46. Nierlich M., *et al, J. Phys. (France)* **40**, 701, (1979); Williams C.E., *et al, J. Polym. Sci. Polym. Lett.* **17**, 379-384, (1979).
47. Kaji K., Urakawa H., Kanaya T. and Kitamaru R., *Macromolecules* **49**, 1835-1839, (1984).
48. Ise N., Okubo T., Kunugi S., Matsuoka H., Yamamoto K., Ishii Y., *J. Chem. Phys.* **81**, 3294, (1984).
49. Nishida K., Kaji K., Kanaya T., *Macromolecules* **31**, 7378-7384, (1998).
50. Netz R., *Eur. Phys. J. E*, to be published.
51. Boué F., Cotton J.P., Lapp A., Jannink G., *J. Chem. Phys.* **101**, 2562-2568, (1994).
52. Spiteri M.N., PhD thesis, Paris XI 1997.
53. Hayter J., Jannink G., Brochard F., de Gennes P.G., *J. Phys. Lett.* **41**, L-451, (1980).
54. The term "aggregate" is misleading as it gives an image of a static cluster of chains and their counterions. However one plausible interpretation is the presence of large concentration fluctuations which are sometimes referred to as "temporal aggregates" or "random inhomogeneities".
55. Brett D.E., Amis E. J., *Macromolecules* **31**, 7378-7384, (1998).
56. Sedlak M., *Macromolecules* **26**, 1158-1162, (1993).
57. Schmitz K.S. 1994 in *Macro-ion Characterization*, Chapter 1, ACS Symposium Series.
58. Odijk T., *J. Polym. Sci. Polym. Phys. Ed.* **15**, 477-483, (1977).
59. Odijk T. , Houwaart A.C., *J. Polym. Sci. Polym. Phys. Ed.* **16**, 627-639, (1978).
60. Skolnick J. , Fixman M., *Macromolecules* **10**, 944-948, (1977).
61. Barrat J.L., Joanny J.-F., *Europhys. Lett.* **24**, 333, (1993); Ha B.Y., Thirumalai D., *Macromolecules* **28**, 577, (1995); Witten T., Pincus P., *Europhys. Lett.* **3**, 315, (1993).
62. Witten T., Li H., *Macromolecules* **28**, 5921, (1995) .
63. Micka U., Kremer K., *J. Phys. Condens. Matter (UK)* **8**, 9463, (1995).
64. Spiteri M.N., Boué F., Cotton J.P., Lapp A., *Phys. Rev. Lett.* **77**, 5418, (1996).
65. Essafi W. PhD thesis, Paris VI 1996.
66. Essafi W., Lafuma F., Williams C.E. 1994 in *Macro-ion Characterization. From Dilute Solutions to Complex Fluids*, K.S. Schmitz, ed., ACS, Washington, **548**, 278.
67. Essafi W., Lafuma F., Williams C.E., *J. de Phys. II* **5**, 1269-1275, (1995).
68. Carbajal-Tinoco M.D., Williams C.E., *Europhysics Lett.* **52**, 284-290, (2000).
69. Carbajal-Tinoco M.D., Ober R., Dolbnya I., Bras W., Williams C.E. 2001, preprint.
70. Khokhlov A.R., *J. Phys. A* **13**, 979, (1980).
71. Dobrynin A.V., Rubinstein M., Obukhov S.P., *Macromolecules* **29**, 2974, (1996).
72. Dobrynin A.V., Rubinstein M., *Macromolecules* **32**, 915-922, (1999).
73. Chatellier X. PhD thesis, Strasbourg 1998.
74. Waigh T.A., Ober R., Williams C.E., Galin J.C., *Macromolecules* **34**, 1973-1980, (2001).
75. Micka U., Holm C., Kremer K., *Langmuir* **15**, 4033-4044, (1999).
76. Batzill St., Luxemburger R., Deike R., Weber R., *Eur. Phys. J. B* **1**, 491-501, (1998).

Index